Monografie Matematyczne

Instytut Matematyczny Polskiej Akademii Nauk (IMPAN)

Volume 66

(New Series)

Founded in 1932 by
S. Banach, B. Knaster, K. Kuratowski,
S. Mazurkiewicz, W. Sierpiński, H. Steinhaus

Volumes 31–62 of the series
Monografie Matematyczne were published by
PWN – Polish Scientific Publishers, Warsaw

Paul F.X. Müller

Isomorphisms between H^1 Spaces

Birkhäuser Verlag
Basel · Boston · Berlin

Author:

Paul F.X. Müller
Institute of Analysis
Johannes Kepler University Linz
Altenbergerstr. 69
Austria
e-mail: pfxm@bayou.uni-linz.ac.at

2000 Mathematics Subject Classification 30D55, 42C10, 46B03, 46B07, 46-99, 47B38, 60G46

A CIP catalogue record for this book is available from the Library of Congress, Washington D.C., USA

Bibliographic information published by Die Deutsche Bibliothek
Die Deutsche Bibliothek lists this publication in the Deutsche Nationalbibliografie; detailed bibliographic data is available in the Internet at <http://dnb.ddb.de>.

ISBN: 3-7643-2431-7 Birkhäuser Verlag, Basel – Boston – Berlin

Part of Springer Science+Business Media
Cover design: Micha Lotrovsky, CH-4106 Therwil, Switzerland
Printed on acid-free paper produced of chlorine-free pulp. TCF ∞
Printed in Germany
ISBN-10: 3-7643-2431-7
ISBN-13: 978-3-7643-2431-5

9 8 7 6 5 4 3 2 1

www.birkhauser.ch

To Joanna

Contents

Preface

The theory of Hardy spaces is firmly rooted in the fields of complex analysis, trigonometric series, and probability. Its history began in the decade between 1906 and 1916. In 1906 appeared P. Fatou's treatise on power series and trigonometric series in Acta Mathematica. G. H. Hardy proved in 1915 that integral means of power series in the unit disk are log-convex, just in time to be included in the first edition of E. Landau's "Darstellung und Begründung einiger neuerer Ergebnisse der Funktionentheorie" (1916). In the same year F. and M. Riesz presented their treatise "Über die Randwerte einer analytischen Funktion" to the Scandinavian congress of mathematicians in Stockholm.

The classical theory of Hardy spaces is concerned with the boundary behavior of analytic functions on the unit disk and with estimates for the Fourier coefficients of the limiting function. Given a power series $f(z) = \sum a_n z^n$, Hardy and Littlewood (1930) proved that on the unit circle the L^1 norm of its non-tangential maximal function is equivalent to the L^1 norm of its boundary values. R. E. A. C. Paley (1933) shows that the sequence of lacunary coefficients is square summable provided that the boundary values are integrable on the unit circle, more precisely that

$$\left(\sum |a_{2^n}|^2\right)^{1/2} \leq \frac{1}{\pi} \int_{-\pi}^{\pi} |f(e^{it})| dt.$$

Kolmogorov's theorem (1925) asserts that the Hilbert transform is of weak type one-one. These three results, together with the F. and M. Riesz theorem, were crucial in establishing the importance of the Hardy space $H^1(\mathbb{T})$ of integrable functions on the unit circle for which the harmonic extension to the unit disk is analytic. Probability, in the 1920's, was confronted with analogous problems of boundary convergence in the work of N. Wiener (construction of Brownian motion), A. Kolmogorov (sums of independent random variables) and J. Khintchine (the law of large numbers) to name just a few. Thus emerged martingales, martingale convergence theorems and J. L. Doob's maximal function.

By 1977 there existed two highly developed classes of H^1 spaces: these were the atomic H^1 spaces which are linked to analytic functions via Fefferman's duality theorem, and the martingale H^1 spaces consisting of martingales for which Doob's maximal function is integrable. A remarkably simple object, called *dyadic* H^1, appears in both of these classes.

Several of the most important theorems for atomic H^1 spaces directly correspond to a theorem for the martingale class (atomic decomposition, equivalence of square functions and maximal functions, H^1 – BMO duality, singular integrals correspond to martingale transforms, etc.) . Given the obvious analogies between the two families of Banach spaces, A. Pełczyński conjectured that there are deeper connections to be uncovered. A. Pełczyński's conjecture asked specifically whether dyadic H^1 and $H^1(\mathbb{T})$ (the L^1 subspace consisting of boundary values of analytic functions) are isomorphic as Banach spaces. For a family of Banach spaces, whose members are constructed by one and the same rule, the question of its isomorphic classification is inevitable. This applied in particular to the class of H^1 spaces, and clarifying the special role played by dyadic H^1 (with its obvious unconditional basis) posed a problem of particular interest. Its solution by B. Maurey opened a new line of investigation for the Banach space H^1. In this treatise we give a thorough analysis of B. Maurey's work and its consequences, including L. Carleson's construction of an unconditional basis for $H^1(\mathbb{T})$.

This monograph is a study of dyadic H^1, its isomorphic invariants and its position within the two classes of martingale and atomic H^1 spaces. Simultaneously, we provide a detailed analysis of the Haar system and of the operators that are built from it. These include rearrangement operators, orthogonal projections, paraproducts and Calderon–Zygmund integral operators.

This book was written under the conviction that the intensive occupation with the Haar system is fully justified by its impact on wide ranging areas of analysis and probability. The following list has been compiled with the aim of substantiating this point of view.

1. The Haar system is clearly the simplest martingale difference sequence, yet we have B. Maurey's theorem to the effect that every martingale is conveniently embedded in the partial sums of a Haar series.

2. Calderon–Zygmund operators can be expressed as paraproducts plus an absolutely converging series of basic operators rearranging and splitting the Haar system. This is the content of Figiel's integral representation.

3. In establishing isomorphic invariants one is clearly free to choose the simplest object among the spaces known to be isomorphic to the Banach space under investigation. For instance, dyadic H^1 is the most unassuming example among the martingale spaces $H^1[(\mathcal{F}_n)]$ and also among the atomic spaces $H^1(X,d,\mu)$. Hence we work in dyadic H^1 to establish isomorphic invariants and simultaneously we search in the classes $H^1[(\mathcal{F}_n)]$ and $H^1(X,d,\mu)$ for the spaces that are isomorphic to dyadic H^1. This line of development was initiated by the outstanding work of B. Maurey. It led to the result that $H^1(X,d,\mu)$ is isomorphic to one of the following spaces, H^1, $(\sum H^1_n)_1$ and ℓ^1 (the same conclusion holds for martingale spaces when the generating σ-algebras are purely atomic).

4. Expressing a problem using Haar functions and exploiting their relative simplicity leads us quite often straight to the point, that is, to the problem's inherent combinatorial difficulty. This occurs for instance in the construction of multipliers into SL^∞, in Johnson's factorization of operators on L^p, or in the construction of a resolving operator to establish the uniform approximation property for H^1.

5. The Haar system is a reliable instrument for separating core ideas from their surrounding technicalities. The best known examples illustrating this point are obtained by specializing, to the dyadic setting, the classical theorems of real and harmonic analysis, such as the atomic decomposition, Fefferman's inequality, Hardy–Littlewood maximal function estimates, good λ inequalities, interpolation and extrapolation of operators etc.

Contents. The first chapter introduces the Haar system.

In Chapter 1 we present the classical inequalities of Khintchine, Burkholder, Fefferman, Hardy–Littlewood. It contains a discussion of large deviation inequalities, Schechtman's sign-embeddings and multipliers into SL^∞. We prove that the unconditionality of the Haar basis yields convergence of the Walsh–Paley expansion and estimates for the dyadic gradient in L^p. We study counterexamples pertaining to the Walsh expansion along increasing multiplicities.

General martingales and biorthogonal systems are analyzed by exploiting their relation to the Haar system. The first chapter concludes with estimates for paraproducts, Figiel's representation of singular integral operators and with a discussion of orthogonal projections in H^1.

In Chapter 2 we review basic concepts of functional analysis. We list the known complemented subspaces of H^1, prove the Banach space decomposition principle of A. Pełczyński and discuss analytic families of operators on H^p spaces.

The analysis of rearrangement operators and orthogonal projections is predominant in this book. It gives rise to hard problems involving combinatorics of colored dyadic intervals. We address these problems by constructing a coherent set of combinatorial techniques, forming the content of Chapter 3.

In Chapter 4 we study martingale H^1 spaces. We present a detailed analysis of Maurey's isomorphism, thus showing the existence of an unconditional basis. Exploiting the unconditional basis, obtained by Maurey's isomorphism, yields the classification theorem for martingale H^1 spaces.

Chapter 5 covers the isomorphic invariants of H^1. We establish dichotomies for complemented subspaces of H^1, prove that H^1 and $H^1(\ell^2)$ are not isomorphic as Banach spaces, and that H^1 satisfies the uniform approximation property.

Chapter 6 contains the presentation of atomic H^1 spaces and their relation to dyadic H^1. We treat the example of H^1_{at} in considerable detail, obtaining equivalent norms using the Hilbert transform, the Lusin area function and the non-tangential maximal function. Chapter 6 contains a careful presentation of L. Carleson's biorthogonal system. We show that it is an unconditional basis for

H^1_{at} and thus prove B. Maurey's theorem that H^1_{at} and H^1 are isomorphic Banach spaces. The chapter closes with the classification theorem for atomic H^1 spaces.

Prerequisites. I assume only that the reader knows basic real, complex and functional analysis, and some probability theory, and refer to [76] by C. Goffman and G. Pedrick, to the first chapters of [163] by Z. Nehari, and to the appendix of [61] by R. Durrett.

Acknowledgment. My research interests were formed by my collaboration with P. Wojtaszczyk at the Polish Academy of Sciences in Warszawa (1982–1984), with G. Schechtman at the Weizmann Institute of Science in Rehovot (1988–1991) and with P. W. Jones at Yale University in New Haven (1995–1998, and 2000). It is with great pleasure that I acknowledge the profound influence of P. Wojtaszczyk, G. Schechtman and P. W. Jones on my mathematical development and on the topics covered in this monograph.

I would also like to thank those mathematicians I discussed with in Linz, M. Blümlinger, J. B. Cooper, A. Kamont, K. Kiener, B. Kirchheim, V. Pillwein, W. Schachermayer, M. Schmuckenschläger, C. Schütt, C. Stegall and A. Wakolbinger. I am especially indebted to B. Kirchheim with whom I organized a series of exciting seminars (1992–1995) which focussed intensely on problems of hard analysis. My records of the seminars were the starting point for the work on this book.

Special thanks go to V. Pillwein who read the entire manuscript, made many valuable suggestions and detected numerous errors. Financial support was provided by the Austrian Science Foundation (FWF) project P15907-N08.

Chapter 1

The Haar System: Basic Facts and Classical Results

This chapter introduces the Haar system. It contains basic inequalities associated with the Haar system and their interpretation as boundedness properties of Haar multipliers, paraproducts, rearrangement operators and orthogonal projections.

1.1 Bases in L^p

In this section we establish that the Haar system is an unconditional basis in L^p $(1 < p < \infty)$. Proving inequalities of Khintchine and Burkholder we obtain a square function characterization of these spaces. We show that unconditionality of the Haar system implies that the Walsh system (in Walsh–Paley order) is a Schauder basis in L^p $(1 < p < \infty)$. We discuss the multiplicity of Walsh functions and study associated approximation problems.

The Haar basis in L^p $(1 \le p < \infty)$

An interval $I \subseteq [0,1]$ is called a dyadic interval if there exists $n \in \mathbb{N} \cup \{0\}$, and $1 \le k \le 2^n$ such that

$$I = \left[\frac{k-1}{2^n}, \ \frac{k}{2^n}\right[.$$

Let I_1 be the left half of I and I_2 be the right half of I. Then I_1, I_2 are again dyadic intervals. Explicitly they are given as

$$I_1 = \left[\frac{2k-2}{2^{n+1}}, \frac{2k-1}{2^{n+1}}\right[\quad \text{and} \quad I_2 = \left[\frac{2k-1}{2^{n+1}}, \frac{2k}{2^{n+1}}\right[.$$

We call I_1, I_2 the dyadic successors of I. Conversely I is called the dyadic predecessor of I_1 and I_2. We denote by $\mathcal{D}$ the collection of all dyadic intervals, and we let $\mathcal{D}_n$ denote the finite sub-collections $\{I \in \mathcal{D} : |I| \geq 2^{-n}\}$. Observe that $\mathcal{D}$ is nested in the following sense. If $I, J \in \mathcal{D}$ are not disjoint, then either $I \subseteq J$ or $J \subseteq I$.

We define the L^∞ normalized Haar system indexed by dyadic intervals as follows. For $I \in \mathcal{D}$ we let

$$h_I = \begin{cases} 1 & \text{on the left half of } I, \\ -1 & \text{on the right half on } I, \\ 0 & \text{otherwise.} \end{cases}$$

The Haar system is biorthogonal in $L^2([0,1])$. Following are several remarks relating the expansion with respect to the Haar system to a sequence of conditional expectation operators. Let $\mathcal{A}_n = \{I \in \mathcal{D} : |I| = 2^{-n}\}$. Denote by $\mathbb{E}_n$ the conditional expectation with respect to the σ-algebra generated by $\mathcal{A}_n$. Explicitly $\mathbb{E}_n(f)$ is given as

$$\mathbb{E}_n(f) = \sum_{I \in \mathcal{A}_n} \left[\int_I f dt\right] \frac{1_I}{|I|}. \tag{1.1.1}$$

Haar's identity. The conditional expectation $\mathbb{E}_n(f)$ and the expansion of f along the Haar system are related through A. Haar's identity,

$$\mathbb{E}_n(f) = \int_0^1 f dt + \sum_{J \in \mathcal{D}_{n-1}} \langle f, \frac{h_J}{|J|} \rangle h_J. \tag{1.1.2}$$

The following remarks verify the basic identity (1.1.2). Let I be a dyadic interval, let I_1 be the left half of I, and let I_2 be the right half of I. These identities are evident,

$$\begin{aligned} 2 \cdot 1_{I_1} &= 1_I + h_I, \\ 2 \cdot 1_{I_2} &= 1_I - h_I. \end{aligned} \tag{1.1.3}$$

Multiplying the equations of (1.1.3) by a function f and integrating gives the relations

$$\begin{aligned} \frac{1}{|I_1|} \int_{I_1} f dt &= \frac{1}{|I|} \int_I f dt + \langle f, \frac{h_I}{|I|} \rangle, \\ \frac{1}{|I_2|} \int_{I_2} f dt &= \frac{1}{|I|} \int_I f dt - \langle f, \frac{h_I}{|I|} \rangle. \end{aligned} \tag{1.1.4}$$

Recall that $h_I = 1$ on I_1 and $h_I = -1$ on I_2. Hence the two equations of (1.1.4) may be compressed into one line as follows. Let $j \in \{1, 2\}$, and let $x \in I_j$. Then,

$$\frac{1}{|I_j|} \int_{I_j} f dt = \frac{1}{|I|} \int_I f dt + \langle f, \frac{h_I}{|I|} \rangle h_I(x). \tag{1.1.5}$$

Equation (1.1.5) relates the average over a dyadic interval, the Haar coefficient of its dyadic predecessor and the average over the dyadic predecessor. A moment's reflection shows that the identity expressed in (1.1.5) may be repeatedly applied. Carrying out the iteration shows the following statement. Let $K \in \mathcal{D}$ and let $\mathcal{C}(K) = \{J \in \mathcal{D} : J \supset K\}$. (We use $J \supset K$ to denote that K is strictly contained in J.) Then for $x \in K$ we obtain

$$\frac{1}{|K|}\int_K f dt = \int_0^1 f dt + \sum_{J \in \mathcal{C}(K)} \langle f, \frac{h_J}{|J|}\rangle h_J(x). \tag{1.1.6}$$

Comparing (1.1.6) and (1.1.1) shows that the identity (1.1.2) holds true.

The linear span of the indicator functions $\{1_I : I \in \mathcal{D}\}$ forms a dense subspace of $L^p([0,1])$, $1 \le p < \infty$. Thus Haar's identity (1.1.2) implies that augmenting the Haar system $\{h_I : I \in \mathcal{D}\}$ by the constant function $1_{[0,1]}$ gives a complete biorthogonal system in $L^2([0,1])$.

Partial sum operators associated to the Haar system. In order to form partial sum operators we need to impose a linear ordering on the Haar system $\{h_I : I \in \mathcal{D}\}$. We do this by associating an integer to the binary expansion of the left endpoint of a dyadic interval. Let $n \in \mathbb{N} \cup \{0\}$, and $1 \le k \le 2^n$. We form $I \in \mathcal{D}$ and $i \in \mathbb{N}$, as

$$I = \left[\frac{k-1}{2^n}, \; \frac{k}{2^n}\right[\qquad \text{and} \qquad i = 2^n + k - 1.$$

Then define

$$h_i = h_I. \tag{1.1.7}$$

By the defining equation (1.1.7) the sequence $\{h_i\}_{i=1}^{\infty}$ is mapped bijectively onto the Haar system indexed by dyadic intervals $\{h_I : I \in \mathcal{D}\}$. We include the constant function by putting $h_0 = 1_{[0,1]}$. Denote by P_ℓ the partial sum operator with respect to the sequence of Haar functions $\{h_i\}_{i=0}^{\infty}$. Thus

$$P_\ell(f) = \sum_{i=0}^{\ell} \langle f, \frac{h_i}{\|h_i\|_2^2}\rangle h_i. \tag{1.1.8}$$

The representation formula (1.1.9) below implies that the partial sum operators $\{P_\ell\}_{\ell=0}^{\infty}$ are given by a sequence of conditional expectations. This statement summarizes A. Haar's original analysis [79] of the orthogonal system that is now named after him.

Proposition 1.1.1. *For every $\ell \in \mathbb{N}$ there exists a collection of pairwise disjoint dyadic intervals $\mathcal{P}_\ell$ so that*

$$P_\ell(f) = \sum_{I \in \mathcal{P}_\ell} \left[\int_I f dt\right] \frac{1_I}{|I|}. \tag{1.1.9}$$

Proof. Let $\ell \in \mathbb{N}$. Let $L \in \mathcal{D}$ be the dyadic interval which is uniquely defined by the relation

$$h_\ell = h_L.$$

Let $n \in \mathbb{N}$ be such that $|L| = 2^{-n}$. Let $\mathcal{A}_n = \{I \in \mathcal{D} : |I| = 2^{-n}\}$.

Next depending on the position of L we will define a splitting of $\mathcal{A}_n$ into $\mathcal{R}_L$ and $\mathcal{L}_L$. Given L we let $\mathcal{R}_L = \{K \in \mathcal{A}_n : \sup K \geq \sup L + 2^{-n}\}$. The collection $\mathcal{R}_L$ contains those dyadic intervals of length 2^{-n} which are strictly to the right of L. Next let $\mathcal{L}_L = \{K \in \mathcal{A}_n : \sup K \leq \sup L\}$. Thus $\mathcal{L}_L$ contains the interval L and the dyadic intervals of length 2^{-n} which lie to the left of L. For $K \in \mathcal{L}_L$ we let K_1 be the left half of K and K_2 be the right half of K. Next define the collection $\mathcal{P}_\ell$ by taking the union

$$\mathcal{P}_\ell = \mathcal{R}_L \cup \bigcup_{K \in \mathcal{L}_L} \{K_1, K_2\}.$$

We now turn to verifying that with $\mathcal{P}_\ell$ the identity (1.1.9) holds true. The starting point is the observation that

$$\{h_i : 1 \leq i \leq \ell\} = \{h_I : I \in \mathcal{D}_{n-1} \cup \mathcal{L}_L\}.$$

Hence with $\mathcal{I}_\ell = \mathcal{D}_{n-1} \cup \mathcal{L}_L$ we rewrite $P_\ell(f)$ as

$$P_\ell(f) = \int_0^1 f\,dt + \sum_{I \in \mathcal{I}_\ell} \langle f, \frac{h_I}{|I|} \rangle h_I. \tag{1.1.10}$$

Comparing the right-hand side of (1.1.10) with the identity in (1.1.6) gives (1.1.9). □

Remarks. 1. The above proof shows that the length of the intervals in $\mathcal{P}_\ell$ is related to the index ℓ by the following two-sided estimate,

$$\frac{1}{2\ell} \leq |I| \leq \frac{2}{\ell} \quad \text{for} \quad I \in \mathcal{P}_\ell. \tag{1.1.11}$$

2. Let $\mathcal{F}_\ell$ be the σ -algebra generated by the collection of pairwise disjoint dyadic intervals $\mathcal{P}_\ell$. Then $\mathcal{F}_\ell \subset \mathcal{F}_{\ell+1}$. Hence the partial sum operators of the Haar system are given by conditional expectation operators associated to an increasing sequence of σ -algebras.

Schauder's theorem. The next theorem states that on $L^p([0,1])$ the partial sum operators $\{P_\ell\}_{\ell=0}^\infty$ are uniformly bounded. This is J. Schauder's contribution to the analysis of the Haar system. The representation formula (1.1.9) provided the starting point for J. Schauder in [185].

Theorem 1.1.2. *Let* $1 \le p \le \infty$, *let* $f \in L^p([0,1])$. *Then for every* $\ell \in \mathbb{N}$,

$$\|P_\ell f\|_{L^p} \le \|f\|_{L^p}. \tag{1.1.12}$$

For $1 \le p < \infty$, *the linear span of* $\{h_i\}_{i=0}^\infty$ *is dense in* $L^p([0,1])$ *and*

$$\lim_{\ell \to \infty} \|P_\ell f - f\|_{L^p} = 0. \tag{1.1.13}$$

Proof. By Proposition 1.1.1 the partial sum operator can be written as

$$P_\ell(f) = \sum_{I \in \mathcal{P}_\ell} \left[\int_I f dt\right] \frac{1_I}{|I|}, \tag{1.1.14}$$

where $\mathcal{P}_\ell$ is a collection of pairwise disjoint dyadic intervals. Applying Hölder's inequality to the right-hand side of (1.1.14) gives that,

$$\begin{aligned} \|P_\ell f\|_{L^p}^p &= \sum_{I \in \mathcal{P}_\ell} \left|\int_I f dt\right|^p |I|^{1-p} \\ &\le \sum_{I \in \mathcal{P}_\ell} \int_I |f|^p dt. \end{aligned}$$

The sum appearing in the last line of the above estimate is just $\|f\|_{L^p}^p$. Thus (1.1.12) holds true for $1 \le p < \infty$. The case of L^∞ follows directly from (1.1.14).

The linear span of the indicator functions of dyadic intervals forms a dense subspace of $L^p([0,1])$, $1 \le p < \infty$. Hence by (1.1.2) the linear span of the Haar system is dense in $L^p([0,1])$, $1 \le p < \infty$.

It remains to verify (1.1.13). Let $f \in L^p([0,1])$ and $1 \le p < \infty$. Given $\epsilon > 0$ there exist $\ell_0 \in \mathbb{N}$ and $g \in \operatorname{span}\{h_i : i \le \ell_0\}$ such that $\|f - g\|_{L^p} < \epsilon$. Then for $\ell \ge \ell_0$ it is easy to see that $g = \sum_{i=0}^{\ell} b_i h_i$. Hence $P_\ell(g) = g$ and the following identity holds,

$$f - P_\ell(f) = f - g + P_\ell(g - f). \tag{1.1.15}$$

The identity (1.1.15) and the norm estimate (1.1.12) imply (1.1.13). □

For $f \in L^p([0,1])$, $1 \le p < \infty$, we call the series

$$\sum_{i=0}^{\infty} a_i h_i \quad \text{where} \quad a_i = \langle f, \frac{h_i}{\|h_i\|_2^2} \rangle,$$

the Haar expansion of f and $\{a_i\}$ the sequence of Haar coefficients. By J. Schauder's Theorem 1.1.2, the Haar expansion converges to f in $L^p([0,1])$ and the Haar coefficients are uniquely determined.

Schauder bases. The preceding theorem leads to the notion of bases in Banach spaces. A sequence $\{x_i\}_{i=1}^{\infty}$ in a Banach space E is called a Schauder basis if for every $x \in E$ there exists a unique sequence of coefficients $\{a_i\}$ so that

$$\lim_{n\to\infty} \|x - \sum_{i=1}^{n} a_i x_i\|_E = 0.$$

Thus associated to a Schauder basis there is the sequence of partial sum operators defined by the equation

$$P_n\left(\sum_{i=1}^{\infty} a_i x_i\right) = \sum_{i=1}^{n} a_i x_i.$$

A theorem of St. Banach asserts that for a Schauder basis $\{x_i\}_{i=1}^{\infty}$, the partial sum operators are uniformly bounded, i.e.,

$$\sup_n \|P_n : E \to E\| < \infty. \tag{1.1.16}$$

The supremum in (1.1.16) is called the basis constant of $\{x_i\}_{i=1}^{\infty}$ in E.

We denote by E^* the dual Banach space to E consisting of continuous linear functionals on E. A Schauder basis gives rise to continuous biorthogonal functionals as follows. For $x = \sum_{i=1}^{\infty} a_i x_i$, define a linear functional $x_i^* : E \to \mathbb{R}$ by putting

$$x_i^*(x) = a_i.$$

Thus $x_i^*(x)x_i = (P_i - P_{i-1})(x)$, hence

$$\|x_i^*\|_{E^*}\|x_i\|_E \leq 2 \sup_n \|P_n : E \to E\| < \infty,$$

and

$$x_i^*(x_j) = \begin{cases} 1 & \text{if } i = j, \\ 0 & \text{if } i \neq j. \end{cases}$$

By Theorem 1.1.2 for $1 \leq p < \infty$, the Haar system $\{h_i\}_{i=0}^{\infty}$ is a Schauder basis in $L^p([0,1])$. The basis constant of $\{h_i\}_{i=0}^{\infty}$ in $L^p([0,1])$ is equal to 1. The biorthogonal functional to h_i is $h_i/\|h_i\|_2^2$.

Khintchine's inequality

The Khintchine inequality asserts that for series of Rademacher functions the norm in L^2 and the norm in L^p, $p < \infty$, are equivalent. For $n \in \mathbb{N}$, we let

$$r_n = \sum_{\{I:|I|=2^{-n+1}\}} h_I, \tag{1.1.17}$$

denote the n-th Rademacher function. The sequence $\{r_n, n \in \mathbb{N}\}$ is the Rademacher system. It consists of statistically independent functions taking the values $\{+1, -1\}$, each with probability one-half.

Following is the Khintchine inequality for L^1.

Theorem 1.1.3. *(Khintchine's inequality). For every sequence of real numbers $\{a_n\}$ and for every $N \in \mathbb{N}$,*

$$\int_0^1 \left| \sum_{n=1}^{N} a_n r_n(y) \right| dy \geq \frac{1}{\sqrt{2}} \left(\sum_{n=1}^{N} a_n^2 \right)^{1/2} . \tag{1.1.18}$$

The constant $\frac{1}{\sqrt{2}}$, appearing on the right-hand side of Khintchine's inequality is best possible. In this sharp form the inequality (1.1.18) is due to S. Szarek [200].

The Walsh system. While the Khintchine inequality itself involves just the Rademacher functions, the proof relies heavily on certain properties of the Walsh system. We recall the definition of the Walsh system. Let A be a finite set of nonnegative integers. Then the Walsh function w_A is defined as a product of Rademacher functions as follows,

$$w_A = \prod_{k \in A} r_k.$$

Using the empty set as index we denote the constant function as $1_{[0,1]} = w_\emptyset$. The Walsh system $\{w_\emptyset\} \cup \{w_A : A \subset \mathbb{N}, A \,\text{finite}\}$ is complete and orthonormal in $L^2([0,1])$. Hence every $X \in L^2([0,1])$ can be expanded in its Walsh series,

$$X = \sum \langle X, w_A \rangle w_A.$$

The Haar system and the Walsh system are related by the basic identity of J.L. Walsh,

$$\operatorname{span}\{w_A : \max A \leq n\} = \operatorname{span}\{h_I \,:\, |I| \geq 2^{-n+1}\}. \tag{1.1.19}$$

Hence if $X \in L^2([0,1])$ is constant on all dyadic intervals of length $= 2^{-n}$, then $X \in \operatorname{span}\{w_\emptyset\} \cup \{w_A : \max A \leq n\}$, and conversely.

The dyadic derivative. Given a finite linear combination of Walsh functions X, we define the dyadic derivative of X by the equation

$$D(X) = \sum |A| \langle X, w_A \rangle w_A.$$

Proof of Theorem 1.1.3. Fix real numbers $\{a_n\}$, let $N \in \mathbb{N}$ and form

$$X(t) = \left| \sum_{n=1}^{N} a_n r_n(t) \right| .$$

Proving Kintchine's Inequality amounts to showing that the following estimate holds,

$$\int_0^1 X^2(t)dt \le 2\left(\int_0^1 X(t)dt\right)^2.$$

Overview. Below we will see that the function $X(t) = |\sum_{n=1}^N a_n r_n(t)|$ is distinguished by satisfying the following two very peculiar properties:

$$\text{If } |A|\langle X, w_A\rangle \neq 0, \text{ then } |A| \ge 2. \tag{1.1.20}$$

$$\text{For every } t \in [0,1], \text{ we have } X(t) \ge D(X)(t). \tag{1.1.21}$$

Let us accept temporarily that $X(t)$ satisfies the properties (1.1.20) and (1.1.21) and let us show how Kintchine's inequality follows.

$$\begin{aligned}\int_0^1 X^2(t)dt &\ge \int_0^1 X(t)D(X)(t)dt \\ &= \sum |A|\langle X, w_A\rangle^2 \\ &\ge \sum_{|A|\ge 2} 2\langle X, w_A\rangle^2 \\ &= 2\int_0^1 X^2(t)dt - 2\left(\int_0^1 X(t)dt\right)^2.\end{aligned}$$

Now subtract $2\int_0^1 X^2(t)dt$ on both sides and multiply the resulting inequality by -1. This gives

$$\int_0^1 X^2(t)dt \le 2\left(\int_0^1 X(t)dt\right)^2,$$

as claimed.

Verification of (1.1.20). We prove now that $\langle X, w_A\rangle = 0$ for $|A| = 1$. The Rademacher functions are those Walsh functions w_A for which $|A| = 1$. Hence (1.1.20) is just stating that

$$\langle X, r_n\rangle = 0,$$

for every Rademacher function. Note that each Rademacher function is odd around the point 1/2. That is for $0 < t \le 1/2$ we have that

$$r_n(-t + \frac{1}{2}) = -r_n(t + \frac{1}{2}).$$

By taking absolute values we find that

$$X(t) = \left|\sum_{n=1}^N a_n r_n(t)\right|$$

is even around the same point 1/2. The interval $[0,1]$ is clearly symmetric around the point 1/2. Hence,

$$\int_0^1 X(t) r_n(t)dt = 0.$$

Verification of (1.1.21). Next we prove the pointwise estimate $X \geq DX$. Note that X is constant on all intervals of length $= 2^{-N}$. In other words, X depends only on (products and sums of) the Rademacher functions $r_1(t), \ldots, r_N(t)$. Consequently, there exists a function F such that

$$X(t) = F(r_1(t), \ldots, r_N(t)).$$

The next identity for the dyadic derivative is easy to check for individual Walsh functions w_A, hence by linear extension it holds in general. It should be thought of as the dyadic chain rule.

$$\begin{aligned} 2 \cdot D(X) = & F(r_1, \ldots, r_N) - F(-r_1, \ldots, r_N) \\ & + F(r_1, r_2, \ldots, r_N) - F(r_1, -r_2, \ldots, r_N) \\ & \vdots \\ & + F(r_1, \ldots, r_N) - F(r_1, \ldots, -r_N). \end{aligned}$$

Clearly the N summands on the left column of the above sum are identical and equal to X. Now we let R_j denote the operator acting on the j-th coordinate of F by reversing the sign. That is

$$(R_j F)(r_1, \ldots, r_j, \ldots, r_N) = F(r_1, \ldots, -r_j, \ldots, r_N).$$

We may rewrite the dyadic chain rule for X in the following way,

$$\begin{aligned} 2 \cdot D(X) &= N \cdot X - \sum_{j=1}^N (R_j F)(r_1, \ldots, r_N) \\ &= N \cdot X - \sum_{j=1}^N R_j(|a_1 r_1 + \cdots + a_N r_N|). \end{aligned}$$

Now we continue by finding a minorization of the sum on the right-hand side. By the triangle inequality we have

$$\sum_{j=1}^N R_j(|a_1 r_1 + \cdots + a_N r_N|) \geq \left| \sum_{j=1}^N R_j\left((a_1 r_1 + \cdots + a_N r_N)\right) \right|.$$

The sum on the right-hand side is clearly a linear combination of the Rademacher functions $r_1, \ldots, r_N$. Observe that each Rademacher function appears $N-2$ times.

Hence

$$\sum_{j=1}^{N} R_j\left((a_1r_1+\cdots+a_Nr_N)\right)=(N-2)(a_1r_1+\cdots+a_Nr_N).$$

Combining we observe the following lower bound,

$$\sum_{j=1}^{N} R_j(|a_1r_1+\cdots+a_Nr_N|)\geq (N-2)|a_1r_1+\cdots+a_Nr_N|.$$

It remains to insert this estimate in the identity expressed by the dyadic chain rule. This gives that,

$$\begin{aligned}2\cdot D(X)&\geq N\cdot X-(N-2)\cdot|a_1r_1+\cdots+a_Nr_N|\\&=2X.\end{aligned}$$

□

Kahane's inequality. We may now review the proof of Theorem 1.1.3. We showed that for any choice of scalar coefficients (a_n), the function

$$X(t)=\left|\sum a_nr_n(t)\right|$$

satisfies the following surprising estimate,

$$\int_0^1 X^2(t)dt\leq 2\left(\int_0^1 X(t)dt\right)^2.$$

J. P. Kahane extended Khintchine's inequality in the following way. For any normed linear space E and any sequence (x_n) in E the non-negative function

$$Z(t)=\left\|\sum x_nr_n(t)\right\|_E$$

satisfies

$$\int_0^1 Z^2(t)dt\leq C\left(\int_0^1 Z(t)dt\right)^2,$$

where C is a universal constant. The preceding estimate is known as Kahane's inequality. We point out that the proof of Theorem 1.1.3 not only shows Kahane's inequality, but also that the constant in Kahane's inequality is the same as in the real case. Indeed, we have that

$$\int_0^1 Z^2(t)dt\leq 2\left(\int_0^1 Z(t)dt\right)^2.$$

Next we prove Khintchine's inequality for L^p, where $p\geq 2$. The constant we obtain in Khintchine's inequality is of order $\sqrt{p}$. For $p\to\infty$, this gives the right asymptotics.

Theorem 1.1.4. *Let $p \geq 2$. Let $\{a_n\}$ be a sequence of real numbers. Then the following inequality holds,*

$$\left(\int_0^1 |\sum_{n=1}^N a_n r_n(t)|^p dt\right)^{1/p} \leq A\sqrt{p}\left(\sum_{n=1}^N a_n^2\right)^{1/2}.$$

Proof. Let $a \in \mathbb{R}$. Then for each Rademacher function r_n the following inequality holds,

$$\begin{aligned}\int_0^1 \exp\{ar_n\}dt &= \frac{1}{2}(\exp\{a\} + \exp\{-a\}) \\ &\leq \exp\left\{\frac{a^2}{2}\right\}.\end{aligned} \tag{1.1.22}$$

Next fix $\{a_n\}$ and normalize such that $\sum_{n=1}^N a_n^2 = 1$. Then we form

$$F(t) = \sum_{n=1}^N a_n r_n(t).$$

Let $\mu > 0$. As the Rademacher functions $\{r_n\}$ are independent we may iterate inequality (1.1.22) to obtain that

$$\int_0^1 \exp\{\mu F\} \leq \prod_{n=1}^N \exp\{\frac{\mu^2 a_n^2}{2}\} = \exp\{\frac{\mu^2}{2}\}.$$

Next observe that $\exp\{\mu|F(t)|\} \leq \exp\{\mu F(t)\} + \exp\{-\mu F(t)\}$. Hence,

$$\int_0^1 \exp\{\mu|F|\} \leq 2\exp\{\frac{\mu^2}{2}\}.$$

By the inequality of Chebyshef, the following estimate holds for the distribution function of $|F|$. For every $\alpha > 0$,

$$\begin{aligned}|\{\mu|F| > \alpha\}| &\leq \frac{\int_0^1 \exp\{\mu|F(t)|\}dt}{\exp\alpha} \\ &\leq 2\exp\{\frac{\mu^2}{2} - \alpha\}.\end{aligned} \tag{1.1.23}$$

We specialize the estimate (1.1.23) by taking $\alpha = \mu^2$, then

$$|\{|F| > \mu\}| \leq 2\exp\{-\frac{\mu^2}{2}\}. \tag{1.1.24}$$

Integrating inequality (1.1.24) gives an upper bound for $||F||_p^p$. After a change of variables we find an estimate for $||F||_p^p$ in terms of the Gamma function, for which standard inequalities exist. Indeed,

$$\begin{aligned}\int_0^1 |F(t)|^p dt &= p\int_0^\infty s^{p-1}|\{|F|>s\}|ds \\ &\le 2p\int_0^\infty s^{p-1}\exp\left\{-\frac{s^2}{2}\right\}ds \\ &= 2p2^{\frac{p-2}{2}}\int_0^\infty r^{\frac{p-2}{2}}\exp\{-r\}dr \\ &= 2p2^{\frac{p-2}{2}}\Gamma\left(\frac{p-2}{2}\right).\end{aligned}$$

At this point we use a simple estimate for the Gamma function. Namely,

$$\Gamma\left(\frac{p-2}{2}\right) \le (Ap)^{\frac{p}{2}}.$$

□

Remark. With a surprising trick we obtain from Theorem 1.1.4 the L^1 version of Khintchine's inequality (albeit with a constant that is larger that $\sqrt{2}$). Define

$$X = \left|\sum_{n=1}^N a_n r_n\right|.$$

Hölder's inequality implies that $\|X\|_2 \le \|X\|_1^{1/3}\|X\|_4^{2/3}$. Apply Theorem 1.1.4 with $p=4$. This gives $\|X\|_4 \le 2A\|X\|_2$. Cancelling $\|X\|_2^{2/3}$ we get $\|X\|_2^{1/3} \le (2A)^{2/3}\|X\|_1^{1/3}$ hence the L^1 version of Khintchine's inequality with constant $4A^2$.

Basic sequences. In a Banach space E, a sequence $\{y_i\}_{i=1}^\infty$ is called a basic sequence if it is a Schauder basis in its closed linear span. Two basic sequences $\{y_i\}_{i=1}^\infty$ in E and $\{z_i\}_{i=1}^\infty$ in F are called equivalent if there exist $c>0$ and $C>0$ so that

$$c\|\sum_{i=1}^n a_i z_i\|_F \le \|\sum_{i-1}^n a_i y_i\|_E \le C\|\sum_{i=1}^n a_i z_i\|_F,$$

for every $n \in \mathbb{N}$ and every choice of scalars $a_1 \dots a_n$.

Let $1 \le p < \infty$. Khintchine's inequality asserts that the Rademacher functions in L^p are equivalent to the unit vector basis in the Hilbert space ℓ^2 (in particular the closed linear span of the Rademacher functions is isomorphic to ℓ^2). If moreover $1<p<\infty$, then

$$R_1(f) = \sum_{n=1}^\infty \langle f, r_n\rangle r_n,$$

the orthogonal projection onto span$\{r_n\}_{n=1}^{\infty}$, is a bounded operator on L^p. Indeed, considering first $p > 2$ and $f \in L^p$ we obtain from Theorem 1.1.4 that

$$\left\| \sum_{n=1}^{\infty} \langle f, r_n \rangle r_n \right\|_p \leq A\sqrt{p} \left(\sum_{n=1}^{\infty} \langle f, r_n \rangle^2 \right)^{1/2}. \tag{1.1.25}$$

Consequently,

$$\|R_1(f)\|_p \leq A\sqrt{p}\|f\|_2 \leq A\sqrt{p}\|f\|_p.$$

Next choose $1 < q < 2$ and let p be the Hölder conjugate exponent defined by $1/p + 1/q = 1$. Then $p > 2$. Let $g \in L^q$ be a finite linear combination of Haar functions. Then $R_1(g)$ is a finite linear combination of Rademacher functions. Determine $f \in L^p$ so that $\|f\|_p = 1$ and $\|R_1(g)\|_q = \langle R_1(g), f \rangle$. Using that R_1 is self adjoint, we obtain from (1.1.25) that

$$\begin{aligned} \langle R_1(g), f \rangle &= \langle g, R_1(f) \rangle \\ &\leq \|g\|_q \|R_1(f)\|_p \\ &\leq A\sqrt{p}\|g\|_q. \end{aligned} \tag{1.1.26}$$

Hence $\|R_1(g)\|_q \leq A\sqrt{p}\|g\|_q$, as claimed.

Burkholder's inequality

We give a proof of Burkholder's inequality, and obtain the square function characterization of L^p by merging the inequalities of Burkholder and Khintchine.

Theorem 1.1.5. *Let $1 < p < \infty$. Let $f \in L^p([0,1])$ with*

$$f = \int_0^1 f dt + \sum_{I \in \mathcal{D}} a_I h_I.$$

Let $\epsilon_I \in \{-1, 1\}$, then

$$g = \pm \int_0^1 f dt + \sum_{I \in \mathcal{D}} \epsilon_I a_I h_I,$$

satisfies the estimate

$$||f||_p \leq (p^* - 1)||g||_p,$$

where $p^* = \max\{p, p/(p-1)\}$.

Proof. By a simple duality argument it suffices to prove the case when $p > 2$. Then $p^* = p$. We assume that $f = \int f + \sum a_I h_I$ is a finite linear combination of Haar functions. We will show that

$$||f||_p \leq (p - 1)||g||_p.$$

With the function

$$v(x,y) = |x|^p - (p-1)^p|y|^p,$$

we have that

$$||f||_p^p - (p-1)^p||g||_p^p = \int_0^1 v(f(t),g(t))dt.$$

Hence we have to show that

$$\int_0^1 v(f(t),g(t))dt \leq 0. \tag{1.1.27}$$

To this end we define Burkholder's most famous function,

$$u(x,y) = p(1-\frac{1}{p})^{p-1}(|x|+|y|)^{p-1}(|y|-(p-1)|x|). \tag{1.1.28}$$

This function became famous because of the following four properties:

(B 1) u dominates v,

$$v(x,y) \leq u(x,y).$$

(B 2) For every $a \in \mathbb{R}$ and $\epsilon \in \{-1,1\}$,

$$u(x+a,y+\epsilon a) + u(x-a,y-\epsilon a) \leq 2u(x,y).$$

(B 3) u is symmetric,

$$u(x,y) = u(-x,-y).$$

(B 4) $u(0,0) = 0$.

We denote the mean value of f over the dyadic interval J by f_J. That is,

$$f_J = \frac{1}{|J|}\int_J f.$$

Correspondingly we denote the mean value of g over J by g_J. Let J_1 be the left half of J and let J_2 be the right half of J. We claim that for Burkholder's function u the following inequality holds:

$$|J_1|u(f_{J_1},g_{J_1}) + |J_2|u(f_{J_2},g_{J_2}) \leq |J|u(f_J,g_J). \tag{1.1.29}$$

Note that for $t \in J_1$ we have that $h_J(t) = 1$ whereas for $t \in J_2$, $h_J(t) = -1$. Hence by (1.1.3) for f the following identities hold,

$$f_{J_1} = f_J + a_J,$$

$$f_{J_2} = f_J - a_J,$$

and analogously we have for g,

$$g_{J_1} = g_J + \epsilon_J a_J,$$

$$g_{J_2} = g_J - \epsilon_J a_J.$$

Inserting the above identities and using property (B 2) of Burkholder's function we have that

$$\begin{aligned} u(f_{J_1}, g_{J_1}) + u(f_{J_2}, g_{J_2}) &= u(f_J + a_J, g_J + \epsilon_J a_J) + u(f_J - a_J, f_J - \epsilon_J a_J) \\ &\leq 2u(f_J, g_J). \end{aligned}$$

It remains to multiply both sides with $|J|/2$ to see that the claimed inequality (1.1.29) holds true. Next fix $k \in \mathbb{N}$ and sum the inequality (1.1.29) over those dyadic intervals for which the length is equal to 2^{-k}. Thus we obtain from inequality (1.1.29) this crucial rescaling estimate,

$$\sum_{\{K:|K|=2^{-k}\}} |K| u(f_K, g_K) \leq \sum_{\{J:|J|=2^{-k+1}\}} |J| u(f_J, g_J). \tag{1.1.30}$$

Note that the right-hand side of the estimate (1.1.30) is a rescaled version of the left-hand side; the difference is that the length of the dyadic intervals we are summing over has increased by a factor of 2. The assertion of (1.1.30) is that by rescaling from intervals of length 2^{-k} to larger intervals of length 2^{-k+1} we are increasing the value of the sum.

Now we turn to verifying (1.1.27). As f and g are finite linear combinations there exists a (large) constant k_0 so that f and g are constant on dyadic intervals of length $\leq 2^{-k_0}$. Hence,

$$\int_0^1 v(f(t), g(t))dt = \sum_{\{K:|K|=2^{-k_0}\}} v(f_K, g_K)|K|.$$

Recall that Burkholder's function u is a pointwise majorant of v, hence,

$$v(f_K, g_K) \leq u(f_K, g_K),$$

and we obtain the following line from which we start an inductive procedure,

$$||f||_p^p - (p-1)^p ||g||_p^p \leq \sum_{\{K:|K|=2^{-k_0}\}} u(f_K, g_K)|K|. \tag{1.1.31}$$

Now we exploit the rescaling estimates (1.1.30). We start with the right-hand side of the inequality (1.1.31) where the summation is taken over 2^{k_0} intervals of length $= 2^{-k_0}$. Call the resulting sum $\sigma(k_0)$. Repeatedly applying the rescaling estimate (1.1.30) shows that the following chain of inequalities holds true,

$$\sigma(k_0) \leq \sigma(k_0 - 1) \leq \cdots \leq \sigma(1) \leq \sigma(0).$$

To form $\sigma(0)$ we take the sum over one interval of length $=1$. Thus $\sigma(0) = u(f_{[0,1]}, g_{[0,1]})$. Finally we recall that the Burkholder function u is symmetric, by (B 3), and that $u(0,0) = 0$. This gives that

$$\begin{aligned}||f||_p^p - (p-1)^p||g||_p^p &\leq \sum_{\{K:|K|=2^{-k_0}\}} u(f_K, g_K)|K| \\ &\leq u(f_{[0,1]}, g_{[0,1]}) \\ &= \frac{1}{2}\left(u(f_{[0,1]}, g_{[0,1]}) + u(-f_{[0,1]}, -g_{[0,1]})\right) \\ &\leq u(0,0) = 0.\end{aligned}$$

□

Given $f \in L^p([0,1])$ we define the square function of f as follows,

$$S(f) = \left(\sum a_I^2 h_I^2\right)^{1/2}, \quad \text{where} \quad a_I = \langle f, \frac{h_I}{|I|}\rangle.$$

Khintchine's inequality together with Burkholder's theorem imply a square function characterization of L^p, when $1 < p < \infty$. The original proof of Theorem 1.1.6, with different constants is due to R. E. A. C. Paley.

Theorem 1.1.6. *Let $1 < p < \infty$. Let $f \in L^p([0,1])$ such that $\int f = 0$. Then*

$$(2\sqrt{2})^{-1}(p^*-1)^{-1}||f||_p \leq ||S(f)||_p \leq \sqrt{2}(p^*-1)||f||_p,$$

where $p^* = \max\{p, p/(p-1)\}$ denotes the constants appearing in Burkholder's inequality.

Proof. Let $f \in L^p([0,1])$, and expand it in its Haar series,

$$f = \sum a_I h_I.$$

Let $\{r_I\}$ be an enumeration of the Rademacher system indexed by dyadic intervals. For $y \in [0,1]$, we define

$$f_y = \sum a_I h_I r_I(y).$$

Then by Burkholder's inequality, Theorem 1.1.5, for each $y \in [0,1]$, the following inequality holds,

$$\int_0^1 |f_y(t)|^p dt \leq (p^*-1)^p \int_0^1 |f(t)|^p dt.$$

Now we pass from the above pointwise estimate, holding for every $y \in [0,1]$, to the average over $y \in [0,1]$. Clearly the average satisfies the same upper bound, that is,

$$\int_0^1 \int_0^1 |f_y(t)|^p dt \leq (p^*-1)^p \int_0^1 |f(t)|^p dt.$$

Next we fix $t \in [0,1]$ and we apply Minkowski's inequality followed by Khintchine's inequality, Theorem 1.1.3,

$$\int_0^1 |f_y(t)|^p dy \geq \left(\int_0^1 |f_y(t)| dy\right)^p \geq \left(\frac{1}{\sqrt{2}} S(f)(t)\right)^p .$$

It remains to use Fubini's theorem and string together the above estimates.

$$\begin{aligned}\left(\frac{1}{\sqrt{2}}\right)^p \int_0^1 S^p(f)(t)dt &\leq \int_0^1 \int_0^1 |f_y(t)|^p dydt \\ &= \int_0^1 \int_0^1 |f_y(t)|^p dtdy \qquad (1.1.32) \\ &\leq (p^* - 1)^p \int_0^1 |f(t)|^p dt.\end{aligned}$$

This proves the right-hand side inequality in Theorem 1.1.6.

We now turn to the left-hand side. Here we distinguish between the cases $p \geq 2$ and $p \leq 2$. We start with the latter case, $p \leq 2$. We apply Burkholder's inequality in the following way. For each $y \in [0,1]$,

$$\int_0^1 |f(t)|^p dt \leq (p^* - 1)^p \int_0^1 |f_y(t)|^p dt. \qquad (1.1.33)$$

Consequently the same estimate holds true if we take the average over $y \in [0,1]$ of the right hand side in (1.1.33). Thus

$$\int_0^1 |f(t)|^p dt \leq (p^* - 1)^p \int_0^1 \int_0^1 |f_y(t)|^p dtdy. \qquad (1.1.34)$$

Next observe that for $p \leq 2$, Minkowski's inequality gives the estimate

$$\int_0^1 |f_y(t)|^p dy \leq \left(\int_0^1 |f_y(t)|^2 dy\right)^{p/2} . \qquad (1.1.35)$$

Finally we combine the estimates (1.1.34) and (1.1.35) and apply Fubini's theorem as follows,

$$\begin{aligned}\int_0^1 |f(t)|^p dt &\leq (p^* - 1)^p \int_0^1 \int_0^1 |f_y(t)|^p dtdy \\ &\leq (p^* - 1)^p \int_0^1 \left(\int_0^1 |f_y(t)|^2 dy\right)^{p/2} dt \\ &= (p^* - 1)^p \int_0^1 S^p(f)(t)dt.\end{aligned}$$

Summing up we showed that for $1 < p \leq 2$ the left-hand side estimate in Theorem 1.1.6 holds, that is,

$$(p^* - 1)^{-1}||f||_p \leq ||S(f)||_p.$$

Next we turn to the case when $p \geq 2$. Here we use duality. Let $q \leq 2$ be the Hölder conjugate index to p, thus $1/p + 1/q = 1$. Let $h \in L^q$ with $||h||_q \leq 2$, and $\int h = 0$ so that

$$||f||_p = \langle f, h \rangle.$$

By the inequalities of Cauchy–Schwarz and Hölder we obtain that

$$\begin{aligned} \langle f, h \rangle &\leq \int_0^1 S(f)(t) S(h)(t) dt \\ &\leq ||S(f)||_p ||S(h)||_q. \end{aligned}$$

We established above (1.1.32). Thereby we proved the right-hand side estimate of Theorem 1.1.6. Now we apply (1.1.32) obtaining $||S(h)||_q \leq \sqrt{2}(q^* - 1)||h||_q$. As $||h||_q \leq 2$, this gives

$$||f||_p \leq 2\sqrt{2}(q^* - 1)||S(f)||_p.$$

Finally we observe that for $p \geq 2$, we have $q \leq 2$, and also $q^* - 1 = p - 1 = p^* - 1$. Summing up for $p \geq 2$,

$$||f||_p \leq 2\sqrt{2}(p^* - 1)||S(f)||_p. \qquad \square$$

Unconditional bases. In a Banach space E let $\{x_i\}_{i=1}^{\infty}$ be a Schauder basis with biorthogonal functionals $\{x_i^*\}_{i=1}^{\infty}$. We say that $\{x_i\}_{i=1}^{\infty}$ is an unconditional basis for E if the following condition holds: There exists $K > 0$ so that for each $x \in E$,

$$\sup_n \sup_{\epsilon_i \in \{-1,1\}} \left\| \sum_{i=1}^n \epsilon_i x_i^*(x) x_i \right\|_E \leq K\|x\|_E. \tag{1.1.36}$$

The infimum over all constants $K > 0$ satisfying (1.1.36) is the unconditional basis constant of $\{x_i\}_{i=1}^{\infty}$ in E. By Schauder's Theorem 1.1.2 the Haar system is a basis in $L^p([0,1]), 1 \leq p < \infty$. Burkholder's inequality asserts that for $1 < p < \infty$ the Haar system is actually an unconditional basis and that the unconditional basis constant of the Haar system in L^p is bounded by $\max\{p, p/(p-1)\} - 1$. The constant $\max\{p, p/(p-1)\} - 1$ is known to be best possible (see for instance Chapter 8 in [165] by I. Novikov and E. M. Semenov).

The UMD property of Banach spaces. Before examining further consequences of Burkholder's Theorem 1.1.5 we will devote a paragraph to the definition of the Banach spaces for which the vector-valued extension of Burkholder's theorem holds true. A Banach space satisfies the UMD property if for $1 < p < \infty$ there exists a constant $C_p > 0$ so that the following estimate holds,

$$\int_0^1 \left\| \sum_{I \in \mathcal{H}} \epsilon_I a_I h_I(t) \right\|_E^p dt \leq C_p^p \int_0^1 \left\| \sum_{I \in \mathcal{H}} a_I h_I(t) \right\|_E^p dt, \tag{1.1.37}$$

for every finite collection $\mathcal{H}$ of dyadic intervals, every $\{a_I \in E : I \in \mathcal{H}\}$ and every choice of signs $\{\epsilon_I \in \{+1, -1\} : I \in \mathcal{H}\}$. Basic properties of UMD spaces are recorded in B. Maurey's seminal paper on the Haar system [139]. These include:

(a) If E has the UMD property, then every closed subspace of E is reflexive.

(b) The observation of G. Pisier that if (1.1.37) holds for some p_0 $(1 < p_0 < \infty)$ then (1.1.37) holds for all $1 < p < \infty$. In particular to verify the UMD property for E it suffices to establish (1.1.37) for L^2.

B. Maurey's work on the Haar system was the starting point for intensive investigations of Banach spaces with the UMD property. We pay close attention to the methods developed for the study of UMD spaces and present in detail the following:

(a) B. Maurey's proof that martingale differences are unconditional in the reflexive L^p spaces (Theorem 1.4.1).

(b) J. Bourgain's proof of E. M. Stein's martingale inequality (Theorem 1.4.2).

(c) T. Figiel's representation of singular integral operators and T. Figiel's expansion of biorthogonal systems (Theorem 1.5.3 and Theorem 1.4.3).

(d) The proof by A. Naor and G. Schechtman of G. Pisier's inequality on dyadic gradients (Theorem 1.1.7).

The Walsh system in L^p $(1 < p < \infty)$

Here we use the Haar basis to study the Walsh system. In particular we establish estimates for Walsh series by applying the unconditionality of the Haar basis in L^p. The best known application is Paley's proof that the Walsh system (in Walsh–Paley order) is a Schauder basis in L^p. A recent connection between the Haar basis and the Walsh system is established in the work of A. Naor and G. Schechtman on L^p estimates for dyadic gradients.

Besides expansion in Walsh–Paley order we test a different and natural method of approximation using Walsh functions. The method is briefly described as follows: First project a given function f onto the span of Walsh functions w_A with $|A| = m$. Denote the result by $R_m(f)$. To recover f use the associated partial sum operators

$$\sum_{m=0}^{n} R_m(f),$$

and let n tend to infinity. Clearly this method works well for the Hilbert space L^2. Below however we will present the surprising theorem of B. Roider to the effect that for any $p \neq 2$ there is $f \in L^p$ which is *not* approximated by the above partial sums. The proof presented here is due to K. Kiener, who builds on core ideas of P. Enflo's construction of a separable Banach space without the approximation property.

The following link between the Walsh system and the Haar system was observed by J. L. Walsh in 1923. Let $g \in L^p([0,1])$ with $\int_0^1 g dt = 0$. Let $k \in \mathbb{N}$, and put

$$d_k = \sum_{\{A \subset \mathbb{N} : \max A = k\}} \langle g, w_A \rangle w_A.$$

By definition, the sequence $\{d_k\}$ is disjointly supported over the Walsh system. Simultaneously, and this is the point of Walsh's observation, $\{d_k\}$ is also disjointly supported over the *Haar system.* Indeed, the following identity holds,

$$d_k = \sum_{\{I : |I| = 2^{-k+1}\}} \langle g, \frac{h_I}{|I|} \rangle h_I.$$

Consequently $g = \sum_{k=1}^{\infty} d_k$, with convergence in L^p. Moreover with Burkholder's theorem the observation of Walsh implies that

$$\left\| \sum_{k=1}^{\infty} \pm d_k \right\|_{L^p} \geq (p^* - 1)^{-1} \|g\|_{L^p}. \tag{1.1.38}$$

The Poincaré inequality for the dyadic gradient

We apply the unconditionality of the Haar basis in L^p to prove a dyadic version of Poincaré's inequality. The dyadic gradient is defined as follows. Let $N \in \mathbb{N}$, and fix f so that $f - \int_0^1 f dt \in \operatorname{span}\{w_A : \max A \leq N\}$. Hence the Walsh expansion of f is of the form

$$f = \int_0^1 f dt + \sum_{A \subseteq \{1,\dots,N\}} c_A w_A, \quad \text{where } c_A \in \mathbb{R}. \tag{1.1.39}$$

Thus f is expressed through the Rademacher functions $\{r_1, \dots, r_N\}$ and can be represented as

$$f(t) = F(r_1(t), \dots, r_N(t)).$$

Then the dyadic partial derivative ∂_i is defined by the equation

$$\partial_i f(t) = \frac{1}{2} \left[F(r_1(t), \dots, r_i(t), \dots, r_N(t)) - F(r_1(t), \dots, -r_i(t), \dots, r_N(t)) \right].$$

In the course of proving Khintchine's inequality we observed that the dyadic derivative $D(f)$, defined through the Walsh expansion of f by

$$Df = \sum |A| \langle w_A, f \rangle w_A,$$

satisfies

$$Df = \sum_{i=1}^{N} \partial_i f.$$

This identity holds true since ∂_i acts on individual Walsh functions by the relation

$$\partial_i w_A = \begin{cases} w_A & \text{if } i \in A, \\ 0 & \text{if } i \notin A. \end{cases} \tag{1.1.40}$$

Finally define the dyadic gradient $\nabla f = (\partial_i f)_{i=1}^N$ and put

$$|\nabla f| = \left(\sum_{i=1}^{N} (\partial_i f)^2 \right)^{1/2}.$$

Theorem 1.1.7. *Let* $1 < p < \infty$. *Let* $f \in L^p$ *have Walsh expansion* (1.1.39). *Then*

$$\|f - \int_0^1 f dt\|_{L^p} \le C_p \|\nabla f\|_{L^p}.$$

Proof. We start by isolating an algebraic identity relating ∂_i to the Walsh expansion of f. Let $a_i \in \mathbb{R}$, and let $i \le N$. Put

$$g = \sum_{i=1}^{N} a_i \partial_i f.$$

Equation (1.1.40), the defining relation for ∂_i, gives the following useful representation of g. For $t \in [0,1]$,

$$g(t) = \sum_{A \subseteq \{1,\dots,N\}} \left(\sum_{i \in A} a_i \right) \langle f, w_A \rangle w_A(t). \tag{1.1.41}$$

Next we split the Walsh expansion of g using the basic observation of Walsh. For $k \le N$, put

$$d_k(t) = \sum_{\{A : \max A = k\}} \left(\sum_{i \in A} a_i \right) \langle f, w_A \rangle w_A(t). \tag{1.1.42}$$

By the identity of Walsh, the sequence $\{d_k\}$ is disjointly supported over the Haar system. Hence Burkholder's Theorem 1.1.5 gives the lower bound

$$\left\| \sum_{k=1}^{N} \pm d_k \right\|_{L^p} \ge (p^* - 1)^{-1} \|g\|_{L^p}. \tag{1.1.43}$$

Now we focus on finding estimates for $\|\nabla f\|_{L^p}$. Recall that $\{r_i\}$ denotes the Rademacher system. Fix $s \in [0,1]$ and $i \le n$. We will apply the above observations using $a_i = r_i(s)$. Displaying the dependence on $s \in [0,1]$ we write

$$d_k(s,t) = \sum_{\{A:\max A = k\}} \left(\sum_{i \in A} r_i(s) \right) \langle f, w_A \rangle w_A(t).$$

Note that the integral identity below holds, since the integrands on the right-hand side and on the left-hand side coincide.

$$\int_0^1 \int_0^1 \left| \sum_{i=1}^N r_i(s) \partial_i f(t) \right|^p dsdt = \int_0^1 \int_0^1 \left| \sum_{k=1}^N d_k(s,t) \right|^p dsdt. \tag{1.1.44}$$

Khintchine's inequality gives pointwise estimates for the length of the dyadic gradient. For $t \in [0,1]$, we have that

$$|\nabla f(t)|^p \ge \frac{1}{(Ap)^{p/2}} \int_0^1 \left| \sum_{i=1}^N r_i(s) \partial_i f(t) \right|^p ds.$$

Integrate over $t \in [0,1]$ and invoke the identity (1.1.44). This gives

$$\|\nabla f\|_{L^p}^p \ge \frac{1}{(Ap)^{p/2}} \int_0^1 \int_0^1 \left| \sum_{k=1}^N d_k(s,t) \right|^p dsdt. \tag{1.1.45}$$

Next we apply Burkholder's inequality in the form (1.1.43). Clearly $r_k(s) = \pm 1$ for each fixed $s \in [0,1]$. Hence the following lower bound holds,

$$\left\| \sum_{k=1}^N d_k(s,\cdot) \right\|_{L^p} \ge (p^* - 1)^{-1} \left\| \sum_{k=1}^N r_k(s) d_k(s,\cdot) \right\|_{L^p}. \tag{1.1.46}$$

Integrate the estimate (1.1.46) over $s \in [0,1]$. By triangle inequality, the integral of the norms $\ge$ norm of the integral. Thus we arrive at the following lower bound for $\|\nabla f\|_{L^p}$.

$$C_p \|\nabla f\|_{L^p} \ge \left\| \int_0^1 \sum_{k=1}^N r_k(s) d_k(s,\cdot) ds \right\|_{L^p}, \tag{1.1.47}$$

where $C_p = (Ap)^{1/2}(p^* - 1)$. It remains to evaluate the expression appearing on the right-hand side of (1.1.47) . A glance at the definition of $d_k(s,t)$ gives that

$$\int_0^1 r_k(s) d_k(s,t) ds = \sum_{\{A:\max A = k\}} \langle f, w_A \rangle w_A(t).$$

Hence for $C_p = (Ap)^{1/2}(p^* - 1)$, we found the lower bound

$$C_p \|\nabla f\|_{L^p} \geq \left\| \sum_{k=1}^{N} \sum_{\{A:\max A = k\}} \langle f, w_A \rangle w_A(t) \right\|_{L^p} .$$

Finally observe that the right-hand side of the above estimate coincides with the norm in L^p of $f - \int_0^1 f$. □

Pisier's inequality. Estimates for the dyadic gradient were established first by G. Pisier [178] in the context of vector-valued coefficients $c_A \in E$, where E is a Banach space. Following is the statement of G. Pisier's original inequality. Let

$$f = \int_0^1 f dt + \sum_{A \subseteq \{1,\dots,N\}} c_A w_A, \quad \text{where } c_A \in E. \tag{1.1.48}$$

Equation (1.1.40) defines the dyadic partial derivatives $\{\partial_i f\}$ by linear extension, and $\{r_i\}$ denotes the Rademacher functions, then

$$\int_0^1 \|f(x) - \int_0^1 f\|_E^p dx \leq (2e \log N)^p \int_0^1 \| \sum_{i=1}^{N} r_i(t)\, \partial_i f(x) \|_E^p dt dx. \tag{1.1.49}$$

In the vector-valued case the constant on the right-hand side depends on N. M. Talagrand [201] shows that $2e \log N$, appearing in (1.1.49) is of the correct order. Moreover in [201] M. Talagrand proves the scalar case with a constant independent of N. The proof of Theorem 1.1.7 presented above is due to A. Naor and G. Schechtman [162]. Verify, by inspection, that their proof carries over to the case where the coefficients in the Walsh expansion of f belong to a Banach space with the UMD property.

Convergence and divergence of Walsh series

So far we were concerned with estimates for finite linear combinations of Walsh functions, hence questions of convergence did not arise. Passing from finite sums to infinite series gives rise to interesting and subtle problems of ordering the Walsh system. For the Walsh–Paley enumeration $\{w_j\}_{j=0}^{\infty}$ we show that for $1 < p < \infty$,

$$\sup_{\|f\|_p \leq 1} \sup_n \left\| \sum_{j=0}^{n} \langle f, w_j \rangle w_j \right\|_p \leq p^* - 1, \tag{1.1.50}$$

where $p^* = \max\{p, p/(p-1)\}$. Clearly this estimate implies convergence of the corresponding partial sums in L^p. On the other hand, ordering the Walsh system along increasing multiplicity leads to divergence of the corresponding partial sums in L^p for $p \neq 2$. We prove below that for $1 < p \neq 2 < \infty$,

$$\sup_{\|f\|_p \leq 1} \sup_n \left\| \sum_{m=0}^{n} R_m(f) \right\|_p = \infty, \tag{1.1.51}$$

where R_m denotes the orthogonal projection onto Walsh functions of multiplicity m. The estimate (1.1.50) is due to R.E.A.C Paley, the particular constant $p^* - 1$ is a consequence of Burkholder's Theorem 1.1.5. The lower bound (1.1.51) is a result of B. Roider.

The Walsh–Paley order. We now define the Walsh–Paley order in $\{w_A\}$, and show Paley's theorem that (thus ordered) the Walsh system is a Schauder basis in $L^p, 1 < p < \infty$. Let $n \in \mathbb{N}$, with dyadic expansion

$$n = \sum_{j=0}^{\infty} \epsilon_j 2^j,$$

where $\epsilon_j \in \{0, 1\}$. (Clearly only finitely many ϵ_j are $\neq 0$.) Then we define

$$w_n = \prod_{j=0}^{\infty} r_{j+1}^{\epsilon_j}. \tag{1.1.52}$$

We also define $w_0 = 1$. Note that $w_1 = r_1$, $w_2 = r_2$ and that in general $w_{2^m} = r_{m+1}$. Moreover the sequence $\{w_n\}_{n=1}^{\infty}$ is an enumeration of the Walsh system $\{w_A : A \subset \mathbb{N},\ A \text{ finite }\}$.

Paley's identity. We prepare the proof of Paley's theorem by stating identities relating the Walsh system to the Haar system. First we have that

$$\begin{aligned} \sum_{j=0}^{2^k-1} w_j &= \prod_{n=1}^{k} (1 + r_n) \\ &= 2^k 1_{[0, 2^{-k}[}. \end{aligned} \tag{1.1.53}$$

Let I be a dyadic interval with $|I| = 2^{-k}$. The identity (1.1.53) leads to a formula for the Walsh expansion of the Haar function h_I. Moreover for $t \in I$ and $x \in [0, 1]$,

$$\sum_{j=2^k}^{2^{k+1}-1} w_j(t) w_j(x) = 2^k h_I(t) h_I(x). \tag{1.1.54}$$

The above formulas were known to (and used by) Walsh in 1923. Following is Paley's identity from 1931 which is the key to his proof that the Walsh system can be ordered to become a Schauder basis in $L^p, 1 < p < \infty$: Let $x, t \in [0,1]$, and let $n = \sum_{k=0}^{\infty} \epsilon_k 2^k$. Then

$$w_n(x)w_n(t)\sum_{j=0}^{n-1} w_j(x)w_j(t) = \sum_{k=0}^{\infty} \epsilon_k \sum_{\{I:|I|=2^{-k}\}} 2^k h_I(x)h_I(t). \tag{1.1.55}$$

Next we show that the partial sum operators with respect to the Walsh system in Walsh–Paley order are convergent in L^p $(1 < p < \infty)$. By the identity of Walsh it suffices to prove that they form a sequence of uniformly bounded operators.

Theorem 1.1.8. *Let $1 < p < \infty$, and let $\{w_j\}_{j=0}^{\infty}$ denote the Walsh system in Walsh–Paley order. Then the partial sum operator,*

$$T_n(f) = \sum_{j=0}^{n-1} \langle f, w_j \rangle w_j,$$

satisfies the norm estimate

$$\|T_n f\|_p \le (p^* - 1)\|f\|_p, \tag{1.1.56}$$

where $p^* = \max\{p, p/(p-1)\}$.

Proof. Fix $n \in \mathbb{N}$. Instead of working directly with T_n we consider the operator

$$\tilde{T_n} : f \mapsto w_n T_n(w_n f).$$

Clearly in L_p the norms of the operators T_n and $\tilde{T_n}$ coincide. Hence it suffices to show the estimate (1.1.56) with T_n replaced by $\tilde{T_n}$. The following observation is crucial to Paley's proof. It identifies $\tilde{T_n}$ as a $\{0,1\}$ multiplier on the Haar(!) system. Precisely, we claim that for a given n there exist coefficients $\delta_I \in \{0,1\}$ so that

$$\tilde{T_n}(f) = \sum_{I \in \mathcal{D}} \delta_I \langle f, \frac{h_I}{|I|} \rangle h_I. \tag{1.1.57}$$

To verify (1.1.57) we observe that

$$\begin{aligned}\tilde{T_n}(f)(x) &= \sum_{j=0}^{n-1} \langle w_n f, w_j \rangle w_j(x) w_n(x) \\ &= \int_0^1 f(t) w_n(x) w_n(t) \sum_{j=0}^{n-1} w_j(x) w_j(t) dt.\end{aligned} \tag{1.1.58}$$

The kernel appearing in the second line of (1.1.58) has been identified in (1.1.55). It is equal to

$$\sum_{j=0}^{\infty} \epsilon_j \sum_{\{I:|I|=2^{-j}\}} 2^j h_I(t) h_I(x),$$

where $\epsilon_j \in \{0,1\}$ is determined by the dyadic expansion of n. Replacing the kernels and carrying out the integration gives that

$$\tilde{T_n}(f) = \sum_{j=0}^{\infty} \epsilon_j \sum_{\{I:|I|=2^{-j}\}} \langle f, \frac{h_I}{|I|} \rangle h_I.$$

Having established (1.1.57) it suffices to apply Theorem 1.1.5 to conclude that

$$\|\tilde{T_n}(f)\|_p \le (p^* - 1)\|f\|_p. \qquad \square$$

The multiplicity of Walsh functions. The multiplicity of a given Walsh function is defined to be the number of Rademacher functions used in its definition. Formally, let A be a finite subset of $\mathbb{N}$. Then the multiplicity of $w_A = \prod_{j\in A} r_j$, is the cardinality of A. For $w_\emptyset = 1_{[0,1]}$ we define the multiplicity to be equal to 0.

Let W_m denote the Walsh functions of multiplicity m. Let $\{w_{m,k}\}_{k=1}^\infty$ be any enumeration of the elements in W_m. Let R_m denote the orthogonal projection onto the span of W_m. Thus for $f \in L^2$,

$$R_m(f) = \sum_{k=1}^{\infty} \langle w_{m,k}, f\rangle w_{m,k}. \tag{1.1.59}$$

For $m \neq m'$ the sets W_m and $W_{m'}$ are disjoint, hence the spaces $\operatorname{span}\{W_m\}$ are orthogonal and

$$\|f\|_2^2 = \sum_{m=0}^{\infty} \|R_m(f)\|_2^2.$$

It follows that the partial sums

$$\sum_{m=0}^{n} \sum_{k=1}^{\infty} \langle w_{m,k}, f\rangle w_{m,k} \tag{1.1.60}$$

form a convergent sequence in L^2 approximating f. Hence elements in L^2 are well approximated by adding Walsh functions of increasing multiplicity. We now study this method of approximation in L^p, when $p \neq 2$. The starting point is a theorem obtained by A. Bonami [15] and K. Kiener [112] independently. It extends Khintchine's inequality to Walsh functions of multiplicity m.

Theorem 1.1.9. *Let $1 < p < \infty$ and let $m \in \mathbb{N}$. There exist constants $c_{m,p} > 0$ and $C_{m,p} > 0$ so that for every sequence $\{a_k\}_{k=1}^\infty \in \ell^2$,*

$$c_{m,p}\left(\sum_{k=1}^{\infty} a_k^2\right)^{1/2} \le \left\|\sum_{k=1}^{\infty} a_k w_{m,k}\right\|_p \le C_{m,p}\left(\sum_{k=1}^{\infty} a_k^2\right)^{1/2}. \tag{1.1.61}$$

(a) *If $p > 2$, then $c_{m,p} = 1$ and $C_{m,p} \le (C\sqrt{p})^m$.*

(b) *If* $p < 2$, *then* $C_{m,p} = 1$ *and* $c_{m,p} \geq C_{m,4}^{-2}$.

Moreover R_m, *the orthogonal projection onto the span of Walsh functions with multiplicity* m, *is a bounded and self-adjoint operator on* $L^p, 1 < p < \infty$.

Proof. First we treat $p \geq 2$. The proof proceeds by induction. For $m = 1$ the theorem holds by Khintchine's inequality. Next assume that the theorem holds for a given $m \in \mathbb{N}$. We will now show that it holds for $m+1$. Let $\mathcal{K}_{m+1} = \{A \subset \mathbb{N} : |A| = m+1\}$. The sets

$$\{A \in \mathcal{K}_{m+1} : \max A = n\}, \quad n \in \mathbb{N},$$

form a partition of $\mathcal{K}_{m+1}$. Let $\{a_A : A \in \mathcal{K}_{m+1}\}$ be a sequence where only finitely many entries are $\neq 0$. Define $h = \sum_{A \in \mathcal{K}_{m+1}} a_A w_A$. Let $n \in \mathbb{N}$ and put

$$d_n = \sum_{\substack{A \in \mathcal{K}_{m+1} \\ \max A = n}} a_A w_{A \setminus \{n\}} r_n. \tag{1.1.62}$$

We get $h = \sum d_n$, and by the identity of Walsh, the sequence $\{d_n\}$ is disjointly supported over the Haar system. The square function characterization of L^p implies that

$$\|h\|_p \leq C_p \left\| \left(\sum d_n^2 \right)^{1/2} \right\|_p. \tag{1.1.63}$$

(We prove in Theorem 1.3.4 below that $C_p \leq C\sqrt{p}$ for $p \geq 2$.) Note that r_n is a common factor of each of the Walsh functions appearing in (1.1.62). Hence

$$d_n^2 = \left(\sum_{\substack{A \in \mathcal{K}_{m+1} \\ \max A = n}} a_A w_{A \setminus \{n\}} \right)^2.$$

Since $p \geq 2$, the triangle inequality in $L^{p/2}$ and the above formula for d_n^2 imply that the right-hand side of (1.1.63) is bounded by

$$C_p \left(\sum_{n \in \mathbb{N}} \left\| \sum_{\substack{A \in \mathcal{K}_{m+1} \\ \max A = n}} a_A w_{A \setminus \{n\}} \right\|_p^2 \right)^{1/2}. \tag{1.1.64}$$

Observe that for $A \in \mathcal{K}_{m+1}$ and $\max A = n$ the Walsh function $w_{A \setminus \{n\}}$ has multiplicity m. In fact $\{w_{A \setminus \{n\}} : A \in \mathcal{K}_{m+1}, \max A = n\}$ equals $\{w_B : B \in \mathcal{K}_m, \max B \leq n-1\}$. Using the induction hypothesis we obtain that

$$\left\| \sum_{\substack{A \in \mathcal{K}_{m+1} \\ \max A = n}} a_A w_{A \setminus \{n\}} \right\|_p^2 \leq C_{m,p}^2 \sum_{\substack{A \in \mathcal{K}_{m+1} \\ \max A = n}} a_A^2.$$

Recall that the collections $\{A \in \mathcal{K}_{m+1}, \max A = n\}$, $n \in \mathbb{N}$ form a partition of $\mathcal{K}_{m+1}$. Thus we obtained that

$$\|h\|_p \leq C_{m+1,p} \left(\sum_{A \in \mathcal{K}_{m+1}} a_A^2 \right)^{1/2} . \tag{1.1.65}$$

Let R_m be the orthogonal projection onto the span of Walsh functions with multiplicity m. Clearly $\|R_m f\|_2 \leq \|f\|_2$, and (since $p \geq 2$) $\|f\|_2 \leq \|f\|_p$. Hence (1.1.65) implies that for $f \in L^p$,

$$\begin{aligned} \|R_m f\|_p &\leq C_{m,p} \|R_m f\|_2 \\ &\leq C_{m,p} \|f\|_p. \end{aligned} \tag{1.1.66}$$

We pass to $p \leq 2$ using a well-known trick based on Hölder's inequality. Clearly it suffices to consider $p = 1$. Define

$$h = \left| \sum_{A \in \mathcal{K}_m} a_A w_A \right| .$$

Hölder's inequality implies that $\|h\|_2 \leq \|h\|_1^{1/3} \|h\|_4^{2/3}$, and by the first part we get $\|h\|_4 \leq C_{m,4} \|h\|_2$. Cancelling $\|h\|_2^{2/3}$ gives $\|h\|_2^{1/3} \leq C_{m,4}^{2/3} \|h\|_1^{1/3}$ which finishes the proof for $p \leq 2$. □

Given Theorem 1.1.9, K. Kiener conjectured that the convergence of the partial sums (1.1.60) could be extended to L^p $(1 < p \neq 2 < \infty)$. This conjecture was disproved by B. Roider (1974) in a letter to K. Kiener. Below we present K. Kiener's proof [113] of B. Roider's result that for $p \neq 2$ there exists $g \in L^p$ which is *not approximated* by its partial sums

$$\sum_{m=0}^{n} \sum_{k=1}^{\infty} \langle w_{m,k}, g \rangle w_{m,k}.$$

The proof of K. Kiener [113] starts with the combinatorial identity (1.1.76) which played a central role in the construction of the first Banach space without the approximation property. Specifically (1.1.76) appears in P. Enflo's classical work [62].

Roider's example. Note that R_m, the orthogonal projection onto the span of Walsh functions with multiplicity m, is a self-adjoint operator. Hence it suffices to prove (1.1.51) for $p < 2$. To this end we fix $n \in \mathbb{N}$, and consider

$$f = 2^{2n} 1_{[0, 2^{-2n}[} \tag{1.1.67}$$

Observe that by (1.1.53) f can be expanded using Walsh functions of multiplicity $\leq 2n$. Thus $f \in \text{span}\{W_0, \dots, W_{2n}\}$. Next project f onto the space of Walsh functions with multiplicity $\leq n$, and denote the resulting function by X. Thus

$$X = \sum_{m=0}^{n} R_m(f). \tag{1.1.68}$$

Clearly in L^2 we have $\|X\|_2 \leq \|f\|_2$. The next theorem shows that for $p \neq 2$ the norm of X in L^p is considerably larger than that of f.

Theorem 1.1.10. *Let* $1 < p < 2$. *Then*

$$\|X\|_p \geq C 2^{n(2/p-1)} n^{-1/2} \|f\|_p, \tag{1.1.69}$$

where $C > 0$ *is a universal constant independent of* n *or* p.

Proof. We begin by defining a partition of the unit interval. (On each set of this partition we will later identify the value of X.) Let $0 \leq r \leq 2n$. Define $P_r \subset [0,1]$ by the following rule: Let $t \in [0,1]$, then put $t \in P_r$ if there exists $A \subseteq \{1, \dots, 2n\}$ so that

$$|A| = r$$

and so that for each $m \in \{1, \dots, 2n\}$

$$r_m(t) = \begin{cases} -1, & \text{if } m \in A, \\ 1, & \text{if } m \notin A. \end{cases} \tag{1.1.70}$$

The proof splits naturally into three parts. First we show that the function X is constant on the sets $\{P_r\}$ and we represent the value of X on $\{P_r\}$ by an elegant complex integral formula. This part of the proof is based on Enflo's identity. Second we parameterize the complex integrals and find closed formulas for their value. Third we use this information to obtain a lower bound for $\|X\|_p$.

The first two parts of the proof consist in identifying the values of $X(t)$ defined by (1.1.68) on the sets $\{P_r\}$. First, by definition, $|X(t)| = 1 + \binom{2n}{1} + \cdots + \binom{2n}{n}$, for $t \in P_0 \cup P_{2n}$. We will show next that for $t \in P_1$, we have $2|X(t)| = \binom{2n}{n}$ and that in general for $k = 1, 2, \dots n-1$,

$$2|X(t)| = \binom{2n}{n}\binom{n}{k}\binom{2n}{2k}^{-1} \quad \text{for} \quad t \in P_{2k} \cup P_{2k+1}. \tag{1.1.71}$$

We obtain the identities (1.1.71) from an elegant formula due to K. Kiener.

$$X(t) = \frac{1}{2\pi i} \oint (1-z)^{r-1}(1+z)^{2n-r} \frac{dz}{z^{n+1}} \quad \text{for} \quad t \in P_r, \tag{1.1.72}$$

where the contour integral is taken over the boundary of the unit disk.

Applications of Cauchy's integral formula. We turn to proving (1.1.72). Recall that expanding $f = 2^{2n}1_{[0,2^{-2n}[}$ in its Walsh series gives $f = \sum_{j=0}^{2^{2n}-1} w_j$. Hence for $m \leq 2n$, we have $R_m(f) = \sum_{w \in W_m} w$. Now fix $t \in [0,1]$ and $z \in \mathbb{D}$. Then define

$$F(z,t) = \sum_{m=0}^{2n} z^m R_m(f)(t). \tag{1.1.73}$$

Verify the following identity by multiplying the factors of the product on the right-hand side of (1.1.74),

$$F(z,t) = \prod_{m=1}^{2n} (1 + z r_m(t)). \tag{1.1.74}$$

The product representation of $F(z,t)$ yields that for fixed $r \leq 2n$ and $z \in \mathbb{D}$ the the function

$$t \mapsto F(z,t)$$

is constant on P_r. Indeed for $t \in P_r$,

$$\prod_{m=1}^{2n} (1 + r_m(t) z) = (1 - z)^r (1+z)^{2n-r}. \tag{1.1.75}$$

We apply the Cauchy integral formula to (1.1.73) and use the equation established in (1.1.75). This gives Enflo's identity that for $t \in P_r$,

$$R_m(f)(t) = \frac{1}{2\pi i} \oint (1-z)^r (1+z)^{2n-r} \frac{dz}{z^{m+1}}, \tag{1.1.76}$$

where the contour integral is taken over the boundary of the unit disk. Take the sum of the equations (1.1.76) using that

$$\sum_{m=0}^{n} z^{-m-1} = \frac{(1 - z^{n+1})}{(1-z) z^{n+1}}.$$

Hence by summing (1.1.76) and Cauchy's integral formula we find that for $t \in P_r$,

$$\sum_{m=0}^{n} R_m(f)(t) = \frac{1}{2\pi i} \oint (1-z)^{r-1} (1+z)^{2n-r} \frac{dz}{z^{n+1}}. \tag{1.1.77}$$

Thus we verified the complex integral representation (1.1.72). We remark that the representation of $X(t)$ in (1.1.72) differs from that of $R_n(f)(t)$ by a factor of $(1-z)$ in the contour integral.

Evaluating complex integrals. Next we determine the complex integral (1.1.77). With the following parametrization we replace (1.1.77) by a standard definite integral whose value is well known. Let $z = e^{i\theta}$. Then (1.1.77) equals

$$\frac{1}{2\pi}\int_0^{2\pi}(1-e^{i\theta})^{r-1}(1+e^{i\theta})^{2n-r}\frac{d\theta}{e^{in\theta}}. \tag{1.1.78}$$

We simplify the integrand of (1.1.78) using elementary trigonometric identities.

$$e^{-i\theta/2}(1-e^{i\theta}) = (-2i)\sin(\theta/2) \quad \text{and} \quad e^{-i\theta/2}(1+e^{i\theta}) = 2\cos(\theta/2). \tag{1.1.79}$$

Consequently,

$$\begin{aligned}
&e^{-in\theta}(1-e^{i\theta})^{r-1}(1+e^{i\theta})^{2n-r}\\
&= (-i)^{r-1}2^{2n-1}e^{-i\theta/2}\sin^{r-1}(\theta/2)\cos^{2n-r}(\theta/2)\\
&= (-i)^{r-1}2^{2n-1}\left\{\sin^{r-1}(\theta/2)\cos^{2n-r+1}(\theta/2) - i\sin^r(\theta/2)\cos^{2n-r}(\theta/2)\right\}.
\end{aligned}$$

Next replace the integrand in (1.1.78) by the last line of the above equation. Then, by the change of variables

$$\varphi = \theta/2 \quad \text{and} \quad d\theta = 2d\varphi,$$

we find that (1.1.78) coincides with the integral

$$\frac{(-i)^{r-1}2^{2n}}{2\pi}\int_0^{\pi}\left\{\sin^{r-1}(\varphi)\cos^{2n-r+1}(\varphi) - i\sin^r(\varphi)\cos^{2n-r}(\varphi)\right\}d\varphi. \tag{1.1.80}$$

To derive the pointwise estimates (1.1.71) we evaluate the trigonometric integrals of (1.1.80). This requires us to distinguish between the cases where r is odd and where r is even.

Observe that $\cos(\varphi)$ is an odd(!) function around the point $\pi/2$ and that $\sin(\varphi)$ is even(!) around $\pi/2$. Clearly if r is an even number, then $2n - r + 1$ is odd. Hence

$$\int_0^{\pi}\sin^r(\varphi)\cos^{2n-r}(\varphi)d\varphi = 2\int_0^{\pi/2}\sin^r(\varphi)\cos^{2n-r}(\varphi)d\varphi,$$

and

$$\int_0^{\pi}\sin^{r-1}(\varphi)\cos^{2n-r+1}(\varphi)d\varphi = 0.$$

Similarly if r is odd, then $2n - r$ is odd, hence

$$\int_0^{\pi}\sin^{r-1}(\varphi)\cos^{2n-r+1}(\varphi)d\varphi = 2\int_0^{\pi/2}\sin^{r-1}(\varphi)\cos^{2n-r+1}(\varphi)d\varphi,$$

and

$$\int_0^{\pi}\sin^r(\varphi)\cos^{2n-r}(\varphi)d\varphi = 0.$$

Next recall that for $x, y > 0$ we have

$$2\int_0^{\pi/2} \sin^{2x-1}(\varphi)\cos^{2y-1}(\varphi) = \frac{\Gamma(x)\Gamma(y)}{\Gamma(x+y)}, \tag{1.1.81}$$

where $\Gamma(x) = \int_0^\infty t^{x-1}e^{-t}dt$. In particular $\Gamma(n+1) = n!$ and $\Gamma(1/2) = \sqrt{\pi}$. In summary it follows from Kiener's integral representation (1.1.80) and the identity (1.1.81) that for $r = 2k$ and $t \in P_{2k}$,

$$|X(t)| = \frac{2^{2n}}{2\pi} \times \frac{\Gamma\left(k+\frac{1}{2}\right)\Gamma\left(n-k+\frac{1}{2}\right)}{n!}. \tag{1.1.82}$$

Similarly, for $r = 2k+1$ and $t \in P_{2k+1}$, we obtain from (1.1.80) and (1.1.81) that

$$|X(t)| = \frac{2^{2n}}{2\pi} \times \frac{\Gamma\left(k+\frac{1}{2}\right)\Gamma\left(n-k+\frac{1}{2}\right)}{n!}. \tag{1.1.83}$$

Thus (for different reasons) the value of $X(t)$ on the set P_{2k} coincides with its value on P_{2k+1}. We replace the expressions involving Gamma functions by more elementary formulas. Recall the so-called duplication formula for the Gamma function. It states that for $x > 0$,

$$\Gamma(x+\frac{1}{2}) = \frac{\sqrt{\pi}}{2^{2x-1}}\frac{\Gamma(2x)}{\Gamma(x)}.$$

Apply the duplication formula with $x = k$ and write $1/2 = k/(2k)$. This gives,

$$\Gamma(k+\frac{1}{2}) = \frac{\sqrt{\pi}}{2^{2k}}\frac{(2k)!}{(k)!}.$$

Apply it to $x = n-k$ and use $1/2 = (n-k)/(2n-2k)$ to obtain that

$$\Gamma(n-k+\frac{1}{2}) = \frac{\sqrt{\pi}}{2^{2n-2k}}\frac{(2n-2k)!}{(n-k)!}.$$

Substitute into (1.1.82) and (1.1.83), then simplify to obtain (1.1.71) by arithmetic.

The lower bound for $\|X\|_p$. Now we turn to the last part of the proof where we apply (1.1.71) to obtain a lower bound for the integral $\int_0^1 |X(t)|^p dt$. Since $X(t)$ is constant on the sets P_r we denote

$$X_r = X(t) \quad \text{for} \quad t \in P_r.$$

Note that the disjoint sets P_r, $0 \le r \le 2n$ are defined so that their measure is given by $|P_r| = 2^{-2n}\binom{2n}{r}$. Hence (1.1.71) gives

$$\begin{aligned} 2^{2n}\int_0^1 |X(t)|^p dt &= \sum_{r=0}^{2n} |X_r|^p \binom{2n}{r} \\ &\ge \sum_{k=0}^{n-1} |X_{2k}|^p \left[\binom{2n}{2k} + \binom{2n}{2k+1}\right]. \end{aligned} \tag{1.1.84}$$

The formula for evaluating consecutive binomial coefficients gives

$$\left[\binom{2n}{2k} + \binom{2n}{2k+1}\right] = \binom{2n+1}{2k+1}.$$

Note that $\binom{2n+1}{2k+1} \geq \binom{2n}{2k}$. Inserting the values (1.1.71) for $|X_{2k}|$ we obtain the identities

$$|2X_{2k}|^p \binom{2n}{2k} = \binom{2n}{n}^p \binom{n}{k}^p \binom{2n}{2k}^{-(p-1)}.$$

Then by (1.1.84) we obtain that

$$2^{2n+p} \int_0^1 |X(t)|^p dt \geq \binom{2n}{n}^p \times \sum_{k=0}^{n-1} \binom{n}{k}^p \binom{2n}{2k}^{-(p-1)}. \tag{1.1.85}$$

Recall Hölder's inequality in the form

$$\sum_{k=0}^{n-1} a_k^p b_k^{-(p-1)} \geq \left(\sum_{k=0}^{n-1} a_k\right)^p \left(\sum_{k=0}^{n-1} b_k\right)^{-(p-1)}. \tag{1.1.86}$$

Apply it with $a_k = \binom{n}{k}$ and $b_k = \binom{2n}{2k}$. We continue by giving closed formulas for the two factors appearing on the right-hand side of (1.1.86).

$$\sum_{k=0}^{n-1} \binom{n}{k} = 2^n - 1 \quad \text{and} \quad \sum_{k=0}^{n-1} \binom{2n}{2k} \leq 2^{2n}.$$

In summary we obtained a lower bound for the right-hand side of (1.1.85). It follows that

$$\int_0^1 |X(t)|^p dt \geq \binom{2n}{n}^p \times 2^{-np} \times 2^{-p-1}.$$

The remaining calculations are based on Stirling's formula, $n! \sim \sqrt{2\pi n} \cdot n^n \cdot e^{-n}$. It allows us to compute $\binom{2n}{n} \sim 2^{2n} n^{-1/2}$. Inserting this value in the previous lower bound for the norm of X in L^p we find

$$\int_0^1 |X(t)|^p dt \geq C n^{-p/2} 2^{np}.$$

Next recall that we put $f = 2^{2n} 1_{[0,2^{-2n}[}$. Hence $\int f^p dt = 2^{2n(p-1)}$. Consequently the following lower estimate holds true.

$$\int |X(t)|^p dt \geq C n^{-p/2} 2^{-np+2n} \int f^p dt. \tag{1.1.87}$$

□

Recall that R_n denotes the orthogonal projection onto the span of Walsh functions with multiplicity n. We pointed out that the integral representations (1.1.72) for X and (1.1.76) for $R_n(f)$ are almost identical. With this observation it is easy to see that the proof of Theorem 1.1.10 gives a lower estimate for the operator norm of R_n on L^p. Theorem 1.1.9, on the other hand, provides an upper estimate. In summary the following estimates hold,

$$c2^{cn(1-2/p)} \leq \|R_n\|_{L^p} \leq (C\sqrt{p})^n, \quad \text{for} \quad p > 2. \tag{1.1.88}$$

Comparing the right and left hand side in (1.1.88) gives rise to the following problem.

Problem 1.1.11. *Narrow the gap between the lower and upper estimates for the operator norm of R_n on L^p. Notice that the test function* (1.1.67) *in the proof of Theorem 1.1.10 is very special. Hence it is likely that the lower estimate in* (1.1.88) *can be improved.*

Since the appearance of Paley's work the Walsh system has been continuously investigated from the point of view of harmonic analysis and approximation theory. Moreover the Walsh system plays a remarkable role in the analysis of classical Banach spaces where it is used to exhibit spaces and operators with extremal (and often unexpected) properties. P. Enflo's original construction of a Banach space without the approximation property is one of the most outstanding examples representing this line of investigation (see the notes for references). Through the work of B. Roider and K. Kiener in 1974 the methods of P. Enflo found an early application to a concrete approximation problem for Walsh functions in L^p.

In 1992 J. Bourgain [25] extended B. Roider's theorem: Denoting by X_n^p the subspace of L^p spanned by the Walsh functions

$$\{w_A : A \subseteq \{1, \ldots, 2n\}, |A| \leq n\},$$

J. Bourgain [25] proves a trace estimate for any operator T on L^p whose range is contained in X_n^p,

$$\mathrm{tr(T)} \leq B2^{2n(1-\gamma)}\|T\|_{L^p}, \tag{1.1.89}$$

where $B = B(p)$ and $\gamma = \gamma(p) > 0$ when $p \neq 2$. The trace of the orthogonal projection onto X_n^p is larger than $c2^{2n}$, hence (1.1.89) gives a lower estimate for the norm of the projection. Thus (1.1.89) implies Roider's estimate (1.1.51).

1.2 Dyadic H^1

In this book we will work intensively with the following dyadic versions of H^1 and BMO. The space BMO consists of all square integrable functions, having mean zero,

$$h = \sum_{J \in \mathcal{D}} b_J h_J$$

for which

$$||h||_{\mathrm{BMO}} = \sup_{I\in\mathcal{D}} \left(\frac{1}{|I|} \sum_{J\subseteq I} b_J^2 |J| \right)^{1/2} < \infty.$$

The space H^1 consists of integrable functions f with integrable square functions. For

$$f = \int_0^1 f dt + \sum_{J\in\mathcal{D}} a_J h_J$$

the square function is given by

$$S(f)(x) = \left(\sum_{J\in\mathcal{D}} a_J^2 1_J(x) \right)^{1/2},$$

and H^1 is the space of integrable functions for which

$$||f||_{H^1} = \left| \int_0^1 f dt \right| + \int_0^1 S(f)(x)\, dx < \infty.$$

Fefferman's inequality

Next we identify BMO as the dual space of H^1. We begin by showing that each $h \in \mathrm{BMO}$ induces a bounded linear functional on H^1. This follows from the next theorem, the content of which is known as Fefferman's inequality. In this section we present two proofs of Fefferman's inequality.

Theorem 1.2.1. *(Fefferman's inequality) Let f be a finite linear combination of Haar functions, and let $h \in BMO$, then*

$$\left| \int fh \right| \leq 2\sqrt{2} ||f||_{H^1} \cdot ||h||_{BMO}.$$

Proof. Let $h = \sum b_J h_J$ be the Haar expansion of $h \in \mathrm{BMO}$. Since

$$\int fhdt = \sum \langle f, h_J \rangle b_J,$$

we may assume that $\{J : b_J \neq 0\}$ is contained in $\{J : \langle f, h_J\rangle \neq 0\}$. In particular we assume that h is a finite linear combination of Haar functions.

Let I be a fixed dyadic interval. We define $S(h \,|\, I)$, the square function localized to I as

$$S(h \,|\, I)(x) = \left(\sum_{J\subseteq I} b_J^2 1_J(x) \right)^{1/2}.$$

Now fix a point x in the unit interval. Note that $S(h \mid I)(x) < \infty$ for every dyadic interval I. Next define $I(x)$ to be the largest dyadic interval containing x such that

$$S^2(h \mid I(x))(x) \leq 2 \sup_{I \ni x} \frac{1}{|I|} \sum_{L \subseteq I} b_L^2 |L|. \tag{1.2.1}$$

Note that, by definition, we have the following upper bound for the square function localized to the interval $I(x)$,

$$S(h \mid I(x))(x) \leq \sqrt{2} ||h||_{BMO}, \quad x \in [0,1]. \tag{1.2.2}$$

Thus Fefferman's inequality is reduced to showing that the following estimate holds true,

$$\left| \int f h dx \right| \leq 2 \int S(f)(x) S(h \mid I(x))(x) dx. \tag{1.2.3}$$

To prove the integral estimate (1.2.3) we verify first that for every dyadic interval J, the following measure estimate holds,

$$|\{x \in J : I(x) \supseteq J\}| \geq |J|/2. \tag{1.2.4}$$

To this end we fix $J \in \mathcal{D}$ and consider the set

$$A = \{x \in J : I(x) \text{ is strictly contained in } J\}.$$

Then using (1.2.1), we estimate

$$\begin{aligned} \int_J S^2(h \mid J)(x)\, dx &\geq \int_A S^2(h \mid J)(x)\, dx \\ &> 2 \int_A \sup_{x \in I} \frac{1}{|I|} \sum_{L \subseteq I} b_L^2 |L|\, dx \\ &\geq 2 \frac{|A|}{|J|} \int_J S^2(h \mid J)(x)\, dx. \end{aligned} \tag{1.2.5}$$

The last inequality in (1.2.5) used the following pointwise bound that holds for any $x \in J$,

$$\sup_{I \ni x} \frac{1}{|I|} \sum_{L \subseteq I} b_L^2 |L| \geq \frac{1}{|J|} \int_J S^2(h \mid J)(x)\, dx.$$

Finally in (1.2.5) we cancel the factor $\int_J S^2(h \mid J)(x)\, dx$ and obtain that $|A| \leq |J|/2$. Taking complements we obtain the measure estimate (1.2.4), that is

$$|\{x \in J : I(x) \supseteq J\}| \geq \frac{1}{2} |J|.$$

We rewrite the estimate (1.2.4), in the following integral form, using indicator functions,

$$\int_0^1 1_J(x) \cdot 1_{\{y : I(y) \supseteq J\}}(x)\, dx \geq \frac{1}{2} |J|.$$

Now fix $f = \sum a_J h_J$ in H^1. With the preceding analysis of h, we estimate as follows.

$$\begin{aligned}
\left|\int fh\,dx\right| &\leq \sum |a_J b_J||J| \\
&\leq 2\sum |a_J b_J| \int 1_J(x) 1_{\{y:I(y)\supseteq J\}}(x)\,dx \\
&= 2\int \sum_{I(x)\supseteq J} |b_J a_J| 1_J(x)\,dx \\
&\leq 2\int S(f)(x) S(h\,|\,I(x))\,dx.
\end{aligned}$$

Thus we verified that (1.2.3) holds. Recalling the pointwise upper bound (1.2.2) completes the proof of Fefferman's inequality. □

Let h be square integrable, with Haar expansion

$$h = \int_0^1 h dx + \sum_{J\in\mathcal{D}} b_J h_J.$$

At a given point $x \in [0,1]$ the sharp function of h is defined to be

$$h^\sharp(x) = \left(\sup_I \frac{1}{|I|} \sum_{J\subseteq I} b_J^2 |J|\right)^{1/2},$$

where the supremum is extended over all dyadic intervals which contain the given point $x \in [0,1]$. Looking back at the proof of Theorem 1.2.1 we observe that the sharp function played an important role in the proof of Fefferman's inequality. It appeared on the right-hand side of the estimate (1.2.1). Even more, the sharp function appears implicitly in the definition of BMO. Indeed, by inspection, the space BMO consists of all functions with uniformly bounded sharp functions, and

$$||h||_{\mathrm{BMO}} = \sup_{x\in[0,1]} h^\sharp(x). \tag{1.2.6}$$

Observe that combining (1.2.1) and the integral estimate (1.2.3) we obtain the following inequality, that immediately gives Fefferman's inequality,

$$\left|\int_0^1 fhdx\right| \leq 2\sqrt{2}\int_0^1 S(f)h^\sharp dx. \tag{1.2.7}$$

It is worth pointing out that the above proof focusses exclusively on the BMO function h. The fact that $f \in H^1$, enters only at the very end when we observe that the right hand side of (1.2.7) is bounded by $2\sqrt{2}||h^\sharp||_\infty \int_0^1 S(f)dx$.

As an application of Khintchine's inequality we next prove a converse to Fefferman's inequality. In that way we are showing that Theorem 1.2.1 is sharp.

Theorem 1.2.2. *For each $f \in H^1$, with $\int f = 0$, there exists $h \in BMO$ such that*

$$\langle f, h \rangle \geq \frac{1}{2\sqrt{2}} ||f||_{H^1} \cdot ||h||_{BMO}.$$

Proof. For the purpose of applying Khintchine's inequality to the Haar coefficients of a given function f, it is more convenient to first re-index the sequence of Rademacher functions using dyadic intervals. Let $\{r_I : I \in \mathcal{D}\}$ be any enumeration of the Rademacher functions. Let $f \in H^1$, and let

$$f = \sum a_I h_I$$

be its Haar expansion. For $y \in [0, 1]$, we define f_y by its Haar expansion as

$$f_y = \sum a_I h_I r_I(y).$$

Note that $f_y \in H^1$. Applying Khintchine's inequality we obtain that

$$\int_0^1 \int_0^1 |f_y| dy\, dx \geq \frac{1}{\sqrt{2}} \int \left(\sum a_I^2 h_I^2(x) \right)^{1/2} dx = \frac{1}{\sqrt{2}} ||f||_{H^1}.$$

Changing the order of integration one finds at least one point $y \in [0, 1]$ such that

$$||f_y||_{L^1} \geq \frac{1}{\sqrt{2}} ||f||_{H^1}.$$

Now we choose $g \in L^\infty$ such that $\langle g, f_y \rangle = ||g||_{L^\infty} \cdot ||f_y||_{L^1}$. Next we expand $g - \int g$ in its Haar series,

$$g - \int g = \sum c_I h_I,$$

and define

$$h = \sum \epsilon_I c_I h_I,$$

where $\epsilon_I = r_I(y)$. By the biorthogonality of the Haar system we have that $\langle f, h \rangle = \langle f_y, g \rangle$, and by the definition of the BMO norm we have also,

$$||h||_{\text{BMO}} = ||g - \int g||_{\text{BMO}} \leq 2||g||_{L^\infty}.$$

Combining these observations gives

$$\begin{aligned} \langle f, h \rangle &= \langle f_y, g \rangle \\ &= ||f_y||_{L^1} ||g||_\infty \\ &\geq \frac{1}{2\sqrt{2}} ||f||_{H^1} ||h||_{\text{BMO}}. \end{aligned}$$

□

By Fefferman's inequality every $h \in \mathrm{BMO}$ defines a continuous linear functional on the space H^1, and moreover the unit ball of BMO norms H^1 by the converse to Fefferman's inequality. We summarize Theorems 1.2.1 and 1.2.2 as follows. Let $f \in H^1$ with $\int f = 0$, then

$$\frac{1}{2\sqrt{2}}||f||_{H^1} \leq \sup\{\int fh\,dx : ||h||_{\mathrm{BMO}} \leq 1\} \leq 2\sqrt{2}||f||_{H^1}. \tag{1.2.8}$$

We now show that every continuous linear functional on H^1, determines a function in BMO. For $n \in \mathbb{N}$ let $f_n : [0,1] \to \mathbb{R}$ be integrable. Recall that the sequence $(f_n)_{n=1}^{\infty}$ belongs to the space $L^1(\ell^2)$ if

$$||(f_n)_{n=1}^{\infty}||_{L^1(\ell^2)} = \int_0^1 \left(\sum_{n=1}^{\infty} f_n^2\right)^{1/2} dx < \infty.$$

The dual Banach space of $L^1(\ell^2)$ is canonically identified with $L^\infty(\ell^2)$, consisting of sequences $(h_n)_{n=1}^{\infty}$ so that

$$||(h_n)_{n=1}^{\infty}||_{L^\infty(\ell^2)} = \sup_{x\in[0,1]} \left(\sum_{n=1}^{\infty} h_n^2(x)\right)^{1/2} < \infty,$$

where "$\sup_{x\in[0,1]}$" denotes the essential supremum over the unit interval, defining the norm in L^∞.

Theorem 1.2.3. *For every continuous linear functional $L : H^1 \to \mathbb{R}$ there exists $h \in BMO$ so that*

$$||h||_{BMO} \leq ||L||, \tag{1.2.9}$$

and so that for every finite linear combination of Haar functions g,

$$L(g) = \int_0^1 hg\,dt. \tag{1.2.10}$$

Proof. The idea of the proof comes with the useful observation that H^1 is canonically represented as a closed subspace of $L^1(\ell^2)$. Hence by the Hahn–Banach theorem every continuous linear functional extends to an element in the dual space of $L^1(\ell^2)$, which is the space $L^\infty(\ell^2)$. The details are as follows. Let $f \in H^1$, with $\int f = 0$. For $n \in \mathbb{N}$ let

$$f_n = \sum_{\{I:|I|=2^{-n-1}\}} \langle f, \frac{h_I}{|I|}\rangle h_I. \tag{1.2.11}$$

Then the square function of f satisfies

$$S(f) = \left(\sum f_n^2\right)^{1/2}.$$

Thus the linear map $f \mapsto (f_n)$ provides the embedding of H^1 to a closed subspace of $L^1(\ell^2)$, and

$$||f||_{H^1} = ||(f_n)_{n=1}^{\infty}||_{L^1(\ell^2)}.$$

With the Hahn Banach theorem extend the linear functional L to a functional on $L^1(\ell^2)$, with the same norm. Let $\tilde{L} : L^1(\ell^2) \to \mathbb{R}$, be an extension satisfying

$$||L|| = ||\tilde{L}||,$$
$$L(f) = \tilde{L}((f_n)),$$

for $f \in H^1$, and $(f_n)_{n=1}^{\infty}$ defined by (1.2.11). Next recall that the dual space of $L^1(\ell^2)$ is identified as $L^\infty(\ell^2)$, in the following sense. There exists a sequence (h_n) such that

$$||\tilde{L}|| = \left(\sup_{x\in[0,1]} \sum h_n^2(x) \right)^{1/2},$$

and for every $(x_n) \in L^1(\ell^2)$,

$$\tilde{L}((x_n)) = \sum_{n-1}^{\infty} \int_0^1 x_n h_n dt. \tag{1.2.12}$$

Next we define h, using the sequence (h_n), by putting

$$h = \sum_{n=1}^{\infty} \sum_{\{I:|I|=2^{-n-1}\}} \langle h_n, \frac{h_I}{|I|} \rangle h_I.$$

We test the equality (1.2.12) on the sequence (f_n) defined by (1.2.11). This gives

$$\begin{aligned}\tilde{L}((f_n)) &= \sum_{n=1}^{\infty} \sum_{\{I:|I|=2^{-n-1}\}} \langle f, \frac{h_I}{|I|} \rangle \langle h_I, h_n \rangle \\ &= \langle f, h \rangle.\end{aligned}$$

Thus when restricted to (f_n) the action of $\tilde{L}$ is given by integration against h,

$$\tilde{L}((f_n)) = \int_0^1 f h dt.$$

This proves (1.2.10). Now we show that $h \in \mathrm{BMO}$, and that

$$||h||_{\mathrm{BMO}} \leq \left(\sup_{x\in[0,1]} \sum h_n^2(x) \right)^{1/2}.$$

Evaluate the BMO norm of h by fixing a dyadic interval J with $|J| = 2^{-m-1}$. Let $(h)_J$ denote the mean value of h over the interval J. By Bessel's inequality we obtain that,

$$\int_J |h-(h)_J|^2 dt = \sum_{n\geq m} \sum_{\{I\subseteq J:|I|=2^{-n-1}\}} \langle h_n, \frac{h_I}{|I|}\rangle^2 |I|$$
$$\leq \int_J \sum_{n=1}^{\infty} h_n^2 dt.$$

Dividing by $|J|$ and then taking the supremum over all dyadic intervals J, we find that

$$\|h\|_{\text{BMO}} \leq \left(\sup_{x\in[0,1]} \sum_{n=1}^{\infty} h_n^2(x) \right)^{1/2},$$

and consequently,

$$||h||_{\text{BMO}} \leq ||\tilde{L}||$$

which proves (1.2.9). □

We will now present a different approach to Fefferman's inequality. In this approach the proof is carried out by analyzing the functions in H^1. We will show that every $f \in H^1$ with small enough norm can be decomposed into a convex combination of dyadic atoms. A function a is a dyadic atom if there exists a dyadic interval I such that

$$\operatorname{supp} a \subseteq I,$$

$$\int_0^1 a^2(t)dt \leq |I|^{-1},$$

and

$$\int_0^1 a(t)dt = 0.$$

We will also say that the constant function $1_{[0,1]}$ is a dyadic atom. For a given $f \in H^1$ with Haar expansion

$$f = \sum_{I\in\mathcal{D}} x_I h_I,$$

we define the Haar support of f to be the following collection of dyadic intervals, $\{I \in \mathcal{D} : x_I \neq 0\}$. In the proof of Theorem 1.2.4 we obtain the decomposition of f into atoms by first decomposing the Haar support of f into pairwise disjoint collections of dyadic intervals. At this point we use, for the first time in this book, a stopping time decomposition of the Haar support. Many more applications of this fundamental technique will be given in later chapters.

Theorem 1.2.4. *For $f \in H^1$, there exists a sequence of dyadic atoms, (a_i) and a sequence of scalars (c_i), such that*

$$f - \int_0^1 f = \sum_{i=1}^{\infty} c_i a_i,$$

and

$$\sum_{i=1}^{\infty} |c_i| \leq 16 \int_0^1 S(f).$$

Conversely, every dyadic atom a belongs to H^1 and satisfies

$$\int_0^1 S(a)dt \leq 1.$$

Before we present the proof we should point out how Theorem 1.2.1 (Fefferman's inequality) can be deduced from the atomic decomposition of H^1. Theorem 1.2.4 may be paraphrased by saying that it suffices to prove Fefferman's inequality only for dyadic atoms $a \in H^1$. Now fix a dyadic atom a and a dyadic interval I such that $\operatorname{supp} a \subseteq I$, and $\int_0^1 a^2(t)dt \leq |I|^{-1}$. Next fix $h \in \mathrm{BMO}$. Let $(h)_I$ denote the mean value of h over the interval I. Then, using $\int_0^1 adt = 0$ and the Cauchy–Schwarz inequality, we estimate as follows,

$$\begin{aligned}\left|\int_0^1 ahdt\right| &= \left|\int_I a(h-(h)_I)dt\right| \\ &\leq \|a\|_{L^2}\|1_I(h-(h)_I)\|_{L^2}.\end{aligned}$$

Finally, observe that the biorthogonality of the Haar system gives that

$$\|1_I(h-(h)_I)\|_{L^2} \leq |I|^{1/2}\|h\|_{\mathrm{BMO}}.$$

Below the following notations are repeatedly used. For a collection of dyadic intervals $\mathcal{F}$ we denote by $\max \mathcal{F}$ the maximal intervals of $\mathcal{F}$. As dyadic intervals are nested, $\max \mathcal{F}$ consists of pairwise disjoint intervals. For a dyadic interval I we denote by $Q(I)$ the collection of all dyadic intervals which are contained in I, thus

$$Q(I) = \{J \in \mathcal{D} : J \subseteq I\}.$$

Proof of Theorem 1.2.4. Let $f \in H^1$, with Haar expansion

$$f = \sum x_J h_J.$$

Thus we assume that $\int_0^1 f = 0$. We will decompose f by splitting its Haar support. We start with the unit interval $[0,1]$. Let $n([0,1])$ be the smallest integer such that

$$\Omega([0,1]) = \{t \in [0,1] : S(f)(t) \geq 2^{n([0,1])}\}$$

satisfies

$$|\Omega([0,1])| \leq \frac{1}{4}.$$

Let $\mathcal{F}([0,1])$ be the collection of dyadic intervals which are contained in $\Omega([0,1])$. Then define

$$\mathcal{C}([0,1]) = Q([0,1]) \setminus \mathcal{F}([0,1]).$$

Note that

$$f_{[0,1]} = \sum_{J \in \mathcal{C}([0,1])} x_J h_J$$

is defined to satisfy the pointwise estimate

$$S(f_{[0,1]}) \leq 2^{n([0,1])}.$$

Next we define $\mathcal{E}_1$ to be the collection of maximal dyadic intervals of $\mathcal{F}([0,1])$, that is,

$$\mathcal{E}_1 = \max \mathcal{F}([0,1]).$$

We continue the construction inductively. Suppose we have already defined collections of pairwise disjoint dyadic intervals $\mathcal{E}_1, \cdots, \mathcal{E}_i$. Fix $I \in \mathcal{E}_i$. Let $n(I)$ be the smallest integer such that

$$\Omega(I) = \{t \in I : S(f)(t) \geq 2^{n(I)}\}$$

satisfies

$$|\Omega(I)| \leq \frac{1}{4}|I|.$$

Let $\mathcal{F}(I)$ be the collection of dyadic intervals which are contained in $\Omega(I)$. Then let

$$\mathcal{C}(I) = Q(I) \setminus \mathcal{F}(I).$$

We define now

$$f_I = \sum_{J \in \mathcal{C}(I)} x_J h_J$$

which satisfies the pointwise estimate

$$S(f_I) \leq 2^{n(I)}.$$

Moreover the support of $S(f_I)$ is contained in the dyadic interval I. Hence

$$\int S(f_I)^2 \leq |I| 2^{2n(I)}.$$

Now we let

$$\mathcal{E}(I) = \max \mathcal{F}(I),$$

and we take the union over $I \in \mathcal{E}_i$ to define

$$\mathcal{E}_{i+1} = \bigcup \{\mathcal{E}(I) : I \in \mathcal{E}_i\}.$$

As $\mathcal{E}_i$ consists of pairwise disjoint dyadic intervals, it is easy to see that $\mathcal{E}_{i+1}$ is a collection of pairwise disjoint dyadic intervals. This completes the induction step. Finally we define $\mathcal{E} = \bigcup \mathcal{E}_i$. Recall that along with the construction of $\mathcal{E}$ we verified that

$$a_I = \frac{f_I}{|I| 2^{n(I)}}$$

is a dyadic atom. This suggests a decomposition of f,

$$\begin{aligned} f &= \sum_{I \in \mathcal{E}} f_I \\ &= \sum_{I \in \mathcal{E}} |I| 2^{n(I)} a_I. \end{aligned}$$

It remains to show that the coefficients in front of the atoms are in fact summable. We do this by proving that

$$\sum_{I \in \mathcal{E}} |I| 2^{n(I)} \leq 16 \int S(f). \tag{1.2.13}$$

We start by observing that the stopping time definition of $n(I)$ implies the measure estimate

$$|\{t \in I : S(f)(t) \geq 2^{n(I)-1}\}| \geq \frac{|I|}{4}.$$

Then we continue with the proof of the convexity relation (1.2.13),

$$\begin{aligned} \sum_{I \in \mathcal{E}} |I| 2^{n(I)} &\leq 4 \sum_{I \in \mathcal{E}} 2^{n(I)} |\{t \in I : S(f) \geq 2^{n(I)-1}\}| \\ &\leq 8 \sum_{I \in \mathcal{E}} 2^{n(I)-1} |\{t \in I : S(f) \geq 2^{n(I)-1}\}| \\ &\leq 8 \sum_{k \in \mathbb{N}} 2^k |\{S(f) \geq 2^k\}| \\ &\leq 16 \int S(f). \end{aligned}$$

We find an atomic decomposition of f by rewriting the equation $f = \sum_{I \in \mathcal{E}} f_I$ as

$$f = \sum_{I \in \mathcal{E}} |I| 2^{n(I)} \frac{f_I}{|I| 2^{n(I)}}.$$

The converse is actually very easy. We fix an atom a, supported in a dyadic interval I such that $\int a^2 \leq |I|^{-1}$. As $\int a = 0$, the dyadic square function $S(a)$ has its support in I as well. Hence by the Cauchy–Schwarz inequality we estimate

$$\int S(a) \leq |I|^{1/2}(\int S^2(a))^{1/2}.$$

The hypothesis that a is an atom implies of course $(\int S(a)^2)^{1/2} \leq |I|^{-1/2}$. Thus we showed that $\int S(a) \leq 1$. □

In the course of the above proof we decomposed f by splitting the Haar support into pairwise disjoint collections of dyadic intervals $\{\mathcal{C}(I) : I \in \mathcal{E}\}$. Being the result of a stopping time decomposition, the collections $\mathcal{C}(I)$ satisfy a very important connectedness property: First, the collection $\mathcal{C}(I)$ contains exactly one maximal interval, namely I. Second, the following holds true. Let $J \in \mathcal{C}(I)$ and let K be contained in the Haar support of f, then

$$J \subseteq K \subseteq I \text{ implies } K \in \mathcal{C}(I).$$

The collection $\mathcal{C}(I)$ is called a block of dyadic intervals in the Haar support of f. Next the index set $\mathcal{E}$ and its construction deserves further analysis, that provides an important link to BMO. Recall that we defined $\mathcal{E} = \bigcup \mathcal{E}_i$. Reviewing the proof we gave of Theorem 1.2.4 shows that the collections $\mathcal{E}_i$ are linearly ordered and shrinking uniformly at a geometric rate. Precisely, the following holds.

1. If $I \in \mathcal{E}_i$, $J \in \mathcal{E}_j$ and $I \subseteq J$, then $i \geq j$.
2. If $J \in \mathcal{E}_j$ and $i \geq j$
$$\sum_{\{I \in \mathcal{E}_i\, :\, I \subseteq J\}} |I| \leq 2^{-i+j}|J|.$$

As a result we obtain that $\mathcal{E}$ satisfies the Carleson packing condition: For any $J \in \mathcal{E}$,

$$\sum_{\{I \in \mathcal{E} : I \subseteq J\}} |I| \leq 2|J|. \tag{1.2.14}$$

Summing up: *The proof of Theorem* 1.2.4 *produced a decomposition of the Haar support into blocks of dyadic intervals* $\{\mathcal{C}(I) : I \in \mathcal{E}\}$, *such that the index set* $\mathcal{E}$ *satisfies the Carleson packing condition* (1.2.14).

Sharp maximal functions in L^p

Renorming L^p with the square function $||S(f)||_p$ and then passing to the limit $p \to 1$ gives H^1, which is thus a natural endpoint of the L^p scale. We will show that in the opposite direction BMO appears as endpoint of the L^p scale when we renorm with the sharp function and then pass to the limit $p \to \infty$. Let $f \in L^2([0,1])$, with Haar expansion

$$f = \int f + \sum b_I h_I.$$

For $x \in [0,1]$, the sharp function is defined as

$$f^\sharp(x) = \left(\sup_{x \in I} \frac{1}{|I|} \sum_{J \subseteq I} b_J^2 |J| \right)^{1/2},$$

where the supremum is taken over all dyadic intervals containing x. We observed in (1.2.6), by comparing the definitions of BMO and $f^\sharp$, that $f \in \mathrm{BMO}$ if $f^\sharp \in L^\infty$.

Now we show that the sharp function provides an equivalent renorming of L^p when $p > 2$.

Theorem 1.2.5. *Let $p > 2$. If $f \in L^p([0,1])$ with $\int f = 0$, then*

$$(8(p-1))^{-1} ||f||_p \le ||f^\sharp||_p \le 2 \left(\frac{p}{p-2} \right)^{1/p} ||f||_p.$$

Proof. Let $q = p/(p-1)$ be the Hölder conjugate index of p. As $p > 2$ we have that $q < 2$. Choose $g \in L^q$ with $||g||_q \le 2$ and $\int g = 1$, so that

$$||f||_p = \int_0^1 fg dt.$$

Now recall the integral estimate (1.2.3) and the pointwise upper bound (1.2.1) established in the course of the proof of Fefferman's inequality. They give that

$$\int_0^1 fg dt \le 2\sqrt{2} \int_0^1 f^\sharp S(g) dt. \tag{1.2.15}$$

Now we apply Hölder's inequality and Theorem 1.1.6, the square function characterization of L^q, when $q < 2$.

$$\begin{aligned} \int_0^1 f^\sharp S(g) dt &\le ||f^\sharp||_p ||S(g)||_q \\ &\le ||f^\sharp||_p 2\sqrt{2}(p-1) ||g||_q \\ &\le 2\sqrt{2}(p-1) ||f^\sharp||_p. \end{aligned}$$

This shows the left-hand side inequality of Theorem 1.2.5, that is,

$$||f||_p \le 8(p-1) ||f^\sharp||_p.$$

We obtain the right-hand side estimate by comparing the sharp function $f^\sharp$ to the following version of the Hardy–Littlewood maximal function. For $x \in [0,1]$ define

$$M_2(f)(x) = \sup_{x \in I} \left(\frac{1}{|I|} \int_I f^2 \right)^{1/2},$$

where we take the supremum of all dyadic intervals which contain the point x. A moments reflection shows that pointwise inequalities hold between the sharp function and the maximal function of Hardy and Littlewood,

$$f^\sharp(x) \le M_2(f)(x). \tag{1.2.16}$$

For $p > 2$, the Hardy–Littlewood maximal theorem gives an upper bound for the L^p norm of $M_2(f)$, in terms of the L^p norm of f. More precisely by Theorem 1.2.6 below, the following holds for $p > 2$,

$$||M_2(f)||_p \le 2 \left(\frac{p}{p-2} \right)^{1/p} ||f||_p.$$

Assuming the validity of Theorem 1.2.6, we obtain from (1.2.16) the norm estimate for the sharp function,

$$||f^\sharp||_p \le ||M_2(f)||_p \le 2 \left(\frac{p}{p-2} \right)^{1/p} ||f||_p. \qquad \square$$

Completing the proof that the sharp function provides an equivalent norm on L^p we show the L^p estimate for the Hardy–Littlewood maximal function. Besides M_2 we will encounter different versions of the Hardy–Littlewood maximal function. Hence we will prove the required estimate in a more general form.

Theorem 1.2.6. *Fix p, q, such that $1 \le q < p$. Let $f \in L^p([0,1])$, then*

$$M_q(f)(x) = \left(\sup_{x \in I} \frac{1}{|I|} \int_I |f|^q \right)^{1/q}$$

satisfies the norm estimate

$$||M_q(f)||_p \le 2 \left(\frac{p}{p-q} \right)^{1/p} ||f||_p.$$

Proof. Fix $f \in L^p([0,1])$. Without loss of generality we assume that $f \ge 0$. Let $t > 0$ and define the following function h by truncating f at the places where $f(x) \le t/2$,

$$h(x) = \begin{cases} f(x) & \text{if } f(x) > t/2, \\ 0 & \text{otherwise.} \end{cases}$$

Put $g = f - h$. Then $g(x) \leq t/2$, hence $f(x) \leq h(x) + t/2$, and $M_q(f) \leq M_q(h) + t/2$. Consequently

$$\{M_q(f) > t\} \subseteq \{M_q(h) > t/2\}. \tag{1.2.17}$$

By Lemma 1.2.7 below, the maximal function M_q satisfies a weak type $q - q$ estimate. We apply Lemma 1.2.7 to h and recall that the support of h is contained in the set $E = \{x : f(x) > t/2\}$. This gives

$$\begin{aligned} t^q |\{M_q(h) > t/2\}| &\leq 2^q ||h||_q^q \\ &\leq 2^q \int_E f^q dx. \end{aligned} \tag{1.2.18}$$

The pieces of information obtained so far are the inclusion relation (1.2.17) and the weak type estimate (1.2.18) Now we combine them and reach an upper bound for the distribution function

$$|\{M_q(f) > t\}| \leq \frac{2^q}{t^q} \int f^q(x) 1_{\{f > t/2\}}(x) dx. \tag{1.2.19}$$

Finally we obtain estimates for $||M_q(f)||_p$ by integration from (1.2.19) in the following way,

$$\begin{aligned} ||M_q(f)||_p^p &= p \int_0^\infty t^{p-1} |\{M_q(f) > t\}| dt \\ &\leq 2^q p \int_0^\infty t^{p-1} t^{-q} \int f^q(x) 1_{\{f > t/2\}}(x) dx dt \\ &= 2^q p \int_0^1 f^q(x) \int_0^{2f(x)} t^{p-q-1} dt dx \\ &= 2^q \frac{p}{p-q} \int_0^1 f^q(x) (2f)^{p-q}(x) dx. \end{aligned} \tag{1.2.20}$$

It remains to collect terms and to take the p-th root to obtain that

$$||M_q(f)||_p \leq 2 \left(\frac{p}{p-q} \right)^{1/p} \|f\|_p. \qquad \square$$

Now we prove the fact that the maximal function M_q satisfies a weak type $q - q$ estimate.

Lemma 1.2.7. *For $h \in L^q$, and $t > 0$ define*

$$\Omega_t = \{x : M_q(h)(x) > t\}.$$

Then

$$t^q |\Omega_t| \leq ||h||_q^q.$$

Proof. For $x \in \Omega_t$ there exists a uniquely determined maximal dyadic interval I_x such that

$$\int_{I_x} |h|^q > t^q |I_x|.$$

Let $\mathcal{M} = \{I_x : x \in \Omega_t\}$. Note that $\mathcal{M}$ consists of pairwise disjoint dyadic intervals which cover the set Ω_t. Hence

$$\sum_{I \in \mathcal{M}} |I| = |\Omega_t|.$$

For $I \in \mathcal{M}$ we have that

$$\int_I |h|^q > t^q |I|.$$

We sum this estimate over all $I \in \mathcal{M}$ and use that $\mathcal{M}$ consists of pairwise disjoint dyadic intervals. This gives the weak type $q - q$ estimate as follows,

$$\begin{aligned} ||h||_q^q &\geq \sum_{I \in \mathcal{M}} \int_I |h|^q \\ &\geq \sum_{I \in \mathcal{M}} t^q |I| \\ &= t^q |\Omega_t|. \end{aligned}$$

□

For $p > 1$ we use H^p to denote L^p equipped with the square function norm

$$\|g\|_{H^p} = \left| \int_0^1 g dt \right| + \left(\int_0^1 (\sum_I a_I^2 h_I^2)^{p/2} \right)^{1/p} \quad \text{where} \quad a_I = \langle g, \frac{h_I}{|I|} \rangle.$$

Assume that $q > 2$ and that $f : [0, 1] \to \mathbb{R}$ satisfies $||f^\sharp||_{L^q} < \infty$. Assume that g is a finite linear combination of Haar functions. The inequality

$$\int_0^1 f g dt \leq 2\sqrt{2} \int_0^1 f^\sharp S(g) dt$$

implies that f induces a bounded linear functional on H^p where $1/p + 1/q = 1$. The next theorem states that the converse holds as well: Every bounded linear functional on H^p induces $f : [0, 1] \to \mathbb{R}$ such that $||f^\sharp||_{L^q} < \infty$.

Theorem 1.2.8. *Let $1 \leq p < 2$, and let q be the Hölder conjugate exponent to p satisfying $1/p + 1/q = 1$. For every continuous linear functional $L : H^p \to \mathbb{R}$ there exists $f : [0, 1] \to \mathbb{R}$ so that $f^\sharp \in L^q$, with*

$$||f^\sharp||_{L^q} \leq \sqrt{2} \left(\frac{q}{q-2} \right)^{1/q} ||L||,$$

and so that for every finite linear combination of Haar functions g,

$$L(g) = \int_0^1 fgdt.$$

Proof. Let g be a finite linear combination of Haar functions. For $n \in \mathbb{N}$ denote by g_n the orthogonal projection of g onto the span$\{h_I : |I| = 2^{-n}\}$. That is,

$$g_n = \sum_{\{I:|I|=2^{-n-1}\}} \langle g, \frac{h_I}{|I|} \rangle h_I. \tag{1.2.21}$$

The following identity holds for the square function of g,

$$S(g) = (\sum g_n^2)^{1/2}.$$

We represent H^p as a closed subspace of $L^p(\ell^2)$. Let $\tilde{g} = (g_n)_{n=1}^{\infty}$. Then

$$||g||_{H^p} = ||\tilde{g}||_{L^p(\ell^2)}.$$

By the Hahn–Banach theorem we extend the linear functional L to a functional on $L^p(\ell^2)$, with the same norm. Let $\tilde{L} : L^p(\ell^2) \to \mathbb{R}$, be an extension satisfying $||L|| = ||\tilde{L}||$, and

$$L(g) = \tilde{L}(\tilde{g}) \quad \text{for} \quad g \in H^p.$$

Next recall that the dual space of $L^p(\ell^2)$ is identified as $L^q(\ell^2)$, in the following sense. There exists a sequence (f_n) such that

$$||\tilde{L}|| = ||(\sum f_n^2)^{1/2}||_{L^q}, \tag{1.2.22}$$

and for every $(h_n) \in L^p(\ell^2)$,

$$\tilde{L}((h_n)) = \sum_{n=1}^{\infty} \langle f_n h_n \rangle. \tag{1.2.23}$$

Next use the sequence (f_n), to define f. We put

$$f = \sum_{n=1}^{\infty} \sum_{\{I:|I|=2^{-n-1}\}} \langle f_n, \frac{h_I}{|I|} \rangle h_I.$$

Testing the identity (1.2.23) on the sequence (g_n) defined by (1.2.21) we obtain that

$$\begin{aligned} \tilde{L}((g_n)) &= \sum_{n=1}^{\infty} \sum_{\{I:|I|=2^{-n-1}\}} \langle g, \frac{h_I}{|I|} \rangle \langle h_I, f_n \rangle \\ &= \langle g, f \rangle. \end{aligned}$$

Thus restricted to (g_n) the action of $\tilde{L}$ is given by integration as

$$\tilde{L}((g_n)) = \int_0^1 gf\,dt.$$

Now we show that $f^\sharp \in L^q$, and that

$$||f^\sharp||_q \leq \sqrt{2}\left(\frac{q}{q-2}\right)^{1/q} ||(\sum f_n^2)^{1/2}||_q.$$

At a given point $x \in [0,1]$, we evaluate the sharp function of f by fixing a dyadic interval J such that $x \in J$. Choose $m \in \mathbb{N}$ such that $|J| = 2^{-m-1}$. By Bessel's inequality, we obtain

$$\begin{aligned}\int_J |f-(f)_J|^2 dx &= \sum_{n\geq m} \sum_{\{I\subseteq J:|I|=2^{-n-1}\}} \langle f_n, \frac{h_I}{|I|}\rangle^2 |I| \\ &\leq \int_J \sum_{n=1}^{\infty} f_n^2 dx.\end{aligned}$$

Divide by $|J|$ and take the supremum over all dyadic intervals containing the fixed point $x \in [0,1]$. Thus we restate the last estimate as a pointwise inequality between the sharp function and the Hardy–Littlewood maximal function M_1. For every $x \in [0,1]$,

$$f^\sharp(x) \leq \left(M_1(\sum_{n=1}^{\infty} f_n^2)\right)^{1/2}(x). \tag{1.2.24}$$

By our assumption $q > 2$, hence by Theorem 1.2.6 the maximal function M_1 acts boundedly on the space $L^{q/2}$. Moreover the following norm estimate holds true,

$$\|M_1(h)\|_{q/2} \leq 2\left(\frac{q}{q-2}\right)^{2/q} \|h\|_{q/2}.$$

This allows us to finish the proof as follows. By (1.2.24) and by Theorem 1.2.6,

$$\begin{aligned}||f^\sharp||_q &\leq \|M_1(\sum_n f_n^2)\|_{q/2}^{1/2} \\ &\leq \sqrt{2}\left(\frac{q}{q-2}\right)^{1/q} ||(\sum_n f_n^2)^{1/2}||_q.\end{aligned}$$

Invoking (1.2.22) gives that $||f^\sharp||_q \leq \sqrt{2}\left(\frac{q}{q-2}\right)^{1/q} \|L\|$, as claimed. □

1.3 Bounded square functions and large deviation inequalities

We return to square functions in $L^p, 1 < p < \infty$. In Theorem 1.1.6 we combined the inequalities of Burkholder, and Khintchine to prove the square function characterization of L^p asserting that for $f \in L^p$, with $\int f = 0$,

$$(2\sqrt{2})^{-1}(p^* - 1)^{-1}||f||_p \leq ||S(f)||_p \leq \sqrt{2}(p^* - 1)||f||_p.$$

In this section we show that for $2 \leq p < \infty$ the following equivalence holds,

$$\frac{1}{16\sqrt{p}}||f||_p \leq ||S(f)||_p \leq \frac{\sqrt{p}}{\sqrt{2}}||f||_p,$$

when $\int f = 0$. The constants of the above inequality which are of order $\sqrt{p}$, are known to represent the correct asymptotic behavior as $p \to \infty$. Along with the proof we introduce important tools of general interest in real analysis. These include a detailed study of the exponential function,

$$\exp\{f - \frac{1}{2}S^2(f)\},$$

large deviation inequalities, stopping time arguments, and the use of so-called good λ inequalities. We apply these methods to study Schechtman's sign-embedding of high-dimensional Euclidean subspaces into ℓ_n^1. We also show how to obtain non-trivial multipliers into SL^∞ (the Banach space consisting of functions with bounded square functions).

Square functions in L^p $(2 < p < \infty)$

We prove now a strong upper bound for the expectation of the exponentials $\exp\{f - \frac{1}{2}S^2(f)\}$. The following theorem has wide ranging applications. These include the law of the iterated logarithms, eigenvalue estimates for the Laplacian, embedding theorems for fractional Sobolev spaces, the existence of high-dimensional Euclidean sections in convex bodies and the square function inequality,

$$||f||_p \leq 16\sqrt{p}||S(f)||_p, \quad \text{for} \quad p \geq 2,$$

where $f \in L^p$ and $\int f = 0$. The notes at the end of the chapter provide the references to the applications of our next theorem.

Theorem 1.3.1. *Let $f \in L^2([0,1])$ with $\int_0^1 f dt = 0$. Then for every $\mu > 0$,*

$$\int_0^1 \exp\{\mu f - \frac{\mu^2}{2}S^2(f)\} \leq 1.$$

Proof. As $\int_0^1 f dt = 0$ we expand f in its Haar series as

$$f = \sum a_I h_I.$$

We fix a dyadic interval I. The following estimate is basic and elementary. It controls the mean values of $\exp\{a_I h_I\}$ over I.

$$\begin{aligned}\int_I \exp\{a_I h_I\} &= \frac{|I|}{2}(\exp(a_I) + \exp(-a_I))\\ &\leq |I| \exp\left\{\frac{a_I^2}{2}\right\}.\end{aligned} \tag{1.3.1}$$

Now we split the Haar series of f according to the levels $\{J : |J| = 2^{-k}\}$. Write

$$d_k = \sum_{\{J:|J|=2^{-k}\}} a_J h_J.$$

We restate the estimate (1.3.1) in the following way. For every dyadic interval I of length $|I| = 2^{-k}$,

$$\begin{aligned}\int_I \exp\{d_k - \frac{d_k^2}{2}\} &= \int_I \exp\{a_I h_I - \frac{a_I^2}{2} h_I^2\}\\ &\leq |I|.\end{aligned}$$

For future reference it is convenient to rewrite the last estimate using conditional expectations. Recall that $\mathbb{E}_k$ denotes the conditional expectation with respect to the σ-algebra generated by the dyadic intervals of length $= 2^{-k}$. We showed so far that

$$\mathbb{E}_k(\exp\{d_k - \frac{d_k^2}{2}\}) \leq 1. \tag{1.3.2}$$

Next let $n \in \mathbb{N}$. For $k \leq n-1$, the conditional expectation $\mathbb{E}_n$ reproduces the functions d_k and d_k^2. Therefore we have the following straightforward identities,

$$\begin{aligned}\int_0^1 \exp\left\{\sum_{k=0}^{n} d_k - \frac{d_k^2}{2}\right\} dt &= \int_0^1 \exp\left\{\sum_{k=0}^{n-1} d_k - \frac{d_k^2}{2}\right\}(\exp\{d_n - \frac{d_n^2}{2}\})dt\\ &= \int_0^1 \exp\left\{\sum_{k=0}^{n-1} d_k - \frac{d_k^2}{2}\right\}\mathbb{E}_n(\exp\{d_n - \frac{d_n^2}{2}\})dt.\end{aligned}$$

The basic inequality (1.3.2) and the above identities imply that

$$\int_0^1 \exp\left\{\sum_{k=0}^{n} d_k - \frac{d_k^2}{2}\right\} dt \leq \int_0^1 \exp\left\{\sum_{k=0}^{n-1} d_k - \frac{d_k^2}{2}\right\} dt. \tag{1.3.3}$$

The net effect of (1.3.3) is that we increased the value of the integral while we reduced the upper limit of summation from n to $n-1$. Now we repeat this procedure to reduce the upper limit of summation from $n-1$ to $n-2$, and we continue in this fashion until we reach the level $k=0$. This gives in summary the inequality

$$\begin{aligned}\int_0^1 \exp\left\{\sum_{k=0}^{n} d_k - \frac{d_k^2}{2}\right\} dt &\leq \int_0^1 \exp\{d_0 - \frac{d_0^2}{2}\}dt \\ &\leq 1.\end{aligned} \tag{1.3.4}$$

By the lemma of Fatou, in the estimate (1.3.4) we may pass to the limit as $n \to \infty$. As

$$\exp\left\{f - \frac{S^2(f)}{2}\right\} = \exp\left\{\sum_{k=0}^{\infty} d_k - \frac{d_k^2}{2}\right\}$$

we proved that $\int \exp\left\{f - \frac{1}{2}S^2(f)\right\} \leq 1$. Finally we fix $\mu > 0$, and replace f by (μf). This gives that

$$\int_0^1 \exp\{\mu f - \frac{\mu^2}{2} S^2(f)\} \leq 1. \qquad \square$$

The space SL^∞ is the class of functions for which the square function is uniformly bounded. Thus $f \in SL^\infty$ iff $S(f) \in L^\infty$. Our first corollary to Theorem 1.3.1 derives a strong upper bound for the distribution of functions $f \in SL^\infty$. We show a square-exponential estimate for the Hardy–Littlewood maximal function $M_1(f)$. To prove this statement we combine Theorem 1.3.1, with the weak type $1:1$ estimate of Lemma 1.2.7 and Jensen's inequality.

Corollary 1.3.2. *Let $f \in L^2([0,1])$ with $\int_0^1 f dt = 0$. If $S(f) \in L^\infty$, then*

$$|\{x : M_1(f)(x) > t\}| \leq 2\exp\left\{-\frac{t^2}{2||S(f)||_\infty^2}\right\}.$$

Proof. We assume that $||S(f)||_\infty = 1$. Write $(|f|)_I$ for the mean value of $|f|$ over the interval I. Analogously we let $(\exp\{|f|\})_I$ denote the mean value of $\exp\{|f|\}$ over I. By Jensen's inequality the mean values are related as follows,

$$\exp\{(|f|)_I\} \leq (\exp\{|f|\})_I. \tag{1.3.5}$$

The inequality (1.3.5) translates into a pointwise estimate for the Hardy–Littlewood maximal function M_1,

$$\exp\{M_1(f)\} \leq M_1(\exp\{|f|\}).$$

By Lemma 1.2.7, the weak type $1:1$ estimate for the maximal function M_1 we have that

$$|\{M_1(\exp\{\mu|f|\}) \geq \exp\alpha\}| \leq \frac{\int \exp\{\mu|f|\}}{\exp\alpha}, \tag{1.3.6}$$

for every $\mu, \alpha > 0$. Theorem 1.3.1 and the hypothesis that $||S(f)||_\infty = 1$ imply that

$$\begin{aligned}\int \exp\{\mu|f|\} &\leq \int \exp\{\mu f\} + \int \exp\{-\mu f\} \\ &\leq 2\exp\{\frac{\mu^2}{2}\}.\end{aligned} \tag{1.3.7}$$

Inserting the integral estimate (1.3.7) into the right-hand side of the measure estimate (1.3.6) gives that

$$|\{M_1(\exp\{\mu|f|\}) \geq \exp\alpha\}| \leq \frac{\int \exp\{\mu|f|\}}{\exp\alpha} \leq 2\exp\left\{\frac{\mu^2}{2} - \alpha\right\}.$$

Summing up we find the following upper bound for the distribution function of $M_1(f)$. Let $\alpha > 0$, then

$$\begin{aligned}|\{\mu M_1(f) > \alpha\}| &= |\{\exp\{\mu M_1(f)\} > \exp\alpha\}| \\ &\leq |\{M_1(\exp(\mu|f|) > \exp\alpha\}| \\ &\leq 2\exp\left\{\frac{\mu^2}{2} - \alpha\right\}.\end{aligned}$$

Specializing this estimate to $\mu = \sqrt{\alpha} = t$, gives that

$$|\{M_1(f) > t\}| \leq 2\exp\left\{-\frac{t^2}{2}\right\},$$

provided that $||S(f)||_\infty = 1$. We get rid of the restriction $||S(f)||_\infty = 1$ by applying the above special case to $f/||S(f)||_\infty$. □

We will next discuss rescaled versions of Corollary 1.3.2. Fix a dyadic interval Q . We now introduce the Hardy–Littlewood maximal function localized to the interval Q. Here we take the supremum only over those intervals which are contained in Q. For $x \in Q$ we put,

$$M_{1,Q}(f)(x) = \sup_{I \subseteq Q,\ x \in I} \frac{1}{|I|} \int_I |f|.$$

Let $f \in L^2$. We denote the mean value of f over the interval Q by f_Q, that is,

$$f_Q = \frac{1}{|Q|} \int_Q f\,dx.$$

Then by rescaling Corollary 1.3.2 from the unit interval to the interval Q, we obtain the following renormalized distributional estimate: If $S(f) \in L^\infty$, then

$$|\{x \in Q : M_{1,Q}(f - f_Q)(x) > t\}| \leq 2|Q|\exp\left\{-\frac{t^2}{2||S(f)||_\infty^2}\right\}. \tag{1.3.8}$$

Our aim is still the square function estimate

$$||f||_p \leq 16p^{1/2}||S(f)||_p. \tag{1.3.9}$$

The essential step in the proof is the next proposition which gives an estimate for the distribution of $M_1(f)$ over the set where $S(f)$ satisfies a pointwise upper bound. The distributional estimates of Proposition 1.3.3 are traditionally called good λ inequalities, or more descriptively, relative distributional estimates. We remark that the appearance of the constant $16p^{1/2}$ in the square function estimate (1.3.9) can be traced to the factor $\exp\left\{-\frac{1}{4\epsilon^2}\right\}$ on the right hand side of the good λ inequality, below. The factor $\exp\left\{-\frac{1}{4\epsilon^2}\right\}$ has in turn its origin in the integral estimate of the exponential

$$\int \exp\{f - \frac{1}{2}S^2(f)\} \leq 1.$$

Proposition 1.3.3. *Let $f \in L^2([0,1])$ with $\int_0^1 f dt = 0$. Then for every $0 < \epsilon \leq 1 - 1/\sqrt{2}$, and $t > 0$ the following distributional inequality holds,*

$$|\{M_1(f) > 2t, S(f) < \epsilon t\}| \leq 2\exp\left\{-\frac{1}{4\epsilon^2}\right\}|\{M_1(f) > t\}|.$$

Proof. Fix $t > 0$, and define $\Omega_t = \{x : M_1(f) > t\}$. For every $x \in \Omega_t$, there exists a maximal dyadic interval I_x such that

$$\frac{1}{|I_x|}\int_{I_x} |f| > t.$$

Let $\mathcal{M} = \{I_x : x \in \Omega_t\}$. The collection $\mathcal{M}$ covers the set Ω_t. By the maximality condition, $\mathcal{M}$ consists of pairwise disjoint dyadic intervals, hence,

$$\sum_{\{Q\in\mathcal{M}\}} |Q| = |\{M_1(f) > t\}|.$$

Maximality gives also the following implication: If $Q \in \mathcal{M}$, and if Q^* denotes the dyadic predecessor of Q, then

$$\frac{1}{|Q^*|}\int_{Q^*} f \leq t.$$

Let $Q \in \mathcal{M}$ and assume that there exists $x \in Q$ such that $M_1(f)(x) > 2t$. Observe that then the relevant contribution to the value of the maximal function $M_1(f)$ at the point x necessarily comes from the intervals which are contained in Q (and not from those containing Q). Precisely, if $Q \in \mathcal{M}$, and $M_1(f)(x) > 2t$, then the following equality holds at the point x,

$$M_1 f(x) = M_{1,Q} f(x). \tag{1.3.10}$$

Expand f in its Haar series,

$$f = \sum a_I h_I.$$

Define $\mathcal{L}$ to be the collection of dyadic intervals I such that the following estimate holds true,

$$\sum_{J \supseteq I} a_J^2 h_J^2(x) \geq \epsilon^2 t^2,$$

for some (or equivalently for all) $x \in I$. Let L be the set of points which is covered by the collection $\mathcal{L}$, thus

$$L = \bigcup_{I \in \mathcal{L}} I.$$

Define $\mathcal{S} = \mathcal{D} \setminus \mathcal{L}$, and put $f_t = \sum_{J \in \mathcal{S}} a_J h_J$. The reason for our choice of $\mathcal{L}$ and $\mathcal{S}$ is that now, by definition, f_t has uniformly bounded square function and satisfies

$$||S(f_t)||_\infty \leq \epsilon t. \tag{1.3.11}$$

Observe that if $x \notin L$, then

$$f_t(x) = f(x), \tag{1.3.12}$$

and furthermore, if J is a dyadic interval containing x we have that

$$\int_J f_t = \int_J f. \tag{1.3.13}$$

So far we defined two important collections of dyadic intervals, $\mathcal{M}$ and $\mathcal{L}$ and we obtained several consequences encoded in their very definitions. Now we analyze the interplay between $\mathcal{M}$ and $\mathcal{L}$. We fix $Q \in \mathcal{M}$, and we assume that there exists $x \in Q$ such that $x \notin L$. Then we claim that the following integral estimate holds,

$$\frac{1}{|Q|} \int_Q f \leq (1 + \epsilon)t. \tag{1.3.14}$$

To verify this claim we consider the dyadic predecessor of Q and call it Q^*. Note that the coefficient a_{Q^*} appearing in the Haar expansion of f is obtained from the mean values of f over Q and over Q^* as follows,

$$\left| \frac{1}{|Q^*|} \int_{Q^*} f - \frac{1}{|Q|} \int_Q f \right| = |a_{Q^*}|. \tag{1.3.15}$$

Recall that for $Q \in \mathcal{M}$, the mean value of $|f|$ over Q^* is bounded by t, hence

$$\frac{1}{|Q^*|} \int_{Q^*} f \leq \frac{1}{|Q^*|} \int_{Q^*} |f| \leq t. \tag{1.3.16}$$

Furthermore if $x \in Q$, and $x \notin L$, then

$$|a_{Q^*}| \leq \epsilon t. \tag{1.3.17}$$

Inserting the estimates (1.3.17) and (1.3.16) into the identity (1.3.15) gives the upper bound

$$\frac{1}{|Q|}\int_Q f \le (1+\epsilon)t,$$

hence inequality (1.3.14) holds true.

Next we claim that for $Q \in \mathcal{M}$, the following estimate holds,

$$|\{x \in Q : M_1(f)(x) > 2t, S(f)(x) < \epsilon t\}| \le 2|Q| \exp\left\{-\frac{1}{4\epsilon^2}\right\}. \tag{1.3.18}$$

We start proving this claim by first rewriting the set whose measure we wish to estimate. Using first (1.3.10) then (1.3.13) we obtain that,

$$\begin{aligned}\{x \in Q : M_1(f)(x) > 2t, S(f)(x) < \epsilon t\} &= \{x \in Q : M_{1,Q}(f)(x) > 2t, x \notin L\}\\ &= \{x \in Q : M_{1,Q}(f_t)(x) > 2t, x \notin L\}.\end{aligned}$$

We write $(f_t)_Q$ for the mean value of f_t over Q. By (1.3.13), for $x \in Q$ and $x \notin L$ the mean values $(f_t)_Q$ and $(f)_Q$ coincide. Invoking (1.3.14) gives that

$$|(f_t)_Q| \le (1+\epsilon)t.$$

By inserting $(f_t)_Q$ we obtain the following inclusion of sets, which incidentally is a one line summary of the proof given so far.

$$\{x \in Q : M_{1,Q}(f_t)(x) > 2t, x \notin L\} \subseteq \{x \in Q : M_{1,Q}\left(f_t - (f_t)_Q\right)(x) > (1-\epsilon)t\}.$$

By definition f_t has uniformly bounded square function with $||S(f_t)||_\infty \le \epsilon t$. Hence (1.3.8) gives the measure estimate

$$|\{x \in Q : M_{1,Q}\left(f_t - (f_t)_Q\right)(x) > (1-\epsilon)t\}| \le 2|Q| \exp\left\{-\frac{(1-\epsilon)^2 t^2}{2\epsilon^2 t^2}\right\}.$$

Now observe that inside the exponential the factors t^2 cancel and that for $0 < \epsilon \le 1 - 1/\sqrt{2}$, we have clearly $(1-\epsilon)^2/2 \ge 1/4$. Thus we verified the claimed inequality (1.3.18).

Finally recall that $\mathcal{M}$ is a collection of pairwise disjoint dyadic intervals, covering the set $\{M_1(f) > t\}$. Now we sum over $Q \in \mathcal{M}$, thereby finishing the proof of Proposition 1.3.3:

$$\begin{aligned}|\{M_1(f) > 2t, S(f) < \epsilon t\}| &= \sum_{Q \in \mathcal{M}} |\{x \in Q : M_1(f)(x) > 2t, S(f)(x) < \epsilon t\}|\\ &\le 2\exp\left\{-\frac{1}{4\epsilon^2}\right\} \sum_{Q \in \mathcal{M}} |Q|\\ &= 2\exp\left\{-\frac{1}{4\epsilon^2}\right\} |\{M_1(f) > t\}|.\end{aligned}$$

□

We now prove a square function characterization of L^p with the correct asymptotic behavior of the constants as $p \to \infty$.

Theorem 1.3.4. *Let $p \geq 2$, and let $f \in L^p([0,1])$ with $\int_0^1 f dt = 0$. Then*

$$||f||_p \leq ||M_1(f)||_p \leq 16\sqrt{p}||S(f)||_p.$$

Proof. We wish to evaluate $||M_1(f)||_p^p$. Fix $t > 0$ and $\epsilon > 0$. We start the proof with the triangle inequality,

$$|\{M_1(f) > 2t\}| \leq |\{M_1(f) > 2t, S(f) < \epsilon t\}| + |\{S(f) \geq \epsilon t\}|. \tag{1.3.19}$$

The effect of the second summand $|\{S(f) \geq \epsilon t\}|$ is easy to evaluate. We have

$$p \int_0^\infty t^{p-1}|\{S(f) \geq \epsilon t\}|dt = \epsilon^{-p}||S(f)||_p^p. \tag{1.3.20}$$

The relative distributional estimate of Proposition 1.3.3 controls the contribution of the first summand. It gives

$$|\{M_1(f) > 2t, S(f) < \epsilon t\}| \leq 2\exp\left\{-\frac{1}{4\epsilon^2}\right\}|\{M_1(f) > t\}|. \tag{1.3.21}$$

We point out that it is the presence of factor $\exp\{-(1/4\epsilon^2)\}$ that allows us to compensate the effect caused by ϵ^{-p} appearing in (1.3.20). Combining the estimates (1.3.19) and (1.3.20) with the relative distributional estimate (1.3.21) gives the following bound for $||M_1(f)||_p^p$.

$$\begin{aligned}\frac{1}{2^p}\int_0^1 M_1(f)^p(x) &= p\int_0^\infty t^{p-1}|\{M_1(f) > 2t\}|dt \\ &\leq p\int_0^\infty t^{p-1}|\{M_1(f) > 2t, S(f) < \epsilon t\}|dt + \epsilon^{-p}||S(f)||_p^p \\ &\leq 2p\exp\{-\frac{1}{4\epsilon^2}\}\int_0^\infty t^{p-1}|\{M_1(f) > t\}|dt + \epsilon^{-p}||S(f)||_p^p \\ &= 2\exp\{-\frac{1}{4\epsilon^2}\}||M_1(f)||_p^p + \epsilon^{-p}||S(f)||_p^p.\end{aligned}$$

Now subtract $\exp\{-(1/\epsilon^2)\}||M_1(f)||_p^p$ on both sides. This gives

$$\left(\frac{1}{2^p} - 2\exp\{-\frac{1}{\epsilon^2}\}\right)||M_1(f)||_p^p \leq \epsilon^{-p}||S(f)||_p^p. \tag{1.3.22}$$

Note that (1.3.22) can become a meaningful estimate only if ϵ is chosen so small that the factor in front of $||M_1(f)||_p^p$ is a positive number. For $p \geq 2$ define

$$\epsilon = \frac{1}{(4\ln 2)^{1/2}(p+2)^{1/2}}.$$

Then $\exp\{-(1/4\epsilon^2)\} \leq 2^{-p-1}$, and $\epsilon^{-p} \leq (8\ln 2)^{p/2} p^{\frac{p}{2}}$. This gives

$$||M_1(f)||_p^p \leq 16^p p^{\frac{p}{2}} ||S(f)||_p^p.$$

It remains to raise both sides to the power $1/p$. □

For $f \in L^p$ with $\int f = 0$ we established the square function estimate

$$||f||_p \leq 16\sqrt{p}||S(f)||_p \quad \text{for} \quad 2 \leq p < \infty.$$

Now we turn to proving the converse inequality. In the proof below we present a method due to A. M. Garsia [75].

Theorem 1.3.5. *Let $p > 2$ and $f \in L^p$. Then,*

$$||S(f)||_p \leq \frac{\sqrt{p}}{\sqrt{2}}||f||_p.$$

Proof. To start the proof we expand f in its Haar series, $f = \int f + \sum a_I h_I$. Then we introduce the following notation to denote partial sum approximations to $S^2(f)$,

$$S_{-1}^2 = 0, \quad \text{and} \quad S_n^2(t) = \sum_{\{I:|I|\geq 2^{-n}\}} a_I^2 h_I^2(t) \quad \text{for} \quad n \in \mathbb{N}_0.$$

We also abbreviate the notation for the square function of f, by writing $S = S(f)$. Below we state two crucial identities, and an inequality which uses that for $p > 2$, the function $r^{p/2-1}$ is increasing on the positive real axis.

$$\begin{aligned} \frac{2}{p} S^p(f)(t) &= \int_0^{S^2(t)} r^{p/2-1} dr \\ &= \sum_{n=0}^{\infty} \int_{S_{n-1}^2(t)}^{S_n^2(t)} r^{p/2-1} dr. \\ &\leq \sum_{n=0}^{\infty} S_n^{p-2}(t) \left\{ S_n^2(t) - S_{n-1}^2(t) \right\}. \end{aligned} \tag{1.3.23}$$

For brevity it is convenient to introduce more notation. Let $n \in \mathbb{N}_0$, then put

$$m_n = S_n^{p-2} - S_{n-1}^{p-2}.$$

Using the summation by parts formula we next rewrite the last line of (1.3.23),

$$\sum_{n=0}^{\infty} S_n^{p-2}(t) \left\{ S_n^2(t) - S_{n-1}^2(t) \right\} = \sum_{n=0}^{\infty} m_n \{ S^2(t) - S_{n-1}^2(t) \}. \tag{1.3.24}$$

Next observe that $\mathbb{E}_n$, the conditional expectation with respect to the σ-algebra generated by the dyadic intervals of length $= 2^{-n}$ reproduces m_n. That is

$$\mathbb{E}_n(m_n) = m_n.$$

Integrating the estimate (1.3.23) over the unit interval and taking into account the identity (1.3.24) we find

$$\frac{2}{p}\int_0^1 S^p(f)(t)dt \leq \sum_{n=0}^{\infty}\int_0^1 m_n\mathbb{E}_n(S^2(t) - S_{n-1}^2(t))dt. \tag{1.3.25}$$

Observe that the conditional expectations $\mathbb{E}_n(S^2 - S_{n-1}^2)(t)$ satisfy the pointwise estimate,

$$\mathbb{E}_n(S^2 - S_{n-1}^2) \leq \mathbb{E}_n(f^2). \tag{1.3.26}$$

Inserting the estimate (1.3.26) into the right-hand side of (1.3.25) and using again that $\mathbb{E}_n(m_n) = m_n$, allows us to continue the right-hand side of (1.3.25) as follows,

$$\sum_{n=0}^{\infty}\int_0^1 m_n\mathbb{E}_n(S^2 - S_{n-1}^2)dt \leq \sum_{n=0}^{\infty}\int_0^1 m_n f^2 dt. \tag{1.3.27}$$

Finally we change the order of integration and summation, and sum the resulting telescoping series $\sum_{k=0}^{\infty} m_k = S^{p-2}$. This gives that

$$\sum_{n=0}^{\infty}\int_0^1 m_n f^2 dt = \int_0^1 S^{p-2} f^2 dt. \tag{1.3.28}$$

Thus starting from (1.3.23) we showed with (1.3.25), (1.3.27) and (1.3.28), this crucial inequality,

$$\frac{2}{p}\int_0^1 S^p(f)(t)dt \leq \int_0^1 S^{p-2} f^2 dt.$$

Now we use Hölder's inequality with $a = p/(p-2)$ and $b = p/2$. This gives

$$\begin{aligned}\int_0^1 S^{p-2} f^2 dt &\leq \left(\int_0^1 S^{(p-2)a}\right)^{1/a}\left(\int_0^1 f^{2b}\right)^{1/b}\\ &= \left(\int_0^1 S^p\right)^{(p-2)/p}\left(\int_0^1 |f^p|\right)^{2/p}.\end{aligned}$$

Summing up we have that

$$\frac{2}{p}\int_0^1 S^p \leq \left(\int_0^1 S^p\right)^{(p-2)/p}\left(\int_0^1 |f^p|\right)^{2/p}.$$

Cancelling the factor $(\int_0^1 S^p)^{(p-2)/p}$ on both sides gives that

$$\frac{2}{p}||S(f)||_p^2 \leq ||f||_p^2.$$

□

In the following paragraphs we will discuss three different applications of the square function characterization of L^p and the SL^∞ estimates that formed the basis of its proof. First, we compare sharp functions with square functions in L^p and let $p \to \infty$, second we use SL^∞ estimates to produce Euclidean subspaces in ℓ_n^1, and third we exploit the $\sqrt{p}$ dependence in Theorem 1.3.5 to prove that the square function of a bounded function is square-exponentially integrable.

Endpoints of the L^p scale. Let $f \in L^p$ with $\int f = 0$. Combining Theorem 1.3.4 and Theorem 1.3.5 gives that for $p \geq 2$,

$$(\sqrt{2}/\sqrt{p})\|S(f)\|_p \leq \|f\|_p \leq 16\sqrt{p}\|S(f)\|_p. \tag{1.3.29}$$

Recall that a function is in SL^∞ if its square function is uniformly bounded. Thus SL^∞ appears as an endpoint of the scale of L^p spaces for $p \to \infty$. Compare this to the renorming obtained in the previous section. For $p \geq 2$,

$$(p-2)/(2p)\|f^\sharp\|_p \leq \|f\|_p \leq 8(p-1)\|f^\sharp\|_p. \tag{1.3.30}$$

This gives a different endpoint to the scale of L^p spaces for $p \to \infty$, namely BMO. A glance at the definitions of square functions and sharp maximal functions shows that SL^∞ is contained in BMO. However much more is true. A result of J. Garnett and P. W. Jones [73] displays the relation between the three spaces L^∞, BMO and SL^∞ that appeared as endpoints of the L^p scale. It asserts that for any $f \in SL^\infty$ with $S(f) \leq 1$ and $\epsilon > 0$ there exists $g \in L^\infty$ satisfying

$$\|f - g\|_{\mathrm{BMO}} \leq \epsilon \quad \text{and} \quad \|g\|_\infty \leq A\epsilon^{-1}, \tag{1.3.31}$$

where $A > 0$ is a universal constant. In particular the closure of L^∞ in BMO contains SL^∞.

Applying the Garnett–Jones result, J. Bourgain [18] obtains a remarkable estimate for the Rademacher projection on H^1. Recall that $\{r_m\}$ denotes the Rademacher system consisting of $\{+1,-1\}$ valued, independent random variables. J. Bourgain's inequality states that

$$\sum_{m=1}^{\infty} \langle h, r_m\rangle^2 \leq C\|h\|_{L^1}\|h\|_{H^1}, \quad \text{for} \quad h \in H^1. \tag{1.3.32}$$

The interesting aspect of J. Bourgain's inequality is the appearance of the L^1 norm on the right-hand side. Replacing $\|h\|_{L^1}\|h\|_{H^1}$ by $\|h\|_{H^1}^2$ would turn it into a trivial estimate. On the other hand replacing $\|h\|_{L^1}\|h\|_{H^1}$ by the smaller number $\|h\|_{L^1}^2$ would lead to a false statement.

To prove (1.3.32) put

$$a_m = \langle h, r_m\rangle\left(\sum \langle h, r_n\rangle^2\right)^{-1/2},$$

and let $f = \sum a_m r_m$. Since $\sum a_m^2 = 1$, we have $f \in SL^\infty$ with $S(f) = 1$. Next define

$$\epsilon = \|h\|_{L^1}^{1/2} \|h\|_{H^1}^{-1/2}.$$

By the Garnett–Jones result there exists $g \in L^\infty$ so that

$$\|f - g\|_{\mathrm{BMO}} \leq \epsilon \quad \text{and} \quad \|g\|_\infty \leq A\epsilon^{-1}. \tag{1.3.33}$$

With the inequalities of Fefferman and Hölder we obtain that

$$\begin{aligned} \left(\sum \langle h, r_m \rangle^2\right)^{1/2} &= \langle h, f \rangle \\ &= \langle h, f - g \rangle + \langle h, g \rangle \\ &\leq 2\sqrt{2}\|h\|_{H^1} \|f - g\|_{\mathrm{BMO}} + \|h\|_{L^1} \|g\|_\infty. \end{aligned} \tag{1.3.34}$$

Use the estimates (1.3.33) in the last line of (1.3.34). Then our choice of $\epsilon > 0$ gives J. Bourgain's inequality (1.3.32).

Schechtman's sign-embedding. In this section we show how SL^∞ estimates lead to Euclidean subspaces of proportional dimension in ℓ_n^1. The resulting embedding — Schechtman's sign-embedding — turns out to be quite explicit.

We write ℓ_n^1 to denote $\mathbb{R}^n$ equipped with the norm of the sequence space ℓ^1. Likewise we denote by ℓ_k^2 the space $\mathbb{R}^k$ equipped with the norm of the Hilbert space ℓ^2. A straightforward exercise with Khintchine's inequality yields a subspace $X \subseteq \ell_n^1$ and a linear bijection $i : \ell_k^2 \to X$ so that

$$k = [\log_2 n] \quad \text{and} \quad \|i\| \cdot \|i^{-1}\| \leq 16.$$

It is a much harder problem to find such an embedding $i : \ell_k^2 \to X$ under the hypothesis that the algebraic dimension of the subspace X is proportional to the dimension of the ambient space ℓ_n^1. That is, under the hypothesis that $k = [\delta n]$. Here, of course, we demand that the factor of proportionality $n/k \sim \delta^{-1}$ is an absolute constant, not depending on n.

In the seminal paper [186] G. Schechtman develops an approach to these problems based integral estimates for the exponentials $\exp\{f - \frac{1}{2}S^2(f)\}$. Specifically, G. Schechtman's construction of the embedding is based on Khintchine's inequality and Corollary 1.3.2 asserting that

$$\left|\left\{\left|f - \int_0^1 f\right| > t\right\}\right| \leq 2\exp\left\{-\frac{t^2}{2\|S(f)\|_\infty^2}\right\}. \tag{1.3.35}$$

Subsequently, martingale methods based on (1.3.35) found numerous applications to related problems in the local theory of Banach spaces. See the notes for references that give an overview of this development.

Next we turn to defining G. Schechtman's embedding. Let $\{e_i : i \leq n\}$ denote the unit vector basis in ℓ_n^1. Thus for coefficients $b_i \in \mathbb{R}$ we get

$$\left\| \sum_{i=1}^{n} b_i e_i \right\|_{\ell_n^1} = \sum_{i=1}^{n} |b_i|.$$

Recall that $\{r_l\}$ denotes the Rademacher system. Fix $k \leq n$ and define

$$x_j(t) = \frac{1}{n} \sum_{i=1}^{n} r_{ik+j}(t) e_i, \quad \text{where} \quad j \leq k \quad \text{and} \quad t \in [0,1].$$

Thus $x_j(t)$ is formed by attaching random signs $\{\pm 1\}$ to the unit vectors of ℓ_n^1 and averaging. Note that for each $t \in [0,1]$ the vectors $x_j(t)$ are of norm 1 in ℓ_n^1. The next theorem asserts that for some $t \in [0,1]$, they are equivalent to the unit vector basis of ℓ_k^2 provided that $k \sim \delta n$.

Theorem 1.3.6. *There exists $\delta > 0$ so that the following holds true. For each $n \in \mathbb{N}$ and $k \leq \delta n$ there exist $t \in [0,1]$ so that*

$$\frac{1}{4} \left(\sum_{j=1}^{k} a_j^2 \right)^{1/2} \leq \left\| \sum_{j=1}^{k} a_j x_j(t) \right\|_{\ell_n^1} \leq 4 \left(\sum_{j=1}^{k} a_j^2 \right)^{1/2}, \tag{1.3.36}$$

for every choice of coefficients $a = (a_j) \in \mathbb{R}^k$.

Proof. Fix coefficients $a = (a_j) \in \mathbb{R}^k$, and assume that $\sum_{j=1}^{k} a_j^2 = 1$. The first part of the proof shows that with $a = (a_j)$ *fixed* the estimates (1.3.36) hold for a very large set of $t \in [0,1]$. The purpose of the second part is to exploit this fact and extend (1.3.36) from *one* set of coefficients to the *entire unit sphere of ℓ_k^2*.

Part 1. For $t \in [0,1]$, we first define the function $f(t)$ and then unwind the definition of the random vectors $x_j(t)$. Thus,

$$\begin{aligned} f(t) &= \left\| \sum_{j=1}^{k} a_j x_j(t) \right\|_{\ell_n^1} \\ &= \frac{1}{n} \sum_{i=1}^{n} \left| \sum_{j=1}^{k} a_j r_{ik+j}(t) \right|. \end{aligned} \tag{1.3.37}$$

Next we estimate the Haar coefficients of f, its square function and its mean value over the interval $[0,1]$. Starting with the Haar coefficients we fix $i \leq n$ and

$m \leq k$. Let I be a dyadic interval. The statistical independence of the Rademacher functions gives the following implication:

$$\text{If} \quad \langle h_I, r_{ik+m}\rangle \neq 0, \quad \text{then} \quad \left|\int \left|\sum_{j=1}^{k} a_j r_{ik+j}(t)\right| h_I(t)dt\right| \leq |I| \cdot |a_m|. \tag{1.3.38}$$

The remaining Haar coefficients vanish. By (1.3.38) the square function of f satisfies the following upper bound.

$$\begin{aligned} S^2(f)(t) &\leq \frac{1}{n^2}\sum_{i=1}^{n}\sum_{m=1}^{k} a_m^2 \\ &\leq \frac{1}{n}\sum_{j=1}^{k} a_m^2. \end{aligned} \tag{1.3.39}$$

Using Khintchine's inequality (for the left-hand side of (1.3.40)) and the biorthogonality of the Rademacher functions (for the right-hand side of (1.3.40)) we obtain that

$$(\sqrt{2})^{-1}\left(\sum_{j=1}^{k} a_j^2\right)^{1/2} \leq \int_0^1 \left|\sum_{j=1}^{k} a_j r_{ik+j}(t)\right| dt \leq \left(\sum_{j=1}^{k} a_j^2\right)^{1/2}. \tag{1.3.40}$$

Recall now that we imposed that $\sum a_j^2 = 1$. In summary we obtained that

$$(\sqrt{2})^{-1} \leq \int_0^1 f(t)dt \leq 1 \quad \text{and} \quad S^2(f)(t) \leq \frac{1}{n}. \tag{1.3.41}$$

It follows from (1.3.41) and (1.3.35) that this deviation inequality holds,

$$\left|\left\{|f - \int_0^1 f| > s\right\}\right| \leq 2\exp\left\{-\frac{s^2 n}{2}\right\}. \tag{1.3.42}$$

By (1.3.41) we get $\{f \geq 2 \quad \text{or} \quad f \leq (2\sqrt{2})^{-1}\} \subseteq \{|f - \int_0^1 f| > (2\sqrt{2})^{-1}\}$. Taking complements and using (1.3.42) with $s = (2\sqrt{2})^{-1}$ gives that

$$\left|\left\{(2\sqrt{2})^{-1} \leq f \leq 2\right\}\right| \geq 1 - 2\exp\left\{-\frac{n}{16}\right\}. \tag{1.3.43}$$

For obvious reasons, (1.3.43) is frequently called the concentration of measure inequality for f. Summing up the first part of the proof, the function f, formed by a single set of coefficients $a = (a_j)$ is bounded (from above and below) on an extremely large subset of the unit interval.

Part 2. Next we pass from a fixed set of coefficients $a = (a_j)$ to a sufficiently large collection of coefficients in the unit sphere of ℓ_k^2. It is at this point of the argument that we will impose a restriction on the size of k.

Let $\epsilon_0 > 0$. Denote by S^{k-1} the unit sphere of ℓ_k^2. Choose a discrete set $M \subseteq S^{k-1}$ — with minimal cardinality — so that for each $b \in S^{k-1}$ there exists $a \in M$ satisfying $|a - b| \leq \epsilon_0$. Verify, by counting, that

$$\text{the cardinality of } M \text{ is bounded by } \exp\{+2k/\epsilon_0\}. \tag{1.3.44}$$

(See, for instance, Proposition 10.II.E in [213].) Now choose $\epsilon_0 > 0$ small enough so that this implication holds: If

$$(2\sqrt{2})^{-1} \leq \left\| \sum_{j=1}^{k} a_j x_j(t) \right\|_{\ell_n^1} \leq 2 \quad \text{for all} \quad (a_j) \in M, \tag{1.3.45}$$

then

$$\frac{1}{4} \leq \left\| \sum_{j=1}^{k} a_j x_j(t) \right\|_{\ell_n^1} \leq 4 \quad \text{for all} \quad (a_j) \in S^{k-1}. \tag{1.3.46}$$

An elementary iteration argument allows one to choose $\epsilon_0 > 0$ satisfying the above property, and ϵ_0 does not depend on the value of k. (The details of the iteration are worked out in [148].) Next apply the first part of the proof to each $a = (a_j) \in M$, and define

$$E_a = \left\{ t : (2\sqrt{2})^{-1} \leq \left\| \sum_{j=1}^{k} a_j x_j(t) \right\|_{\ell_n^1} \leq 2 \right\} \quad \text{and} \quad E = \bigcap_{a \in M} E_a.$$

Inequality (1.3.43) states that $|E_a| > 1 - 2\exp\{-n/16\}$. By (1.3.44), the measure of the intersection E is therefore larger than

$$1 - 2\exp\left\{\frac{-n}{16}\right\}\exp\left\{\frac{+2k}{\epsilon_0}\right\}.$$

Hence, for $k \leq (n/32 - 2)\epsilon_0$ say , the set E is non-empty. For each $t \in E$, by its definition, the estimates (1.3.45) hold true. Passing from (1.3.45) to (1.3.46) completes the proof. □

We emphasize that (1.3.44) and (1.3.43) are the central and competing inequalities of the preceding proof.

The square function of bounded functions. We exploit the constants $\sqrt{p}$ appearing in Theorem 1.3.5 to prove that the square function of a bounded function is square-exponentially integrable.

Corollary 1.3.7. *Let $f \in L^\infty$. Then for $0 < \alpha < 1/e$,*

$$\int_0^1 \exp\{\alpha \frac{S^2(f)}{||f||^2_{L^\infty}}\} dt < \infty.$$

Proof. We use the normalization $||f||_{L^\infty} = 1$. Expanding the exponential in its Taylor series gives

$$\exp\{\alpha S^2(f)\} = \sum_{n=0}^{\infty} \frac{\alpha^n}{n!} S(f)^{2n}.$$

Apply Theorem 1.3.5, with $p = 2n$ to f. This gives

$$\int_0^1 S(f)^{2n} dt \le n^n ||f||^{2n}_{2n}.$$

Note that $n^n/n! \le e^n$, since $n^n/n!$ is just one of the summands in the Taylor expansion of e^n. Then we have

$$\begin{aligned}\int_0^1 \exp\{\alpha S^2(f)\} dt &= \sum_{n=0}^{\infty} \frac{\alpha^n}{n!} \int_0^1 S(f)^{2n} dt \\ &\le \sum_{n=0}^{\infty} \frac{\alpha^n}{n!} n^n \\ &\le \sum_{n=0}^{\infty} \alpha^n e^n.\end{aligned}$$

For $\alpha < 1/e$ the geometric series converges. □

Quadratic exponential integrability is the best estimate available for $S(f)$ when f is uniformly bounded. To see this consider the following example. Let $I_n = [0, 4^{-n})$. The sequence I_n is decreasing, $I_n \supset I_{n+1}$. Next define the following sequence of pairwise disjoint intervals ,

$$K_n = [\frac{3}{4} 4^{-n}, 4^{-n}),$$

and let,

$$f = \sum_{n=0}^{\infty} 1_{K_n}. \tag{1.3.47}$$

As the intervals K_n are pairwise disjoint, f is bounded by 1. Now we compute the Haar coefficients. We have

$$\int h_{I_n} f dt = \frac{-|I_n|}{6};$$

the other coefficients in the Haar expansion of f vanish. Hence,

$$S^2(f)(t) = \sum_{n=0}^{\infty} \frac{1}{36} 1_{I_n}.$$

It follows that

$$S^2(f)(t) = \frac{n}{36} \quad \text{for } t \in I_n \setminus I_{n+1}.$$

This shows that $S(f)$ is unbounded and the distribution function of $S(f)$ satisfies the lower estimate

$$|\{S(f) > t\}| \geq A_0 \exp\{-a_0 t^2\}.$$

Clearly this shows that quadratic exponential integrability is the best one can get for the square function of f, even when f is uniformly bounded. Observe however that for the example (1.3.47) and $m(t) = t$ the product mf is non-trivial in the sense that $\int mf > 0$ and $mf \in SL^\infty$.

In the next subsection we will find such a multiplier for every bounded non-negative function f.

Multipliers into SL^∞

To prove the good-λ inequality in Proposition 1.3.3 we produced functions in SL^∞ by deleting certain coefficients in the Haar expansion of a given function. Thereby we contracted the Haar support and lost any information about its pointwise support. In this subsection we present a method to produce elements in SL^∞ by contracting simultaneously the Haar coefficients and the pointwise support of a given function. Let $f \in L^\infty([0,1])$ be non-negative. Let its Haar expansion be

$$f = \int_0^1 f + \sum a_I h_I.$$

In the next theorem we define a multiplier $m : [0,1] \to [0,1]$, such that

$$h = m \cdot f,$$

is non-trivial and has uniformly bounded square function. We determine m so that the operation

$$f \longmapsto mf$$

is a pointwise contraction on f that also contracts each individual Haar coefficient of f to the extent that h has a bounded square function. The product mf is non-trivial in the sense that $\int mf \geq \exp(-\|f\|_\infty) \int f$. Replacing f with $f/\|f\|_\infty$ allows us to restrict ourselves to the case where f is normalized in $L^\infty([0,1])$.

Theorem 1.3.8. *Let $f \in L^\infty([0,1])$, and assume that $0 \le f \le 1$. Let f_I be the mean value of f over the interval I. Then there are constants $m_I \in [1/6, 6]$ so that the multiplier $m : [0,1] \to [0,1]$, defined by*

$$m = \exp(f - \|f\|_\infty) \exp\left(- \sum a_I^2 m_I (\frac{1}{2} + \frac{1}{f_I}) h_I^2 \right)$$

satisfies the following conditions: The product $h = m \cdot f$ is non-trivial, and

$$\int_0^1 h \ge \exp(-\|f\|_\infty) \int_0^1 f.$$

The Haar coefficients of $h = m \cdot f$ satisfy the upper bound

$$|\langle h, h_I \rangle| \le A|I| \exp\left(- \sum_{J \supseteq I} a_J^2 \right) |a_I|.$$

Consequently the dyadic square function of h is uniformly bounded,

$$\sum_I \left\langle h, \frac{h_I}{|I|} \right\rangle^2 h_I^2 \le CA^2.$$

Proof. Clearly $e^{-S^2(f)}$ is the first candidate for a multiplier to satisfy the conclusions of our theorem. However there are serious obstacles that prevent one from getting satisfactory estimates for the Haar coefficients $\langle f e^{-S^2(f)}, h_I \rangle$. The construction of the multiplier φ below is in direct response to these difficulties; indeed the combinatorial identity (1.3.48) combined with the estimates (1.3.51) are crucial in proving the main inequality (1.3.52) for the Haar coefficients of $f e^{\varphi}$.

Overview. Given $0 \le f \le 1$, we write down its Haar expansion,

$$f = \int_0^1 f + \sum a_I h_I.$$

First we show that for every dyadic interval there exists a constant m_I satisfying

$$\frac{1}{6} \le m_I \le 6,$$

and such that the following identity holds,

$$\int_I f \exp\left(a_I h_I - a_I^2 \left(\frac{1}{2} + \frac{1}{f_I} \right) m_I h_I^2 \right) dt = \int_I f. \tag{1.3.48}$$

Once the sequence $\{m_I\}$ satisfying (1.3.48) is determined we put

$$\varphi = f - \|f\|_\infty - \sum_I a_I^2 m_I \left(\frac{1}{2} + \frac{1}{f_I} \right) h_I^2,$$

and define the multiplier as

$$m = e^{\varphi}.$$

We will verify below that the function $h = mf$ satisfies the conclusions of Theorem 1.3.8.

Verification of (1.3.48). Let I be a fixed dyadic interval. We begin by evaluating the integral

$$\int_I f e^{a_I h_I} dt.$$

Define $R(t)$ to be the remainder term of order 3 in the Taylor expansion of the exponential $\exp(a_I h_I(t))$, that is

$$R(t) = \exp(a_I h_I(t)) - 1 - a_I h_I(t) - \frac{1}{2} a_I^2 h_I^2(t).$$

Define also

$$Q = \int_I f(t) R(t) dt.$$

Then, $|Q| \leq |a_I^3| \int_I f(t)dt$. Next we rewrite ,

$$\begin{aligned}\int_I f(t) e^{a_I h_I(t)} dt &= \int_I f(t) \left(1 + a_I h_I(t) + \frac{1}{2} a_I^2 h_I^2(t) + R(t)\right) dt \\ &= \int_I f(t)dt + a_I^2 |I| + \frac{a_I^2}{2} \int_I f(t)dt + Q \\ &= \int_I f(t)dt \left(1 + \frac{a_I^2 |I|}{\int_I f dt} + \frac{a_I^2}{2} + \frac{Q}{\int_I f dt}\right).\end{aligned}$$

Now recall that we write f_I for the mean value of f over the interval I. Then, clearly, we have the identity

$$1 + \frac{a_I^2 |I|}{\int_I f dt} + \frac{a_I^2}{2} + \frac{Q}{\int_I f dt} = 1 + a_I^2 \left(\frac{1}{2} + \frac{1}{f_I}\right) + \frac{Q}{|I| f_I}.$$

Next we determine constants m_I so that the following equation holds.

$$1 + a_I^2 \left(\frac{1}{2} + \frac{1}{f_I}\right) + \frac{Q}{|I| f_I} = \exp\left(a_I^2 \left(\frac{1}{2} + \frac{1}{f_I}\right) m_I\right). \tag{1.3.49}$$

As $|Q| \leq |a_I^3| |I| f_I$ and $|a_I| \leq f_I$, the solutions m_I of (1.3.49) satisfy the estimates

$$\frac{1}{6} < m_I < 6.$$

Summing up we showed that

$$\int_I f e^{a_I h_I} dt = \int_I f dt \exp\left(a_I^2 \left(\frac{1}{2} + \frac{1}{f_I}\right) m_I\right). \tag{1.3.50}$$

Dividing (1.3.50) by $\exp(a_I^2(\frac{1}{2}+\frac{1}{f_I})m_I)$ gives the identity (1.3.48). Thus we chose a constant $m_I \in [1/6, 6]$ so that the functions f and $f\exp(a_I h_I - a_I^2(\frac{1}{2}+\frac{1}{f_I})m_I h_I^2)$, have identical mean value over the interval I.

Estimates for $\langle f e^{a_I h_I}, h_I\rangle$. Next we show that the Haar coefficient $\langle f e^{a_I h_I}, h_I\rangle$, admits the upper bound

$$\left|\int_I f e^{a_I h_I} h_I\right| \leq 3|a_I|\,|I|. \tag{1.3.51}$$

To verify this we start again with the Taylor expansion of the exponential,

$$\exp(a_I h_I(t)) = 1 + a_I h_I(t) + E(t).$$

Then $|E(t)| \leq a_I^2$. In the integral (1.3.51) we replace $e^{a_I h_I(t)}$ by $1 + a_I h_I(t) + E(t)$ and estimate as follows, using $|E(t)| \leq a_I^2$.

$$\begin{aligned}\left|\int_I f e^{a_I h_I} h_I\right| &= \left|\int_I f\,(1 + a_I h_I + E)\,h_I\right| \\ &\leq \left|\int_I f h_I\right| + |a_I| \int_I f + a_I^2 \int_I f \\ &\leq 2|a_I|\cdot|I| + a_I^2|I|.\end{aligned}$$

As $2|a_I|\cdot|I| + a_I^2|I| \leq 3|a_I|\,|I|$, we have (1.3.51).

The main inequality. We are now ready to prove a crucial estimate for the Haar coefficient of $e^\varphi f$,

$$\left|\int e^\varphi f h_I dt\right| \leq 3|a_I|\,|I| \exp\left(-\sum_{J\supseteq I} a_J^2(\frac{1}{2}+\frac{1}{f_J})m_J\right). \tag{1.3.52}$$

To this end we split φ into three pieces. For $x \in I$ we write

$$\varphi(x) = v_I + \left(a_I h_I(x) - a_I^2(\frac{1}{2}+\frac{1}{f_I})m_I h_I^2(x)\right) + z_I(x),$$

where v_I is defined to be the number

$$v_I = -\|f\|_\infty + \int_0^1 f dt + \sum_{J\supset I} a_J h_J - \sum_{J\supset I} a_J^2 m_J \left(\frac{1}{2}+\frac{1}{f_J}\right) h_J^2. \tag{1.3.53}$$

Recall that by writing $J \supset I$ we denote the summation over all dyadic intervals J that are strictly containing I. Note that z_I is the function given by

$$z_I = \sum_{J\subset I} a_J h_J - \sum_{J\subset I} a_J^2 m_J \left(\frac{1}{2}+\frac{1}{f_J}\right) h_J^2,$$

here $J \subset I$ denotes all dyadic intervals that are strictly contained in I. Now let I_1 be the left half of I and let I_2 be the right half of I. Inductively applying (1.3.48) we obtain the identities

$$\begin{aligned}\int_{I_1} e^{z_I} f &= \int_{I_1} f, \\ \int_{I_2} e^{z_I} f &= \int_{I_2} f.\end{aligned} \tag{1.3.54}$$

The identities (1.3.54) are crucial in reducing estimates for the Haar coefficient $\langle \epsilon^{\varphi} f, h_I \rangle$, to the estimate for $\langle f e^{a_I h_I}, h_I \rangle$ in (1.3.51). Indeed as $e^{a_I h_I}$ is a constant function on both intervals I_1 and I_2, we observe that with (1.3.54),

$$\begin{aligned}\int e^{a_I h_I} e^{z_I} f h_I dt &= \int_{I_1} e^{a_I} e^{z_I} f dt - \int_{I_2} e^{-a_I} e^{z_I} f dt \\ &= \int_{I_1} e^{a_I} f dt - \int_{I_2} e^{-a_I} f dt \\ &= \int e^{a_I h_I} f h_I dt.\end{aligned} \tag{1.3.55}$$

By (1.3.53) we observe that

$$e^{v_I} \leq \exp\left(- \sum_{J \supseteq I} a_J^2 (\frac{1}{2} + \frac{1}{f_J}) m_J \right). \tag{1.3.56}$$

The identity (1.3.55) and the estimates (1.3.51) and (1.3.56) imply the following upper bound for the coefficient $\langle e^{\varphi} f, h_I \rangle$,

$$\begin{aligned}\left| \int e^{\varphi} f h_I dt \right| &= e^{v_I} \exp\left(-a_I^2 (\frac{1}{2} + \frac{1}{f_I}) m_I \right) \int e^{a_I h_I} e^{z_I} h_I f dt \\ &= \exp\left(- \sum_{J \supseteq I} a_J^2 (\frac{1}{2} + \frac{1}{f_J}) m_J \right) \left| \int e^{a_I h_I} f h_I dt \right| \\ &\leq 3|I| \exp\left(- \sum_{J \supseteq I} a_I^2 (\frac{1}{2} + \frac{1}{f_J}) m_J \right) |a_I|.\end{aligned} \tag{1.3.57}$$

Upper bounds for the square function of $e^{\varphi} f$. Next we shorten notation and abbreviate,

$$p_I = \exp\left(- \sum_{J \supseteq I} \frac{a_J^2}{12} \right).$$

Note that p_I is clearly bounded by 1. The sum $\sum_{\{I : x \in I\}} p_I a_I^2$ is a lower Riemann sum for $\int_0^{\infty} \exp(-\frac{s}{12}) ds$, where the partition is given by the partial sums of the series $\sum_{\{I:x\in I\}} a_I^2$. Thus for every $x \in [0,1]$,

$$\sum_{\{I:x\in I\}} p_I a_I^2 \le \int_0^\infty \exp(-\frac{s}{12})ds = 12.$$

It is now quite easy to estimate the square function of $e^\varphi f$. Expand

$$e^\varphi f = \int e^\varphi f + \sum c_I h_I$$

in its Haar series. Then by (1.3.57) and (1.3.56),

$$|c_I| \le p_I a_I.$$

At each point $x \in [0,1]$ we have therefore the uniform bound,

$$\begin{aligned} S^2(e^\varphi f)(x) &\le 9 \sum_{\{I:x\in I\}} p_I a_I^2 \\ &\le 108. \end{aligned}$$

The mean value of $e^\varphi f$. Next we observe that by iteratively applying equation (1.3.48) we arrive at the identity

$$\int_0^1 e^\varphi f dt = \left\| \exp(-f + \int_0^1 f) \right\|_\infty \int_0^1 f dt,$$

that implies the lower bound

$$\int_0^1 e^\varphi f dt \ge \|\exp(-f)\|_\infty \int_0^1 f dt.$$

Finally, as $\varphi \le 0$ the multiplier $m = e^\varphi$, satisfies

$$0 \le m \le 1.$$

□

1.4 Martingales and biorthogonal systems

The Haar basis is at the same time the simplest martingale difference sequence and the simplest biorthogonal system. In this section we extend the content of Burkholder's Theorem 1.1.5 to all martingale difference sequences and to a large class of biorthogonal systems, including the Franklin system and the wavelet family. The results of this section illustrate the use of the Haar system in the solution of problems arising in probability and analysis.

The martingale inequalities of D. Burkholder and E. M. Stein

We define first conditional expectations and recall the notions of martingales and martingale difference sequences. The following summary is based on the appendix in [61] by R. Durrett. We refer to this book for the proofs.

Let $(\Omega, \mathcal{F}, |\,.\,|)$ be a probability space. Let $\mathcal{G} \subseteq \mathcal{F}$ be a sub σ-algebra of $\mathcal{F}$. The theorem of Radon and Nikodym implies that for every $f \in L^1(\Omega, \mathcal{F}, |\,.\,|)$ there exists an (essentially unique) $g : \Omega \to \mathbb{R}$ which is measurable with respect to $\mathcal{G}$ and satisfies

$$\int_A f = \int_A g \quad \text{for } A \in \mathcal{G}.$$

The function g is then called the conditional expectation of f with respect to $\mathcal{G}$, for which we use the notation

$$\mathbb{E}(f|\mathcal{G}) = g.$$

Taking conditional expectations defines a contraction on $L^p(\Omega)$ $(1 \le p \le \infty)$.

Let $\{\mathcal{F}_n\}$ be a sequence of increasing σ-algebras contained in $\mathcal{F}$. Assume that $\mathcal{F}$ coincides with the σ-algebra generated by $\mathcal{F}_0 \cup \cdots \cup \mathcal{F}_n \cup \ldots$. Let $\{f_n\}$ be a sequence in $L^1(\Omega, \mathcal{F})$, such that

$$f_n = \mathbb{E}(f_{n+1}|\mathcal{F}_n).$$

Then $\{f_n\}$ is called a martingale and $d_n = f_n - f_{n-1}$ is called a martingale difference sequence. Examples of martingales are most easily obtained by fixing $f \in L^1(\Omega, \mathcal{F})$ and forming the conditional expectations,

$$f_n = \mathbb{E}(f|\mathcal{F}_n).$$

The partial sum operators of the Haar basis $\{h_i\}_{i=0}^{\infty}$ are given by conditional expectation with respect to an increasing sequence of σ-algebras. See Proposition 1.1.1 and its proof. The same assertion holds for the partial sums of the Rademacher series $\sum_{i=1}^{\infty} a_i r_i$, when $\sum a_i^2 < \infty$.

Burkholder's martingale inequality

For $f \in L^p(\Omega)$ the conditional expectations $f_n = \mathbb{E}(f|\mathcal{F}_n)$ form a convergent sequence in $L^p(\Omega)$ $(1 \le p < \infty)$. Indeed we have

$$\|f - f_n\|_{L^p(\Omega)} \to 0 \quad \text{for} \quad n \to \infty.$$

Hence with $d_n = f_n - f_{n-1}$ the martingale difference series $\sum_{n=1}^{\infty} d_n$ is convergent in $L^p(\Omega)$. The next result extends Burkholder's theorem to martingales. It asserts that in $L^p(\Omega)$, $1 < p < \infty$, the decomposition of f in its martingale differences

$$f = \mathbb{E}(f|\mathcal{F}_0) + \sum_{n=1}^{\infty} d_n$$

is unconditionally converging. We present the proof of B. Maurey [139], which consists of reducing estimates for general martingale difference sequences to the case of Haar functions.

Theorem 1.4.1. *Let $f \in L^p(\Omega, \mathcal{F})$, $1 < p < \infty$. Let $\{\mathcal{F}_n\}$ be a sequence of increasing σ-algebras contained in $\mathcal{F}$. Let*

$$d_n = \mathbb{E}(f|\mathcal{F}_n) - \mathbb{E}(f|\mathcal{F}_{n-1}).$$

Then for every choice of $\epsilon_n \in \{+1, -1\}$, the following estimate holds,

$$\left\| \mathbb{E}(f|\mathcal{F}_0) + \sum_{n=1}^{\infty} \epsilon_n d_n \right\|_{L^p(\Omega)} \le (p^* - 1)\|f\|_{L^p(\Omega)}. \tag{1.4.1}$$

Proof. In this proof, we find it convenient to use systematically the following notation. For a collection of pairwise disjoint dyadic intervals $\mathcal{C}$ we denote

$$Q(\mathcal{C}) = \bigcup_{K \in \mathcal{C}} \{J \in \mathcal{D} : J \subseteq K\}.$$

Thus $Q(\mathcal{C})$ is the collection of intervals which are contained in one of the intervals of $\mathcal{C}$. For a single dyadic interval I we write

$$Q(I) = \{J \in \mathcal{D} : J \subseteq I\}.$$

Overview. Given $\epsilon > 0$ we will construct a sequence of functions $h_n : [0,1] \to \mathbb{R}$ with the following three properties. First, their Haar support, given by

$$\mathcal{H}_n = \{I : \langle h_n, h_I \rangle \neq 0\},$$

forms a sequence of pairwise disjoint collections of dyadic intervals, second,

$$\left\| h_0 + \sum_{n=1}^{\infty} h_n \right\|_{L^p([0,1])} \le (1+\epsilon)\|f\|_{L^p(\Omega)}, \tag{1.4.2}$$

and third, for every choice of $\epsilon_n \in \{+1, -1\}$ the following estimate holds,

$$\left\| \mathbb{E}(f|\mathcal{F}_0) + \sum_{n=1}^{\infty} \epsilon_n d_n \right\|_{L^p(\Omega)} \le (1+\epsilon) \left\| h_0 + \sum_{n=1}^{\infty} \epsilon_n h_n \right\|_{L^p([0,1])}. \tag{1.4.3}$$

Once this sequence is obtained we get (1.4.1) from Burkholder's inequality. Indeed, using (1.4.3), Theorem 1.1.5 and (1.4.2) gives that

$$\begin{aligned}
\left\| \mathbb{E}(f|\mathcal{F}_0) + \sum_{n=1}^{\infty} \epsilon_n d_n \right\|_{L^p(\Omega)} &\le (1+\epsilon) \left\| h_0 + \sum_{n=1}^{\infty} \epsilon_n h_n \right\|_{L^p([0,1])} \\
&\le (p^*-1)(1+\epsilon) \left\| h_0 + \sum_{n=1}^{\infty} h_n \right\|_{L^p([0,1])} \\
&\le (p^*-1)(1+\epsilon)^2 \|f\|_{L^p(\Omega)}.
\end{aligned}$$

The basic observation. Below we will construct the sequence $\{h_n\}$ satisfying (1.4.2) and (1.4.3) by iterating an elementary observation.

Let $\mathcal{C}$ be a collection of pairwise disjoint dyadic intervals covering the unit interval. Thus $\bigcup_{I\in\mathcal{C}} I = [0,1[$. Assume that $\mathcal{C}$ is decomposed into pairwise disjoint collections $\mathcal{C}_1,\ldots,\mathcal{C}_m$. Let C_i be the point-set covered by the collection $\mathcal{C}_i$. The sets C_i are pairwise disjoint and cover the unit interval. Choose now scalars $a_i \in \mathbb{R}$ so that

$$\sum_{i=1}^{m} a_i|C_i| = 0, \tag{1.4.4}$$

and form the function $h : [0,1] \to \mathbb{R}$ by putting

$$h = \sum_{i=1}^{m} a_i 1_{C_i}. \tag{1.4.5}$$

Thus we defined the function h to be measurable with respect to the σ-algebra generated by the collection $\mathcal{C}$. By (1.4.4) we have $\int h = 0$. Next put

$$\mathcal{H} = Q([0,1]) \setminus Q(\mathcal{C}). \tag{1.4.6}$$

Since h is measurable with respect to the σ-algebra generated by $\mathcal{C}$ and $\int h = 0$, the Haar support of h is contained in $\mathcal{H}$. Thus

$$h = \sum_{J\in\mathcal{H}} \langle h, \frac{h_J}{|J|}\rangle h_J. \tag{1.4.7}$$

Summing up for a function h, given by (1.4.5) and (1.4.4), the Haar support is contained in $\mathcal{H}$, defined by (1.4.6).

Perturbation, preparation. Now, let $f \in L^p(\Omega,\mathcal{F})$, $1 < p < \infty$. Let $\{\mathcal{F}_n\}$ be a given sequence of increasing σ-algebras contained in $\mathcal{F}$. We assume for simplicity that $\mathcal{F}_0 = \{\Omega, \emptyset\}$. Let

$$d_n = \mathbb{E}(f|\mathcal{F}_n) - \mathbb{E}(f|\mathcal{F}_{n-1}).$$

Furthermore we assume that each of the martingale differences d_n is a step function. Measured in L^p, the error of approximating by a step function can be made arbitrarily small. Consequently we may assume that each of the σ-algebras $\{\mathcal{F}_n\}$ is generated by a finite set of atoms. Let $\mathcal{A}_n$ denote the atoms of $\{\mathcal{F}_n\}$. The collections $\{\mathcal{A}_n\}$ are linearly ordered in the following sense. If $A \in \mathcal{A}_n$ and $B \in \mathcal{A}_{n+1}$ so that $A\cap B \neq \emptyset$, then $B \subseteq A$. Taking the union and forming $\mathcal{A} = \bigcup \mathcal{A}_n$, gives a collection of sets which satisfies the following condition.

$$\text{If } A, B \in \mathcal{A}, \text{and } A\cap B \neq \emptyset, \text{then } A \subseteq B \text{ or } B \subseteq A. \tag{1.4.8}$$

Next we will reproduce the intersection pattern of $\mathcal{A}$, by using collections of dyadic intervals. Precisely we claim that for each $A \in \mathcal{A}$ there exists a collection of pairwise disjoint dyadic intervals $\mathcal{C}_A$ so that with $C_A = \bigcup_{I\in\mathcal{C}_A} I$ the following conditions hold.

1. $C_\Omega = [0,1]$, and $|C_B| = |B|$ for $B \in \mathcal{A}$.
2. If $A, B \in \mathcal{A}$, and $C_A \cap C_B \neq \emptyset$, then $C_A \subseteq C_B$ or $C_B \subseteq C_A$.
3. $C_B \subseteq C_A$ iff $B \subseteq A$.
4. If $B \subseteq A$, and $I \in \mathcal{C}_A$, then
$$\frac{|I \cap C_B|}{|I|} = \frac{|B|}{|A|}. \tag{1.4.9}$$
5. If $I \in \mathcal{C}_B$, $J \in \mathcal{C}_A$ and $I \subseteq J$ then $B \subseteq A$.

Collections $\{\mathcal{C}_A : A \in \mathcal{A}\}$ with the above properties are constructed inductively in the course of the proof of Proposition 2.1.4 in Chapter 2.

Before we continue we should comment on condition (1.4.9): Let $A \in \mathcal{A}_n$ and $I \in C_A$. Then by (1.4.9), conditioned to I the measures of the sets $\{I \cap \mathcal{C}_B : B \in \mathcal{A}_{n+1}, B \subseteq A\}$ are the same as the measures of $\{B \in \mathcal{A}_{n+1} : B \subseteq A\}$ conditioned to A.

The induction argument. Define $\mathcal{C}_n = \bigcup_{B \in \mathcal{A}_n} \mathcal{C}_B$. Note that $\mathcal{C}_n$ is a collection of pairwise disjoint dyadic intervals. We are going to construct now a sequence of functions $\{h_n\}$ with Haar support contained in
$$\mathcal{H}_n = Q(\mathcal{C}_{n-1}) \setminus Q(\mathcal{C}_n)$$
so that the joint distribution of $\{h_n\}$ is the same as the joint distribution of the martingale difference sequence $\{d_n\}$.

Recall that $\mathcal{F}_0 = \{\Omega, \emptyset\}$. Hence $\mathbb{E}(f|\mathcal{F}_0)$ is a constant function such that $\mathbb{E}(f|\mathcal{F}_0) = \int f$. We let $h_0 : [0,1] \to \mathbb{R}$ be the constant function defined by $h_0(t) = \int f$.

Now we define $h_1 : [0,1] \to \mathbb{R}$. For $d_1 : \Omega \to \mathbb{R}$ there exist coefficients $d(B), B \in \mathcal{A}_1$ so that
$$d_1 = \sum_{B \in \mathcal{A}_1} d(B) 1_B \quad \text{and} \quad \sum_{B \in \mathcal{A}_1} d(B)|B| = 0.$$
Next we modify d_1 replacing the indicator function $1_B : \Omega \to \{0,1\}$ by $1_{C_B} : [0,1] \to \{0,1\}$. By doing this we obtain a copy of d_1 with the same distribution function. We put
$$h_1 = \sum_{B \in \mathcal{A}_1} d(B) 1_{C_B}.$$
Note that h_1 is measurable with respect to the σ-algebra generated by the collection $\mathcal{C}_1$, and $\int h_1 = 0$. Consequently with
$$\mathcal{H}_1 = Q([0,1]) \setminus Q(\mathcal{C}_1),$$

the Haar expansion of h_1 assumes the form

$$h_1 = \sum_{I \in \mathcal{H}_1} \langle h_1, \frac{h_I}{|I|} \rangle h_I.$$

By construction the martingale differences $\{d_0, d_1\}$ and the newly constructed functions $\{h_0, h_1\}$ are identically distributed.

Suppose now that $\{h_0, h_1, \dots h_n\}$ are defined so that for $i \le n$,

$$h_i = \sum_{I \in \mathcal{H}_i} \langle h_i, \frac{h_I}{|I|} \rangle h_I, \quad \text{where} \quad \mathcal{H}_i = Q(\mathcal{C}_{i-1}) \setminus Q(\mathcal{C}_i),$$

and so that the joint distribution of $\{d_0, d_1, \dots, d_n\}$ and $\{h_0, h_1, \dots, h_n\}$ are identical. Next we determine h_{n+1}.

Let $I \in \mathcal{C}_n$. There exists a uniquely determined $A \in \mathcal{A}_n$ such that $I \in \mathcal{C}_A$. Observe that restricting $h_0, h_1, \dots, h_n$ to the interval I gives a constant function. Similarly the restriction of $d_0, d_1, \dots, d_n$ to A is a constant. Next fix $B \in \mathcal{A}_{n+1}$ such that $B \subseteq A$. Then form the collection of dyadic intervals

$$\mathcal{C}_{B,I} = \{J \subseteq I : J \in \mathcal{C}_B\},$$

and let $C_{B,I}$ denote the point set covered by $\mathcal{C}_{B,I}$.

Next we consider the restriction of d_{n+1} to the atom $A \in \mathcal{A}_n$. For $B \in \mathcal{A}_{n+1}$, with $B \subseteq A$, there exist coefficients $d(B)$ such that

$$1_A d_{n+1} = \sum_{B \in \mathcal{A}_{n+1}, B \subseteq A} d(B) 1_B.$$

Note that $\mathbb{E}(d_{n+1}|\mathcal{F}_n) = 0$ implies that $\int_A d_{n+1} = 0$. Next we replace the indicator functions $1_B : A \to \{0,1\}$ by $1_{C_{B,I}} : I \to \{0,1\}$. By (1.4.9), we obtain a copy of $1_A d_{n+1}$, by defining

$$h_{n+1,I} = \sum_{B \in \mathcal{A}_{n+1}, B \subseteq A} d(B) 1_{C_{B,I}}.$$

Observe that $h_{n+1,I}$ is measurable with respect to the σ-algebra generated by $\{J \in \mathcal{C}_{n+1} : J \subseteq I\}$. Moreover $\int_I h_{n+1,I} = 0$. It follows that the Haar support of $h_{n+1,I}$ is contained in

$$\mathcal{H}_{n+1,I} = Q(I) \setminus Q(\{J \in \mathcal{C}_{n+1} : J \subseteq I\}).$$

Finally define h_{n+1} by summing over $I \in \mathcal{C}_n$. Thus,

$$h_{n+1} = \sum_{I \in \mathcal{C}_n} h_{n+1,I}.$$

With this definition of h_{n+1} the functions $\{d_0, \ldots, d_{n+1}\}$ and $\{h_0, \ldots, h_{n+1}\}$ are identically distributed. Finally for $\mathcal{H}_{n+1} = Q(\mathcal{C}_n) \setminus Q(\mathcal{C}_{n+1})$, the Haar expansion of h_{n+1} is given by

$$h_{n+1} = \sum_{I \in \mathcal{H}_{n+1}} \langle h_{n+1}, \frac{h_I}{|I|} \rangle h_I.$$

This completes the inductive construction of the sequence $\{h_n\}$. □

Consequences. Combining Theorem 1.4.1 with Khintchine's inequality gives a large class of square function characterizations for $L^p(\Omega, \mathcal{F}), 1 < p < \infty$. We let $\{\mathcal{F}_n\}$ be a sequence of increasing σ-algebras contained in $\mathcal{F}$. Assume that $\mathcal{F}$ is the σ-algebra generated by $\bigcup \mathcal{F}_n$. For $f \in L^p(\Omega, \mathcal{F})$ define

$$d_0 = \mathbb{E}(f|\mathcal{F}_0), \quad \text{and} \quad d_n = \mathbb{E}(f|\mathcal{F}_n) - \mathbb{E}(f|\mathcal{F}_{n-1}), \quad \text{for} \quad n \in \mathbb{N}.$$

We call $(d_0^2 + \sum_{n=1}^{\infty} d_n^2)^{1/2}$ the martingale square function of f. By Theorem 1.4.1 the decomposition of f into martingale differences

$$f = d_0 + \sum_{n=1}^{\infty} d_n$$

converges unconditionally in $L^p, 1 < p < \infty$, and by Khintchine's inequality we have the following characterization of $L^p(\Omega)$ by martingale-square-functions

$$c_p \|f\|_{L^p(\Omega)} \le \left\| \left(d_0^2 + \sum_{n=1}^{\infty} d_n^2 \right)^{1/2} \right\|_{L^p(\Omega)} \le C_p \|f\|_{L^p(\Omega)}. \tag{1.4.10}$$

For $1 < p < \infty$ we denote by $H^p[(\mathcal{F}_n)]$ the space of all $f \in L^p(\Omega, \mathcal{F})$ for which

$$\|f\|_{H^p[(\mathcal{F}_n)]} = \left\| \left(\mathbb{E}(f|\mathcal{F}_0)^2 + \sum_{n=1}^{\infty} (\mathbb{E}(f|\mathcal{F}_n) - \mathbb{E}(f|\mathcal{F}_{n-1}))^2 \right)^{1/2} \right\|_{L^p(\Omega)} < \infty.$$

By (1.4.10) the identity operator $\mathrm{Id} : L^p(\Omega) \to H^p[(\mathcal{F}_n)]$ is an isomorphism. The importance of this result lies in its generality: One is completely free to choose the σ-algebras $\{\mathcal{F}_n\}$ to decompose a given $f \in L^p(\Omega)$.

Stein's martingale inequality

A very surprising application of Burkholder's martingale inequality is J. Bourgain's proof of E. M. Stein' s martingale inequality.

Theorem 1.4.2. *Let $(\Omega, \mathcal{F}, |\cdot|)$ be a probability space. Let $\mathcal{F}_1 \subseteq \cdots \subseteq \mathcal{F}_m \subseteq \mathcal{F}$ be σ algebras in Ω. Then, for any choice of $f_1, \ldots, f_m \in L^p(\Omega)$ the following estimate holds true,*

$$\int_0^1 \left\| \sum_{n=1}^m r_n(t)\mathbb{E}(f_n|\mathcal{F}_n) \right\|_{L^p(\Omega)}^p dt \leq (p^* - 1)^p \int_0^1 \left\| \sum_{n=1}^m r_n(t) f_n \right\|_{L^p(\Omega)}^p dt, \quad (1.4.11)$$

where $p^* = \max\{p, p/(p-1)\}$, and where $r_1, \ldots, r_m$ are Rademacher functions.

Proof. For $t \in [0,1]$, and $x \in \Omega$ define

$$F(t,x) = \sum_{n=1}^m r_n(t) f_n(x).$$

On the product space $[0,1] \times \Omega$ define the increasing σ-algebras

$$\mathcal{G}_n = \sigma\{r_1, \ldots, r_n\} \otimes \mathcal{F}_n, \quad n \leq m.$$

Observe that taking the conditional expectation with respect to $\mathcal{G}_n$ corresponds to forming partial sums as follows,

$$\mathbb{E}(F \,|\, \mathcal{G}_n)(t,x) = \sum_{i \leq n} r_i(t)\mathbb{E}(f_i|\mathcal{F}_n)(x). \quad (1.4.12)$$

Let $F_0 = 0$, put $F_n = \mathbb{E}(F \,|\, \mathcal{G}_n)$, for $1 \leq n \leq m$, and $d_n(f_i) = \mathbb{E}(f_i \,|\, \mathcal{F}_n) - \mathbb{E}(f_i \,|\, \mathcal{F}_{n-1})$, for $2 \leq n \leq m$. Note that by (1.4.12), we get

$$F_n - F_{n-1} = r_n \, \mathbb{E}(f_n|\mathcal{F}_n) + \sum_{i<n} r_i \, d_n(f_i), \quad \text{for} \quad 1 \leq n \leq m.$$

(In the above formula we read $\sum_{i<1} = 0$, which is consistent with the standard convention that the value of a sum taken over the empty set equals 0.) Hence,

$$\sum_{n=1}^m F_n - F_{n-1} = \sum_{n=1}^m r_n \, \mathbb{E}(f_n|\mathcal{F}_n) + \sum_{n=1}^m \sum_{i<n} r_i \, d_n(f_i).$$

Next apply Theorem 1.4.1 to the martingale differences $F_n - F_{n-1}$ defined on the probability space $[0,1] \times \Omega$. Put $C_p = \max\{p, p/(p-1)\} - 1$. Then for every $s \in [0,1]$, the following estimate holds.

$$\begin{aligned}
&C_p^p \int_\Omega \int_0^1 |F(t,x)|^p dt dx \\
&\geq \int_\Omega \int_0^1 \left| \sum_{n=1}^m r_n(s)(F_n - F_{n-1})(t,x) \right|^p dt dx \\
&= \int_\Omega \int_0^1 \left| \sum_{n=1}^m r_n(s) r_n(t)\mathbb{E}(f_n|\mathcal{F}_n)(x) + \sum_{n=1}^m \sum_{i<n} r_i(s) r_i(t) d_n(f_i)(x) \right|^p dt dx.
\end{aligned} \quad (1.4.13)$$

For $s \in [0,1]$ and $x \in \Omega$, we next rewrite the integrand in the last line of the above estimate. Observe that (by multiplication with $r_n(s)$) the following identity holds.

$$\begin{aligned} &\int_0^1 \left| \sum_{n=1}^m r_n(s) r_n(t) \mathbb{E}(f_n|\mathcal{F}_n)(x) + \sum_{n=1}^m \sum_{i<n} r_i(s) r_i(t) d_n(f_i)(x) \right|^p dt \\ &= \int_0^1 \left| \sum_{n=1}^m r_n(t) \mathbb{E}(f_n|\mathcal{F}_n)(x) + \sum_{n=1}^m \sum_{i<n} r_i(s) r_n(s) r_i(t) d_n(f_i)(x) \right|^p dt. \end{aligned} \tag{1.4.14}$$

Next combine (1.4.14) with (1.4.13) and apply the triangle inequality. Then we exploit that for $i < n$, $\int_0^1 r_i(s) r_n(s) ds = 0$. Thus we obtain the following estimates.

$$\begin{aligned} &C_p^p \int_\Omega \int_0^1 |F(t,x)|^p dt dx \\ &\geq \sup_{s\in[0,1]} \int_\Omega \int_0^1 \left| \sum_{n=1}^m r_n(t) \mathbb{E}(f|\mathcal{F}_n)(x) + \sum_{n=1}^m \sum_{i<n} r_i(s) r_n(s) r_i(t) d_n(f_i)(x) \right|^p dt dx \\ &\geq \int_\Omega \int_0^1 \left| \sum_{n=1}^m r_n(t) \mathbb{E}(f_n|\mathcal{F}_n)(x) + \sum_{n=1}^m \sum_{i<n} [\int_0^1 r_i(s) r_n(s) ds] r_i(t) d_n(f_i)(x) \right|^p dt dx \\ &= \int_\Omega \int_0^1 \left| \sum_{n=1}^m r_n(t) \mathbb{E}(f_n|\mathcal{F}_n)(x) \right|^p dt dx. \end{aligned}$$

□

E.M. Stein in [193] proved that for $f_n \in L^p(\Omega)$, the following martingale inequality holds true,

$$\left\| \left(\sum_{n=1}^m \mathbb{E}(f_n|\mathcal{F}_n)^2 \right)^{1/2} \right\|_{L^p(\Omega)} \leq A_p \left\| \left(\sum_{n=1}^m f_n^2 \right)^{1/2} \right\|_{L^p(\Omega)}, \tag{1.4.15}$$

where $A_p \leq Cp^2/(p-1)$. By Khintchine's inequality we obtain (1.4.11) — with different constants — from (1.4.15), and conversely (1.4.11) implies that E. M. Stein's martingale inequality holds true. The original proof in [193] applies just to scalar valued L^p-functions $f_n : \Omega \to \mathbb{R}$. Above we presented J. Bourgain's proof since it is solely based on Burkholder's theorem and the triangle inequality in the space L^p. Consequently the estimate (1.4.11) carries over to L^p functions $f_n : \Omega \to E$ taking values in a Banach space E with the UMD property.

T. Figiel's approach to biorthogonal systems

Next we turn to biorthogonal systems. We introduce a powerful method of T. Figiel [65] showing that a large class of biorthogonal systems is equivalent to the Haar basis in $L^p([0,1])$. As a corollary we obtain the theorem of S.V. Bočkarev [14] that the Franklin functions form an unconditional basis in $L^p([0,1])$. We fix a family of functions $f_I : [0,1] \to \mathbb{R}$, indexed by dyadic intervals $I \in \mathcal{D}$. In this section we determine conditions on the family $\{f_I\}$ which imply that the linear extension of the map

$$T : h_I \mapsto f_I,$$

defines a bounded operator on the L^p spaces when $1 < p < \infty$. First we impose a set of three conditions on the mean values and on the Haar coefficients of $\{f_I\}$ which imply that the map $T : h_I \mapsto f_I$, extends to a bounded operator on L^p. Later we give conditions on the pointwise behavior of $\{f_I\}$ to the same effect.

We assume that

$$\int_0^1 f_I dt = 0. \tag{1.4.16}$$

Using mean values and Haar coefficients we define what it means for $\{f_I\}$ to be dominated by the Haar system. Let I be a dyadic interval with $|I| = 2^{-i}$. Let $m \in \mathbb{Z}$ such that $I + m|I| \subseteq [0,1]$. Let J be a dyadic interval such that,

$$J \subseteq I + m|I|,$$
$$|J| = 2^{-l}|I|,$$

where $l \in \mathbb{N}$. Assume that there exist $A > 0$ and $\alpha > 0$ independent of I or J so that the mean values and the Haar coefficients of f_I satisfy the estimates

$$\left| \int_{I+m|I|} f_I(y)dy \right| \leq A|I|\{1 + |m|\}^{-1-\alpha}, \tag{1.4.17}$$

and

$$\left| \int_0^1 f_I(y)h_J(y)dy \right| \leq A|J|2^{-l}\{1 + |m|\}^{-1-\alpha}. \tag{1.4.18}$$

We say that the family of functions $\{f_I : I \in \mathcal{D}\}$ is dominated by the Haar system if the three conditions (1.4.16), (1.4.17) and (1.4.18) hold true for every dyadic interval $I \in \mathcal{D}$.

Theorem 1.4.3. *Let $1 < p < \infty$. Let $\{f_I : I \in \mathcal{D}\}$ be a family of functions that is dominated by the Haar system, that is, it satisfies* (1.4.16), (1.4.17) *and* (1.4.18). *Let (x_I) be sequence of scalars. Then the following estimate holds,*

$$\left\| \sum x_I f_I \right\|_{L^p([0,1])} \leq C_p(A,\alpha) \left\| \sum x_I h_I \right\|_{L^p([0,1])},$$

where $C_p = C(A,\alpha) \max\{p, p/(p-1)\}$.

Proof. The idea of the proof is encoded in Figiel's expansion (1.4.28). Its derivation is contained in the first three steps of the proof. In Steps 4 and 5 we show how (1.4.28) implies the conclusion of the theorem.

Step 1. The starting point is the following non-standard expansion of f_I. First cut f_I using the indicator functions of the shifted intervals $I + m|I|$.

$$f_I(x) = \sum_{\{m:I+m|I|\subseteq[0,1]\}} 1_{I+m|I|}(x) f_I(x). \tag{1.4.19}$$

Fix m such that $I+m|I| \subseteq [0,1]$. Then we expand further each summand appearing in (1.4.19). We would like to use Haar series in such a way that only those Haar functions h_J appear for which $J \subseteq I+m|I|$. To make this possible we first take out the mean value of f_I over $I + m|I|$. Thus we derived an identity with important consequences,

$$\begin{aligned} 1_{I+m|I|}(x) f_I(x) &= \frac{1}{|I|} \int_{I+m|I|} f_I(y)dy 1_{I+m|I|}(x) \\ &+ \sum_{l=0}^{\infty} \sum_{\{J\subseteq I+m|I|,\, |J|=2^{-l}|I|\}} \langle f_I, \frac{h_J}{|J|} \rangle h_J(x). \end{aligned} \tag{1.4.20}$$

The mean values, $|I|^{-1} \int_{I+m|I|} f_I(y)dy$, and the remaining series of Haar functions in (1.4.20) will be treated quite differently below.

Step 2. We exploit now that the integral of f_I over the unit interval vanishes. Observe that $\int f_I = 0$ can be written as

$$\sum_{\{m:I+m|I|\subseteq[0,1]\}} \left[\int_{I+m|I|} f_I(y)dy \right] = 0. \tag{1.4.21}$$

Next we take the sum of the mean values appearing in the first line of the equation (1.4.20). The identity (1.4.21) is used below to replace the characteristic functions $1_{I+m|I|}$ by the difference $1_{I+m|I|}(x) - 1_I(x)$ which is much better behaved. Before we start doing this we introduce a convenient abbreviation. Let

$$b_m^I = \frac{1}{|I|} \int_{I+m|I|} f_I(y)dy, \tag{1.4.22}$$

and for $I + m|I| \subseteq [0,1]$ define

$$U_m(h_I) = 1_{I+m|I|} - 1_I.$$

With (1.4.21) we find $\sum_m b_m^I = 0$, hence

$$\sum_{\{m:I+m|I|\subseteq[0,1]\}} b_m^I 1_{I+m|I|}(x) = \sum_{\{m:I+m|I|\subseteq[0,1]\}} b_m^I U_m(h_I)(x). \tag{1.4.23}$$

The mean value estimate of (1.4.17) implies that for each fixed m and I,

$$\sup_I |b_m^I| \leq A\,\{1 + |m|\}^{-1-\alpha}. \tag{1.4.24}$$

Step 3. Here we analyze the Haar series appearing in (1.4.20). Let

$$d_{(m,l)}(f_I) = \sum_{\{J\subseteq I+m|I|,\,|J|=2^{-l}|I|\}} \langle f_I, \frac{h_J}{|J|}\rangle h_J. \tag{1.4.25}$$

The Haar support of $d_{(m,l)}(f_I)$ is a collection of pairwise disjoint dyadic intervals. The union of the intervals appearing in the Haar support of $d_{(m,l)}(f_I)$ equals $I + m|I|$. Hence the square function of $d_{(m,l)}(f_I)$ is supported in the interval $I + m|I|$. Invoking the coefficient estimates (1.4.18) gives that

$$S\left(d_{(m,l)}(f_I)\right) \le A\left\{1+m\right\}^{-1-\alpha} 2^{-l} 1_{I+m|I|}. \tag{1.4.26}$$

Note also that the indicator function $1_{I+m|I|}$ just coincides with $S(T_m(h_I))$, where T_m is the rearrangement operator defined as the linear extension of the map

$$T_m\ :\ h_I \mapsto h_{I+m|I|}, \quad \text{for} \quad I+m|I| \subseteq [0,1].$$

Moreover, we observe that for each m, l fixed, the set of functions $\{d_{(m,l)}(f_I) : I \in \mathcal{D}\}$ is disjointly supported over the Haar system. In summary the following pointwise estimate between the square functions holds true,

$$S\left(\sum_I x_I d_{(m,l)}(f_I)\right) \le A2^{-l}\left\{1+m\right\}^{-1-\alpha} S\left(T_m(\sum_I x_I h_I)\right), \tag{1.4.27}$$

where (x_I) is any given sequence of scalars.

Step 4. Let us summarize the non-standard expansion of f_I developed in the previous three steps. Using that $\int f_I = 0$ we obtained the formula

$$f_I = \sum_m b_m^I U_m(h_I) + \sum_m \sum_{l=0}^{\infty} d_{(m,l)}(f_I). \tag{1.4.28}$$

By the hypothesis this expansion is convergent. Next we take a finite linear combination $g = \sum_I x_I f_I$, where $x_I \in \mathbb{R}$. We apply the representation (1.4.28) to each of the functions f_I. This gives the representation for g as

$$g = \sum_m \sum_I x_I b_m^I U_m(h_I) + \sum_m \sum_{l=0}^{\infty} \sum_I x_I d_{(m,l)}(f_I). \tag{1.4.29}$$

Recall that our aim is to estimate the norm of g in L^p by the norm of $f = \sum x_I h_I$ in L^p. We exploit (1.4.29) by simply applying the triangle inequality. Thus for $\|g\|_{L^p([0,1])}$ we obtain an upper bound by the following series,

$$\sum_m \left\|\sum_I x_I b_m^I U_m(h_I)\right\|_{L^p([0,1])} + \sum_m \sum_{l=0}^{\infty} \left\|\sum_I x_I d_{(m,l)}(f_I)\right\|_{L^p([0,1])}. \tag{1.4.30}$$

In Step 2, we saw that the mean value estimates (1.4.17) imply (1.4.24). Thus using the norm estimate of Haar multipliers (Theorem 1.5.1 below) we find that

$$\left\| \sum_I x_I b^I_m U_m(h_I) \right\|_{L^p([0,1])} \le A_p \{1+|m|\}^{-1-\alpha} \|U_m\|_{L^p([0,1])} \|f\|_{L^p([0,1])} .$$

The analysis of the operators $d_{(m,l)}(f_I)$ is summarized in the estimate (1.4.27) which relates the square function to the basic rearrangement operators T_m. So we get that,

$$\begin{aligned} \left\| \sum_I x_I d_{(m,l)}(f_I) \right\|_{L^p([0,1])} &\le A \{1+|m|\}^{-1-\alpha} 2^{-l/2} \left\| \sum_I x_I T_m(h_I) \right\|_{L^p([0,1])} \\ &\le A \{1+|m|\}^{-1-\alpha} 2^{-l/2} \|T_m\|_{L^p([0,1])} \|f\|_{L^p([0,1])} . \end{aligned}$$

Step 5. In Theorem 3.3.9 below we will prove that the norm of $\|T_m\|_{L^p([0,1])}$, and $\|U_m\|_{L^p([0,1])}$ admit an upper bound of the form $C_p \log_2(4+|m|)$ where C_p depends on the value of p only. Hence the estimates for g, established in Step 4, imply that $\|g\|_{L^p([0,1])} \le C_p \|f\|_{L^p([0,1])}$. □

Before we continue we should try to understand one particular aspect about the non-standard expansion that led to the formula (1.4.28). Namely the role of the hypothesis (1.4.16), that is $\int f_I = 0$. Let $\int f_I = a_I$. In the case where $a_I \neq 0$ we can, of course, still carry out the expansion (1.4.28). The changes resulting from $a_I \neq 0$ are displayed in the identity

$$f_I = a_I 1_I + \sum_m b^I_m U_m(h_I) + \sum_m \sum_{l=0}^{\infty} d_{(m,l)}(f_I). \tag{1.4.31}$$

The question is now if we can also have a result like Theorem 1.4.3 without using that $a_I = 0$, or without making strong assumptions like $\sum_{I\in\mathcal{D}} |a_I| < \infty$. The answer to this question is quite surprising and leads to an interesting relation between L^p and BMO : If $\{f_I\}$ satisfies (1.4.17) and (1.4.18) and if

$$\sum a_I h_I \in \text{BMO},$$

then the conclusion of Theorem (1.4.3) holds true, that is the map

$$T : h_I \mapsto f_I$$

extends linearly to a bounded operator on L^p, when $1 < p < \infty$. How this works and how the BMO condition enters here, will be one of the topics discussed in the next section.

Assume that $\{f_I : I \in \mathcal{D}\}$ is dominated by the Haar system. Incidentally this implies the norm estimates

$$\|f_I\|_2 \le C|I|^{1/2}.$$

Now we want to isolate further conditions on $\{f_I : I \in \mathcal{D}\}$ so that the map

$$T : h_I \mapsto f_I$$

extends to an operator satisfying *two-sided* estimates in L^p. That is, we want to know when $T : h_I \mapsto f_I$ extends to an isomorphism between L^p. A minimal requirement is, of course, that the system $\{f_I : I \in \mathcal{D}\}$ (when augmented with the constant function) is a basis in L^2. So $\{f_I : I \in \mathcal{D}\}$ has to have a well-defined system of biorthogonal functions attached to it. Call it $\{g_I : I \in \mathcal{D}\}$. Let the defining equations for $\{g_I : I \in \mathcal{D}\}$ be given as follows,

$$\int_0^1 g_J dt = 0 \quad \text{and} \quad \int_0^1 f_I g_J dt = \begin{cases} |I| & \text{if } I = J, \\ 0 & \text{if } I \neq J. \end{cases} \tag{1.4.32}$$

Note that then

$$\sup \|g_I\|_2 |I|^{-1/2} < \infty.$$

Our next theorem asserts that if $\{f_I : I \in \mathcal{D}\}$ and $\{g_I : I \in \mathcal{D}\}$ are both dominated by the Haar system, then the linear extension of the map

$$T : h_I \mapsto f_I$$

defines an *isomorphism* on L^p. Theorem 1.4.4 below gives a very useful criterion for equivalence to the Haar system. It also raises the following significant problem: Assume that $\{f_I : I \in \mathcal{D}\}$ is an unconditional basis in L^2 and dominated by the Haar system. Is it true that its biorthogonal system is dominated by the Haar system as well? In many cases the answer is known and positive. Hence Theorem 1.4.4 covers a wide range of examples; it applies, for instance, to the Franklin system, to wavelets and to spline bases.

Theorem 1.4.4. *Let $1 < p < \infty$. Assume that the family $\{f_I : I \in \mathcal{D}\}$ and its biorthogonal system $\{g_I : I \in \mathcal{D}\}$ are both dominated by the Haar system, then for every choice of scalars (x_I) the following estimate holds,*

$$c \left\| \sum x_I h_I \right\|_{L^p} \leq \left\| \sum x_I f_I \right\|_{L^p} \leq C \left\| \sum x_I h_I \right\|_{L^p},$$

where $c = c(A, \alpha)(\max\{p, p/(p-1)\})^{-1}$, and $C = C(A, \alpha) \max\{p, p/(p-1)\}$.

Proof. Let $1 < p < \infty$. By hypothesis, the family $\{f_I : I \in \mathcal{D}\}$ is dominated by the Haar system. Hence the right-hand side estimate follows from Theorem 1.4.3. The left-hand side estimate is equivalent to saying that the linear extension of the map

$$S : f_I \mapsto h_I$$

extends to a bounded operator on L^p. However a simple calculation shows that the transposed operator to S is the linear extension of the map

$$S^* : h_I \mapsto g_I.$$

By our assumption that $\{g_I : I \in \mathcal{D}\}$ is dominated by the Haar system it follows that S^* is a bounded operator on L^p, by Theorem 1.4.3. $\square$

It remains to give a set of pointwise estimates on a family $\{f_I : I \in \mathcal{D}\}$, that guarantees that it is dominated by the Haar system. This is provided by the next proposition. We will see that the mean value condition (1.4.17) corresponds to a pointwise decay estimate (1.4.34) and that condition (1.4.18) corresponds to a decay estimate for the derivative of f_I, expressed in (1.4.35).

Proposition 1.4.5. *Let $A > 0, \alpha > 0$. Let $\{f_I : I \in \mathcal{D}\}$ be a family of functions. Let $I \in \mathcal{D}$ such that $|I| = 2^{-i}$. Assume that*

$$\int_0^1 f_I dt = 0 \tag{1.4.33}$$

and for every $x \in [0, 1]$,

$$|f_I(x)| \leq A \left\{1 + 2^i \operatorname{dist}(x, I)\right\}^{-1-\alpha}, \tag{1.4.34}$$

and every $x, y \in [0, 1]$,

$$|f_I(x) - f_I(y)| \leq A2^i|x - y| \left\{1 + 2^i \operatorname{dist}([x, y], I)\right\}^{-1-\alpha}, \tag{1.4.35}$$

where $A > 0$ and $\alpha > 0$ are independent of $I \in \mathcal{D}$. Then the system $\{f_I : I \in \mathcal{D}\}$ is dominated by the Haar system.

Proof. Fix a dyadic interval I, with $|I| = 2^{-i}$. Let $m \in \mathbb{Z}$. Assume that $I + m|I|$ is contained in the unit interval. By (1.4.34) every $x \in I + m|I|$ satisfies

$$|f_I(x)| \leq A \{1 + |m|\}^{-1-\alpha}.$$

Integrating over $x \in I + m|I|$ gives the claimed estimate of the mean value of f_I over the shifted interval $I + m|I|$,

$$\left| \int_{I+m|I|} f_I(x)dx \right| \leq A|I|\{1 + |m|\}^{-1-\alpha}.$$

Next we turn to estimating the Haar coefficients of f_I. Let J be a dyadic interval such that, $J \subseteq I + m|I|$ and $|J| = 2^{-l}|I|$. If $x, y \in J$, then $|x - y| \leq 2^{-i-l}$. Hence the Lipschitz estimate of (1.4.35) implies that for $x, y \in J$ and $J \subseteq I + m|I|$ we have

$$|f_I(x) - f_I(y)| \leq A2^{-l} \{1 + |m|\}^{-1-\alpha}. \tag{1.4.36}$$

Let m_J denote the midpoint of the interval J. By (1.4.36), we obtain the following estimate for the Haar coefficients,

$$\begin{aligned}\left| \int_0^1 f_I(y)h_J(y)dy \right| &= \left| \int_0^1 \{f_I(y) - f_I(m_J)\}h_J(y)dy \right| \\ &\leq |J| \sup_{y \in J} |f_I(y) - f_I(m_J)| \\ &\leq A|J|2^{-l} \{1 + |m|\}^{-1-\alpha}.\end{aligned}$$

$\square$

The books by B. Kashin and A. Saakyan [111], by Y. Meyer and R.R. Coifman [147], or by G. David [58] discuss in detail classes of orthonormal systems for which the hypothesis of Theorem 1.4.4 and Proposition 1.4.5 are satisfied.

1.5 Basic operators

We present Figiel's representation of integral operators using paraproducts, rearrangements and Haar multipliers. We discuss orthogonal (averaging) projections acting on the Haar system, and introduce a basic criterion implying their boundedness on H^1.

Multipliers and paraproducts

First we discuss multiplier operators acting on the Haar system. Let (x_I) be a sequence in $\mathbb{R}$. We define the multiplier operator X to be the linear extension of the map

$$h_I \mapsto x_I h_I.$$

Thus for f with finite Haar expansion,

$$f = \sum_I a_I h_I,$$

$X(f)$ is defined by

$$X(f) = \sum_I x_I a_I h_I. \tag{1.5.1}$$

So far the multiplier operator X has been defined just algebraically. Now we study its boundedness properties. We show that Burkholder's inequality implies that for bounded sequences (x_I) the induced multiplier is a bounded operator on L^p, $1 < p < \infty$. The proof given below works for any unconditional basis in any Banach space.

Theorem 1.5.1. *Let $1 < p < \infty$. Let (x_I) be a bounded sequence. Then the multiplier defined by* (1.5.1) *satisfies the following norm estimate on L^p,*

$$\|X(f)\|_p \leq (p^* - 1)\left(\sup_I |x_I|\right)\|f\|_p,$$

where $p^* = \max\{p, p/(p-1)\}$.

Proof. Fix $1 < p < \infty$ and let q denote the Hölder conjugate index to p. Choose $g \in L^q$ satisfying $\|g\|_q = 1$, and

$$\|X(f)\|_p = \langle X(f), g\rangle. \tag{1.5.2}$$

With the Haar system we expand $g = \int g dt + \sum b_I h_I$. Then by biorthogonality,

$$\begin{aligned} \langle X(f), g\rangle &= \sum_I x_I a_I b_I |I| \\ &\leq \sup_I |x_I| \sum_I |a_I b_I|\, |I|. \end{aligned} \tag{1.5.3}$$

Next choose $\epsilon_I \in \{\pm 1\}$ such that $\epsilon_I a_I b_I = |a_I b_I|$. Again by biorthogonality,

$$\sum_I \epsilon_I a_I b_I \, |I| = \langle \sum_I \epsilon_I a_I h_I, g\rangle. \tag{1.5.4}$$

With (1.5.3) and (1.5.4) Hölder's inequality gives that

$$\langle X(f), g\rangle \leq \sup_I |x_I| \left\| \sum_I \epsilon_I a_I h_I \right\|_p \|g\|_q. \tag{1.5.5}$$

Burkholder's inequality provides the upper bound

$$\| \sum_I \epsilon_I a_I h_I \|_p \leq (p^* - 1) \|f\|_p.$$

With (1.5.2) and (1.5.5) the last line finishes the proof. □

Next we define paraproduct operators and study their boundedness on the L^p spaces. Let $a \in \mathrm{BMO}$ and let $f \in L^p$ with $1 < p < \infty$. The paraproduct P_a is then defined by,

$$P_a(f) = \sum_I \langle a, \frac{h_I}{|I|}\rangle \langle f, \frac{h_I}{|I|}\rangle 1_I. \tag{1.5.6}$$

The operator P_a is thus defined to be the linear extension of the map

$$h_I \mapsto \langle a, \frac{h_I}{|I|}\rangle 1_I.$$

Paraproducts appear quite naturally in connection with Figiel's method considered in the previous section. There we fixed a family of functions $\{f_I\}$ and asked for conditions which imply that the map

$$T : h_I \to f_I$$

extends to a bounded linear operator on L^p. Following Figiel's approach we demanded that the functions f_I satisfy the mean-value conditions (1.4.17) and estimates for their Haar coefficient expressed in (1.4.18). Recall the following expansion which marked the starting point for Figiel's work:

$$f_I = a_I 1_I + \sum_m b^I_m U_m(h_I) + \sum_m \sum_{l=0}^{\infty} d_{(m,l)}(f_I), \tag{1.5.7}$$

where b^I_m is defined in (1.4.22), $d_{(m,l)}(f_I)$ is defined in (1.4.25) and where

$$a_I = \int_0^1 f_I dt.$$

For a finite linear combination $g = \sum_I x_I f_I$, with $x_I \in \mathbb{R}$ the representation (1.5.7) allows us to estimate the norm of g as follows,

$$\begin{aligned} \|g\|_{L^p} \leq & \left\| \sum_I x_I a_I 1_I \right\|_{L^p} \\ & + \left\| \sum_m \sum_I x_I b^I_m U_m(h_I) \right\|_{L^p} + \left\| \sum_m \sum_{l=0}^{\infty} \sum_I x_I d_{(m,l)}(f_I) \right\|_{L^p} . \end{aligned} \tag{1.5.8}$$

We saw in the proof of Theorem 1.4.3 that (1.4.17) gives a nice upper bound for the second series appearing on the right-hand side of (1.5.8). The hypothesis and (1.4.18) takes care of the third sum in (1.5.8). So, by reviewing the proof of Theorem 1.4.3 we made the following observation. If the family $\{f_I\}$ satisfies (1.4.17) and (1.4.18), then

$$T : h_I \to f_I$$

extends boundedly to L^p, provided that

$$\| \sum_I x_I a_I 1_I \|_{L^p} \leq C_p \| \sum_I x_I h_I \|_{L^p} \tag{1.5.9}$$

where

$$a_I = \int_0^1 f_I dt.$$

But clearly the estimate (1.5.9) states that the paraproduct P_a is bounded on L^p. Simultaneously, the paraproduct is one of the basic operators of Calderón–Zygmund theory, since it plays a crucial role in the proof of the famous $T(1)$ theorem of G. David and J.L. Journé [59]

Theorem 1.5.2. *Let $a \in BMO$, then P_a defines a bounded operator on L^p, for $1 < p < \infty$, and satisfies the estimate*

$$||P_a(f)||_p \leq C_p ||a||_{BMO} ||f||_p,$$

for every $f \in L^p$ with $1 < p < \infty$. The constant C_p can be chosen as follows

$$C_p = \begin{cases} 32(p-1)^{-1} & \text{if } 1 < p \leq 2, \\ 4p^{3/2} & \text{if } 2 \leq p < \infty. \end{cases} \tag{1.5.10}$$

Proof. Let $f \in L^p$ and let $g \in L^q$, where q denotes the Hölder conjugate index to p. We will show that

$$|\langle P_a(f), g\rangle| \leq C_p ||a||_{\mathrm{BMO}} ||f||_p ||g||_q. \tag{1.5.11}$$

The essential step towards the proof is the estimate

$$|\langle P_a(f), g\rangle| \leq 2\sqrt{2}||a||_{\mathrm{BMO}} \int_0^1 M_1(g)(x) S(f)(x) dx, \tag{1.5.12}$$

where $M_1(g)$ denotes the Hardy–Littlewood maximal function introduced in Section 1.1 in Theorem 1.2.6. We use the notation g_I to denote the mean value of g over the interval I,

$$g_I = \frac{1}{|I|} \int_I g.$$

We start with the identity

$$\begin{aligned} \langle P_a(f), g\rangle &= \sum_I \langle a, h_I\rangle \langle f, \frac{h_I}{|I|}\rangle g_I \\ &= \int_0^1 a(x) \left(\sum_I \langle f, \frac{h_I}{|I|}\rangle g_I h_I(x) \right) dx. \end{aligned} \tag{1.5.13}$$

To the integral appearing in the last line of (1.5.13) we apply Fefferman's inequality. This gives that

$$|\langle P_a(f), g\rangle| \leq 2\sqrt{2}||a||_{\mathrm{BMO}} \left\| \sum_I \langle f, \frac{h_I}{|I|}\rangle g_I h_I \right\|_{H^1}. \tag{1.5.14}$$

We estimate the H^1 factor by analyzing its square function pointwise. For $x \in [0,1]$ we have that

$$\sum_I \langle f, \frac{h_I}{|I|}\rangle^2 g_I^2 h_I^2(x) \leq \sup_{I \ni x} g_I^2 \cdot \sum_I \langle f, \frac{h_I}{|I|}\rangle^2 h_I^2(x),$$

where the supremum is taken over all dyadic intervals containing the point x. Note that $\left(\sup_{I\ni x} g_I^2\right)^{1/2} \leq M_1(g)(x)$, where $M_1(g)$ denotes the Hardy–Littlewood maximal function introduced in Theorem 1.2.6. Thus we obtain that

$$\left(\sum_I \langle f, \frac{h_I}{|I|}\rangle^2 g_I^2 h_I^2(x) \right)^{1/2} \leq M_1(g)(x) S(f)(x). \tag{1.5.15}$$

Integrating (1.5.15) gives (1.5.12), and by Hölder's inequality we obtain that

$$|\langle P_a(f), g\rangle| \leq 2\sqrt{2}||a||_{\mathrm{BMO}} ||M_1(g)||_q ||S(f)||_p. \tag{1.5.16}$$

The factors on the right-hand side of (1.5.16) are easily estimated with the upper bound for the Hardy–Littlewood maximal function, Theorem 1.2.6, and the square function characterization of Theorem 1.1.6. Thus (1.5.16) implies that

$$|\langle P_a(f), g\rangle| \le C_p ||a||_{\mathrm{BMO}} ||g||_q ||f||_p, \tag{1.5.17}$$

where $C_p = 8p(p^* - 1)$ and $p^* = \max\{p, p/(p-1)\}$. Taking the supremum (1.5.17) over all $g \in L^q$ with $||g||_q \le 1$ gives that

$$||P_a(f)||_p \le C_p ||a||_{\mathrm{BMO}} ||f||_p.$$

A more careful use of square function estimates is possible by distinguishing between the cases $1 < p < 2$ and $2 < p < \infty$. If we use Theorem 1.1.6 for $1 < p < 2$, and Theorem 1.3.5 for $2 < p < \infty$, then we obtain improved constants C_p as follows,

$$C_p = \begin{cases} 32(p-1)^{-1} & \text{if } 1 < p < 2, \\ 4p^{3/2} & \text{if } 2 < p < \infty. \end{cases} \tag{1.5.18}$$

□

Rearrangement operators and Calderón–Zygmund kernels

Next we introduce rearrangement operators. In combination with paraproducts they are an important tool in the study of integral operators. Most notably in the analysis of Calderón–Zygmund integral operators. In this section we present Figiel's representation of integral operators, to illustrate this point. For $m \in \mathbb{Z}$, we define the operators T_m and U_m by their action on the Haar basis. Let I be a dyadic interval such that $I + m|I| \subseteq [0,1]$. Then we put

$$T_m(h_I) = h_{I+m|I|},$$
$$U_m(h_I) = 1_{I+m|I|} - 1_I.$$

For the remaining intervals we simply put

$$T_m(h_I) = U_m(h_I) = 0.$$

Defined that way, the operators T_m and U_m are bounded operators on L^p when $1 < p < \infty$. We will give the precise estimates later. Now we discuss their connection to integral operators in detail. We start with integral operators for which the kernel, a priorily, belongs to the space $L^2([0,1] \times [0,1])$. To analyze the operator we expand the kernel using the following biorthogonal system in $L^2([0,1] \times [0,1])$.

$$\{h_{I+m|I|}(x) \cdot h_I(y) : I + m|I| \subseteq [0,1],\ m \in \mathbb{Z}, I \in \mathcal{D}\}$$
$$\cup$$
$$\{1_{I+m|I|}(x) \cdot h_I(y) : I + m|I| \subseteq [0,1],\ m \in \mathbb{Z}, I \in \mathcal{D}\}$$
$$\cup$$
$$\{h_{I+m|I|}(x) \cdot 1_I(y) : I + m|I| \subseteq [0,1],\ m \in \mathbb{Z}, I \in \mathcal{D}\}.$$

This system is complete in $L^2([0,1]\times[0,1])$, provided that we augment it with the constant function 1.

Now we fix an integral kernel in $K \in L^2([0,1]\times[0,1])$. We will now expand K using the basis introduced above. This is done in three steps corresponding to the three groups of basis functions introduced. We fix dyadic intervals I, J, with $|I| = |J|$, then we define the Haar coefficients

$$\begin{aligned}(K, h_J \otimes h_I) &= \int_0^1 \int_0^1 K(x,y) h_I(x) h_J(y)\,dx\,dy,\\ (K, 1_J \otimes h_I) &= \int_0^1 \int_0^1 K(x,y) h_I(x) 1_J(y)\,dx\,dy,\\ (K, h_J \otimes 1_I) &= \int_0^1 \int_0^1 K(x,y) 1_I(x) h_J(y)\,dx\,dy.\end{aligned} \tag{1.5.19}$$

Next we introduce three kernels which also correspond to the three groups of the basis.

$$\begin{aligned}K_1(x,y) &= \sum_{m\in\mathbb{Z}} \sum_{\{I:I+m|I|\subseteq[0,1]\}} (K, h_{I+m|I|}\otimes h_I) \frac{h_{I+m|I|}(x)}{|I|}\frac{h_I(y)}{|I|},\\ K_2(x,y) &= \sum_{m\in\mathbb{Z}} \sum_{\{I:I+m|I|\subseteq[0,1]\}} (K, 1_{I+m|I|}\otimes h_I) \frac{1_{I+m|I|}(x)}{|I|}\frac{h_I(y)}{|I|},\\ K_3(x,y) &= \sum_{m\in\mathbb{Z}} \sum_{\{I:I+m|I|\subseteq[0,1]\}} (K, h_{I+m|I|}\otimes 1_I) \frac{h_{I+m|I|}(x)}{|I|}\frac{1_I(y)}{|I|}.\end{aligned}$$

By definition, our original kernel K can be recovered from K_1, K_2, K_3, as follows,

$$K = \int_0^1 \int_0^1 K(x,y)\,dx\,dy + K_1 + K_2 + K_3. \tag{1.5.20}$$

Following is the definition of three Haar multipliers. Their defining sequences come from the Haar coefficients (1.5.19). We let $m \in \mathbb{Z}$ and let I be a dyadic interval, then we define

$$a_m(I) = \frac{(K, h_{I+m|I|}\otimes h_I)}{|I|}. \tag{1.5.21}$$

For fixed m the sequence $(a_m(I))$ defines the multiplier

$$A_m(f) = \sum_{\{I:I+m|I|\subseteq[0,1]\}} a_m(I)\langle f, \frac{h_I}{|I|}\rangle h_I. \tag{1.5.22}$$

Theorem 1.5.1 asserts that if $(a_m(I))$ is bounded, then the Haar multiplier A_m is a bounded operator on L^p and satisfies

$$\|A_m(f)\|_p \le (p^* - 1)\left(\sup_I |a_m(I)|\right)\|f\|_p.$$

The Haar multiplier A_m corresponds to the kernel K_1. Now we define the multiplier corresponding to K_2. Let I be a dyadic interval, then we define

$$b_m(I) = \frac{(K, 1_{I+m|I|} \otimes h_I)}{|I|}.$$

For fixed m the sequence $(b_m(I))$ defines the multiplier

$$B_m(f) = \sum_{\{I:I+m|I|\subseteq[0,1]\}} b_m(I)\langle f, \frac{h_I}{|I|}\rangle h_I.$$

By Theorem 1.5.1, the Haar multiplier B_m satisfies the norm estimates

$$\|B_m(f)\|_p \leq (p^* - 1)\left(\sup_I |b_m(I)|\right)\|f\|_p.$$

It remains to define the Haar multiplier associated to the kernel K_3. Let I be a dyadic interval, then define the coefficients

$$c_m(I) = \frac{(K, h_I \otimes 1_{I+m|I|})}{|I|},$$

which in turn define the multiplier C_m,

$$C_m(f) = \sum_{\{I:I+m|I|\subseteq[0,1]\}} c_m(I)\langle f, \frac{h_I}{|I|}\rangle h_I.$$

Theorem 1.5.1 gives that

$$\|C_m(f)\|_p \leq (p^* - 1)\left(\sup_I |c_m(I)|\right)\|f\|_p.$$

The above discussion shows that in L^p the operator norms of the multipliers A_m, B_m, C_m, are bounded by the quantity

$$|K|_m = (p^* - 1)\sup_I\{|a_m(I)| + |b_m(I)| + |c_m(I)|\},$$

where the supremum is extended over all dyadic intervals I for which $I + m|I|$ is contained in the unit interval.

We are now prepared to formulate Figiel's representation of integral operators. This is a powerful theorem which expresses a given integral operator as a sum of paraproducts, Haar multipliers and rearrangement operators. The classical $T(1)$ theorem of G. David and J.L Journé [59] follows easily from Figiel's representation, thereby showing the importance of these basic operators.

Theorem 1.5.3. *Let $K \in L^2([0,1] \times [0,1])$. Let R be the integral operator defined by the kernel K,*

$$R(f)(x) = \int_0^1 K(x,y)f(y)dy.$$

Let $a = R(1)$, $b = R^*(1)$ and put $K_0 = \int_0^1 \int_0^1 K(x,y)dxdy$. Then R admits the representation

$$R(f) = K_0 + P_a(f) + P_b^*(f) + \sum_{m \in \mathbb{Z}} T_m(A_m(f)) + U_m(B_m(f)) + (U_m C_m)^*(f),$$

and its norm on L^p satisfies the estimate

$$||R||_p \leq K_0 + 8p(p^* - 1)\,(||a||_{BMO} + ||b||_{BMO}) + C_p \sum_{m \in \mathbb{Z}} \log_2(4 + |m|)|K|_m.$$

Here $p^* = \max\{p, p/(p-1)\}$.

We formulated Figiel's representation theorem for integral operators under the initial hypothesis that $K \in L^2([0,1] \times [0,1])$. However the L^2 norm of the kernel does not appear in the conclusion of Theorem 1.5.3, that is the norm estimate of the integral operator

$$R(f)(x) = \int_0^1 K(x,y)f(y)dy.$$

It is therefore possible to extend the scope of the representation Theorem 1.5.3 to singular integral operators, using simple limiting procedures.

Recall from (1.5.20), that the kernel K appearing in Theorem 1.5.3 is decomposed as

$$K = \int_0^1 \int_0^1 K(x,y)dxdy + K_1 + K_2 + K_3.$$

For $i \in \{1,2,3\}$ let R_i denote the integral operator corresponding to the kernel K_i,

$$R_i(f)(x) = \int_0^1 K_i(x,y)f(y)dy.$$

The following three propositions give the respective representations for the integral operators R_i, $i \in \{1,2,3\}$. Put together they give Theorem 1.5.3.

We begin with the simplest case which is the operator R_1.

Proposition 1.5.4. *Let $f \in L^p$. Let A_m be the multiplier given by (1.5.21) and (1.5.22). Then*

$$R_1(f) = \sum_{m \in \mathbb{Z}} T_m(A_m(f)), \tag{1.5.23}$$

and

$$||R_1||_p \leq \sum_{m \in \mathbb{Z}} ||T_m||_p ||A_m||_p,$$

where

$$\|A_m\|_p \leq (p^* - 1)\left(\sup_I |a_m(I)|\right).$$

Proof. Once we have shown the representation (1.5.23), the norm estimate for R_1 follows easily. Indeed by (1.5.22), Theorem 1.5.1 implies that

$$||A_m||_p \leq (p^* - 1)\left(\sup_I |a_m(I)|\right).$$

We turn to the representation of R_1. Note that the action of T_m on the Haar function h_I is given by $T_m(h_I) = h_{I+m|I|}$. The claimed representation for the integral operator R_1 we find by unwinding the definitions involved. The first identity below follows from the definition of the kernel K_1, for the second one we use the definition of the rearrangement T_m, for the last identity we invoke the definition of the multiplier A_m. In summary we find that

$$\begin{aligned}\int_0^1 K_1(x,y)f(y)dy &= \sum_m \sum_I (K, h_{I+m|I|} \otimes h_I)\langle f, \frac{h_I}{|I|}\rangle \frac{h_{I+m|I|}(x)}{|I|} \\ &= \sum_m \sum_I (K, h_{I+m|I|} \otimes h_I)\langle f, \frac{h_I}{|I|}\rangle \frac{T_m(h_I)(x)}{|I|} \\ &= \sum_m T_m(A_m(f))(x).\end{aligned}$$

□

Proposition 1.5.5. *Let $f \in L^p$. Let $a(x) = \int_0^1 K(x,y)dy$, and let P_a be the paraproduct defined by a. Then R_2 has the representation*

$$R_2(f) = P_a(f) + \sum_{m\in\mathbb{Z}} U_m(B_m(f)).$$

Its operator norm admits an upper bound,

$$||R_2||_p \leq 8p(p^* - 1)||a||_{BMO} + \sum_{m\in\mathbb{Z}} ||U_m||_p \|B_m\|_p,$$

and

$$\|B_m\|_p \leq (p^* - 1)\left(\sup_I |b_m(I)|\right).$$

Proof. Here we start by evaluating the integral operator R_2. The definition of the kernel K_2 gives that

$$\int_0^1 K_2(x,y)f(y)dy = \sum_{m\in\mathbb{Z}} \sum_I \left(K, 1_{I+m|I|} \otimes h_I\right) \langle f, \frac{h_I}{|I|}\rangle \frac{1_{I+m|I|}(x)}{|I|}, \qquad (1.5.24)$$

where the inner sum is extended over all dyadic intervals I for which $I + m|I| \subseteq [0,1]$. Next we focus on the function $1_{I+m|I|}(x)$ appearing in the above sum. Using the operator U_m we rewrite $1_{I+m|I|}(x)$ as

$$1_{I+m|I|}(x) = U_m(h_I)(x) + 1_I(x). \tag{1.5.25}$$

Inserting the identity (1.5.25) into (1.5.24) defines a splitting of the operator R_2 into two parts. A first part which involves the operators U_m, and a remaining part which leads to the appearance of the paraproduct P_a.

We isolate first the paraproduct. By changing the order of summation we obtain the identity

$$\begin{aligned}
&\sum_{m\in\mathbb{Z}} \sum_{\{I: I+m|I|\subseteq[0,1]\}} \left(K, 1_{I+m|I|} \otimes h_I\right) \frac{1_I}{|I|} \langle f, \frac{h_I}{|I|}\rangle \\
&= \sum_I \left(\sum_{\{m: I+m|I|\subseteq[0,1]\}} \left(K, 1_{I+m|I|} \otimes h_I\right) \right) \frac{1_I}{|I|} \langle f, \frac{h_I}{|I|}\rangle .
\end{aligned}$$

We continue by identifying the sum appearing inside the brackets. We will show that it is just the Haar coefficient $\langle a, h_I\rangle$, where $a(x) = \int_0^1 K(x,y)dy$. We begin by evaluating

$$\sum_{\{m: I+m|I|\subseteq[0,1]\}} \left(K, 1_{I+m|I|} \otimes h_I\right) = \int_0^1 \int_0^1 K(x,y) h_I(x) dx dy. \tag{1.5.26}$$

Next observe that the double integral on the right-hand side is a Haar coefficient, namely with $a(x) = \int_0^1 K(x,y)dy$,

$$\int_0^1 \int_0^1 K(x,y) h_I(x) dx dy = \langle a, h_I\rangle. \tag{1.5.27}$$

Recall next the definition of the paraproduct operator P_a. It is given by the equation

$$P_a(f)(x) = \sum_I \langle a, h_I\rangle \langle f, \frac{h_I}{|I|}\rangle \frac{1_I(x)}{|I|}.$$

Combining (1.5.26) with (1.5.27), and inserting the value of $\langle a, h_I\rangle$ we arrived at the following identity for the paraproduct,

$$P_a(f)(x) = \sum_{m\in\mathbb{Z}} \sum_{\{I: I+m|I|\subseteq[0,1]\}} \left(K, 1_{I+m|I|} \otimes h_I\right) \langle f, \frac{h_I}{|I|}\rangle \frac{1_I}{|I|}. \tag{1.5.28}$$

Next we unwind the definition of the multipliers B_m, and thus we find that

$$\begin{aligned}
&\sum_{m\in\mathbb{Z}} U_m(B_m(f)) \\
&= \sum_{m\in\mathbb{Z}} U_m \left(\sum_I \langle B_m(f), \frac{h_I}{|I|} \rangle h_I \right) \\
&= \sum_{m\in\mathbb{Z}} \sum_I \left(K, 1_{I+m|I|} \otimes h_I \right) \langle f, \frac{h_I}{|I|} \rangle \frac{U_m(h_I)}{|I|},
\end{aligned}$$

where we extended the inner sum over all dyadic intervals I such that $I + m|I| \subseteq [0,1]$. It remains to combine the last identity with (1.5.24), (1.5.25) and (1.5.28). This gives that

$$\int_0^1 K_2(x,y) f(y) dy = P_a(f)(x) + \sum_{m\in\mathbb{Z}} U_m(B_m(f))(x),$$

which is the claimed representation for R_2. The resulting upper bound for the norm comes from Theorem 1.5.2 that controls the norm of the paraproduct,

$$||P_a||_p \le 8p(p^* - 1)||a||_{\mathrm{BMO}},$$

and Theorem 1.5.1 that gives norm estimates of the multiplier B_m. Hence

$$||B_m||_p \le (p^* - 1) \left(\sup_I |b_m(I)| \right). \qquad \square$$

Following is the representation of R_3. Up to transposition and duality the the proof below is the same as the preceding proof of Proposition 1.5.5.

Proposition 1.5.6. *Let $g \in L^p$. Let $b(y) = \int_0^1 K(x,y)dx$, and let P_b be the paraproduct defined by b. Then the adjoint operator to R_3 has the representation*

$$R_3^*(g) = P_b(g) + \sum_{m\in\mathbb{Z}} U_m(C_m(g)),$$

and

$$||R_3|| \le 8p(p^* - 1)||b||_{BMO} + \sum_{m\in\mathbb{Z}} ||U_m||_p ||C_m||_p,$$

where

$$||C_m||_p \le (p^* - 1) \left(\sup_I |c_m(I)| \right).$$

Proof. Fix $f \in L^p$, and $g \in L^q$. Then we evaluate $\langle R_3(f), g\rangle$. We invoke the definition of the kernel K_3, then we change the order of summation and relabel,

$$\begin{aligned}&\int_0^1\int_0^1 K_3(x,y)f(y)g(x)dxdy\\&=\sum_m\sum_I \left(K, h_{I+m|I|}\otimes 1_I\right)\langle g, \frac{h_{I+m|I|}}{|I|}\rangle\langle f, \frac{1_I(y)}{|I|}\rangle\\&=\sum_J\sum_n \left(K, h_J\otimes 1_{J+n|J|}\right)\langle g, \frac{h_J}{|J|}\rangle\langle f, \frac{1_{J+n|J|}}{|J|}\rangle.\end{aligned}$$

Next we look carefully at the last factor in the above sum. That is $\langle f, \frac{1_{J+n|J|}}{|J|}\rangle$. We rewrite it using the operator U_n as

$$\langle f, \frac{1_{J+n|J|}}{|J|}\rangle = \langle f, \frac{U_n(h_J)}{|J|}\rangle + \langle f, \frac{1_J}{|J|}\rangle.$$

Now the proof proceeds parallel to that of Proposition 1.5.5. We identify the paraproduct first. Recall that we put $b(y) = \int_0^1 K(x,y)dx$. One finds as in the proof of Proposition 1.5.5 that

$$\langle b, h_J\rangle = \sum_n \left(K, h_J\otimes 1_{J+n|J|}\right),$$

where we extend the sum over those $n\in\mathbb{Z}$ for which $J+n|J|\subseteq[0,1]$. The paraproduct $P_b(g)$ is defined by

$$P_b(g) = \sum_J \langle b, h_J\rangle\langle g, \frac{h_J}{|J|}\rangle\frac{1_J}{|J|}.$$

Consequently,

$$\langle f, P_b(g)\rangle = \sum_J\sum_n \left(K, h_J\otimes 1_{J+n|J|}\right)\langle g, \frac{h_J}{|J|}\rangle\langle f, \frac{1_J}{|J|}\rangle.$$

Next we recall the definition of the multiplier C_n, and note that

$$\begin{aligned}&\sum_n\langle f, U_n(C_n(g))\rangle\\&=\sum_J\sum_n \left(K, h_J\otimes 1_{J+n|J|}\right)\langle g, \frac{h_J}{|J|}\rangle\langle f, \frac{U_n(h_J)}{|J|}\rangle.\end{aligned}$$

It follows that

$$\begin{aligned}\langle f, R_3^*(g)\rangle &= \langle R_3(f), g\rangle\\&=\langle f, P_b(g)\rangle + \sum_{n\in\mathbb{Z}}\langle f, U_n(C_n(g))\rangle,\end{aligned}$$

which proves the representation of the operator R_3. The norm estimates are obtained as in the proof of Propositions 1.5.4 and 1.5.5. □

Proof of Theorem 1.5.3. We apply Propositions 1.5.4–1.5.6, and use the following upper bound for the norms of the operators T_m and U_m in L^p.

$$\begin{aligned} ||T_m||_p \leq & \ C_p \log_2(4+|m|), \\ ||U_m||_p \leq & \ C_p \log_2(4+|m|). \end{aligned}$$

These estimates will be proven in Chapter 3, Theorem 3.3.9. With this remark the proof of Theorem 1.5.3 is complete. □

The Hilbert transform is our first example illustrating Theorem 1.5.3. For $x, y \in [0,1]$ and $x \neq y$ we define

$$H(x,y) = \cot \frac{(x-y)}{\pi}.$$

Let $\epsilon > 0$ and put

$$H_\epsilon(x,y) = \begin{cases} H(x,y) & \text{if } |x-y| > \epsilon, \\ 0 & \text{otherwise.} \end{cases}$$

The kernel H_ϵ is a bounded function defined on $[0,1] \times [0,1]$. It gives rise to an integral operator

$$R_\epsilon f(x) = \int_0^1 H_\epsilon(x,y) f(y) dy.$$

Applying Theorem 1.5.3 we will show that the integral operator R_ϵ is bounded in L^p $(1 < p < \infty)$ with norm independent of $\epsilon > 0$.

Corollary 1.5.7. *Let $1 < p < \infty$. The Hilbert transform defined as the limit*

$$R(f) = \lim_{\epsilon \to 0} R_\epsilon(f)$$

is a bounded operator on L^p.

Proof. We will show that for $f \in L^p([0,1])$,

$$||R_\epsilon f||_p \leq A_p ||f||_p, \tag{1.5.29}$$

and that $\lim_{\epsilon\to 0} R_\epsilon(f)$ exists in L^p. To prepare for the application of Theorem 1.5.3 we recall that the kernel H_ϵ belongs to $L^2([0,1]\times[0,1])$. Furthermore

$$\int_0^1 H_\epsilon(x,y)dy = 0 \quad \text{and} \quad \int_0^1 H_\epsilon(x,y)dx = 0.$$

Consequently, the paraproducts in Figiel's representation of R_ϵ are vanishing operators. Using integration by parts we easily obtain that there exists $A > 0$ so that for $m \in \mathbb{Z}$, the following estimates hold,

$$\begin{aligned} |(H_\epsilon, h_I \otimes h_{I+m|I|})| &\leq A(1+|m|)^{-2}|I|, \\ |(H_\epsilon, 1_I \otimes h_{I+m|I|})| &\leq A(1+|m|)^{-2}|I|, \\ |(H_\epsilon, h_I \otimes 1_{I+m|I|})| &\leq A(1+|m|)^{-2}|I|. \end{aligned}$$

The constant A does not depend on I, m or $\epsilon > 0$. Hence Theorem 1.5.3 gives

$$\begin{aligned}||R_\epsilon||_p &\leq C_p A \sum_{m\in\mathbb{Z}} (1+|m|)^{-2}\log_2(4+|m|)\\ &\leq A_p.\end{aligned}$$

In this upper bound the L^2 norm of the kernel does not appear. In particular, the norm of R_ϵ in L^p is bounded independent of ϵ.

Next we show convergence of $R_\epsilon(f)$ when f has bounded derivative. As $\int H_\epsilon(x,y)dy = 0$, we may insert the value $f(x)$ in the integral defining $R_\epsilon f(x)$. Hence

$$R_\epsilon f(x) = \int_0^1 H_\epsilon(x,y)\{f(y)-f(x)\}dy.$$

Fix $0 < \delta < \epsilon$, and estimate the difference

$$\begin{aligned}|R_\delta f(x) - R_\epsilon f(x)| &= \left|\int_0^1 \{H_\delta(x,y) - H_\epsilon(x,y)\}\{f(y)-f(x)\}dy\right|\\ &\leq \int_{\delta\leq|x-y|\leq\epsilon} \left|\cot\frac{(x-y)}{\pi}\right| \cdot |f(y)-f(x)|dy.\end{aligned}$$

Recall that f is assumed to have bounded derivative, hence $|f(y) - f(x)| \leq ||f'||_\infty|x-y|$. Furthermore, the product $|(x-y)\cot(x-y)|$ is bounded. It follows that

$$|R_\delta f(x) - R_\epsilon f(x)| \leq C||f'||_\infty(\epsilon - \delta).$$

Summing up, $R_\epsilon f$ is a Cauchy sequence in L^p, when f has bounded derivative. By (1.5.29) and the Banach–Steinhaus theorem, the limit defined by

$$R(f) = \lim_{\epsilon\to 0} R_\epsilon(f)$$

induces a bounded linear operator on L^p. □

Now we apply Theorem 1.5.3 to a wide range of singular kernels, and prove a version of the $T(1)$ theorem of G. David and J. Journé [59]. A kernel $K(x,y)$ defined for $x \neq y$ is called a Calderón–Zygmund kernel if there exists $\delta > 0$ and $C > 0$ such that for $x \neq y$ the following estimates hold,

$$\begin{aligned}|K(x,y)| &\leq C|x-y|^{-1},\\ |\partial_x K(x,y)| + |\partial_y K(x,y)| &\leq C|x-y|^{-1-\delta},\\ K(x,y) &= -K(y,x).\end{aligned}$$

We will now determine when a Calderón–Zygmund kernel is associated to a bounded linear operator $T : L^p \to L^p$, in the following sense:

$$\int_0^1\int_0^1 K(x,y)f(y)g(x)dxdy = \langle T(f), g\rangle,$$

whenever $f \in L^p([0,1])$, $g \in L^q([0,1])$ and $\operatorname{supp}(f) \cap \operatorname{supp}(g) = \emptyset$.

Fix a Calderón–Zygmund kernel K. We introduce now a canonical approximation procedure for K. For a dyadic interval I define the following collection of dyadic intervals,

$$\mathcal{A}_I = \{J : |J| = |I|,\ \operatorname{dist}(I,J) \geq |I|\}.$$

Fix $n \in \mathbb{N}$. We define the approximating kernel $K^{(n)}$ by setting

$$K^{(n)}(x,y) = \sum_{\{I:|I|=2^{-n}\}} \sum_{J\in\mathcal{A}_I} K(x,y)1_I(x)1_J(y).$$

Note that the approximation is obtained by deleting a thin strip around the diagonal $x = y$, hence $K^{(n)} \in L^2([0,1]\times[0,1])$. By the defining properties for the Calderón–Zygmund kernel K there exists a constant $A_\delta > 0$ such that for every $n \in \mathbb{N}$ the following estimates hold,

$$\left|K^{(n)}\right|_m \leq A_\delta |m|^{-1-\delta}.$$

The L^2 kernel $K^{(n)}$ defines the integral operator

$$R^{(n)}(f)(x) = \int_0^1 K^{(n)}(x,y)f(y)dy.$$

Let $a_n = R^{(n)}(1)$, and $b_n = (R^{(n)})^*(1)$. Theorem 1.5.3 gives the following estimates for the operator norm of $R^{(n)} : L^p \to L^p$.

$$\begin{aligned}||R^{(n)}||_p &\leq C_p(||a_n||_{\mathrm{BMO}} + ||b_n||_{\mathrm{BMO}}) + C_p \sum_{m\in\mathbb{Z}} A_\delta(\log_2(4+|m|))|m|^{-1-\delta}\\ &\leq C_p(||a_n||_{\mathrm{BMO}} + ||b_n||_{\mathrm{BMO}}) + A_{p,\delta}.\end{aligned}$$

We point out that the constants $A_{p,\delta}$ appearing in the norm estimate for $R^{(n)}$ are actually independent of n.

Theorem 1.5.8. *If* $\sup_n ||a_n||_{BMO} < \infty$ *and* $\sup_n ||b_n||_{BMO} < \infty$, *then there exists a bounded linear operator* $T : L^p \to L^p$, $1 < p < \infty$ *such that*

$$\langle T(f), g\rangle = \int_0^1 \int_0^1 K(x,y)f(y)g(x)dxdy,$$

whenever $f \in L^p([0,1])$, $g \in L^q([0,1])$ *and* $\operatorname{supp}(f) \cap \operatorname{supp}(g) = \emptyset$.

Proof. First observe that we defined the approximating kernels by truncating so that for $x \neq y$,

$$K(x,y) = \lim_{n\to\infty} K^{(n)}(x,y).$$

As $\sup_n ||a_n||_{\mathrm{BMO}} < \infty$ there exists a subsequence of (a_n) which converges, weak* in BMO. Denote it again by (a_n). Let f have bounded derivative. Let $g \in L^q$. We show next that $(\langle g, R^{(n)}(f)\rangle)$ is a Cauchy sequence. Let $n \in \mathbb{N}$ and let $m \geq n$. Then

$$\begin{aligned}
&\langle g, (R^{(n)} - R^{(m)})(f)\rangle \\
&= \int_0^1 \int_0^1 (K^{(n)}(x,y) - K^{(m)}(x,y))f(y)g(x)dxdy \\
&= \int_0^1 \int_0^1 \left(K^{(n)}(x,y) - K^{(m)}(x,y)\right)(f(y) - f(x))g(x)dxdy \\
&+ \langle a_n - a_m, (gf)\rangle.
\end{aligned}$$

The definition of the approximating kernels give that

$$\begin{aligned}
|(K^{(n)}(x,y) - K^{(m)}(x,y))| \cdot |f(y) - f(x)| &\leq C|K^{(n)}(x,y) - K^{(m)}(x,y)| \cdot |x-y| \\
&\leq C.
\end{aligned}$$

Moreover $\operatorname{supp}\{(K^{(n)} - K^{(m)})\} \subseteq \{(x,y) : |x-y| \leq 2^{-n}\}$. Thus for each $x \in [0,1]$,

$$\int_0^1 |(K^{(n)}(x,y) - K^{(m)}(x,y))(f(y) - f(x))|dy \leq C2^{-n}.$$

It follows that $\langle g, (R^{(n)} - R^{(m)})(f)\rangle$ is a Cauchy sequence, and for a Lipschitz function f, the following limit exists in L^p,

$$\lim_{n\to\infty} R^{(n)} f.$$

By Theorem 1.5.3 and the hypothesis we find

$$\begin{aligned}
\sup_n ||R^{(n)}||_p &\leq \sup_n C_p(||a_n||_{\mathrm{BMO}} + ||b_n||_{\mathrm{BMO}}) + A_{p,\delta} \\
&< \infty.
\end{aligned}$$

As Lipschitz functions are dense in L^p, the theorem of Banach and Steinhaus implies the existence of a bounded linear operator $T : L^p \to L^p$, such that

$$T(f) = \lim_{n\to\infty} R^{(n)}(f),$$

where the limit exists in L^p; moreover,

$$||T||_p \leq \sup_n ||R^{(n)}||_p.$$

Finally note that if $\operatorname{supp}(f) \cap \operatorname{supp}(g) = \emptyset$, then $\operatorname{dist}(\operatorname{supp}(f), \operatorname{supp}(g)) > 0$. Hence the identity

$$K(x,y) = \lim_{n\to\infty} K^{(n)}(x,y),$$

holding for $x \neq y$, implies that

$$\int_0^1 \int_0^1 K(x,y)f(y)g(x)dxdy = \lim_{n\to\infty} \int_0^1 \int_0^1 K^{(n)}(x,y)f(y)g(x)dxdy,$$

provided that $f \in L^p([0,1])$, $g \in L^q([0,1])$ and $\operatorname{supp}(f) \cap \operatorname{supp}(g) = \emptyset$. Summing up we found a bounded linear operator $T : L^p \to L^p$, such that,

$$\begin{aligned}\langle T(f), g\rangle &= \lim_{n\to\infty} \langle R^{(n)}(f), g\rangle \\ &= \lim_{n\to\infty} \int_0^1 \int_0^1 K^{(n)}(x,y) f(y) g(x) dx dy \\ &= \int_0^1 \int_0^1 K(x,y) f(y) g(x) dx dy,\end{aligned}$$

whenever $f \in L^p([0,1])$, $g \in L^q([0,1])$ and $\operatorname{supp}(f) \cap \operatorname{supp}(g) = \emptyset$. □

Orthogonal projections

We introduce now a class of orthogonal projections which plays a very important role in this book. We will use them for three main reasons. First we construct several families of complemented subspaces in H^1 and BMO. Second their boundedness properties will be crucial in finding isomorphic invariants of H^1, such as the uniform approximation property, the dimension conjecture or dichotomies for complemented subspaces. Third, orthogonal projections yield canonical examples of operators by which we will test (and refute) several natural conjectures. From now on it is convenient to use systematically the following notation. Let $\mathcal{H}$ be a collection of dyadic intervals. Let $f = \sum a_I h_I$. We define $S(f|\mathcal{H})$ by the equation

$$S^2(f \,|\, \mathcal{H}) = \sum_{I \in \mathcal{H}} a_I^2 h_I^2,$$

and call it the square function of f restricted to $\mathcal{H}$. If J is a dyadic interval and if we put $Q(J) = \{I : I \subseteq J\}$, then

$$\int S^2(f \,|\, Q(J)) dt = \sum_{I \subseteq J} a_I^2 |I|.$$

Hence

$$\|f\|_{\mathrm{BMO}}^2 = \sup_J \int S^2(f \,|\, Q(J)) \frac{dt}{|J|}.$$

We let $\mathcal{C}_1, \ldots, \mathcal{C}_N$ be disjoint collections of dyadic intervals. We form linear combinations of Haar functions by putting

$$b_i = \sum \{h_I : I \in \mathcal{C}_i\}, \quad i \leq N, \tag{1.5.30}$$

and we normalize in L^2,

$$k_i = b_i / ||b_i||_2. \tag{1.5.31}$$

The orthogonal projection onto $\text{span}\{b_i\}_{i=1}^N$ is given by

$$Pf = \sum_{i=1}^{N} \langle f, k_i \rangle k_i. \tag{1.5.32}$$

Clearly, by the disjointness of the collections $\mathcal{C}_1, \dots, \mathcal{C}_N$ the operator P is of norm 1 in $L^2([0,1])$.

Now we introduce a condition on $\mathcal{C}_1, \dots, \mathcal{C}_N$ which implies that the operator norm of P on BMO is bounded independent of N. We let C_i denote the point-set covered by the collection $\mathcal{C}_i$, that is, we let

$$C_i = \bigcup_{I \in \mathcal{C}_i} I.$$

Disjoint collections of dyadic intervals $\mathcal{C}_1, \dots, \mathcal{C}_N$ are said to satisfy Jones's compatibility condition (J) if the following conditions (J 1)–(J 4) are satisfied.

(J 1) Each $\mathcal{C}_i$ is a collection of pairwise disjoint dyadic intervals.

(J 2) There exists a constant B, so that for each $J \in \mathcal{C}_j$ and each $i \geq j$ the following estimate holds,

$$\sum_{I \in \mathcal{C}_i, I \subseteq J} \frac{|I|}{|J|} \leq B \frac{|C_j \cap C_i|}{|C_j|}.$$

(J 3) The collections $\{\mathcal{C}_i : 1 \leq i \leq N\}$ are ordered in the following sense: If there exists $I \in \mathcal{C}_i$ and $J \in \mathcal{C}_j$, such that $I \subseteq J$, then $i \geq j$. Conversely, if $i \geq j$, if $I \in \mathcal{C}_i$ and $J \in \mathcal{C}_j$, such that $I \cap J \neq \emptyset$, then $I \subseteq J$.

(J 4) If $j \leq i$ and if $C_j \cap C_i \neq \emptyset$, then $C_i \subseteq C_j$. In particular the sets $\{C_i : 1 \leq i \leq N\}$ are nested; if $C_j \cap C_i \neq \emptyset$, then either $C_i \subseteq C_j$ or $C_j \subseteq C_i$.

If the compatibility condition (J) is satisfied we will show that the BMO norm of the orthogonal projection defined by (1.5.32) is bounded independent of N. Precisely we prove below that the norm of P in BMO is bounded by $\sqrt{B}$, where B is the constant appearing in condition (J 2). Following is a basic criterion for the boundedness of orthogonal projections in BMO.

Theorem 1.5.9. *If $\mathcal{C}_1, \dots, \mathcal{C}_N$ are pairwise disjoint collections of dyadic intervals satisfying Jones's compatibility condition (J), then, on BMO, the orthogonal projection*

$$Pf = \sum_{i=1}^{N} \langle f, k_i \rangle k_i$$

has norm bounded by $\sqrt{B}$.

Proof. Let $f \in \mathrm{BMO}$ with Haar expansion $f = \sum c_I h_I$. We expand Pf in its Haar series

$$Pf = \sum \alpha_I h_I,$$

where $\alpha_I = \langle f, k_i \rangle |\mathcal{C}_i|^{-1/2}$, for $I \in \mathcal{C}_i$, and where $\alpha_I = 0$ for $I \notin \bigcup_{i=1}^N \mathcal{C}_i$.

Now we fix $k \leq N$ and a dyadic interval $J_0 \in \mathcal{C}_k$. Then the following identity holds,

$$\frac{1}{|J_0|} \int S^2(Pf \,|\, Q(J_0)) = \frac{1}{|J_0|} \sum_{I \subseteq J_0} \alpha_I^2 |I|.$$

We will now show that the following estimates holds,

$$\frac{1}{|J_0|} \sum_{I \subseteq J_0} \alpha_I^2 |I| \leq B ||f||_{\mathrm{BMO}}^2.$$

By condition (J 3), if $I \subseteq J_0$ and if $I \in \mathcal{C}_i$, then $i \geq k$. Therefore by changing the order of summation we have the identity

$$\sum_{I \subseteq J_0} \alpha_I^2 |I| = \sum_{i=k}^{N} \langle f, k_i \rangle^2 \sum_{I \in \mathcal{C}_i, I \subseteq J_0} \frac{|I|}{|C_i|}. \tag{1.5.33}$$

We will now estimate the right-hand side of (1.5.33) using the conditions (J 1)–(J 4). Recall that condition (J 2) states that for $J \in \mathcal{C}_k$ and $i \geq k$, the following estimate holds,

$$\sum_{I \in \mathcal{C}_i, I \subseteq J_0} \frac{|I|}{|J_0|} \leq B \frac{|C_i \cap C_k|}{|C_k|}. \tag{1.5.34}$$

Thus conditions (J 2) and (J 3) give the following upper bound for E,

$$\frac{1}{|J_0|} \int S^2(Pf \,|\, Q(J_0)) \leq \frac{B}{|C_k|} \sum_{i=k}^{N} \langle f, k_i \rangle^2 \frac{|C_i \cap C_k|}{|C_i|}. \tag{1.5.35}$$

Now we use condition (J 4) to continue. It follows from (J 4) that if $C_i \cap C_k \neq \emptyset$ and $i \geq k$, then $C_i \subseteq C_k$, and

$$|C_i \cap C_k| = |C_i|.$$

Moreover if $i \geq k$ and C_i is not contained in C_k, then $C_i \cap C_k = \emptyset$. Hence the estimate (1.5.35) implies

$$\frac{1}{|J_0|} \int S^2(Pf \,|\, Q(J_0)) \leq \frac{B}{|C_k|} \sum_{\{i : C_i \subseteq C_k\}} \langle f, k_i \rangle^2. \tag{1.5.36}$$

Next we use the converse part of condition (J 3), Bessel's inequality and the definition of BMO to obtain

$$\begin{aligned}\sum_{\{i:C_i\subseteq C_k\}} \langle f, k_i\rangle^2 &\le \sum_{J\in\mathcal{C}_k}\sum_{K\subseteq J} c_K^2|K| \\ &\le ||f||^2_{\mathrm{BMO}} \sum_{J\in\mathcal{C}_k} |J|.\end{aligned} \tag{1.5.37}$$

Finally recall condition (J 1) that $\mathcal{C}_k$ is a collection of pairwise disjoint dyadic intervals. Thus

$$\sum_{J\in\mathcal{C}_k} |J| = |C_k|. \tag{1.5.38}$$

Inserting identity (1.5.38) and the estimate (1.5.37) into (1.5.36) gives

$$\frac{1}{|J_0|}\int S^2(Pf\,|\,Q(J_0)) \le B||f||^2_{\mathrm{BMO}},$$

as claimed. □

By duality the orthogonal projection P is bounded on H^1 if and only if it is bounded on BMO.

Theorem 1.5.10. *Let $\mathcal{C}_1,\dots,\mathcal{C}_N$ be collections of dyadic intervals satisfying Jones's compatibility condition (J). Let k_i be defined by (1.5.31). Then the orthogonal projection*

$$Pf = \sum_{i=1}^{N}\langle f, k_i\rangle k_i$$

defines a bounded operator on H^1 with norm $\le 8\sqrt{B}$.

Proof. We use Fefferman's duality theorem, the orthogonality of P and Theorem 1.5.9. This gives the following bound for the norm of P,

$$\begin{aligned}||Pf||_{H^1} &\le 2\sqrt{2}\sup\{\langle Pf, g\rangle : ||g||_{\mathrm{BMO}}\le 1\}\\ &= 2\sqrt{2}\sup\{\langle f, Pg\rangle : ||g||_{\mathrm{BMO}}\le 1\}\\ &\le 2\sqrt{2}\sqrt{B}\sup\{\langle f, h\rangle : ||h||_{\mathrm{BMO}}\le 1\}\\ &\le 8\sqrt{B}||f||_{H^1}.\end{aligned}$$

□

The glueing process. Below we need to be able to reduce the rank of a given orthogonal projection. For this purpose we glue together families of orthogonal projections acting locally, into one large orthogonal projection acting globally. We describe now the glueing process in detail, and study boundedness properties of orthogonal projections obtained by this method. We recall first the notion of a block of intervals. It was introduced in Section 1.2. Let $\mathcal{K}$ be a collection of dyadic intervals, and let $\mathcal{B}\subseteq\mathcal{K}$. We say that $\mathcal{B}$ is a block of intervals in $\mathcal{K}$, if the following conditions are satisfied.

1. $\mathcal{B}$ contains exactly one maximal interval, say I.
2. For every $J \in \mathcal{B}$ and $K \in \mathcal{K}$ the following "connectedness" condition holds true:
$$\text{If } J \subseteq K \subseteq I, \text{ then } K \in \mathcal{B}.$$

Given a block of intervals $\mathcal{B}$ we will systematically use the notation $\mathcal{B} = \mathcal{B}(I)$ to indicate that I is the unique maximal interval of the block $\mathcal{B}$.

Assume now that $\{\mathcal{B}(I) : I \in \mathcal{A}\}$ is a collection of disjoint blocks in $\mathcal{K}$. Fix $N, B \in \mathbb{N}$. Assume that for each $I \in \mathcal{A}$ the block $\mathcal{B}(I)$ contains collections of dyadic intervals, $\mathcal{C}_i(I)$, $i \leq N$ so that

$$\mathcal{C}_1(I), \ldots, \mathcal{C}_N(I) \text{ satisfy Jones's compatibility condition (J).} \tag{1.5.39}$$

Let the constant appearing in the compatibility condition be the fixed number $B \in \mathbb{N}$. For $i \leq N$ define

$$b_i = \sum_{I \in \mathcal{A}} \{h_K : K \in \mathcal{C}_i(I)\}, \tag{1.5.40}$$

and let P denote the orthogonal projection onto $\operatorname{span}\{b_i\}$, that is,

$$P(f) = \sum_{i=1}^{N} \langle f, \frac{b_i}{||b_i||_2} \rangle \frac{b_i}{||b_i||_2}. \tag{1.5.41}$$

Applying Theorem 1.5.9 to the projection defined by (1.5.41) gives an upper bound for $\|P\|_{\mathrm{BMO}}$ by $\sqrt{B} \times$ (the cardinality of $\mathcal{A}$). For the applications to follow we now isolate a further structural condition ensuring that the norm of P is bounded by a function depending only on B and $[\![\mathcal{A}]\!]$, where

$$[\![\mathcal{A}]\!] = \sup_{I \in \mathcal{A}} \frac{1}{|I|} \sum_{J \in \mathcal{A},\, J \subseteq I} |J|.$$

This is the purpose of Theorem 1.5.11. For $m \leq N$ and $J \in \mathcal{A}$ let

$$C_m(J) = \bigcup_{K \in \mathcal{C}_m(J)} K.$$

Loosely speaking the hypothesis of Theorem 1.5.11 states that for $I, J \in \mathcal{A}$ the collections $\{C_m(J) : m \leq N\}$ and $\{C_m(I) : m \leq N\}$ are rescaled versions of each other. Precisely, assume that $\{C_m(J) : m \leq N\}$ and $\{C_m(I) : m \leq N\}$ are linked through the following conditions: For $I, J \in \mathcal{A}$ and $m, n \leq N$,

$$C_m(I) \subseteq C_n(I) \text{ if } C_m(J) \subseteq C_n(J), \tag{1.5.42}$$

and

$$\frac{|C_m(I)|}{|C_n(I)|} \leq 2 \frac{|C_m(J)|}{|C_n(J)|}. \tag{1.5.43}$$

Remarks. 1. Note that condition (1.5.42) states that for $I, J \in \mathcal{A}$, the intersection pattern of $\{C_m(J) : m \le N\}$ coincides with that of $\{C_m(I) : m \le N\}$.

2. Below we will apply also the following averaged version of condition (1.5.43). Multiply both sides of (1.5.43) by $|C_n(J)|$, and sum the resulting estimate over $J \in \mathcal{A}$. This gives that

$$\left[\sum_{J\in\mathcal{A}} |C_n(J)|\right] \frac{|C_m(I)|}{|C_n(I)|} \le 2 \sum_{J\in\mathcal{A}} |C_m(J)|.$$

Dividing further by $\sum_{J\in\mathcal{A}} |C_n(J)|$ yields the averaged version of condition (1.5.43): For every $I \in \mathcal{A}$,

$$\frac{|C_m(I)|}{|C_n(I)|} \le 2 \frac{\sum_{J\in\mathcal{A}} |C_m(J)|}{\sum_{J\in\mathcal{A}} |C_n(J)|}.$$

Theorem 1.5.11. *Assume that* (1.5.39) *and the rescaling conditions* (1.5.42) *and* (1.5.43) *are satisfied. Then for the projection P, defined by* (1.5.41) *the following estimates hold: Let $I_0 \in \mathcal{A}$, and $J_0 \in \mathcal{B}(I_0)$, then*

$$\int S^2(P(f)|\mathcal{L}) \le 2B\|f\|^2_{BMO}|J_0|, \tag{1.5.44}$$

where $\mathcal{L} = \{J \in \mathcal{B}(I_0) : J \subseteq J_0\}$. Consequently,

$$||P||_{BMO} \le (2B[\![\mathcal{A}]\!])^{\frac{1}{2}}. \tag{1.5.45}$$

Proof. The proof is divided into two separate parts. First we show how the norm estimate for P follows from (1.5.44). The second part consists in proving (1.5.44).

The norm estimate (1.5.45) **follows from** (1.5.44). Let $f \in \mathrm{BMO}$. Let $\mathcal{H}$ be the Haar support of $P(f)$. For $J_0 \in \mathcal{H}$ there exists $I_0 \in \mathcal{A}$, such that $J_0 \in \mathcal{B}(I_0)$. Next observe that the following inclusion holds,

$$\{J \in \mathcal{H} : J \subseteq J_0\} \subseteq \{J \in \mathcal{B}(I_0) : J \subseteq J_0\} \cup \bigcup_{I\in\mathcal{A},\, I\subset I_0} \mathcal{B}(I).$$

Introducing the following abbreviations simplifies considerably our notation.

$$J_0 \cap \mathcal{H} = \{J \in \mathcal{H} : J \subseteq J_0\},$$
$$J_0 \cap \mathcal{B}(I_0) = \{J \in \mathcal{B}(I_0) : J \subseteq J_0\}.$$

Now we rewrite the square function $S^2(P(f)|J_0 \cap \mathcal{H})$ as

$$S^2(P(f)|J_0 \cap \mathcal{H}) = S^2(P(f)|J_0 \cap \mathcal{B}(I_0)) + \sum_{I\in\mathcal{A},\, I\subset I_0} S^2(P(f)|\mathcal{B}(I)).$$

We conclude that the norm estimate (1.5.45) for P follows from the inequalities

$$\int S^2(P(f)|J_0 \cap \mathcal{B}(I_0)) \leq 2||f||^2_{\mathrm{BMO}} B|J_0|,$$

$$\int S^2(P(f)|\mathcal{B}(I)) \leq 2||f||^2_{\mathrm{BMO}} B|I|.$$

In order to prove (1.5.44), it suffices to verify the first of the above estimates; once this is done we specialize to $I_0 = J_0$ and observe that I_0 was an arbitrary interval from $\mathcal{A}$.

Verification of (1.5.44). Recall that $\mathcal{H}$ is the Haar support of $P(f)$. Restricted to the block $\mathcal{B}(I_0)$ the Haar support of $P(f)$ can be written as

$$\mathcal{B}(I_0) \cap \mathcal{H} = \mathcal{C}_1(I_0) \cup \cdots \cup \mathcal{C}_N(I_0).$$

Hence there exists $m_0 \leq N$ such that $J_0 \in \mathcal{C}_{m_0}(I_0)$. Now define the set of integers

$$U_{m_0} = \{m : C_m(I_0) \subseteq C_{m_0}(I_0)\}.$$

By the hypothesis (1.5.42) the set U_{m_0} does not depend on the choice of the interval I_0. Indeed, if for m_0 and m we have that $C_m(I_0) \subseteq C_{m_0}(I_0)$, then for any other $I \in \mathcal{A}$ the corresponding inclusion holds, namely, $C_m(I) \subseteq C_{m_0}(I)$. (For U_{m_0} to be independent of I_0, we used the hypothesis that the intersection pattern of $\{C_m(I) : m \leq N\}$ is independent of I.)

Now we rewrite, using the index set U_{m_0},

$$S^2(P(f)|J_0 \cap \mathcal{B}(I_0)) = \sum_{m \in U_{m_0}} \langle f, \frac{b_m}{||b_m||_2} \rangle^2 \frac{S^2(b_m|J_0 \cap \mathcal{B}(I_0))}{||b_m||_2^2}. \tag{1.5.46}$$

Fix $m \in U_{m_0}$, and focus on the factor $S^2(b_m|J_0 \cap \mathcal{B}(I_0))$. We claim that for $m \in U_{m_0}$, the following uniform estimate holds,

$$\int \frac{S^2(b_m|J_0 \cap \mathcal{B}(I_0))}{|J_0|} \leq 2B \frac{||b_m||_2^2}{||b_{m_0}||_2^2}. \tag{1.5.47}$$

Observe that the right-hand side of (1.5.47) depends on m_0 and m only. In particular it is independent of $I_0 \in \mathcal{A}$ and $J_0 \in \mathcal{B}(I_0)$.

We turn to verifying (1.5.47). Observe that (1.5.40), the definition of b_m, gives

$$\int S^2(b_m|J_0 \cap \mathcal{B}(I_0)) = \sum_{K \in J_0 \cap \mathcal{C}_m(I_0)} |K|. \tag{1.5.48}$$

Next we exploit that Jones's compatibility condition holds for $\{\mathcal{C}_m(I_0) : m \leq N\}$. Condition $(J2)$ states that

$$\sum_{K \in J_0 \cap \mathcal{C}_m(I_0)} \frac{|K|}{|J_0|} \leq B \frac{|C_m(I_0)|}{|C_{m_0}(I_0)|}. \tag{1.5.49}$$

Note that the upper bound on the right-hand side of (1.5.49) is (now) independent of J_0. Next we remove also the dependence on the particular interval I_0. By the rescaling hypothesis (1.5.43) the value of the ratios $|C_m(I)|/|C_{m_0}(I)|$ is independent of $I \in \mathcal{A}$. Hence by averaging we obtain the inequality

$$\frac{|C_m(I_0)|}{|C_{m_0}(I_0)|} \leq 2\frac{\sum_{I\in\mathcal{A}}|C_m(I)|}{\sum_{I\in\mathcal{A}}|C_{m_0}(I)|}. \tag{1.5.50}$$

The identity (1.5.48) followed by the estimates (1.5.49) and (1.5.50) proves the claim made in (1.5.47) since,

$$||b_m||_2^2 = \sum_{I\in\mathcal{A}} |C_m(I)|.$$

In summary, using locally the compatibility assumption (J) together with the rescaling hypotheses (1.5.42) and (1.5.43) we passed from a small scale quantity to a large scale estimate; we showed that integrating (1.5.46) and invoking (1.5.47) gives that

$$\int \frac{S^2(P(f)|J_0 \cap \mathcal{B}(I_0))}{|J_0|} \leq \frac{2B}{||b_{m_0}||_2^2} \sum_{m\in U_{m_0}} \langle f, \frac{b_m}{||b_m||_2}\rangle^2. \tag{1.5.51}$$

Finally we use Bessel's inequality, and the fact that index set U_{m_0} is actually independent of I_0 to conclude that

$$\begin{aligned}\sum_{m\in U_{m_0}} \langle f, \frac{b_m}{||b_m||_2}\rangle^2 &\leq \sum_{I\in\mathcal{A}} |C_{m_0}(I)| \cdot ||f||_{\mathrm{BMO}}^2 \\ &= ||b_{m_0}||_2^2 ||f||_{\mathrm{BMO}}^2.\end{aligned} \tag{1.5.52}$$

Combining (1.5.51) and (1.5.52) implies (1.5.44). □

The pigeon hole principle. Often we use a very simple pigeon hole principle to realize the hypothesis of Theorem 1.5.11. As before we are given disjoint blocks of dyadic intervals, $\{\mathcal{B}(I) : I \in \mathcal{A}\}$. Let $\mathcal{B}(I)$ contain collections of dyadic intervals

$$\mathcal{C}_1(I), \ldots, \mathcal{C}_N(I),$$

which satisfy Jones's compatibility condition (J). Moreover, we assume that the number of families N, and the constant B appearing in the compatibility condition remain fixed, as I ranges over $\mathcal{A}$. We let

$$C_i(I) = \bigcup_{J\in\mathcal{C}_i(I)} J.$$

Now we assume that a lower bound holds for the density of $C_i(I)$ in I : There exists $\delta > 0$ such that

$$|C_i(I)| > \delta|I|, \qquad \text{for } i \leq N \text{ and } I \in \mathcal{A}. \tag{1.5.53}$$

Then, by the pigeon hole principle, there exists $K = K(\delta)$ and a decomposition of $\mathcal{A}$,

$$\mathcal{A} = \mathcal{A}_1 \cup \cdots \cup \mathcal{A}_K,$$

so that each $\mathcal{A}_k$ satisfies the hypothesis of Theorem 1.5.11. That is, if $I, J \in \mathcal{A}_k$ and $m, n \leq N$, then

$$C_m(I) \subseteq C_n(I) \text{ if } C_m(J) \subseteq C_n(J),$$

and

$$\frac{|C_m(I)|}{|C_n(I)|} \leq 2\frac{|C_m(J)|}{|C_n(J)|}.$$

Let $k \leq K$. Define P_k to be the orthogonal projection onto the span of

$$b_i = \sum_{I \in \mathcal{A}_k} \{h_K : K \in \mathcal{C}_i(I)\}, \quad i \leq N.$$

Then, by Theorem 1.5.11 each of the projections P_k has norm on BMO bounded by $4(B[\![\mathcal{A}_k]\!])^{1/2}$. The sum satisfies the norm estimate

$$\left\| \sum_{k=1}^{K} P_k \right\|_{\mathrm{BMO}} \leq 4K(B[\![\mathcal{A}]\!])^{\frac{1}{2}}, \tag{1.5.54}$$

where $\mathcal{A} = \mathcal{A}_1 \cup \cdots \cup \mathcal{A}_K$, and where $K = K(\delta)$.

1.6 Notes

The work of A. Haar [79] and J. Schauder [185] is represented by Proposition 1.1.1 and Theorem 1.1.2. See [76] by C. Goffman and G. Pedrick for a presentation of the basic properties of Schauder bases, in particular for a proof of St. Banach's theorem mentioned in the text.

Theorem 1.1.3 is the result of S. Szarek [200]. The text presents the method introduced by R. Latała and K. Oleszkiewicz in [124], with simplifications given in [180]. The proof of Theorem 1.1.4 is taken from the appendix D of [192] by E. M. Stein which also contains the trick by which we transferred Khintchine's inequality from L^4 to L^1. Proofs of Burkholder's Theorem 1.1.5 are contained in [33] and [35]. Theorem 1.1.6, the square function characterization of L^p, is originally due to R.E.A.C. Paley [167]. The constants of equivalence in Paley's proof are different from the ones obtained in Theorem 1.1.6.

Applying the unconditionality of the Haar system in L^p $(1 < p < \infty)$ we obtain estimates for the Walsh system. The proof given for Pisier's inequality in the scalar case (that is Theorem 1.1.7 referred to as the dyadic version of Poincaré's inequality) is due to A. Naor and G. Schechtman [162]. It extends without changes to UMD valued L^p spaces. For a different proof see M. Talagrand [201]. Theorem 1.1.8, asserting that the Walsh system in Walsh–Paley order forms a Schauder basis for L^p $(1 < p < \infty)$ is due to R. E. A. C. Paley [167]. Theorem 1.1.9, the extension of Khintchine's inequality to Walsh functions of finite multiplicity, is obtained by A. Bonami [15] and by K. Kiener [112] independently. Theorem 1.1.10 is a result of B. Roider. The proof in the text is based on K. Kiener's work in [113] and on additional, unpublished manuscripts of K. Kiener. In [25] J. Bourgain gives an extension of B. Roider's Theorem 1.1.10.

For a systematic treatment of the Walsh system consult the monograph [77] by B. Golubov, A. Efimov and V. Skvortsov. There, in particular, one finds the identities of Walsh and Paley used in the text.

The Walsh system is used as a model providing insight into problems concerning the trigonometric system. The work of R. Hunt in [87] and [88] that analyzes the Walsh model is an excellent preparation for the study of L. Carleson's treatise [44] on almost everywhere convergence of Fourier series. In [203] Ch. Thiele forms the notion of a quartile operator in the context of the Walsh system. Subsequently quartile operators played a crucial role in the work of M. Lacey and Ch. Thiele [121], [122] on the boundedness of the bilinear Hilbert transform.

In Banach space theory Walsh functions are used for the dual purpose of constructing operators (Walsh matrices) and spaces (spanned by subsets of the Walsh system). Typical applications include:

(a) The construction of operators with unexpected or extremal properties in the work of P. Enflo [62], A. Szankowski [197], [199], A. Pełczyǹski and C. Schütt [173], C. Schütt [189], V.I. Gurarii, M. Kadec and V.I. Macaev [78].

(b) The construction of complemented subspaces in $L^p(1 < p < \infty)$ by J. Bourgain, H. Rosenthal and G. Schechtman [29], and subspaces in L^1 with extremal properties obtained by J. Bourgain [16].

(c) The proof of the existence of a fundamental, total and bounded biorthogonal sequence in every separable Banach space by R. Ovsepian and A. Pełczyǹski [166], [170].

(d) The study of type, cotype and K-convexity in the work of G. Pisier [177], [178].

The proof of Theorem 1.2.1 (Fefferman's Inequality) is taken from the classic paper [64] by C. Fefferman and E. M. Stein. Theorem 1.2.2 is probably a folk result, Theorem 1.2.3 is from A.M. Garsia [75]. The direct approach to the atomic decomposition of H^1 is due to R. Coifman [55]. The proof of Theorem 1.2.4 given in the text is due to S. Janson and P.W. Jones [90]. The sharp function characterization of Theorem 1.2.5 is due to C. Fefferman and E. M. Stein [64]. The proof in the text follows A.M. Garsia [75]. Theorem 1.2.6 is the dyadic version of the Hardy–Littlewood maximal theorem, for its proof we used the presentation in [192]. For Theorem 1.2.8 see again A.M. Garsia [75].

Large deviation inequalities as considered in Section 1.3 originate with the work of W. Hoeffding [85] and K. Azuma [5]. The proof of Theorem 1.3.1 and

Corollary 1.3.2 are taken from [49] by S. Y. A. Chang, M. Wilson and Th. Wolff. The proof of Theorem 1.3.4 based on Proposition 1.3.3 was compiled using [6] by R. Banuelos together with [49] by S. Y. A. Chang, M. Wilson and Th. Wolff and [7] by R. Banuelos and C. Moore. A different approach to Theorem 1.3.4 was developed by J. Pipher [174].

The proof of Theorem 1.3.5 and Corollary 1.3.7 are from Garsia's book [75].

Theorem 1.3.6 is due to G. Schechtman [186]. Its proof introduces martingale methods, in particular SL^∞ estimates, to obtain Euclidean subspaces of proportional dimension in ℓ^1_n. For an overview of the development that followed, we refer to the survey articles by G. Schechtman [187] and W. B. Johnson and G. Schechtman [97] as well as to the books by V. Milman and G. Schechtman [148] and G. Pisier [179] . Historically the first application of large deviation inequalities to a problem in Banach spaces appears in the note [141] by B. Maurey.

T. Figiel, J. Lindenstrauss and V. Milman [68], B. Kashin [110], and G. Schechtman [186] present distinct solutions to the problem of finding Euclidean subspaces of proportional dimension in ℓ^1_n.

Multipliers into SL^∞ as presented in Theorem 1.3.8 are constructed in [104].

The monograph [148] by V. Milman and G. Schechtman contains far-reaching applications to Banach space theory of large deviation inequalities such as Theorem 1.3.1. In a completely different direction N. Makarov [133] applies Theorem 1.3.1 to study the boundary behavior of conformal maps, thereby N. Makarov obtains important estimates of harmonic measure. Recently, J. Bourgain, H. Brezis and P. Mironescu [26], obtained surprising, new applications of large deviation inequalities and the $\sqrt{p}$ dependence in Theorem 1.3.4 to embedding theorems of fractional Sobolev spaces. See also the survey article [30] by H. Brezis.

Theorem 1.3.1 provides the basis for an algorithm that determines the rectifiable pieces of a given data set in $\mathbb{R}^n$. Thus the work of G. Lerman [126] — that builds on the travelling salesman theorem of P. Jones [103] — gives unexpected applications of SL^∞ estimates to geometric measure theory.

The theorem of J. Garnett and P. W. Jones [73] is presented with a stochastic proof in Durrett's book [61]. Both the original proof in [73] and the stochastic proof in [61] are easily specialized to the context of dyadic martingales. The estimate (1.3.32) for the Rademacher projection H^1 is a result of J. Bourgain [18]. J. Bourgain's estimate (1.3.32), is related (in spirit) to R. E. A. C. Paley's inequality on lacunary coefficients [168] and to the interpolatory inequalities for Fourier multipliers, appearing in the work of S. Kislyakov [115] and [116]. (The basis of the relationship is the Lusin square function characterization presented in Section 6.1.)

The proof of Theorem 1.4.1 is due to B. Maurey [139] (see also [144]). The proof of Stein's inequality (Theorem 1.4.2) is due to J. Bourgain and applied in 1983/4 [24]. It extends without changes to the case of L^p spaces of functions with values in a UMD space. The only published account of J. Bourgain's proof seems to be in [69] by T. Figiel and P. Wojtaszczyk. Theorem 1.4.3, Theorem 1.4.4 and Proposition 1.4.5 are results of T. Figiel [65].

The proof of Theorem 1.5.1 is obtained (by specialization) from [130], Proposition 2.c.17. Theorem 1.5.2 is taken from [147] by Y. Meyer and R. Coifman. The proof in the text is strictly limited to the case of real-valued coefficients and not extending to coefficients in a UMD space. However there exist paraproduct estimates due to J. Bourgain that extend to the UMD case. See [69] for the only published proof.

Theorem 1.5.3 and the proof presented in the text are due to T. Figiel [66]. The survey [69] by T. Figiel and P. Wojtaszczyk contains a very informative summary of Figiel's work on singular integral operators. Our presentation incorporates also lecture notes of talks given by T. Figiel at a Conference on the Geometry of Banach spaces in Strobl (1989), and at the Weizmann Institute in Rehovot (1988). Z. Ciesielski [53] gives a detailed presentation of the Haar system on $[0,1] \times [0,1]$ indexed by dyadic squares. J. O. Stromberg's construction [196] of a modified Franklin system, is based on the same principles.

T. Figiel's use of the square indexed Haar system in the expansion of a kernel on $[0,1] \times [0,1]$ resulted in the representation of integral operators as paraproducts, (simple) rearrangements and multipliers. Using the standard Haar system on $[0,1] \times [0,1]$ indexed by dyadic rectangles leads to a different representation of integral operators. It was invented by Y. Meyer [146] to give the first wavelet proof of the $T(1)$ theorem of G. David and J.L Journé [59] . Comparing Y. Meyer's proof [146] of the $T(1)$ theorem to that of T. Figiel [66] the following attracts attention: The norm estimates for rearrangement operators in [66] are replacing the applications of Cotlar's lemma in [146]. Having freed the proof of the $T(1)$ theorem from Cotlar's lemma, T. Figiel [66] extended the $T(1)$ theorem beyond Hilbert spaces to the UMD class (being able to do this provided the main stimulus for T. Figiel's work).

Simultaneously and independent of the work of T. Figiel the advantages of expanding integral kernels along the Haar system indexed by dyadic squares were understood by G. Beylkin, R. Coifman and V. Rokhlin [10] while developing numerical algorithms for the evaluation of integral operators.

The boundedness criteria for orthogonal projections as formulated in Theorem 1.5.9–Theorem 1.5.11 are due to P. W. Jones [102]. The use of orthogonal projections has led to deep results in function spaces other than H^1 and BMO. For instance, the results on complemented subspaces of rearrangement invariant function spaces in the monograph [93] by W. Johnson, B. Maurey, G. Schechtman, L. Tzafriri.

Chapter 2

Projections, Isomorphisms and Interpolation

This chapter reviews three basic concepts of functional analysis. We discuss complemented subspaces (mostly in H^1), the decomposition method of A. Pełczyński, and elements of complex interpolation theory.

2.1 Complemented subspaces

In this section we study Banach spaces which are isomorphic to a complemented subspace of H^1. To this end we introduce three families of Banach spaces closely related to H^1. These are atomic H^1 over spaces of homogeneous type, martingale H^1 associated to an increasing sequence of σ algebras, and Banach spaces spanned by nested, three-valued martingale difference sequences. Furthermore we discuss the Rosenthal space and related Banach spaces spanned by independent random variables.

Standard concepts and terminology

We recall standard concepts associated with the study of Banach spaces. We define projections, isomorphisms, well isomorphic sequences of finite-dimensional spaces, complemented copies, infinite direct sums, and the finite-dimensional building blocks of the spaces H^1, L^p, and BMO.

Projections. A projection P on a Banach space E is a bounded linear operator $P : E \to E$ such that $P^2 = P$. The range of a projection is called a complemented subspace. Clearly the orthogonal projections discussed in Section 1.5 satisfy $P^2 = P$.

Isomorphisms. Two Banach spaces E and F are isomorphic if there exists a bijective linear operator $T : E \to F$ so that

$$\|T\| \cdot \|T^{-1}\| < \infty. \tag{2.1.1}$$

In that case $T : E \to F$ is called an isomorphism between E and F. The infimum of (2.1.1) taken over all isomorphisms between E and F is called the Banach–Mazur distance between E and F. If two Banach spaces E and F are isomorphic, then we say that E is a copy of F and we write $E \sim F$. A property of a Banach space which is preserved by isomorphisms is called an isomorphic invariant.

Well isomorphic sequences. Given two sequences of finite-dimensional Banach spaces $\{E_n\}$ and $\{F_n\}$. We say that $\{E_n\}$ is well isomorphic to $\{F_n\}$ if there exist bijective linear operators $T_n : E_n \to F_n$ such that

$$\sup_{n\in\mathbb{N}} \|T_n\| \cdot \|T_n^{-1}\| < \infty.$$

Complemented copies. Below we will be concerned with Banach spaces that are isomorphic to a complemented subspace of H^1. Observe that a Banach space E is isomorphic to a complemented subspace of a Banach space F if there exist linear operators $M : E \to F$ and $N : F \to E$ so that

$$\begin{array}{ccc} E & \xrightarrow{\mathrm{Id}} & E \\ M \searrow & & \nearrow N \\ & F & \end{array} \quad \text{and} \quad \|M\| \cdot \|N\| < \infty. \tag{2.1.2}$$

Indeed, $M(E)$ is a subspace of F that is isomorphic to E, and $P = M \circ N$ is a projection on F whose range equals $M(E)$. If (2.1.2) holds, then we say that F contains a complemented copy of E.

Infinite direct sums. Among the simplest procedures to generate new Banach spaces from old ones is the formation of infinite direct sums. Let $E_n, n \in \mathbb{N}$ be a sequence of Banach spaces, let $1 \le p \le \infty$. Then $(\sum E_n)_p$ is defined as the space of sequences $(x_n)_{n=1}^\infty$, where $x_n \in E_n$, equipped with the norm

$$\begin{array}{lll} \left(\sum ||x_n||_{E_n}^p\right)^{1/p} & \text{when} & 1 \le p < \infty, \\ \sup ||x_n||_{E_n} & \text{when} & p = \infty. \end{array} \tag{2.1.3}$$

For $1 \le p < \infty$ let q be the Hölder conjugate index to p, satisfying $1/p + 1/q = 1$. The dual space to $(\sum E_n)_p$ is naturally identified as

$$\left(\sum E_n\right)_p^* = \left(\sum E_n^*\right)_q,$$

where E_n^* denotes the dual space to E_n.

Finite-dimensional building blocks. We encounter infinite direct sums most frequently in connection with the following examples. Define the algebraic vector space

$$V_n = \operatorname{span}\{h_J : |J| \geq 2^{-n}\}.$$

The algebraic dimension of V_n is equal to $2^{n+1}-1$. Equipping V_n with the norm of H^1 , L^p or BMO gives the finite-dimensional building blocks of H^1, L^p or BMO,

$$H_n^1 = (V_n, ||\cdot||_{H^1}), \quad L_n^p = (V_n, ||\cdot||_{L^p}), \quad \mathrm{BMO}_n = (V_n, ||\cdot||_{\mathrm{BMO}}).$$

Clearly H_n^1 is complemented in H^1; it is the range of the norm 1 projection,

$$P(f) = \sum_{\{J:|J|\geq 2^{-n}\}} \langle f, h_J \rangle \frac{h_J}{|J|}.$$

The projection P is also a norm 1 operator on L^p and on BMO. Hence L_n^p and BMO_n are one-complemented subspaces of L^p respectively BMO.

The following proposition is simple and useful. Its proof shows how infinite direct sums are represented as subspaces of H^1.

Proposition 2.1.1. *The space $(\sum H_n^1)_1$ is isomorphic to a complemented subspace of H^1.*

Proof. Let (I_n) be a sequence of pairwise disjoint dyadic intervals. Let $f_n : [0,1] \to I_n$ be an affine function mapping $[0,1]$ bijectively onto the interval I_n. The slope of f_n is $|I_n|$. Define the following collections of dyadic intervals contained in I_n,

$$\mathcal{C}_n = \{J \subseteq I_n : |J| \geq 2^{-n}|I_n|\}.$$

Let Y_n be the algebraic vector space $Y_n = \operatorname{span}\{h_J : J \in \mathcal{C}_n\}$. The composition operator T_n defined by

$$T_n : g \mapsto |I_n| g \circ f_n,$$

is an isometry from $(Y_n, ||\cdot||_{H^1})$ onto H_n^1. Note that for $y_n \in Y_n$ we have the identity

$$||\sum y_n||_{H^1} = \sum ||y_n||_{H^1}. \tag{2.1.4}$$

Let Y be the closure in H^1 of $\operatorname{span}\{Y_n : n \in \mathbb{N}\}$. By the above identity $(Y, ||\cdot||_{H^1})$ is isometric to $(\sum H_n^1)_1$. Moreover Y is a complemented subspace of H^1. Indeed let P be the orthogonal projection defined by

$$P(f) = \sum_{n=1}^{\infty} \sum_{J\in\mathcal{C}_n} \langle f, h_J \rangle \frac{h_J}{|J|}.$$

Then $||P||_{H_1} = 1$ and Y is the range of P. $\square$

Let ℓ_n^2 denote the n-dimensional Euclidean space. With obvious modifications the proof of Proposition 2.1.1 shows that $(\sum \ell_n^2)_1$ is isomorphic to a complemented subspace of H^1. By duality the spaces $(\sum \mathrm{BMO}_n)_\infty$ and $(\sum \ell_n^2)_\infty$ are isomorphic to complemented subspaces of BMO. In the next section we will considerably strengthen this observation and prove a theorem of P. Wojtaszczyk that $(\sum \mathrm{BMO}_n)_\infty$ is isomorphic to BMO. This provides us with a useful alternative representation of the space BMO. Consequences of this isomorphism will be drawn in Chapter 5 where we prove isomorphic invariants of H^1 and BMO.

Martingale H^1 spaces

Martingale H^1 spaces provide a class of complemented subspaces of H^1. Let $(\Omega, \mathcal{F}, |\cdot|)$ be a probability space and let

$$\mathcal{F}_0 \subseteq \cdots \subseteq \mathcal{F}_n \subseteq \cdots \mathcal{F}$$

be a sequence of increasing σ-algebras on Ω so that $\mathcal{F}$ is generated by $\bigcup \mathcal{F}_n$. For an integrable function $f : \Omega \to \mathbb{R}$, the induced square function is defined to be

$$\left(\sum_n |\mathbb{E}(f|\mathcal{F}_n) - \mathbb{E}(f|\mathcal{F}_{n-1})|^2 \right)^{1/2},$$

where $\mathbb{E}(f|\mathcal{F}_n)$ denotes the conditional expectation of f with respect to $\mathcal{F}_n$. Then $H^1[(\mathcal{F}_n)]$ is the space of integrable functions $f : \Omega \to \mathbb{R}$, with integrable square function. The norm on $H^1[(\mathcal{F}_n)]$ is defined by

$$||f||_{H^1[(\mathcal{F}_n)]} = \int_\Omega \left(\mathbb{E}(f|\mathcal{F}_0)^2 + \sum_n |\mathbb{E}(f|\mathcal{F}_n) - \mathbb{E}(f|\mathcal{F}_{n-1})|^2 \right)^{1/2}.$$

The space $\mathrm{BMO}[(\mathcal{F}_n)]$ consists of all functions $g \in L^2(\Omega)$ so that

$$\|g\|_{\mathrm{BMO}[(\mathcal{F}_n)]} = \sup_{n \in \mathbb{N}} \mathbb{E}(|g - \mathbb{E}(g|\mathcal{F}_{n-1})|^2 | \mathcal{F}_n) < \infty.$$

The space $\mathrm{BMO}[(\mathcal{F}_n)]$ is (identifiable with) the dual space of $H^1[(\mathcal{F}_n)]$ as follows. Let $f, g \in L^2(\Omega)$, then

$$|\mathbb{E}(f \cdot g)| \leq \sqrt{2} ||f||_{H^1[(\mathcal{F}_n)]} \|g\|_{\mathrm{BMO}[(\mathcal{F}_n)]}.$$

Thus every $g \in \mathrm{BMO}[(\mathcal{F}_n)]$ induces a bounded linear functional on $H^1[(\mathcal{F}_n)]$, by

$$L(f) = \lim_{n \to \infty} \mathbb{E}(f_n \cdot g_n), \quad f_n = \mathbb{E}(f|\mathcal{F}_n) \text{ and } g_n = \mathbb{E}(f|\mathcal{F}_n).$$

Conversely for every bounded linear functional $L : H^1[(\mathcal{F}_n)] \to \mathbb{R}$ there exists $g \in \mathrm{BMO}[(\mathcal{F}_n)]$ so that

$$L(f) = \lim_{n \to \infty} \mathbb{E}(f_n \cdot g_n) \quad \text{for } f \in H^1[(\mathcal{F}_n)].$$

The book of A. M. Garsia [75] is the standard reference to martingale inequalities, $H^1[(\mathcal{F}_n)]$ and $\mathrm{BMO}[(\mathcal{F}_n)]$.

Dyadic H^1 is in the class of $H^1[(\mathcal{F}_n)]$ spaces. To see this let $\mathcal{F}_n$ be the σ-algebra generated by dyadic intervals of length $\geq 2^{-n}$. This observation, and many other analogies between general martingale H^1 space and the dyadic H^1, suggest strong ties between these spaces. On the other hand also the Banach spaces ℓ^1 and $L^1([0,1])$ are in this class. To see how $L^1([0,1])$ arises, let $\mathcal{F}_0 = \{\emptyset, [0,1]\}$ and define $\mathcal{F}_1 = \mathcal{F}_2 \cdots$ to be the Lebesgue σ-algebra on the unit interval $[0,1]$. For this filtration $H^1[(\mathcal{F}_n)]$ coincides with $L^1([0,1])$. The Banach spaces L^1, ℓ^1, and H^1 share very few isomorphic invariants. In particular, by a result of A. Pełczyński, $L^1([0,1])$ is not isomorphic to a subspace of a Banach space with an unconditional basis. We conclude that without putting restrictions on the filtration $(\mathcal{F}_n)$ the ties between the spaces in the class $H^1[(\mathcal{F}_n)]$ and dyadic H^1 are not very strong. However, after imposing the simplest condition ruling out the case of L^1, B. Maurey [144] obtained this truly remarkable theorem:

Theorem 2.1.2. *If the σ-algebras $\mathcal{F}_n$ are purely atomic, then $H^1[(\mathcal{F}_n)]$ is isomorphic to a complemented subspace of H^1.*

It takes up almost all of Chapter 4 to prove B. Maurey's result. In Chapter 4 we present also two applications to Maurey's theorem. First we use it to obtain the isomorphic classification of martingale H^1 spaces when the underlying filtration consists of purely atomic σ-algebras. Separately we apply its proof to show that the span of finitely many Haar functions in L^p is well isomorphic to ℓ^p_n.

Atomic H^1 spaces

The next examples of complemented subspaces in H^1 are the atomic H^1 spaces appearing in analysis. On a set X let $d(.,.)$ be a quasi-metric, that is,

$$\begin{aligned} d(x,y) &= 0 \text{ if and only if } x = y, \\ d(x,y) &= d(y,x), \\ d(x,z) &\leq K(d(x,y) + d(y,z)) \end{aligned}$$

for $x, y, z \in X$. The balls of radius r in X are the following sets,

$$B(x,r) = \{y \in X : d(x,y) < r\}.$$

We let μ be a probability measure on X satisfying the doubling condition,

$$\mu(B(x,2r)) \leq A\mu(B(x,r)),$$

for some A and all $x \in X$, $r > 0$. Following R. Coifman and G. Weiss [56] we call the triple (X, d, μ) a space of homogeneous type. An atom for (X, d, μ) is either a constant function on X or a function $a : X \to \mathbb{R}$ such that

$$\int a d\mu = 0,$$

and

$$\int a^2 d\mu \leq \mu(B)^{-1},$$

where B is a ball in X such that

$$\operatorname{supp} a \subseteq B.$$

We define the associated atomic H^1 space to be the space of functions $f : X \to \mathbb{R}$ which admit a decomposition into atoms,

$$f = \sum c_i a_i,$$

where a_i are atoms for (X, d, μ), and where

$$\sum |c_i| < \infty.$$

We denote this space by $H^1(X, d, \mu)$. The norm of $f \in H^1(X, d, \mu)$ is

$$||f||_{H^1(X,d,\mu)} = \inf\{\sum |c_i|\},$$

where the infimum is taken over all decompositions of f into atoms. The following two examples appear most frequently.

1. Historically the first example of a space of homogeneous type is the boundary of the unit disk $\mathbb{T} = \{z \in \mathbb{D} : |z| = 1\}$, where the metric is given by the Euclidean distance, and the measure is the Lebesgue measure on $\mathbb{T}$. The space $H^1(\mathbb{T}, \mathrm{dt}, |\cdot|)$ is denoted H^1_{at}. In Chapter 6 we present a thorough discussion of the space H^1_{at}. Section 6.1 covers the basic theorems about H^1_{at}, relating it to functions for which the Hilbert transform and the Lusin area function are integrable; Section 6.2 contains L. Carleson's proof of B. Maurey's theorem that H^1_{at} is isomorphic to H^1.

2. By Theorem 1.2.4 we identified the space of homogeneous type (X, d, μ) for which the associated atomic H^1 space coincides with dyadic H^1. Indeed, let X be the unit interval $[0, 1]$, let μ be Lebesgue measure on $[0, 1]$, and let $d(x, y)$ be the length of the smallest dyadic interval containing both x and y. Then (X, d, μ) is a space of homogeneous type. The following example sheds some light on this metric. We fix numbers $n,\ q \in \mathbb{N}$ such that $n \geq q$. Consider I_{q-1} a dyadic interval of length 2^{-q+1}. Let m be the midpoint of I_{q-1}. For each $q \leq j \leq n$ define

$$J_j = [m - 2^{-j}, m), \quad \text{and} \quad I_j = [m, m + 2^{-j}). \tag{2.1.5}$$

Observe that J_j and I_j are adjacent dyadic intervals, and that

$$J_q \cup I_q = I_{q-1}.$$

The interval I_{q-1} is then the smallest dyadic interval containing I_j and J_j. Hence for $q \leq j \leq n$, $x \in I_j$ and $y \in J_j$ we obtain

$$d(x,y) = |I_{q-1}|.$$

Given $x \in [0,1]$, the ball $B(x,r)$ of radius r in (X,d,μ) is just the dyadic interval I for which $x \in I$ and

$$r/2 \leq |I| < r.$$

Hence a function a is an atom for (X,d,μ) if there exists a dyadic interval I such that

$$\operatorname{supp} a \subseteq I,$$

$$\int a^2(t)dt \leq |I|^{-1},$$

and

$$\int a(t)dt = 0.$$

By Theorem 1.2.4 for $f \in H^1$, there exists a decomposition

$$f - \int f = \sum c_i a_i,$$

where a_i are atoms for (X,d,μ), and c_i are scalars such that

$$\sum |c_i| \leq 16 \int S(f).$$

Conversely, any atom a for (X,d,μ) satisfies

$$\int S(a) \leq 1.$$

In Section 6.5 we will show that atomic H^1 spaces are isomorphic to complemented subspaces in H^1, for each choice of (X,d,μ). More precisely we will prove the following Theorem [155].

Theorem 2.1.3. *For every space of homogeneous type, $H^1(X,d,\mu)$ is isomorphic to a complemented subspace of H^1. Moreover $H^1(X,d,\mu)$ and H^1 are isomorphic if (and only if)*

$$\mu\{x \in X : \mu(\{x\}) = 0\} > 0.$$

We close this section by describing the dual space of $H^1(X,d,\mu)$. For $x \in X$ and $r > 0$ let $B(x,r) = \{y \in X : d(x,y) < r\}$. Let $h \in L^2(X)$ and $B = B(x,r)$, then we write $m_B(h)$ to denote the mean value of h over B, thus,

$$m_B(h) = \int_B h \frac{d\mu}{\mu(B)}.$$

A function $h \in L^2(X)$ belongs to $\mathrm{BMO}(X, d, \mu)$ if

$$\|h\|_{\mathrm{BMO}(X,d,\mu)} = \sup_B \left[\int_B |h - m_B(h)|^2 \frac{d\mu}{\mu(B)} \right]^{1/2} < \infty, \tag{2.1.6}$$

where the supremum is taken over all balls $B = B(x, r)$, $x \in X$, and $r > 0$. The classical survey article [56] by R. R. Coifman and G. Weiss is our standard reference to spaces of homogeneous type, $H^1(X, d, \mu)$ and $\mathrm{BMO}(X, d, \mu)$. The duality results described below are proved there in detail.

The dual space of $H^1(X, d, \mu)$ is characterized as $\mathrm{BMO}(X, d, \mu)$. This characterization comprises two statements. First, each $h \in \mathrm{BMO}(X, d, \mu)$ induces a bounded linear functional on $H^1(X, d, \mu)$. For $f \in H^1(X, d, \mu)$, with atomic decomposition

$$f = \sum \lambda_i a_i \quad \text{and} \quad \sum |\lambda_i| < \infty,$$

the mapping

$$f \to \lim_{N \to \infty} \sum_{i=1}^{N} \lambda_i \left[\int_X h a_i d\mu \right], \tag{2.1.7}$$

is a well-defined linear functional on $H^1(X, d, \mu)$. Conversely, if x^* is in the dual space of $H^1(X, d, \mu)$ then there exists a unique $h \in \mathrm{BMO}(X, d, \mu)$ so that $\|h\|_{\mathrm{BMO}(X,d,\mu)} \leq C\|x^*\|$ and so that for $f \in H^1(X, d, \mu)$ the value of $x^*(f)$ is given by the mapping (2.1.7).

Three-valued martingale differences

Dyadic intervals are distinguished by the following properties: First for a dyadic interval I there exists $n \in \mathbb{N}$ such that $|I| = 2^{-n}$. Second, if I, J are dyadic intervals and $I \cap J \neq \emptyset$, then either $I \subseteq J$ or $J \subseteq I$. We now consider arbitrary collections of measurable sets which satisfy the second property but not necessarily the first. Let $\mathcal{E}$ be a countable collection of measurable subsets in the measure space $[0, 1]$, equipped with Lebesgue measure. We say that $\mathcal{E}$ is a nested collection (of measurable sets) if the elements of $\mathcal{E}$ satisfy the following condition,

$$\text{if } A, B \in \mathcal{E} \text{ and } A \cap B \neq \emptyset, \text{ then } A \subseteq B \text{ or } B \subseteq A.$$

We let $\mathcal{E}$ play the role of dyadic intervals and define spaces of three-valued martingale difference sequences. As soon as $\mathcal{E}$ is fixed the construction of these spaces is in complete analogy to that of H^1 and BMO.

A nested collection $\mathcal{E}$ gives rise to spaces $X[\mathcal{E}]$ and $Y[\mathcal{E}]$ of three-valued martingale difference sequences. We start with an analog to the Haar system. Let $\{r_A : A \in \mathcal{E}\}$ be any enumeration of the sequence of Rademacher functions. Then for $A \in \mathcal{E}$, put

$$d_A(w, t) = 1_A(w) r_A(t).$$

Note that the functions $\{d_A\}$ assume only the values $\{+1,-1,0\}$, they form a martingale difference sequence. The support of d_A is the set $A \times [0,1]$. Hence the support sets of the functions $\{d_A : A \in \mathcal{E}\}$ form a nested collection of measurable sets. We define the space $X[\mathcal{E}]$ to consist of those $f = \sum_{A\in\mathcal{E}} a_A d_A$ for which

$$||f||_{X[\mathcal{E}]} = \int \left(\sum_{A\in\mathcal{E}} a_A^2 1_A \right)^{1/2} < \infty.$$

This norm makes $X[\mathcal{E}]$ a Banach space, and $\{d_A : A \in \mathcal{E}\}$ an unconditional basis for $X[\mathcal{E}]$. We define the space $Y[\mathcal{E}]$ to consist of all functions $h = \sum_{A\in\mathcal{E}} b_A d_A$, for which

$$||h||_{Y[\mathcal{E}]} = \sup_{A\in\mathcal{E}} \left(\frac{1}{|A|} \sum_{B\in\mathcal{E},\, B\subseteq A} b_B^2 |B| \right)^{1/2} < \infty.$$

Repeating the proof of Theorem 1.2.1 (Fefferman's inequality) gives the integral estimate

$$\left| \int f h \, dx \right| \le 2\sqrt{2} ||f||_{X[\mathcal{E}]} ||h||_{Y[\mathcal{E}]} \tag{2.1.8}$$

for any finite linear combination $f = \sum a_A d_A$. Moreover $Y[\mathcal{E}]$ can be identified with the dual space of $X[\mathcal{E}]$.

By choosing different collections $\mathcal{E}$ we should be able to produce several different spaces in the class $X[\mathcal{E}]$. Below we find that the following list of Banach spaces can be obtained by choosing $\mathcal{E}$ appropriately,

$$\begin{gathered} \ell^1,\ \ell^2,\ \ell^1 \oplus \ell^2,\ (\sum \ell_n^2)_1,\ (\sum \ell_n^2)_1 \oplus \ell^2,\ (\sum \ell^2)_1, \\ (\sum H_n^1)_1,\ (\sum H_n^1)_1 \oplus \ell^2,\ (\sum H_n^1)_1 \oplus (\sum \ell^2)_1, H^1. \end{gathered} \tag{2.1.9}$$

It was pointed out by P. Wojtaszczyk [211] that the spaces in (2.1.9) are all different. By that we mean that no two of them are isomorphic. We will now show how to select nested collections $\mathcal{E}$ so that $X[\mathcal{E}]$ is isomorphic to a given space from the list (2.1.9).

1. Let $\mathcal{C}$ be a collection of dyadic intervals. Then clearly $\mathcal{C}$ is a nested collection of measurable sets and for $\mathcal{E} = \mathcal{C}$, the unconditional basis for $X[\mathcal{E}]$,

   $$d_I(w,t) = 1_I(w) r_I(t), \quad I \in \mathcal{C},$$

   is equivalent to the Haar functions $\{h_I : I \in \mathcal{C}\}$ in H^1. Thus the class of spaces $X[\mathcal{E}]$ contains the class of all subspaces in H^1 which are spanned by subsequences of the Haar system. Clearly H^1 can be obtained in that way. In the course of proving Proposition 2.1.1 we showed how to choose a subsequence of the Haar system so that its closed linear span is isometric to $(\sum H_n^1)_1$. Finally we obtain the space ℓ^1 by letting $\mathcal{C}$ be a collection of pairwise disjoint dyadic intervals.

2. Here we construct $\mathcal{E}$ such that $X[\mathcal{E}]$ is isomorphic to ℓ^2. Let $I_0 = [0,1]$ and for $n \in \mathbb{N}$ let $I_n = [0, 1 - 4^{-n}]$. Then the collection $\mathcal{E} = \{I_n : n \in \mathbb{N}_0\}$ is a nested collection and $X[\mathcal{E}]$ is isomorphic to ℓ^2. Indeed the unconditional basis for $X[\mathcal{E}]$ is
$$d_n(w,t) = 1_{I_n}(w) r_n(t).$$
For this system we have
$$\frac{1}{2}\left(\sum_{n=0}^{\infty} a_n^2\right)^{1/2} \le \left\| \sum_{n=0}^{\infty} d_n a_n \right\|_{X[\mathcal{E}]} \le \left(\sum_{n=0}^{\infty} a_n^2\right)^{1/2}.$$

3. Now we define a collection $\mathcal{E}$ such that $X[\mathcal{E}]$ is isomorphic to $(\sum \ell^2)_1$. For $n \in \mathbb{N}$, let I_n be a sequence of pairwise disjoint dyadic intervals. Then for $m \ge 1$ choose $I_{n,m} \subseteq I_n$ such that $|I_{m,n}| = |I_n|(1 - 4^{-m})$. Now consider $\mathcal{E} = \{I_{n,m} : m, n \in \mathbb{N}\}$, and define the normalized unconditional basis for $X[\mathcal{E}]$,
$$d_{n,m}(w,t) = |I_{n,m}|^{-1} 1_{I_{n,m}}(w) r_{n,m}(t),$$
where $r_{n,m}$ denotes independent Rademacher functions. Let $\{a_{n,m}\}$ be a sequence of scalars. For each n fixed we have an ℓ^2 estimate,
$$\frac{1}{2}\left(\sum_{m=1}^{\infty} a_{n,m}^2\right)^{1/2} \le \left\| \sum_{m=1}^{\infty} a_{m,n} d_{n,m} \right\|_{X[\mathcal{E}]} \le \left(\sum_{m=1}^{\infty} a_{n,m}^2\right)^{1/2}.$$
Next we observe that for $n \ne n'$ and $m, m' \in \mathbb{N}$, the intervals $I_{n,m}$ and $I_{n',m'}$ are disjoint. Hence,
$$\begin{aligned}\frac{1}{2}\sum_{n=1}^{\infty}\left(\sum_{m=1}^{\infty} a_{n,m}^2\right)^{1/2} &\le \left\| \sum_{n=1}^{\infty}\sum_{n=1}^{\infty} a_{n,m} d_{n,m} \right\|_{X[\mathcal{E}]} \\ &\le \sum_{n=1}^{\infty}\left(\sum_{m=1}^{\infty} a_{n,m}^2\right)^{1/2}.\end{aligned}$$
In other words the unit vectors of $(\sum \ell^2)_1$ are equivalent to $\{d_{n,m} : n, m \in \mathbb{N}\}$.

4. In our final example we describe an entire class of collections $\mathcal{E}$ for which $X[\mathcal{E}]$ is isomorphic to ℓ^1. We consider now collections $\mathcal{E}$ which satisfy the Carleson packing condition,
$$[\![\mathcal{E}]\!] = \sup_{A \in \mathcal{E}} \frac{1}{|A|} \sum_{B \subseteq A,\, B \in \mathcal{E}} |B| < \infty.$$
We claim that then $X[\mathcal{E}]$ is isomorphic to ℓ^1. To see this we consider the unconditional basis,
$$d_A(w,t) = 1_A(w) r_A(t), \; A \in \mathcal{E}.$$

We will show that $\{|A|^{-1}d_A : A \in \mathcal{E}\}$ is equivalent to the unit vector basis of ℓ^1. To this end, we fix $a_A \in \mathbb{R}$. Let

$$f = \sum a_A d_A(w,t),$$

and define

$$h = \sum \operatorname{sign}(a_A) d_A(w,t).$$

A moment's reflection shows that $||h||_{Y[\mathcal{E}]} \leq [\![\mathcal{E}]\!]^{1/2}$, and that $\sum |a_A||A| = \int hf$. Applying (2.1.8), that is Fefferman's inequality for nested martingale differences, we obtain the upper bound

$$\sum |a_A||A| \leq 2\sqrt{2}[\![\mathcal{E}]\!]^{1/2}||f||_{X[\mathcal{E}]}. \tag{2.1.10}$$

Next we show that the converse estimate holds true as well. Indeed, note that $||d_A||_{X[\mathcal{E}]} = |A|$, and apply the triangle inequality to f. This gives the corresponding lower bound,

$$\sum |a_A||A| \geq ||f||_{X[\mathcal{E}]}. \tag{2.1.11}$$

Combining (2.1.10) and (2.1.11) we find that $\{|A|^{-1}d_A : A \in \mathcal{E}\}$ is equivalent to the unit vector basis of ℓ^1.

It is now easy to combine these constructions and to obtain the remaining members of the list (2.1.9). We remarked already that the spaces appearing in (2.1.9) are known to be isomorphically different (see [211]). Next we show that each of the spaces $X[\mathcal{E}]$ is isomorphic to a complemented subspace of H^1. Thus we have at once ten different examples of complemented subspaces of H^1.

Nested collections can be decomposed naturally in a sequence of generations. This is done as follows. Given $A \in \mathcal{E}$ we denote by $G_1(A,\mathcal{E}) \subseteq \mathcal{E}$ the collection of maximal elements of $\mathcal{E}$ which are strictly contained in A. This is called the first generation of A. As $\mathcal{E}$ is nested, $G_1(A,\mathcal{E})$ is a collection of pairwise disjoint sets. Now we define the following generations by induction, using the first step as a model. Suppose that the first $p-1$ generations, $G_1(A,\mathcal{E}),\ldots,G_{p-1}(A,\mathcal{E})$, are already defined. Then we put

$$\mathcal{H} = \{B \in \mathcal{E} : B \subset A\} \setminus \{G_1(A,\mathcal{E}) \cup \cdots \cup G_{p-1}(A,\mathcal{E})\},$$

and define the p-th generation of A as

$$G_p(A,\mathcal{E}) = G_1(A,\mathcal{H}).$$

Let $G_0(\mathcal{E})$ denote the collection of maximal elements $\mathcal{E}$. Inductively, for $p \in \mathbb{N}$, we define

$$G_p(\mathcal{E}) = \bigcup_{A \in G_{p-1}(\mathcal{E})} G_1(A,\mathcal{E}).$$

Proposition 2.1.4. *For any nested family of measurable sets $\mathcal{E}$, the space $X[\mathcal{E}]$ is isometric to a complemented subspace of H^1.*

Proof. To simplify the presentation of the proof we make the convenient but inessential assumption that for each $A \in \mathcal{E}$, $G_1(A, \mathcal{E})$ is a finite set.

We will construct collections $\{\mathcal{C}_A : A \in \mathcal{E}\}$ of pairwise disjoint dyadic intervals which provide a model of the collection $\mathcal{E}$, since $\{\mathcal{C}_A\}$ will be defined to satisfy the following properties.

1. Let C_A be the point-set covered by the collection $\mathcal{C}_A$. Then $\{C_A : A \in \mathcal{E}\}$ is a nested family of subsets of $[0,1]$ and $|A| = |C_A|$.
2. $C_B \subseteq C_A$ iff $B \subseteq A$.
3. If $B \subseteq A$, and $I \in \mathcal{C}_A$, then
$$\frac{|I \cap C_B|}{|I|} = \frac{|B|}{|A|}.$$
4. If $I \in \mathcal{C}_B$, $J \in \mathcal{C}_A$ and $I \subseteq J$, then $B \subseteq A$.

Now we construct these collections by induction. We fix a maximal element in $\mathcal{E}$, say A. As $|A| \leq 1$, there exists a collection $\mathcal{C}_A$ of pairwise disjoint dyadic intervals so that
$$C_A = \bigcup_{I \in \mathcal{C}_A} I$$
satisfies $|C_A| = |A|$. It is easy to ensure that for $A, A' \in G_0(\mathcal{E})$, with $A \neq A'$ we have $C_A \cap C_{A'} = \emptyset$. Thus we define $\mathcal{C}_A$ for each A in the generation $G_0(\mathcal{E})$. Next we turn to $\mathcal{C}_B$ for the sets B in the first generation of $\mathcal{E}$. Let $A \in G_0(\mathcal{E})$. Recall that $G_1(A, \mathcal{E})$ consists of pairwise disjoint elements from $\mathcal{E}$. For each $I \in \mathcal{C}_A$ and each $B \in G_1(A, \mathcal{E})$ we find easily a collection $\mathcal{C}_{B,I}$ of pairwise disjoint dyadic intervals with the following properties: Let $C_{B,I}$ be the point-set covered by the collection $\mathcal{C}_{B,I}$, then $C_{B,I} \subseteq I$ and
$$\frac{|C_{B,I}|}{|I|} = \frac{|B|}{|A|}.$$
Furthermore we arrange also for the following condition to hold: If $B_1, B_2 \in G_1(A, \mathcal{E})$ are different (and hence disjoint sets), then
$$C_{B_1,I} \cap C_{B_2,I} = \emptyset.$$
Now we define $\mathcal{C}_B$ for $B \in G_1(A, \mathcal{E})$ by taking the union,
$$\mathcal{C}_B = \bigcup_{I \in \mathcal{C}_A} \mathcal{C}_{B,I}.$$
Having defined $\mathcal{C}_B$ for $B \in G_0(\mathcal{E}), \dots G_p(\mathcal{E})$ we continue as follows. Fix $A \in G_p(\mathcal{E})$. Then $G_1(A, \mathcal{E})$ consists of pairwise disjoint elements from $\mathcal{E}$. Hence for each $I \in \mathcal{C}_A$

and each $B \in G_1(A, \mathcal{E})$ there exists a collection of pairwise disjoint dyadic intervals $\mathcal{C}_{B,I}$ such that $\mathcal{C}_{B,I} \subseteq I$, and

$$\frac{|\mathcal{C}_{B,I}|}{|I|} = \frac{|B|}{|A|}.$$

Now we obtain $\mathcal{C}_B$ for $B \in G_1(A, \mathcal{E})$ by taking unions,

$$\mathcal{C}_B = \bigcup_{I \in \mathcal{C}_A} \mathcal{C}_{B,I}.$$

The resulting collections $\{\mathcal{C}_A : A \in \mathcal{E}\}$ satisfy the above conditions 1) – 4). Consequently they satisfy Jones' compatibility condition (J). Now we form the linear combinations

$$b_A = \sum \{h_I : I \in \mathcal{C}_A\},$$

and put $k_A = b_A/||b_A||_2$. By Theorem 1.5.9, the orthogonal projection

$$Pf = \sum_{A \in \mathcal{E}} \langle f, k_A \rangle k_A$$

is a bounded operator on H^1 with norm ≤ 1. The range of P is the closed linear span of $\{b_A : A \in \mathcal{E}\}$. Thus it remains to prove that in H^1, the closure of span$\{b_A : A \in \mathcal{E}\}$ is isomorphic to $X[\mathcal{E}]$. To show this let

$$d_A(\omega, t) = 1_A(\omega) r_A(t), \quad A \in \mathcal{A},$$

be the unconditional basis for $X[\mathcal{E}]$ and fix a sequence $a_A \in \mathbb{R}$. Observe that the distribution of $\sum_{A \in \mathcal{E}} a_A^2 d_A^2$ is the same as the distribution of $\sum_{A \in \mathcal{E}} a_A^2 b_A^2$. Thus it follows that

$$\left\| \sum_{A \in \mathcal{E}} a_A d_A \right\|_{X[\mathcal{E}]} = \left\| \sum_{A \in \mathcal{E}} a_A b_A \right\|_{H^1}.$$

□

So far we used the spaces $X[\mathcal{E}]$ to produce examples of complemented subspaces in H^1. They play a second, even more important role in the isomorphic classification of martingale H^1 spaces. The link to the martingale H^1 spaces is provided by Maurey's theorem. Following is a more detailed formulation of Theorem 2.1.2.

Theorem 2.1.5. *Let $(\mathcal{F}_n)$ be a sequence of increasing and purely atomic σ-algebras. There exists a nested collection $\mathcal{E}$, satisfying*

$$A, B \in \mathcal{E} \text{ and } A \subset B \text{ implies } |A| \leq |B|/2,$$

such that $H^1[(\mathcal{F}_n)]$ is isomorphic to $X[\mathcal{E}]$. Consequently $H^1[(\mathcal{F}_n)]$ is a space with an unconditional basis.

Maurey's result is the main step in the classification of martingale H^1 spaces. We will prove Theorem 2.1.5 in Chapter 4 and complement it with the following result.

Theorem 2.1.6. *Let $\mathcal{E}$ be a nested collection of measurable sets satisfying*

$$A, B \in \mathcal{E} \text{ and } A \subset B \text{ implies } |A| \leq |B|/2.$$

Then $X[\mathcal{E}]$ is isomorphic to one of the spaces $\ell^1, (\sum H_n^1)_1$, or H^1.

Above we found ten different Banach spaces in the class $X[\mathcal{E}]$. It is not possible to obtain more spaces by this method, since the isomorphic types of the family $X[\mathcal{E}]$ are classified in [151]: For every choice of $\mathcal{E}$, the space $X[\mathcal{E}]$, is isomorphic to one of the ten spaces listed in (2.1.9) above.

The Rosenthal space and independent random variables

Next we use independent random variables to construct complemented subspaces in H^1, respectively BMO. We present the Rosenthal space, its predual, the independent sum of BMO_n and the intersection of weighted L^2 with BMO.

The Rosenthal space. We define the intersection of ℓ^∞ with a weighted ℓ^2 space. More precisely, let (w_i) be a bounded sequence of positive real numbers such that

$$\sum w_i = \infty, \qquad \text{and} \qquad \lim w_i = 0.$$

Then the sequence space $R = R(\{w_i\})$ consists of all bounded sequences (a_i) for which the following norm is finite,

$$||(a_i)||_R = \left(\sup a_i^2 + \sum a_i^2 w_i\right)^{1/2}.$$

The space R that is now called the Rosenthal space was introduced and studied first by H. Rosenthal in [181]. Rosenthal showed in particular that the isomorphic type of $R = R(\{w_i\})$ is determined by the following two conditions on the weights,

$$\sum w_i = \infty, \text{ and } \lim w_i = 0.$$

Precisely, if $\{v_i\}$ is a sequence of positive real numbers so that $\sum v_i = \infty$ and $\lim v_i = 0$, then $R' = R(\{v_i\})$ is isomorphic to $R = R(\{w_i\})$. The proof showing the existence of an isomorphism between R and R' is based on A. Pełczyński's decomposition method. It was observed in [158] that the Rosenthal space is isomorphically different from the spaces $Y[\mathcal{E}]$ that are generated by nested three-valued martingale difference sequences.

Proposition 2.1.7. *The Rosenthal space R is isomorphic to a complemented subspace of BMO.*

Proof. The proof is divided into three basic steps. We first define a subspace of BMO, by constructing the sequence (b_i) in (2.1.12). Then we show that in BMO the sequence (b_i) is equivalent to the unit vector basis of the Rosenthal space. In a third step we verify that the orthogonal projection onto the linear span of (b_i) is bounded on BMO. Combined, these three steps imply that the Rosenthal space is isomorphic to a complemented subspace of BMO.

Step 1. We are given a sequence (w_i) of positive real numbers ≤ 1. Without loss of generality we assume that the sequence (w_i) is sorted in decreasing order. First we group them according to their value, as follows. For $n \geq 1$ define the index set

$$I_n = \{i : 2^{-n} \leq w_i \leq 2^{-n+1}\}.$$

Let i_n denote the cardinality of the set I_n. Next we select pairwise disjoint intervals of natural numbers K_n, and M_n such that $|K_n| = n$, $|M_n| = i_n$, and so that

$$\sup K_n < \inf M_n \leq \sup M_n < \inf K_{n+1}.$$

We use the index set K_n to define the indicator functions

$$X_n = \prod_{s \in K_n} \frac{1}{2}(1 + r_s).$$

The supports of the functions X_n form a sequence of independent sets, the measure of the support of X_n equals 2^{-n}. Next we attach the following frequencies to the indicator function X_n. For $j \in M_n$, we form

$$b_{n,j} = X_n r_j.$$

Now transform the lexicographic order on the set

$$\{(n,j) : n \in \mathbb{N}, j \in M_n\}$$

into the linear order on the natural numbers, and accordingly we relabel the sequence $\{b_{n,j}\}$, as

$$b_i = \sum_{I \in \mathcal{C}_i} h_I, \quad i \in \mathbb{N}. \tag{2.1.12}$$

Let C_i be the point-set covered by the collection $\mathcal{C}_i$. Then by construction the measure of C_i is related to the weights w_i as follows,

$$w_i/2 \leq |C_i| \leq w_i.$$

Step 2. We will now show that in BMO the sequence $\{b_i\}$ spans the Rosenthal space. Fix a bounded sequence $a_i \in \mathbb{R}$ and put

$$f = \sum a_i b_i.$$

We verify the equivalence of norms

$$||f||^2_{\mathrm{BMO}} \le \max_{i\in\mathbb{N}} a_i^2 + \sum_{i=1}^{\infty} a_i^2 w_i \le 2||f||^2_{\mathrm{BMO}}. \tag{2.1.13}$$

We show first the left-hand side estimate in (2.1.13). To this end we expand f in its Haar series and let let $\sum \alpha_I h_I$ be the Haar expansion of f. Next we fix $n \in \mathbb{N}$ and $I \in \mathcal{C}_n$. Note that for $i \ge n+1$ the sets I and C_i are independent. Hence

$$\int_I |b_i|^2 = |I \cap C_i| = |I| \cdot |C_i|.$$

Therefore we observe the identities

$$\begin{aligned}\sum_{J\subseteq I} \alpha_J^2 |J| &= |I| a_n^2 + \sum_{i=n+1}^{\infty} a_i^2 \int_I |b_i|^2 \\ &= |I| a_n^2 + \sum_{i=n+1}^{\infty} a_i^2 |I||C_i|.\end{aligned}$$

Dividing by $|I|$, and taking the supremum over all dyadic intervals I gives

$$\begin{aligned}\sup_I \frac{1}{|I|} \sum_{J\subseteq I} \alpha_J^2 |J| &\le \sup_n \left\{ a_n^2 + \sum_{i=n+1}^{\infty} a_i^2 |C_i| \right\} \\ &\le \sup_i a_i^2 + \sum_{i=1}^{\infty} a_i^2 |C_i|.\end{aligned}$$

As $|C_i| \le w_i$ we have verified the left-hand side of (2.1.13).

To prove the reverse estimate, observe that $\{b_i\}$ is disjointly supported over the Haar system. Hence for $f = \sum a_i b_i$ we have $||f||_{\mathrm{BMO}} \ge \sup |a_i|$. The norm of f in L^2 is bounded by its BMO norm, hence,

$$\begin{aligned}||f||^2_{\mathrm{BMO}} &\ge ||f||^2_{L^2} \\ &= \sum_{i=1}^{\infty} a_i^2 |C_i|.\end{aligned}$$

Combining this with $|C_i| \ge w_i/2$ we obtain the right-hand side of (2.1.13).

Step 3. The orthogonal projection onto span$\{b_i\}$ is defined by

$$P(f) = \sum_{i=1}^{\infty} \langle f, \frac{b_i}{||b_i||_2} \rangle \frac{b_i}{||b_i||_2}.$$

By (2.1.13) boundedness of P on BMO is easy to check. Indeed (2.1.13) gives that

$$||P(f)||_{\mathrm{BMO}}^2 \leq \sup_i \langle f, \frac{b_i}{||b_i||_2^2} \rangle^2 + \sum_{i=1}^{\infty} \langle f, \frac{b_i}{||b_i||_2^2} \rangle^2.$$

Next observe that $||b_i||_2^2 = |C_i|$. Hence by Bessel's inequality

$$\langle f, \frac{b_i}{||b_i||_2^2} \rangle^2 \leq ||f||_{\mathrm{BMO}}^2.$$

Moreover we have the L^2 estimate

$$\begin{aligned} \sum_{i=1}^{\infty} \langle f, \frac{b_i}{||b_i||_2^2} \rangle^2 &\leq ||f||_2^2 \\ &\leq ||f||_{\mathrm{BMO}}^2. \end{aligned}$$

In summary we showed that $||P(f)||_{\mathrm{BMO}}^2 \leq 2||f||_{\mathrm{BMO}}^2$, thus proving that the Rosenthal space is isomorphic to a complemented subspace of BMO. □

The predual of Rosenthal's space. Let R_* be the space of all sequences (a_i) for which

$$\|(a_i)\|_{R_*} = \left(\sum_{i=1}^{\infty} \min\{|a_i|, a_i^2 w_i^{-1}\} \right)^{1/2} < \infty.$$

This space can be identified — canonically — with the predual of Rosenthal's space. By Proposition 2.1.7 and duality the space R_* is isomorphic to a complemented subspace of H^1. A direct proof of this fact, not using duality, follows from work of W. B. Johnson and G. Schechtman [95]. It was verified in [158] that R_* is not isomorphic to one of the spaces listed in (2.1.9).

The independent sum of BMO$_n$. The next example is very close to Rosenthal's space. Recall that $L_n^2 \subset L^2$ and $\mathrm{BMO}_n \subset \mathrm{BMO}$ denote the subspaces spanned by the Haar functions $\{h_J : |J| \geq 2^{-n}\}$. Define the intersection

$$\left(\sum \mathrm{BMO}_n\right)_\infty \cap \left(\sum L_n^2\right)_2$$

equipped with the norm

$$||(x_n)|| = \left(\max_{n \in \mathbb{N}} ||x_n||_{\mathrm{BMO}}^2 + \sum_{n=1}^{\infty} ||x_n||_2^2 \right)^{1/2}.$$

This space is commonly called the independent sum of the spaces BMO_n, and denoted by $(\sum \mathrm{BMO}_n)_{\mathrm{ind}}$. The name comes from the embedding defined below. There we use statistically independent copies of BMO_n as building blocks of a subspace isomorphic to $(\sum \mathrm{BMO}_n)_\infty \cap (\sum L^2_n)_2$. We point out that implicitly we actually determine a subsequence of the Walsh system that spans a copy of the independent sum $(\sum \mathrm{BMO}_n)_{\mathrm{ind}}$.

Theorem 2.1.8. *The independent sum $(\sum BMO_n)_{ind}$ is isomorphic to a complemented subspace of BMO.*

Proof. We start by defining a subspace of BMO which is isomorphic to the independent sum $(\sum \mathrm{BMO}_n)_{\mathrm{ind}}$. First we fix $i \in \mathbb{N}$ and let $N = \{i, \dots, i+n\}$. Now we define a copy of BMO_n using the Rademacher functions $\{r_j : j \in N\}$. We start by defining

$$b_{[0,1]} = r_i.$$

Now we suppose that the functions b_I have been constructed for all dyadic intervals of length $\geq 2^{-\ell}$. Then fix a dyadic interval J of length $2^{-\ell}$, and split J into its left half J_1, and its right half J_2. Then we put

$$\begin{aligned} b_{J_1} &= r_{i+\ell+1} 1_{\{b_J = 1\}}, \\ b_{J_2} &= r_{i+\ell+1} 1_{\{b_J = -1\}}. \end{aligned}$$

In this way we defined b_I for every dyadic interval I with $|I| \geq 2^{-n}$. We let $G(N) = \operatorname{span}\{b_I : |I| \geq 2^{-n}\}$. Clearly, when equipped with the BMO norm, the space $G(N)$ is isometric to BMO_n. Incidentally we defined $G(N)$ in such a way that it coincides with the span of Walsh functions $\{w_A : A \subseteq \{i, \dots, i+n\}\}$.

Moreover $G(N)$ is the range of a norm 1 projection. To see this, we renormalize in L^2, and put $k_I = b_I / ||b_I||_2$. Then by Theorem 1.5.9 the orthogonal projection

$$P_N(f) = \sum \langle f, k_I \rangle k_I$$

is bounded on BMO with norm ≤ 1. The range of P_N is $G(N)$.

Next we produce a sequence of independent copies of BMO_n. We fix a sequence of disjoint intervals $N_n \subset \mathbb{N}$, such that $|N_n| = n$ and

$$\sup N_n < \inf N_{n+1}.$$

Define $G_n = G(N_n)$ using the above construction. Note that if $x_n \in G_n$ and if $x_m \in G_m$, then x_n and x_m are independent functions, for $m \neq n$. Repeating the proof of the estimates (2.1.13) shows that for any sequence of functions $x_n \in G_n$, the following equivalence of norms holds true,

$$\max ||x_n||^2_{\mathrm{BMO}} + \sum ||x_n||_2^2 \leq \left\| \sum x_n \right\|^2_{\mathrm{BMO}} \leq 2 \max ||x_n||^2_{\mathrm{BMO}} + 2 \sum ||x_n||_2^2. \tag{2.1.14}$$

Note that the orthogonal projection onto span$\{G_n,\ n \in \mathbb{N}\}$ is given by the operator

$$P = \sum P_{N_n}.$$

By the equivalence of norms (2.1.14) one obtains that P is a bounded operator on BMO. Indeed, let $f \in \mathrm{BMO}$, then $P_{N_n}(f) \in G_n$, hence

$$\begin{aligned}\|P(f)\|_{\mathrm{BMO}}^2 &= \|\sum P_{N_n}(f)\|_{\mathrm{BMO}}^2 \\ &\leq 2\max \|P_{N_n}(f)\|_{\mathrm{BMO}}^2 + 2\sum \|P_{N_n}(f)\|_2^2 \\ &\leq 2\|f\|_{\mathrm{BMO}}^2 + 2\|f\|_2^2.\end{aligned}$$

As $\|f\|_2 \leq \|f\|_{\mathrm{BMO}}$, this shows that $\|P(f)\|_{\mathrm{BMO}} \leq 2\|f\|_{\mathrm{BMO}}$. Summing up we proved our claim that $(\sum \mathrm{BMO}_n)_{\mathrm{ind}}$ is isomorphic to a complemented subspace of BMO. □

It is not known whether the Rosenthal space and the independent sum of BMO_n are isomorphic Banach spaces.

Problem 2.1.9. *Prove that R is not isomorphic to $(\sum BMO_n)_{ind}$.*

In Chapter 5 we study structural results on complemented subspaces of H^1. One of them states the following dichotomy: A complemented subspace of H^1 either contains a copy of ℓ^2, or it is isomorphic to a complemented subspace of $(\sum H_n^1)_1$. To arrive at that result we use that the members of the following family of subspaces are well complemented in BMO. We fix $p < q$ and let

$$S(p,q,M) = \mathrm{span}\{h_J : 2^{-q} \leq |J| < 2^{-p}\}.$$

We define the unit ball in $S(p,q,M)$ as the intersection of the unit ball of BMO and $1/M$ times the unit ball of L^2. This is achieved using the norm

$$\max\{M||x||_{L^2}, ||x||_{\mathrm{BMO}}\}.$$

The dual space of $S(p,q,M)$ consists of

$$T(p,q,M) = \mathrm{span}\{h_J : 2^{-q} \leq |J| < 2^{-p}\},$$

where the norm of $x \in T(p,q,M)$ is defined as

$$|x|_{T(p,q,M)} = \inf\{||y||_{H^1} + ||z||_{L^2}M^{-1} : x = y + z\}.$$

Proposition 2.1.10. *If $M \leq 2^{p/2}$, then the space $S(p,q,M)$ is isomorphic to a complemented subspace of BMO. Embedding and projection have norm bounded by 8. By duality $T(p,q,M)$ is isomorphic to a complemented subspace of H^1, again embedding and projection have norm bounded by 8.*

Proof. By Theorem 2.1.8 it suffices to construct embedding and projection between $S(p,q,M)$ and $(\sum \mathrm{BMO}_n)_{\mathrm{ind}}$. We let $\mathcal{A} = \{J : J \text{ dyadic } |J| = 2^{-p}\}$. Then for $x \in S(p,q,M)$ we have the simple and useful identity

$$\max\{M||x||_{L^2}, ||x||_{\mathrm{BMO}}\} = \max\left\{\left(\sum_{J\in\mathcal{A}} M^2||x\cdot 1_J||_{L^2}^2\right)^{1/2}, \quad \max_{J\in\mathcal{A}} ||x1_J||_{\mathrm{BMO}}\right\}.$$

Recall that we chose $M^2 \le 2^p$, and that the support of $x \cdot 1_J$ has measure $\le 2^{-p}$. Therefore we may find a linear map of slope $\le M^2$ mapping J into $[0,1]$. This linear change of variables maps the function $x\cdot 1_J$ into an element $z_J \in \mathrm{BMO}_n$ such that,

$$\begin{aligned} ||z_J||_{L^2} &= M||x\cdot 1_J||_{L^2}, \\ ||z_J||_{\mathrm{BMO}} &= ||x\cdot 1_J||_{\mathrm{BMO}}. \end{aligned}$$

In this way we obtain an embedding

$$T : S(p,q,M) \to \left(\sum \mathrm{BMO}_n\right)_{\mathrm{ind}}, \qquad x \mapsto (z_J)_{J\in\mathcal{A}}$$

for which

$$\max\{M||x||_{L^2}, ||x||_{\mathrm{BMO}}\} \le ||Tx||_{(\sum \mathrm{BMO}_n)_{\mathrm{ind}}} \le 2\max\{M||x||_{L^2}, ||x||_{\mathrm{BMO}}\}.$$

Note that the orthogonal projection onto the range of T is (trivially) bounded on $(\sum \mathrm{BMO}_n)_{\mathrm{ind}}$ with norm $= 1$. Thus by Theorem 2.1.8, $S(p,q,M)$ is isomorphic to a complemented subspace of BMO. The norm of the embedding is $\le 2\sqrt{2}$ the norm of the projection is bounded by 2. □

Sequences of uniformly complemented subspaces. Given a sequence of finite dimensional Banach spaces $\{E_n\}$, we say that $\{E_n\}$ is well isomorphic to uniformly complemented subspaces of F if there exist linear operators $A_n : E_n \to F$ and $B_n : F \to E_n$ so that

$$\begin{array}{ccc} E_n & \xrightarrow{\mathrm{Id}} & E_n \\ A_n \searrow & & \nearrow B_n \\ & F & \end{array} \qquad \text{and} \qquad \sup_{n\in\mathbb{N}} \|A_n\|\cdot\|B_n\| < \infty.$$

Equivalently we say that F contains uniformly complemented copies of $\{E_n\}$.

Let $\{p_n\}$, $\{q_n\}$ and $\{M_n\}$ be sequences of natural numbers so that $p_n \le q_n$ and $M_n \le 2^{p_n/2}$. Then, by Proposition 2.1.10 the spaces $\{T(p_n,q_n,M_n)\}$ are well isomorphic to uniformly complemented subspaces of H^1. (Or equivalently, H^1 contains uniformly complemented copies of $\{T(p_n,q_n,M_n)\}$.)

Weighted intersections

The next class of examples is obtained by intersecting ℓ^∞ with a weighted version of BMO. The weight is given by a fixed sequence of scalars, $0 \leq x_I \leq 1$. We define the space $\mathcal{S}^\infty$ to consist of all bounded sequences (a_I) for which the following norm is finite,

$$||(a_I)||_{\mathcal{S}^\infty} = ||\sum x_I a_I h_I||_{\mathrm{BMO}} + \sup |a_I|.$$

The norm in $\mathcal{S}^\infty$ depends of course explicitly on the chosen weight. For example, when $x_I = 1$ for all dyadic intervals I, then the norm of $\mathcal{S}^\infty$ is equivalent to the norm of BMO. When the weight disappears, that is when $x_I = 0$, for all I, then the norm of $\mathcal{S}^\infty$ is equivalent to that of ℓ^∞. When $x_I \in \{0, 1\}$, then $\mathcal{S}^\infty$ is the direct sum of ℓ^∞ and the span of a subsequence of the Haar basis. Our notation, however, suppresses this dependence. In Chapter 4 we will give the isomorphic classification of the spaces $\mathcal{S}^\infty$. It turns out that BMO and ℓ^∞ are the only isomorphic types which appear in this class. The first step in this direction is the following result.

Theorem 2.1.11. *The space $\mathcal{S}^\infty$ is isomorphic to a complemented subspace of BMO. The norm of embedding and projection are independent of the choice of scalars, $0 \leq x_I \leq 1$.*

Proof. The proof consists of two steps. It starts by showing that for every choice of scalars $0 \leq x_I \leq 1$, there exists a sequence $g_I \in \mathrm{BMO}$ which spans a subspace of BMO that is isomorphic to $\mathcal{S}^\infty$. Here the construction of the functions g_I reflects the given sequence of weights $\{x_I\}$, so that the norm of the isomorphism is independent of the given weight. In the second step we show that the orthogonal projection

$$P(f) = \sum_I \langle f, \frac{g_I}{||g_I||_2} \rangle \frac{g_I}{||g_I||_2}$$

is bounded in BMO.

For convenience we assume that each of the numbers x_I is a negative power of 2. Then depending on the weights $\{x_I\}$ we choose a rapidly increasing sequence of natural numbers $1 << m_0 << m_1 << m_2 \cdots$. Below it will become clear how rapidly increasing this sequence has to be. Let $\mathcal{A}_i$ be the dyadic intervals of length 2^{-m_i}. First define the collection $\mathcal{C}_{[0,1]} = \mathcal{A}_0$. Then select a collection $\mathcal{B}_{[0,1]} \subseteq \mathcal{C}_{[0,1]}$ such that

$$x_{[0,1]}^2 = |B_{[0,1]}|,$$

where $B_{[0,1]}$ is the point-set covered by $\mathcal{B}_{[0,1]}$. We can make this choice, since we are assuming that $x_{[0,1]}$ is a negative power of 2, and that $1 << m_0$. Denote by I_1 the left half of the unit interval and by I_2 the right half of the unit interval. Next fix $J \in \mathcal{C}_{[0,1]}$, and select $\mathcal{C}_{I_1}(J) \subset \mathcal{A}_1$ and $\mathcal{C}_{I_2}(J) \subset \mathcal{A}_1$ so that the point-set covered by $\mathcal{C}_{I_i}$ is contained in J and has measure equal to $|J|/2$. Moreover we select the intervals in such a way that the relative density of $\mathcal{C}_{I_i}(J)$ is the same on the left half of J and on the right half of J, that is,

$$|C_{I_i}(J) \cap J_j| = \frac{1}{2}|J_j|, \quad i,j \in \{1,2\},$$

where $C_{I_i}(J)$ is the point-set covered by $\mathcal{C}_{I_i}(J)$, and where J_1 is the left half of J, and J_2 is the right half of J. Next we select $\mathcal{B}_{I_1}(J) \subseteq \mathcal{C}_{I_1}(J)$, and $\mathcal{B}_{I_2}(J) \subseteq \mathcal{C}_{I_2}(J)$ such that

$$\frac{1}{2}x_{I_1}^2 = \frac{|B_{I_1}(J)|}{|J|},$$

and

$$\frac{1}{2}x_{I_2}^2 = \frac{|B_{I_2}(J)|}{|J|},$$

where $B_{I_i}(J)$ is the point-set covered by the collection $\mathcal{B}_{I_i}(J)$. Now we take the union over $J \in \mathcal{C}_{[0,1]}$. This gives for $i \in \{1,2\}$, the collections

$$\mathcal{C}_{I_i} = \bigcup_{J \in \mathcal{C}_{[0,1]}} \mathcal{C}_{I_i}(J) \quad \text{and} \quad \mathcal{B}_{I_i} = \bigcup_{J \in \mathcal{C}_{[0,1]}} \mathcal{B}_{I_i}(J).$$

We continue by induction following the pattern of the first step. Thus for each I we construct two families, $\mathcal{B}_I$ and $\mathcal{C}_I$, consisting of pairwise disjoint dyadic intervals satisfying the following properties: The collection $\mathcal{B}_I$ is a subcollection of $\mathcal{C}_I$. Let C_J be the point-set covered by the collection $\mathcal{C}_J$. Then $|C_J| = |J|$, and moreover the family $\{C_I\}$ is a nested family of sets, so that $C_I \subseteq C_J$ if $I \subseteq J$. Note that if $K \in \mathcal{C}_I$ and $L \in \mathcal{C}_J$ and $K \subseteq L$, then we have $C_I \subseteq C_J$, or equivalently, $I \subseteq J$. Moreover we observe the following: Fix a dyadic interval J, and let

$$g_J = \sum \{h_K : K \in \mathcal{B}_J\}.$$

Suppose that I is a dyadic interval which is strictly contained in J . Then for $K \in \mathcal{B}_J$, we have the following identities expressing a strong degree of self similarity,

$$\frac{1}{|K|}\int_K g_I^2 = ||g_I||_2^2 \frac{1}{|J|} = x_I^2 \frac{|I|}{|J|}. \tag{2.1.15}$$

Having made these observations it is easy to show that the orthogonal projection P defined by

$$P(f) = \sum_I \langle f, \frac{g_I}{||g_I||_2}\rangle \frac{g_I}{||g_I||_2}$$

is bounded in BMO. To prove boundedness of P, we fix $f \in \mathrm{BMO}$ and let $g = P(f)$. Then we fix a dyadic interval J, and we let $K \in \mathcal{B}_J$. We calculate using the observation in (2.1.15),

$$\begin{aligned} \frac{1}{|K|}\int_K |g - \frac{1}{|K|}\int_K g|^2 &= \sum_{I \subseteq J} \langle f, \frac{g_I}{||g_I||_2}\rangle^2 \frac{1}{|K|}\int_K \frac{g_I^2}{||g_I||_2^2} \\ &= \langle f, \frac{g_J}{||g_J||_2}\rangle^2 \frac{1}{||g_J||_2^2} + \sum_{I \subset J} \langle f, \frac{g_I}{||g_I||_2}\rangle^2 \frac{1}{|J|}. \end{aligned} \tag{2.1.16}$$

We continue estimating the first summand of the second line in (2.1.16). Note that $||g_J||_2^2 = |B_J|$, and that the Haar support of g_J is the collection $\mathcal{B}_J$. Hence by Bessel's inequality and Cauchy–Schwarz we estimate as follows.

$$\begin{aligned}\langle f, \frac{g_J}{||g_J||_2}\rangle^2 \frac{1}{||g_J||_2^2} &= \left[\frac{1}{|B_J|}\sum_{K\in\mathcal{B}_J}\langle f, h_K\rangle\right]^2 \frac{1}{|B_J|}\\ &\le \frac{1}{|B_J|}\left[\sum_{K\in\mathcal{B}_J}|K|\right]\cdot\left[\frac{1}{|B_J|}\sum_{K\in\mathcal{B}_J}\langle f, h_K\rangle^2|K|^{-1}\right]\\ &\le C||f||^2_{\mathrm{BMO}}.\end{aligned}$$

Now we turn to estimating the second summand in the second line of (2.1.16). Recall that for a given interval J the support of the function g_J is contained in the set C_J. Recall also that $|C_J| = |J|$, and that the collection $\{C_I\}$ is a nested family of sets. Now we estimate using Bessel's inequality and the definition of BMO.

$$\begin{aligned}\sum_{I\subset J}\langle f, \frac{g_I}{||g_I||_2}\rangle^2 \frac{1}{|J|} &\le \frac{1}{|J|}\sum_{K\subseteq C_J}\langle f, h_K\rangle^2|K|^{-1}\\ &\le ||f||^2_{\mathrm{BMO}}.\end{aligned}$$

Thus we showed that the orthogonal projection P is bounded on BMO.

With a similar calculation we evaluate now the norm in BMO of a linear combination $g = \sum a_I g_I$. To this end we fix a dyadic interval J and $K \in \mathcal{B}_J$. Then we calculate using the observation (2.1.15),

$$\begin{aligned}\frac{1}{|K|}\int_K |g - \frac{1}{|K|}\int_K g|^2 &= \sum_{I\subseteq J}\frac{1}{|K|}\int_K a_I^2 g_I^2\\ &= a_J^2 + \sum_{I\subset J} a_I^2 x_I^2 \frac{|I|}{|J|}.\end{aligned}$$

From this identity we deduce easily the following equivalence of norms,

$$||g||_{\mathrm{BMO}} \sim ||\sum a_I x_I h_I||_{\mathrm{BMO}} + \sup|a_I|.$$

Summing up, we showed that the weak *-closed linear span of $\{g_I\}$ is isomorphic to $\mathcal{S}^\infty$, and complemented in BMO. □

2.2 Pełczyński's decomposition method

We now present an extremely flexible and powerful tool to prove that two Banach spaces are isomorphic: the decomposition method of A. Pełczyński. In this monograph we will encounter several examples where the decomposition method permits us to show that two given Banach spaces are isomorphic, even when the explicit determination of the isomorphism is unknown.

Let X be a Banach space and let $1 \leq p \leq \infty$. The ℓ^p direct sum of X is the Banach space of all sequences $\{f_i : i \in \mathbb{N}\}$ with $f_i \in X$ such that $\{||f_i|| : i \in \mathbb{N}\} \in \ell^p$. We denote it by $\left(\sum X\right)_p$. The norm of a given sequence $\{f_i : i \in \mathbb{N}\}$ in $\left(\sum X\right)_p$ is defined as follows,

$$\begin{array}{lll} \left(\sum ||f_i||_X^p\right)^{1/p} & \text{when} & 1 \leq p < \infty, \\ \sup ||f_i||_X & \text{when} & p = \infty. \end{array}$$

With these ingredients we can formulate and prove the decomposition method.

Theorem 2.2.1. *(A. Pełczyński.)* $1 \leq p \leq \infty$. *Let X and Y be Banach spaces and suppose that X is isomorphic to its ℓ^p direct sum. Then the following holds. If X is isomorphic to a complemented subspace of Y, and Y is isomorphic to a complemented subspace of X, then X is isomorphic to Y.*

Proof. If X is isomorphic to a complemented subspace of Y, then (by definition) there exists a Banach space E such that $Y \sim E \oplus X$. Now we show that Y is isomorphic to $X \oplus Y$. Indeed

$$\begin{aligned} Y \sim E \oplus Y \sim E \oplus \left(\sum X\right)_p \\ \sim E \oplus X \oplus \left(\sum X\right)_p \\ \sim Y \oplus X. \end{aligned}$$

Next we show that X is isomorphic to $Y \oplus X$. We write $X \sim F \oplus Y$ and then,

$$\begin{aligned} X &\sim \left(\sum X\right)_p \\ &\sim \left(\sum F \oplus Y\right)_p \\ &\sim \left(\sum F\right)_p \oplus \left(\sum Y\right)_p \oplus Y \\ &\sim \left(\sum F \oplus Y\right)_p \oplus Y \\ &\sim X \oplus Y. \end{aligned}$$

□

Infinite direct sums. The development of Banach space theory, as recorded for instance in [130], testifies to the flexibility and extreme power of A. Pełczyński's decomposition method. In this book we use it frequently to show that a given space is isomorphic to H^1. The hypothesis of the decomposition method requires then that we know that H^1 is isomorphic to its ℓ^1 direct sum, denoted $(\sum H^1)_1$. This motivates our next theorem.

Theorem 2.2.2. *H^1 is isomorphic to $(\sum H^1)_1$.*

Proof. Clearly, $(\sum H^1)_1$ is isomorphic to its own ℓ^1 direct sum, and contains a complemented copy of H^1. We proceed by constructing a complemented subspace of H^1 isomorphic to $(\sum H^1)_1$. Fix a sequence of pairwise disjoint dyadic intervals $\{I_n : n \in \mathbb{N}\}$. Then we let $\mathcal{C}_n = \{J \in \mathcal{D} : J \subseteq I_n\}$. The closure of $\text{span}\{h_J : J \in \mathcal{C}_n\}$ is isomorphic to H^1. Moreover for $f_n \in \text{span}\{h_J : J \in \mathcal{C}_n\}$ we have the identity

$$\sum ||f_n||_{H^1} = \left\|\sum f_n\right\|_{H^1}.$$

Hence the closure of $X = \text{span}\{h_J : J \in \bigcup \mathcal{C}_n\}$ in H^1 is isomorphic to $(\sum H^1)_1$. The restriction operator

$$R : \sum_{I\in\mathcal{D}} a_I h_I \to \sum_n \sum_{I\in\mathcal{C}_n} a_I h_I$$

is of norm 1, and provides a projection onto X. By Pełczyński's decomposition principle, this implies that $(\sum H^1)_1$ is isomorphic to H^1. □

The following result of P. Wojtaszczyk provides a surprising representation of BMO.

Theorem 2.2.3. *BMO is isomorphic to $(\sum BMO_n)_\infty$.*

Proof. By duality Theorem 2.2.2 implies that BMO is isomorphic to $(\sum \text{BMO})_\infty$. Hence $(\sum \text{BMO}_n)_\infty$ is isomorphic to a complemented subspace of BMO. By the decomposition principle of A. Pełczyński it remains to show that BMO is isomorphic to a complemented subspace of $(\sum \text{BMO}_n)_\infty$.

We start by defining the embedding. Let $\mathcal{D}_n = \{I \in \mathcal{D} : |I| \geq 2^{-n}\}$. For $f \in \text{BMO}$, we let

$$f_n = \sum_{J\in\mathcal{D}_n} \langle f, \frac{h_J}{|J|}\rangle h_J.$$

The sequence (f_n) converges in the weak-* topology of BMO and the limit is f. Moreover $||f|| = \sup_n ||f_n||$. Thus the operator

$$E : \text{BMO} \to (\sum \text{BMO}_n)_\infty$$
$$f \mapsto (f_n),$$

is bounded with $||E|| = 1$. Next we define an operator

$$R : (\sum \text{BMO}_n)_\infty \to \text{BMO}$$
$$(x_n) \mapsto R((x_n))$$

satisfying the following property: If (x_n) is a weak-* convergent sequence with $x_n \in \text{BMO}_n$, and if

$$x_n \to x \text{ in } \sigma(\text{BMO}, H^1),$$

then

$$R((x_n)) = x.$$

Any operator with this property inverts the action of E. Indeed, for $f \in \text{BMO}$, $(Ef)_n \to f$ in the $\sigma(\text{BMO}, H^1)$ topology. Therefore

$$\begin{aligned} REf &= R((Ef)_n) \\ &= f. \end{aligned}$$

It remains to show the existence of the operator R. For that we use a compactness argument. We let B be the unit ball of $(\sum \text{BMO}_n)_\infty$ and let C be the unit ball of BMO. Both sets are equipped with their respective weak-* topologies, thus they are compact. Now consider the maps

$$\begin{aligned} R_m : B &\to C \\ (x_n) &\mapsto x_m. \end{aligned}$$

By Tychonov's theorem the product space C^B is compact and therefore the maps R_m have a cluster point in C^B. We let $R : B \to C$ be such a cluster point. Clearly additivity and homogeneity of the maps R_m transfer to the cluster point R which can therefore be extended linearly to a bounded linear operator $R : (\sum \text{BMO}_n)_\infty \to \text{BMO}$. Moreover as R is a cluster point of the sequence $\{R_m\}$, it satisfies

$$R((x_n)) = x,$$

if $x_n \in \text{BMO}_n$, and

$$x_n \to x \text{ in } \sigma(\text{BMO}, H^1). \qquad \square$$

H^1 with values in ℓ^2_n. Next we introduce the Banach space of H^1 functions taking values in ℓ^2_n (the n-dimensional Euclidean space). An integrable $f : [0,1] \to \ell^2_n$ with Haar series

$$f(t) = \sum h_I(t) x_I, \quad x_I \in \ell^2_n$$

belongs to $H^1(\ell^2_n)$ iff its square function

$$\left(\sum h_I^2(t) \|x_I\|_2^2\right)^{1/2}$$

is integrable. With the norm

$$\int_0^1 \left(\sum h_I^2(t) \|x_I\|_2^2\right)^{1/2} dt,$$

the space $H^1(\ell^2_n)$ is clearly a Banach space.

We will show now that H^1 and $H^1(\ell^2_n)$ are isomorphic in such a way that the norm of the isomorphism and its inverse is bounded independently of n. By contrast H^1 and $H^1(\ell^2)$ are two Banach spaces which are not isomorphic. This important theorem of J. Bourgain will be presented in Chapter 5.

Theorem 2.2.4. *For every $n \in \mathbb{N}$, there exists an isomorphism $T : H^1 \to H^1(\ell_n^2)$ such that*

$$||T|| \cdot ||T^{-1}|| < C,$$

where C is independent of n.

Proof. Clearly H^1 is isomorphic to a complemented subspace of $H^1(\ell_n^2)$, and by Theorem 2.2.2, H^1 is isomorphic to its ℓ^1 sum. Hence by Pełczyński's decomposition method, the existence of the isomorphism is guaranteed when we show that $H^1(\ell_n^2)$ is isomorphic to a complemented subspace of H^1, and the norm of embedding and projection are bounded independently of n.

We will now present the construction. We fix $n \in \mathbb{N}$. Then we pick a sequence $s_m \in \mathbb{N}$ and a sequence of intervals $R_m \subseteq \mathbb{N}$ such that the cardinality of R_m equals n, and

$$s_{m-1} < \inf R_m < \sup R_m < s_m.$$

We first use the sequence $\{s_m\}$ and the Rademacher functions $\{r_{s_m}\}$, to construct a block basis equivalent to the Haar system as follows. Put

$$k_{[0,1]} = r_{s_1},$$

and suppose that k_J has been defined for all dyadic intervals of length $\geq 2^{-m}$, then we take a dyadic interval J of length equal to 2^{-m}. Split J into dyadic intervals J_1, J_2 of length 2^{-m-1}, and define

$$\begin{aligned} k_{J_1} &= 1_{\{k_J=1\}} r_{s_{m+1}}, \\ k_{J_2} &= 1_{\{k_J=-1\}} r_{s_{m+1}}. \end{aligned}$$

In this way we construct k_J for each dyadic interval J.

Next we attach to k_J a set of Rademacher functions to produce copies of ℓ_n^2. Fix $i \leq n$, and denote the i-th element in the set (of integers) R_m by $i(m)$. Then, for a dyadic interval J of length 2^{-m} we define

$$x(J,i) = k_J r_{i(m)}.$$

Our first claim is that the orthogonal projection onto $\operatorname{span}\{x(J,i) : J \in \mathcal{D}, i \leq n\}$ is bounded on H^1 with norm ≤ 8. The orthogonal projection is given explicitly by

$$P(f) = \sum_{J \in \mathcal{D}} \sum_{i=1}^{n} \left\langle f, \frac{x(J,i)}{||x(J,i)||_2} \right\rangle \frac{x(J,i)}{||x(J,i)||_2}.$$

To prove the norm estimate we only have to observe that the collections

$$\mathcal{C}_{J,i} = \{K \in \mathcal{D} : \langle h_K, x(J,i) \rangle \neq 0\}, \quad J \in \mathcal{D}, i \leq n,$$

(can be ordered to) satisfy Jones' compatibility condition (J).

Our second claim is that in H^1 the closure of span$\{x(J,i) : J \in \mathcal{D}, i \leq n\}$ is isomorphic to $H^1(\ell_n^2)$. We fix a sequence of scalars $a(J,i)$, and observe the identity

$$\begin{aligned}\int \left(\sum_{J\in\mathcal{D}} \sum_{i=1}^{n} x(J,i)^2 a(J,i)^2 \right)^{1/2} &= \int \left(\sum_{J\in\mathcal{D}} k_J^2(s) \sum_{i=1}^{n} a(J,i)^2 \right)^{1/2} ds \\ &= \int \left(\sum_{J\in\mathcal{D}} k_J^2(s) ||a_J||^2 \right)^{1/2} ds,\end{aligned}$$

where $||a_J||_2^2 = \sum_{i=1}^n a(J,i)^2$. Finally we observe that $\{k_J : J \in \mathcal{D}\}$ has the same joint distribution as the Haar system. Hence

$$\left\| \sum_{J\in\mathcal{D}} \sum_{i=1}^{n} x(J,i) a(J,i) \right\|_{H^1} = \left\| \sum_{J\in\mathcal{D}} h_J a_J \right\|_{H^1(\ell_n^2)}$$

where $a_J = (a(J,i), \dots, a(J,n))$. This completes the proof that $H^1(\ell_n^2)$ is isomorphic to a complemented subspace of H^1. The norm of embedding and projection are bounded independent of n. □

2.3 Interpolation of operators

In this section we present powerful methods of complex interpolation. We give applications to averaging projections and rearrangement operators.

The Riesz convexity theorem is a widely used interpolation result. In the most basic form its statement involves a single operator T, bounded and defined both on L^1 and L^∞. With

$$||T||_{L^p} = ||T : L^p \to L^p||$$

the conclusion of the Riesz convexity theorem is that for $1 \leq p \leq \infty$ and $\theta = 1/p$,

$$||T||_{L^p} \leq ||T||_{L^1}^{1-\theta} ||T||_{L^\infty}^{\theta}.$$

To illustrate this result we prove it for finite-dimensional spaces ℓ_n^p, consisting of $\mathbb{R}^n$ equipped with the norm

$$\|x\|_p = \left(\sum_{i=1}^{n} |x_i|^p \right)^{1/p}.$$

Let the linear operator T on ℓ_n^p be given by the matrix $(t_{j,k})_{j,k=1}^n$. For $x \in \mathbb{R}^n$ the j-th component of Tx is just $\sum_{k=1}^n t_{j,k} x_k$. Hence

$$||Tx||_p^p = \sum_{j=1}^{n} |\sum_{k=1}^{n} t_{j,k} x_k|^p.$$

Rewrite the right-hand side, multiplying and dividing by the factor $(\sum_{l=1}^{n}|t_{j,l}|)^p$. Thus we prepare for the application of Minkowski's inequality. We obtain the identity

$$\|Tx\|_p^p = \sum_{j=1}^{n}\left[\sum_{l=1}^{n}|t_{j,l}|\right]^p |\sum_{k=1}^{n}\frac{t_{j,k}}{\sum_{l=1}^{n}|t_{j,l}|}x_k|^p.$$

For each fixed $j \leq n$ apply Minkowski's inequality in the form

$$|\sum_{k=1}^{n}\frac{t_{j,k}}{\sum_{l=1}^{n}|t_{j,l}|}x_k|^p \leq \sum_{k=1}^{n}\frac{|t_{j,k}|}{\sum_{l=1}^{n}|t_{j,l}|}|x_k|^p.$$

We obtain this upper bound,

$$\begin{aligned}\|Tx\|_p^p &\leq \sum_{j=1}^{n}\left[\sum_{l=1}^{n}|t_{j,l}|\right]^{p-1}\sum_{k=1}^{n}|t_{j,k}||x_k|^p \\ &\leq \sup_{j\leq n}\left[\sum_{l=1}^{n}|t_{j,l}|\right]^{p-1}\left[\sup_{k\leq n}\sum_{j=1}^{n}|t_{j,k}|\right]\sum_{k=1}^{n}|x_k|^p.\end{aligned}$$

Taking the p-th root on both sides gives that

$$\|Tx\|_p \leq M_1^{\frac{p-1}{p}} M_\infty^{\frac{1}{p}}\|x\|_p,$$

where $M_\infty = \sup_{k\leq n}\sum_{j=1}^{n}|t_{j,k}|$ and $M_1 = \sup_{j\leq n}\sum_{l=1}^{n}|t_{j,l}|$. To complete the proof of the Riesz convexity theorem it remains to observe that $\|T : \ell_n^1 \to \ell_n^1\| = M_1$, and $\|T : \ell_n^\infty \to \ell_n^\infty\| = M_\infty$. If the endpoint L^∞ is replaced by L^2, then the theorem of Riesz asserts that for $1 \leq p \leq 2$ and $\theta = 2 - 2/p$,

$$||T||_{L^p} \leq ||T||_{L^1}^{1-\theta}||T||_{L^2}^{\theta}.$$

In this section we prove interpolation theorems parallel to the Riesz convexity theorem, where we use H^1 in place of L^1. The averaging projections

$$P(f) = \sum\langle f, \frac{k_i}{||k_i||_2}\rangle\frac{k_i}{||k_i||_2},$$

where

$$k_i = \sum\{h_I : I \in \mathcal{C}_i\},$$

were studied in Section 1.5 as operators on H^1. There we showed (Theorem 1.5.9), that P is bounded on H^1 if the defining collections of dyadic intervals $\mathcal{C}_i$ satisfy Jones's compatibility condition. On the other hand being an orthogonal projection, P is bounded on the space L^2 with norm ≤ 1, as long as the collections $\{\mathcal{C}_i\}$ are pairwise disjoint. With interpolation theorems for single operators acting between

Hardy spaces we find that P is bounded on H^p, for $1 < p < 2$, satisfying the norm estimate

$$||P(f)||_{H^p} \leq C||P||_{H^1}^{1-\theta}||f||_{H^p}, \quad \text{where} \quad \theta = 2 - 2/p.$$

(Recall that for $f = \sum a_I h_I$, we denote $S(f) = (\sum a_I^2 h_I^2)^{1/2}$ and $\|f\|_{H^p} = \|S(f)\|_{L^p}$.)

The family of rearrangement operators of the L^p normalized Haar system is a basic example where interpolation theorems for a single operator are not quite sufficient; more refined tools of complex interpolation are used instead. A rearrangement operator is determined by an injective map τ acting on dyadic intervals. By linear extension it induces a rearrangement of the L^p normalized Haar system,

$$T_p : \frac{h_I}{|I|^{1/p}} \mapsto \frac{h_{\tau(I)}}{|\tau(I)|^{1/p}}. \tag{2.3.1}$$

On a given Haar function h_I the operator T_p acts as follows,

$$T_p(h_I) = h_{\tau(I)} \left(\frac{|I|}{|\tau(I)|}\right)^{1/p}.$$

The definition of T_p depends on both the underlying rearrangement and the value of p. We use results of complex interpolation to prove that for the operators $\{T_p : 1 < p < 2\}$ the boundedness on L^p is actually determined by the rearrangement alone. Precisely, we will prove that if there exists $1 < p_0 < 2$, such that T_{p_0} is bounded on L^{p_0}, then for all $1 < p < 2$, the operator T_p is already bounded on L^p. The methods presented in this section are based on ideas from two distant areas of real and complex analysis. These include:

(a) Stopping time arguments and a procedure for effectively decomposing functions. This idea already appeared in the proof of the atomic decomposition of H^1. See Theorem 1.2.4.

(b) A maximum principle for subharmonic functions valid in unbounded domains such as the vertical strip

$$V = \{x + iy : x \in (0, 1), y \in \mathbb{R}\}.$$

We summarize now the results of this section.

1. We identify the Calderón product of H^1 and H^2. Let $1 < p \leq 2$ and $\theta = 2 - 2/p$. For $u \in H^p$ with $\int u = 0$ the norm satisfies

$$c||u||_{H^p} \leq \inf\{||x||_{H^1}^{1-\theta}||y||_2^{\theta} : u = \sum x(I)^{1-\theta}y(I)^{\theta}h_I\} \leq C||u||_{H^p}, \tag{2.3.2}$$

   where $x = \sum x(I)h_I$ and $y = \sum y(I)h_I$. In the inequalities (2.3.2) the constants $c > 0$ and $C > 0$ are independent of p. By definition the Calderón product of H^1 and H^2 is given by the central expression in (2.3.2).

2. We present G. Pisier's norm on H^1. Let $g \in H^1$ have Haar expansion $g = \sum g(I)h_I$. Let $1 < p < 2$ be fixed and let $\theta = 2 - 2/p$. Then the norm of g in H^1 can be evaluated via $H^p(1 < p < 2)$ and L^2 in the following way.

$$c||g||_{H^1}^{1-\theta} \leq \sup || \sum g(I)^{1-\theta} w(I)^{\theta} h_I ||_{H^p} \leq ||g||_{H^1}^{1-\theta}, \tag{2.3.3}$$

where the supremum is extended over all $w \in L^2$ with $||w||_2 \leq 1$, and Haar expansion $w = \sum w(I)h_I$.

3. Let $1 < p < 2$. For T_p, defined by (2.3.1), denote $||T_p||_p = ||T_p : H^p \to H^p||$. Then
$$c||T_p||_p \leq ||T_1||_1^{1-\theta} \leq C||T_p||_p, \text{ where } \theta = 2 - 2/p.$$

4. Let $1 < p < 2$ and $\theta = 2 - 2/p$. On H^p the norm of an analytic family of operators $\{J_z\}$ satisfies the convexity estimate
$$||J_\theta(f))||_{H^p} \leq CR_0^{1-\theta} R_1^{\theta} ||f||_{H^p},$$
where $R_0 = \sup_t ||J_{it}||_{H^1}$, and $R_1 = \sup_t ||J_{1+it}||_{L^2}$.

A crucial link between the renormings of H^p ($1 < p < 2$) and G. Pisier's norm on H^1 is given by Fefferman's inequality as formulated in Theorem 2.3.1 below. For instance, both, the inequalities (2.3.24) and the norm (2.3.3) express converse estimates to (2.3.4) thereby showing its sharpness.

Theorem 2.3.1. *Let $x \in H^1$, $v \in BMO$, and let $w, y, \in L^2$, be finitely supported over the Haar system. For $\theta \in (0,1)$, form the functions*

$$u = \sum x(I)^{1-\theta} y(I)^{\theta} h_I \quad \text{and} \quad f = \sum v(I)^{1-\theta} w(I)^{\theta} h_I,$$

where $x(I) = \langle x, h_I \rangle / |I|$, $y(I) = \langle y, h_I \rangle / |I|$, $v(I) = \langle v, h_I \rangle / |I|$ and $w(I) = \langle w, h_I \rangle / |I|$. Then,

$$\left| \int_0^1 uf \right| \leq \left(2\sqrt{2} ||x||_{H^1} ||v||_{BMO} \right)^{1-\theta} (||y||_{L^2} ||w||_{L^2})^{\theta}. \tag{2.3.4}$$

Proof. First we apply Hölder' s inequality with the parameters $p = 1/(1-\theta)$ and $q = 1/\theta$. Then we use Fefferman's inequality and the Cauchy–Schwarz inequality as follows.

$$\begin{aligned} \left| \int_0^1 uf \right| &= \left| \sum x(I)^{1-\theta} v(I)^{1-\theta} w(I)^{\theta} y(I)^{\theta} |I| \right| \\ &\leq \left(\sum |x(I)v(I)||I| \right)^{1-\theta} \left(\sum |w(I)y(I)||I| \right)^{\theta} \\ &\leq \left(2\sqrt{2} ||x||_{H^1} ||v||_{\mathrm{BMO}} \right)^{1-\theta} (||y||_{L^2} ||w||_{L^2})^{\theta}. \end{aligned}$$

□

Calderón's product of H^1 and H^2

Recall that on H^p $(1 \le p < \infty)$ the norm is defined as follows. Given an integrable function f with $\int f = 0$, let

$$S(f) = (\sum_I f(I)^2 h_I^2)^{1/2} \quad \text{where} \quad f(I) = \langle f, \frac{h_I}{|I|} \rangle,$$

and put

$$\|f\|_{H^p} = \left(\int_0^1 S(f)^p \right)^{1/p}.$$

In this subsection we will identify the Calderón product of H^1 and H^2 by proving the estimates stated in (2.3.2). We start with the left-hand side of (2.3.2). Its proof consists of two applications of Hölder's inequality.

Proposition 2.3.2. *Let $1 < p \le 2$, and $\theta = 2 - 2/p$. Then for $x \in H^1$, and $y \in L^2$, with Haar expansion $x = \sum x(I) h_I$ and $y = \sum y(I) h_I$, the following estimate holds,*

$$\left\| \sum x(I)^{1-\theta} y(I)^\theta h_I \right\|_{H^p} \le ||x||_{H^1}^{1-\theta} ||y||_2^\theta.$$

Proof. Fix $x \in H^1$, and let $y \in L^2$, be finitely supported over the Haar system. Let

$$u = \sum x(I)^{1-\theta} y(I)^\theta h_I.$$

First use Hölder's inequality for sequence spaces with the conjugate exponents $\frac{1}{\theta}$ and $\frac{1}{1-\theta}$. This shows that the following estimate holds pointwise for the square function of u,

$$S(u) \le S(x)^{1-\theta} S(y)^\theta. \tag{2.3.5}$$

Next apply Hölder's inequality for integrals to the conjugate exponents $\frac{1}{p(1-\theta)}$ and $\frac{2}{p\theta}$. This gives

$$\int_0^1 S(x)^{p(1-\theta)} S(y)^{p\theta} \le \left(\int_0^1 S(x) \right)^{p(1-\theta)} \left(\int_0^1 S(y)^2 \right)^{p\theta/2}. \tag{2.3.6}$$

Combining the estimates (2.3.6) and (2.3.5) we obtain that

$$||u||_{H^p} \le ||x||_{H^1}^{1-\theta} ||y||_2^\theta.$$

□

Next we verify that the right-hand side estimate of (2.3.2) holds true. We will show that for $u \in H^p$ the Haar coefficient, $u(I) = \langle u, h_I/|I| \rangle$, admits a factorization,

$$u(I) = x(I)^{1-\theta} y(I)^\theta,$$

so that the norm of $x = \sum x(I) h_I$ and $y = \sum y(I) h_I$ satisfy

$$||x||_{H^1}^{1-\theta} ||y||_2^\theta \le 2^5 ||u||_{H^p}.$$

We constructively determine $x \in H^1$ and $y \in L^2$ with these properties using the following decomposition theorem that is based on stopping time arguments.

Theorem 2.3.3. *Let $1 \leq s < \infty$. Let $g \in H^s$ with $\int g = 0$. There exists a sequence of functions $\{g_i\}$ disjointly supported over the Haar system such that*

$$g = \sum_{i=1}^{\infty} g_i,$$

and so that with $B_i = ||S(g_i)||_\infty$, the series

$$x = \sum g_i B_i^{-1+s} \quad \text{and} \quad y = \sum g_i B_i^{-1+s/2},$$

are converging respectively in H^1, and L^2. They satisfy the norm estimates,

$$||x||_{H^1} \leq 8 \cdot 2^{2s} ||g||_{H^s}^s \quad \text{and} \quad ||y||_2 \leq \sqrt{8} \cdot 2^s ||g||_{H^s}^{s/2}.$$

Proof. Let $1 \leq s < \infty$, and pick $g \in H^s$ with $\int g = 0$. We will now define the decomposition of g using a stopping time argument based on the size of the square function $S(g)$.

We start with the unit interval $[0,1]$. Let $n([0,1])$ be the smallest integer such that

$$\Omega([0,1]) = \{t \in [0,1] : S(g)(t) \geq 2^{n([0,1])}\}$$

satisfies

$$|\Omega([0,1])| \leq \frac{1}{4}.$$

Let $\mathcal{F}([0,1])$ be the collection of dyadic intervals which are contained in $\Omega([0,1])$. Then define

$$\mathcal{C}([0,1]) = \mathcal{Q}([0,1]) \setminus \mathcal{F}([0,1]),$$

and

$$g_{[0,1]} = \sum_{J \in \mathcal{C}([0,1])} \langle g, \frac{h_J}{|J|} \rangle h_J.$$

Note that the square function of $g_{[0,1]}$ satisfies the pointwise estimate

$$S(g_{[0,1]}) \leq 2^{n([0,1])}. \tag{2.3.7}$$

Next define the collection of pairwise disjoint dyadic intervals,

$$\mathcal{E}_1 = \max \mathcal{F}([0,1]).$$

We continue the construction inductively. Suppose we have already defined collections of pairwise disjoint dyadic intervals $\mathcal{E}_1, \ldots, \mathcal{E}_i$. Fix $I \in \mathcal{E}_i$. Let $n(I)$ be the smallest integer such that the set

$$\Omega(I) = \{t \in I : S(g)(t) \geq 2^{n(I)}\}$$

satisfies

$$|\Omega(I)| \leq \frac{1}{4}|I|.$$

Let $\mathcal{F}(I)$ be the collection of dyadic intervals which are contained in $\Omega(I)$. Then define

$$\mathcal{C}(I) = Q(I) \setminus \mathcal{F}(I).$$

Note that

$$g_I = \sum_{J \in \mathcal{C}(I)} \langle g, \frac{h_J}{|J|} \rangle h_J$$

satisfies the pointwise estimate,

$$S(g_I) \leq 2^{n(I)}, \tag{2.3.8}$$

and that the support of $S(g_I)$ is contained in I. Now we put

$$\mathcal{E}(I) = \max \mathcal{F}(I),$$

and taking unions over $I \in \mathcal{E}_i$ we define

$$\mathcal{E}_{i+1} = \bigcup \{\mathcal{E}(I) : I \in \mathcal{E}_i\}.$$

Thus the collection $\mathcal{E}_{i+1}$ consists of pairwise disjoint dyadic intervals. This completes the construction part of the proof. We now analyze consequences resulting from this decomposition. We put $\mathcal{E} = \bigcup \mathcal{E}_i$. First notice that for any $I \in \mathcal{E}$,

$$\sum_{J \in \mathcal{E} \cap I, J \neq I} |J| \leq \frac{1}{2}|I|. \tag{2.3.9}$$

We will next verify that the following convexity relation holds,

$$\sum_{I \in \mathcal{E}} |I| 2^{sn(I)} \leq 2^{2s+3} ||S(g)||_s^s. \tag{2.3.10}$$

We start the verification of (2.3.10) with this crucial observation: The stopping time definition of $n(I)$ implies the measure estimate

$$|\{t \in I : S(g)(t) \geq 2^{n(I)-1}\}| \geq \frac{|I|}{4}.$$

Multiplying the above line with $2^{sn(I)}$, and taking the sum over all $I \in \mathcal{E}$, gives

$$\begin{aligned}
\sum_{I \in \mathcal{E}} |I| 2^{sn(I)} &\leq 4 \sum_{I \in \mathcal{E}} 2^{sn(I)} |\{t \in I : S(g) \geq 2^{n(I)-1}\}| \\
&\leq 4 \cdot 2^s \sum_{I \in \mathcal{E}} 2^{s(n(I)-1)} |\{t \in I : S(g) \geq 2^{n(I)-1}\}| \\
&\leq 4 \cdot 2^s \sum_{k \in \mathbb{N}} 2^{ks} |\{S(g) \geq 2^k\}| \\
&\leq 8 \cdot 2^{2s} \int S^s(g).
\end{aligned}$$

Hence (2.3.10) is verified.

The pieces $\{g_I : I \in \mathcal{E}\}$ are disjointly supported over the Haar system. Therefore

$$S^2(g) = \sum_{I\in\mathcal{E}} S^2(g_I),$$

and the decomposition $g = \sum g_I$ converges unconditionally in H^s. We relabel $\{g_I : I \in \mathcal{E}\}$ as $\{g_i : i \in \mathbb{N}\}$, and accordingly we rename $\{2^{n(I)} : I \in \mathcal{E}\}$ as $\{B_i : i \in \mathbb{N}\}$.

Next we verify the L^2 norm estimate for $y = \sum g_i B_i^{-1+s/2}$. By (2.3.8) we have that $S(g_i) \le B_i$. Hence the convexity relation (2.3.10) gives the estimates

$$\begin{aligned} \|\sum g_i B_i^{-1+s/2}\|_2^2 &= \sum \int S^2(g_i) B_i^{-2+s} \\ &\le \sum_{I\in\mathcal{E}} |I| 2^{sn(I)} \qquad (2.3.11) \\ &\le 8 \cdot 2^{2s} \|S(g)\|_s^s. \end{aligned}$$

To obtain the estimate for the H^1 norm of $x = \sum g_i B_i^{-1+s}$, we use triangle inequality, (2.3.8) and the convexity relation (2.3.10),

$$\begin{aligned} \|\sum g_i B_i^{-1+s}\|_{H^1} &\le \sum \int S(g_i) B_i^{-1+s} \\ &\le \sum_{I\in\mathcal{E}} |I| 2^{sn(I)} \qquad (2.3.12) \\ &\le 8 \cdot 2^{2s} \|S(g)\|_s^s. \end{aligned}$$

□

Remark. Let $z = \sum g_i B_i^{-1}$, then $\|z\|_{\mathrm{BMO}} \le 2$. This follows from the pointwise estimate $S(g_i) \le B_i$ and the fact that $\mathcal{E}$ satisfies 2-Carleson condition (2.3.9).

$$\begin{aligned} \|\sum g_i B_i^{-1}\|_{\mathrm{BMO}} &\le [\![\mathcal{E}]\!]^{1/2} \sup_i B_i^{-1} \|g_i\|_{\mathrm{BMO}} \qquad (2.3.13) \\ &\le 2. \end{aligned}$$

□

By simple arithmetic we will verify below that the decomposition of Theorem 2.3.3 yields the right-hand side estimate of (2.3.2).

Theorem 2.3.4. *Let $1 \le p \le 2$ and $\theta = 2 - 2/p$. For every $u \in H^p$ with $\int u = 0$, there exist $x \in H^1$ and $y \in L^2$, such that*

$$\|x\|_{H^1}^{1-\theta} \|y\|_2^\theta \le 2^5 \|u\|_{H^p},$$

and

$$u = \sum x(I)^{1-\theta} y(I)^\theta h_I,$$

where $x(I)$ and $y(I)$ denote the Haar coefficients of x, respectively y.

Proof. Let $1 < p < 2$ and $\theta = 2 - 2/p$. To $u \in H^p$ we apply the decomposition Theorem 2.3.3. Thus u can be decomposed as

$$u = \sum u_i,$$

and with $B_i = ||S(u_i)||_\infty$ we form

$$x = \sum u_i B_i^{-1+p} \quad \text{and} \quad y = \sum u_i B_i^{-1+p/2}. \tag{2.3.14}$$

To prove the theorem we show that the functions $u \in H^p$, $y \in L^2$, and $x \in H^1$ are related through their Haar coefficients and by a convexity estimate for their norms. We claim that the following holds,

$$\langle x, h_I\rangle^{1-\theta}\langle y, h_I\rangle^{\theta} = \langle u, h_I\rangle \quad \text{and} \quad ||x||_{H^1}^{1-\theta}||y||_2^{\theta} \leq 2^5||u||_{H^p}.$$

We start by verifying the convexity estimate $||x||_{H^1}^{1-\theta}||y||_2^{\theta} \leq 2^5||u||_{H^p}$. Recall the upper bounds established in Theorem 2.3.3,

$$||x||_{H^1} \leq 2^{2p+3}||u||_{H^p}^{p} \quad \text{and} \quad ||y||_2 \leq 2^{p+3/2}||u||_{H^p}^{p/2}.$$

Multiplying we find that

$$||x||_{H^1}^{1-\theta}||y||_2^{\theta} \leq A||u||_{H^p}^{p(1-\theta)}||u||_{H^p}^{p\theta/2},$$

where $A = 2^{(2p+3)(1-\theta)}2^{(p+3/2)\theta}$. Using that $\theta = 2 - 2/p$ we calculate,

$$p(1-\theta) + p\theta/2 = 1 \quad \text{and} \quad A \leq 2^5.$$

Next we check that the Haar coefficients satisfy the identity

$$\langle x, h_J\rangle^{1-\theta}\langle y, h_J\rangle^{\theta} = \langle u, h_J\rangle. \tag{2.3.15}$$

To this end let J be a dyadic interval in the Haar support of u. The construction of Theorem 2.3.3 defined the functions $\{u_i\}$, by splitting the Haar support of u. Hence for the given J there exists a unique i such that J is in the Haar support of u_i. Then,

$$\langle x, h_J\rangle = B_i^{-1+p}\langle u, h_J\rangle, \quad \text{and} \quad \langle y, h_J\rangle = B_i^{-1+p/2}\langle u, h_J\rangle.$$

Raise the coefficients to the respective powers $1-\theta$ and θ, and multiply them. This gives

$$\langle x, h_J\rangle^{1-\theta}\langle y, h_J\rangle^{\theta} = B_i^{(-1+p)(1-\theta)}B_i^{(-1+p/2)\theta}\langle u, h_J\rangle.$$

With our choice of $\theta = 2 - 2/p$ we have

$$(-1+p)(1-\theta) + (-1+\frac{p}{2})\theta = 0.$$

Inserting gives

$$B_i^{(-1+p)(1-\theta)}B_i^{(-1+p/2)\theta} = 1,$$

hence the claimed relation (2.3.15) between the Haar coefficients. □

The renorming of (2.3.2) follows from Theorem 2.3.4 and Proposition 2.3.2.

Pisier's norm on H^1

Here we specialize to H^1 the norm invented by G. Pisier in [176]. To this end we carry further the ideas introduced in the course of proving the renorming (2.3.2). Our treatment is guided by the interpretation of G. Pisier's norm as a uniqueness result for the endpoint of complex interpolation scales of Banach lattices, due to M. Cwikel, P. Nilsson and G. Schechtman [57].

Theorem 2.3.5. *Let $g \in H^1$ have Haar expansion $g = \sum g(I)h_I$. For $1 < p < 2$, and $\theta = 2 - 2/p$,*

$$c||g||_{H^1}^{1-\theta} \leq \sup \left\{ \left\| \sum g(I)^{1-\theta} w(I)^\theta h_I \right\|_{H^p} : ||w||_2 \leq 1 \right\} \leq ||g||_{H^1}^{1-\theta}, \qquad (2.3.16)$$

where $c = 2^{-5}$.

Proof. The ingredients of the proof are Proposition 2.3.2 for the right-hand side of (2.3.16), and the stopping time decomposition of Theorem 2.3.3 for the left-hand side.

Fix $g \in H^1$ with Haar expansion $g = \sum g(I)h_I$, and let $1 < p < 2$. We show first that the right-hand side inequality holds. Choose $w \in L^2$ with Haar expansion $w = \sum w(I)h_I$. Then put

$$u = \sum g(I)^{1-\theta} w(I)^\theta h_I.$$

Proposition 2.3.2 gives an upper estimate for the norm of $u \in H^p$,

$$||u||_{H^p} \leq ||g||_{H^1}^{1-\theta} ||w||_2^\theta.$$

This proves that the right-hand side of (2.3.16) holds true.

To prove the left-hand side of (2.3.16) we will constructively determine $w \in L^2$ so that $||w||_2 \leq 2^{5/2}$ and

$$||g||_{H^1}^{1-\theta} \leq 4 \left\| \sum g(I)^{1-\theta} w(I)^\theta h_I \right\|_{H^p}. \qquad (2.3.17)$$

Assume first that $||g||_{H^1} = 1$. Apply Theorem 2.3.3 to g. Let $\{g_i\}$ be the resulting decomposition so that $g = \sum g_i$. By definition $\operatorname{supp} S(g_i)$ is a dyadic interval, say I_i. By (2.3.8) and (2.3.10) we have

$$\sum \|S(g_i)\|_\infty |I_i| \leq 2^5 ||g||_{H^1}. \qquad (2.3.18)$$

Define

$$w_i = g_i Y_i^{1/2}, \quad \text{where} \quad Y_i = \frac{\|S(g_i)\|_\infty |I_i|}{\|S(g_i)\|_2^2},$$

and form $w = \sum w_i$. The choice of Y_i gives $\|w_i\|_2^2 = \|S(g_i)\|_\infty |I_i|$. Hence by (2.3.18),

$$\|w\|_2^2 \leq 2^5 ||g||_{H^1}. \qquad (2.3.19)$$

Let $\{w_i(J)\}$, respectively $\{g_i(J)\}$, denote the Haar coefficients of w_i, respectively g_i. Then the following identities hold,

$$\begin{aligned}\sum g(J)^{1-\theta} w(J)^{\theta} h_J &= \sum_i \sum_J g_i(J)^{1-\theta} w_i(J)^{\theta} h_J \\ &= \sum_i Y_i^{\theta/2} g_i.\end{aligned}$$

The stopping time decomposition of g implies that

$$\sum_i Y_i^{p\theta/2} \|g_i\|_{H^p}^p \leq 4 \left\| \sum g(J)^{1-\theta} w(J)^{\theta} h_J \right\|_{H^p}^p . \tag{2.3.20}$$

A detailed verification of (2.3.20) is given in Proposition 3.1.5 below. Observe that by Hölder's inequality,

$$\left(\int_0^1 S(g_i) \right) \left(\int_0^1 S^2(g_i) \right)^{p-1} \leq \left(\int_0^1 S(g_i)^p \right) \|S(g_i)\|_\infty^{p-1} |I_i|^{p-1}. \tag{2.3.21}$$

Indeed, first we pull out $\|S(g_i)\|_\infty$ from the second factor on the left-hand side of (2.3.21). Thus it is bounded by

$$\left(\int_0^1 S(g_i) \right)^p \|S(g_i)\|_\infty^{p-1}.$$

Next using Hölder's inequality and the fact that $S(g_i)$ is supported on the interval I_i one gets

$$\left(\int_0^1 S(g_i) \right)^p \leq \left(\int_0^1 S(g_i)^p \right) |I_i|^{p-1}.$$

Collecting terms gives (2.3.21).

Next recall that $p\theta/2 = p - 1$. Hence we rewrite (2.3.21) as

$$\|g_i\|_{H^1} \leq Y_i^{p\theta/2} \|g_i\|_{H^p}^p .$$

Taking the sum over i gives that

$$\sum_i \|g_i\|_{H^1} \leq \sum_i Y_i^{p\theta/2} \|g_i\|_{H^p}^p . \tag{2.3.22}$$

Combining the estimates (2.3.22) and (2.3.20) gives (2.3.17) under the assumption that $\|g\|_{H^1} = 1$. The homogeneities in these estimates and in (2.3.19) imply the general case. □

Illustrating how interpolation methods are applied we prove a theorem relating norms of rearrangement operators. Recall that a rearrangement operator is determined by an injective map τ acting on dyadic intervals. We let T_p be the linear extension of the map

$$T_p : \frac{h_I}{|I|^{1/p}} \mapsto \frac{h_{\tau(I)}}{|\tau(I)|^{1/p}}.$$

For the operator norm of T_p on H^p we write as before $||T_p||_p = ||T_p : H^p \to H^p||$.

Theorem 2.3.6. *For $1 < p < 2$ let $\theta = 2 - 2/p$. Then*

$$c||T_p||_p \leq ||T_1||_1^{1-\theta} \leq C||T_p||_p,$$

where $c = 2^{-5}$, and $C = 2^5$.

Proof. We start with the right-hand side inequality showing that T_1 is bounded on H^1 provided that T_p is bounded on H^p. The idea is to apply the renorming of Theorem 2.3.5 to the range of the rearrangement operator. We fix $g \in H^1$, and $w \in L^2$. Let $g(I)$, $w(I)$ denote their respective Haar coefficients. Next form

$$u = \sum g(I)^{1-\theta} w(I)^\theta h_I.$$

Recall that $1 - \theta + \theta/2 = 1/p$ to observe that the following identity holds between the Haar coefficients of $T_1 g$, $T_2 w$ and $T_p u$,

$$(T_1 g)(J)^{1-\theta}(T_2 w)(J)^\theta = (T_p u)(J). \tag{2.3.23}$$

After these preparatory remarks we start estimating. First we express the norm of $T_1 g$ in H^1 utilizing Pisier's renorming. Let $c = 2^{-5}$, then by Theorem 2.3.5,

$$c||T_1 g||_{H^1}^{1-\theta} \leq \sup \left\{ \left\| \sum_{J \in \tau(\mathcal{D})} (T_1 g)(J)^{1-\theta} w(J)^\theta h_J \right\|_{H^p} : ||w||_2 \leq 1 \right\}.$$

Next we observe that on L^2 the operator T_2 is of norm 1, and invertible on its image. Hence the preceding supremum coincides with

$$\sup \left\{ \left\| \sum_{J \in \tau(\mathcal{D})} (T_1 g)(J)^{1-\theta}(T_2 w)(J)^\theta h_J \right\|_{H^p} : ||T_2 w||_2 \leq 1 \right\}.$$

Observe that we know the value of the coefficients, $(T_1 g)(J)^{1-\theta}(T_2 w)(J)^\theta$. (They were identified in (2.3.23) above.) Inserting we rewrite the above supremum as

$$\sup \left\{ \left\| T_p \left[\sum_{I \in \mathcal{D}} g(I)^{1-\theta} w(I)^\theta h_I \right] \right\|_{H^p} : ||w||_2 \leq 1 \right\}.$$

As T_p is bounded on H^p the last line admits the upper estimate

$$||T_p||_p \sup \left\{ \left\| \sum_{I \in \mathcal{D}} g(I)^{1-\theta} w(I)^{\theta} h_I \right\|_{H^p} : ||w||_2 \leq 1 \right\}.$$

By Theorem 2.3.5 the second factor on the right-hand side of the preceding inequality is bounded by $||g||_{H^1}^{1-\theta}$. Thus we showed that the following estimate holds,

$$||T_1||_1^{1-\theta} \leq 2^5 ||T_p||_p.$$

Next we prove the converse estimate. Let $u \in H^p$ with Haar expansion $u = \sum u(I) h_I$. By Theorem 2.3.4, there exist $x \in H^1$, and $y \in L^2$, such that

$$u(I) = x(I)^{1-\theta} y(I)^{\theta} \quad \text{and} \quad ||x||_{H^1}^{1-\theta} ||y||_2^{\theta} \leq 2^5 ||u||_{H^p}.$$

Note that the Haar coefficients of $T_p u$ satisfy

$$(T_p u)(J) = (T_1 x)(J)^{1-\theta} (T_2 y)(J)^{\theta}.$$

Hence we estimate using first Proposition 2.3.2, and then $||T_2||_2 = 1$.

$$\begin{aligned} ||T_p u||_{H^p} &\leq ||T_1 x||_{H^1}^{1-\theta} ||T_2 y||_2^{\theta} \\ &\leq ||T_1||_1^{1-\theta} ||x||_{H^1}^{1-\theta} ||y||_2^{\theta} \\ &\leq 2^5 ||T_1||_1^{1-\theta} ||u||_{H^p}. \end{aligned}$$

□

Dual estimates

Here we show the dual estimates to the renorming (2.3.2) and give a second proof of Theorem 2.3.5. Let $1 < p < 2$, $\theta = 2 - 2/p$ and $u \in H^p$ with $\int u = 0$. Then

$$c||u||_{H^p} \leq \sup \left\{ \int u \left(\sum v(I)^{1-\theta} w(I)^{\theta} h_I \right) dt \right\} \leq C ||u||_{H^p}, \tag{2.3.24}$$

where the supremum is taken over $v = \sum v(I) h_I$ and $w = \sum w(I) h_I$, satisfying $||v||_{\mathrm{BMO}} \leq 1$, and $||y||_2 \leq 1$. The constants $c > 0$ and $C > 0$ in (2.3.24) are independent of p.

The following theorem asserts that the left-hand side estimate in (2.3.24) holds true.

Theorem 2.3.7. *Let $1 \leq p \leq 2$ and $\theta = 2 - 2/p$. For $u \in H^p$ with $\int u = 0$, there exist $v \in BMO$, $w \in L^2$, with $||v||_{BMO} \leq 1$ and $||w||_2 \leq 1$, so that their Haar coefficients $v(I), w(I)$ give*

$$||u||_{H^p} \leq C \left| \int u \left(\sum v(I)^{1-\theta} w(I)^{\theta} h_I \right) dt \right|,$$

where $C > 0$ is independent of p.

The proof of Theorem 2.3.7 uses the stopping time argument of Theorem 2.3.3 applied to the sharp function in place of the square function. Assume first that $1 \leq p < 4/3$ and let q denote the Hölder conjugate index to p. In Section 1.2 we proved that for $u \in H^p$ there exists $f \in L^q$ so that

$$||u||_{H^p} \leq C\left|\int_0^1 uf dt\right| \quad \text{and} \quad \|f^\sharp\|_q \leq 1.$$

Next let $4/3 < p \leq 2$. By the square function characterization of L^p there exists $f \in L^q$ so that

$$||u||_{H^p} \leq C\left|\int_0^1 uf dt\right| \quad \text{and} \quad \|S(f)\|_q \leq 1.$$

In either case f admits a representation as

$$f = \sum v(I)^{1-\theta} w(I)^\theta h_I,$$

where $v = \sum v(I)h_I$ and $w = \sum w(I)h_I$ satisfy $||v||_{\text{BMO}} \leq C$ and $||w||_2 \leq C$. The reader is invited to supply the details by adapting the proof of Theorem 2.3.3.

The right-hand side estimate in (2.3.24) is a corollary to Theorem 2.3.4 and to Fefferman's inequality as stated in Theorem 2.3.1.

Corollary 2.3.8. *Let $1 < p < 2$, and $\theta = 2 - 2/p$. For $v \in BMO$, $w \in L^2$, and $u \in H^p$, the following estimate holds,*

$$\left|\int_0^1 u\left(\sum v(I)^{1-\theta} w(I)^\theta h_I\right)\right| \leq 2^6\sqrt{2}||u||_{H^p}||v||_{BMO}^{1-\theta}||w||_2^\theta.$$

Proof. By the decomposition Theorem 2.3.4, $u \in H^p$ can be represented as

$$u = \sum x(I)^{1-\theta} y(I)^\theta h_I,$$

where

$$||x||_{H^1}^{1-\theta}||y||_2^\theta \leq 2^5||u||_{H^p}.$$

Inserting the representation of u and applying Theorem 2.3.1 we obtain that

$$\left|\int_0^1 u\left(\sum v(I)^{1-\theta} w(I)^\theta h_I\right)\right| \leq \sqrt{2}2^6||u||_{H^p}||v||_{\text{BMO}}^{1-\theta}||w||_2^\theta,$$

as claimed. □

Summing up, the renorming of H^p (2.3.2) follows from Theorem 2.3.4 and Proposition 2.3.2. Its dual version (2.3.24) we obtain by combining Theorem 2.3.7 and Corollary 2.3.8. In the next sub-section we combine (2.3.2) and (2.3.24) to prove a basic boundedness principle for analytic families of operators.

The duality proof of Theorem 2.3.5. It is instructive to see how G. Pisier's original proof [176] — via its interpretation by M. Cwikel, P. Nilsson and G. Schechtman [57] — specializes to the context of H^1. To this end we give a second proof for the left-hand side of (2.3.16) using these sources.

Fix $g \in H^1$, $w \in L^2$ and consider

$$u = \sum g(I)^{1-\theta} w(I)^{\theta} h_I.$$

By Corollary 2.3.8 the following lower bound holds,

$$\begin{aligned} 2^6\sqrt{2}||u||_{H^p} &\geq \sup\left\{\int u\left(\sum z(I)^{1-\theta}y(I)^{\theta}h_I\right)dt : ||y||_2 \leq 1,\ ||z||_{\mathrm{BMO}} \leq 1\right\} \\ &= \sup\left\{\sum g(I)^{1-\theta}z(I)^{1-\theta}w(I)^{\theta}y(I)^{\theta}|I| : ||y||_2 \leq 1,\ ||z||_{\mathrm{BMO}} \leq 1\right\}. \end{aligned}$$

Inserting the preceding lower bound gives an expression displaying a high degree of symmetry,

$$\begin{aligned} &2^6\sqrt{2}\sup\left\{\left\|\sum g(I)^{1-\theta}w(I)^{\theta}h_I\right\|_{H^p} : ||w||_2 \leq 1\right\} \\ &\geq \sup\left\{\sum g(I)^{1-\theta}z(I)^{1-\theta}w(I)^{\theta}y(I)^{\theta}|I| : ||y||_2 \leq 1, ||w||_2 \leq 1,\ ||z||_{\mathrm{BMO}} \leq 1\right\}. \end{aligned}$$

Expressing the self duality of Hilbert spaces we write

$$\left\{\sum w(I)^{\theta}y(I)^{\theta}|I| : ||y||_2 \leq 1, ||w||_2 \leq 1\right\} = \left\{\sum \mu(I)^{\theta}|I| : \sum|\mu(I)||I| \leq 1\right\}.$$

Inserting this identity and thereby reintroducing a different form of asymmetry, we find that

$$\begin{aligned} &\sup\left\{\sum g(I)^{1-\theta}z(I)^{1-\theta}w(I)^{\theta}y(I)^{\theta}|I| : ||y||_2 \leq 1, ||w||_2 \leq 1,\ ||z||_{\mathrm{BMO}} \leq 1\right\} \\ &= \sup\left\{\sum g(I)^{1-\theta}z(I)^{1-\theta}\mu(I)^{\theta}|I| : \sum|\mu(I)||I| \leq 1,\ ||z||_{\mathrm{BMO}} \leq 1\right\}. \end{aligned}$$

Finally we invoke duality between the sequence spaces $\ell^{\frac{1}{\theta}}$ and $\ell^{\frac{1}{1-\theta}}$ in the following way. For any sequence of scalars $a(I)$,

$$\sup\left\{\sum a(I)\mu(I)^{\theta}|I| : \sum|\mu(I)||I| \leq 1\right\} = \left(\sum |a(I)|^{\frac{1}{1-\theta}}|I|\right)^{1-\theta}.$$

With $a(I) = g(I)^{1-\theta}z(I)^{1-\theta}$, the preceding identity gives

$$\begin{aligned} &\sup\left\{\sum g(I)^{1-\theta}z(I)^{1-\theta}\mu(I)^{\theta}|I| : \sum|\mu(I)||I| \leq 1,\ ||z||_{\mathrm{BMO}} \leq 1\right\} \\ &= \sup\left\{\left(\sum |g(I)z(I)||I|\right)^{1-\theta} : ||z||_{\mathrm{BMO}} \leq 1\right\} \\ &= \sup\left\{\int gz : ||z||_{\mathrm{BMO}} \leq 1\right\}^{1-\theta}. \end{aligned}$$

In summary we showed the following minorizations, thereby completing the proof of the theorem.

$$\begin{aligned}&\sup\left\{\left\|\sum g(I)^{1-\theta}w(I)^{\theta}h_I\right\|_{H^p} : \|w\|_2 \le 1\right\}\\ &\ge (2^6\sqrt{2})^{-1}\sup\left\{\int gz : \|z\|_{\mathrm{BMO}} \le 1\right\}^{1-\theta}\\ &\ge 2^{-8}\|g\|_{H^1}^{1-\theta}.\end{aligned}$$

□

Analytic families of operators

Now we open a second source of strong estimates. Following is the three lines theorem. This is a maximum principle for analytic functions in the vertical strip V. Its applications form the core results of complex interpolation theory.

Theorem 2.3.9. *Let $F(z)$ be analytic in $V = \{x + iy : x \in (0,1),\ y \in \mathbb{R}\}$. Assume that F extends continuously to the boundary of V, and that there exists a polynomial P such that*

$$\log|F(x+iy)| \le P(y),\ \ x + iy \in V.$$

If $\sup_t |F(it)| \le R_0$ and $\sup_t |F(1+it)| \le R_1$, then for every $\theta \in (0,1)$,

$$|F(\theta)| \le R_0^{1-\theta}R_1^{\theta}.$$

First we slightly change the appearance of Theorem 2.3.9. Thus the above maximum principle becomes directly applicable to show boundedness of operators. For $z \in V$ we let J_z be a linear operator defined a priorily on the set of finite linear combinations of Haar functions. For f, g finitely supported over the Haar system, we assume that the map

$$F(z) = \int J_z(g)f$$

is analytic in the vertical strip V, and extends continuously the boundary. Furthermore we assume that there exists a polynomial P, depending on the choice of f, g, such that

$$\log|F(x+iy)| \le P(y), \qquad \text{where} \qquad x + iy \in V.$$

On the boundary of V we have J_{it} and J_{1+it} implicitly defined as follows. We put

$$J_{it}(g,f) = \lim_{z\to it}\int J_z(g)f,$$

$$J_{1+it}(g,f) = \lim_{z\to 1+it}\int J_z(g)f,$$

thereby defining linear operators J_{it}, J_{1+it}, which act initially on finite linear combinations of Haar functions. Under these conditions we call $\{J_z : z \in V\}$ an analytic family of operators.

Throughout this sub-section we assume that for $t \in \mathbb{R}$, the operators J_{it}, and J_{1+it} have bounded extensions to H^1 and L^2 respectively, and that they satisfy the norm estimates

$$||J_{it}||_{H^1} \leq R_0 \quad \text{and} \quad ||J_{1+it}||_{L^2} \leq R_1. \tag{2.3.25}$$

Below we prove that for analytic families of operators the hypothesis (2.3.25) implies that for $p \in (1,2)$ and $\theta = 2 - 2/p$, the operator J_θ satisfies

$$||J_\theta g||_{H^p} \leq C R_0^{1-\theta} R_1^\theta ||g||_{H^p},$$

where $C > 0$ is independent of p. The first step in this direction is the following boundedness principle for analytic families of operators. Its power comes from the three lines theorem and the Fefferman inequality.

Theorem 2.3.10. *Let $x \in H^1$, $v \in BMO$, and let $w, y, \in L^2$. Assume that x, v, w, y, are finitely supported over the Haar system. For $\theta \in (0,1)$, form the functions*

$$g = \sum x(I)^{1-\theta} y(I)^\theta h_I, \quad \text{and} \quad f = \sum v(I)^{1-\theta} w(I)^\theta h_I.$$

Then, the operator J_θ satisfies the estimate

$$\left| \int_0^1 (J_\theta g)\, f \right| \leq \left(2\sqrt{2} R_0 ||x||_{H^1} ||v||_{BMO} \right)^{1-\theta} \left(R_1 ||y||_2 ||w||_2 \right)^\theta.$$

Proof. Let $x \in H^1$, $v \in$ BMO, and let $w, y, \in L^2$. For $z \in V$ we form the functions

$$g_z = \sum x(I)^{1-z} y(I)^z h_I, \text{ and } f_z = \sum v(I)^{1-z} w(I)^z h_I.$$

For $z \in V$ the map

$$F(z) = \int_0^1 (J_z g_z)\, f_z$$

is analytic in V with continuous boundary values. Since the functions x, y, v, w are finite linear combinations of Haar functions there exists a constant $C = C(x, y, v, w)$ such that for $z \in V$,

$$\log |F(z)| \leq C \log(4 + |z|).$$

On the boundary of V the following identities hold for the norm of f_z,

$$||f_{it}||_{\mathrm{BMO}} = ||v||_{\mathrm{BMO}}, \quad \text{and} \quad ||f_{1+it}||_2 = ||w||_2.$$

Moreover by assumption we have for g_z,

$$||J_{it}(g_{it})||_{H^1} \leq R_0 ||x||_{H^1}, \quad \text{and} \quad ||J_{1+it}(g_{1+it})||_2 \leq R_1 ||y||_2.$$

Fefferman's inequality implies that

$$\begin{aligned}|F(it)| &= \left|\int_0^1 (J_{it}g_{it})\, f_{it}\right| \\ &\leq 2\sqrt{2}R_0||x||_{H^1}||v||_{\mathrm{BMO}}.\end{aligned}$$

The inequality of Cauchy–Schwarz gives

$$\begin{aligned}|F(1+it)| &= \left|\int_0^1 (J_{1+it}g_{1+it})\, f_{1+it}\right| \\ &\leq R_1||w||_2||y||_2.\end{aligned}$$

Applying the three lines theorem we obtain that

$$|F(\theta)| \leq \left(2\sqrt{2}R_0||x||_{H^1}||v||_{\mathrm{BMO}}\right)^{1-\theta} (R_1||y||_2||w||_2)^{\theta}.$$

This completes the proof, since

$$\left|\int_0^1 (J_\theta g)\, f\right| = |F(\theta)|. \qquad \square$$

The next results are derived by combining the decomposition Theorems 2.3.4 and 2.3.7 with the estimate of Theorem 2.3.10. These tools complement each other very well, despite their origin in well separated branches of analysis. Following is the main step towards the proof of the boundedness theorem for analytic families of operators.

Corollary 2.3.11. *Let $1 \leq p \leq 2$, and $\theta = 2 - 2/p$. Let $x \in H^1$ and $y \in L^2$ with Haar expansion $x = \sum x(I)h_I$ and $y = \sum y(I)h_I$; then*

$$\left\|J_\theta\left(\sum x(I)^{1-\theta}y(I)^\theta h_I\right)\right\|_{H^p} \leq CR_0^{1-\theta}R_1^\theta||x||_{H^1}^{1-\theta}||y||_2^\theta,$$

where $C > 0$ is independent of p.

Proof. Fix $x \in H^1$, and let $y \in L^2$, be finitely supported over the Haar system. Let

$$u = J_\theta\left(\sum x(I)^{1-\theta}y(I)^\theta h_I\right). \tag{2.3.26}$$

By Theorem 2.3.7 there exist $v \in \mathrm{BMO}$ and $w \in L^2$, such that $||v||_{\mathrm{BMO}} \leq 1$, $||w||_2 \leq 1$, and

$$||u||_{H^p} \leq C\left|\int_0^1 u\left(\sum v(I)^{1-\theta}w(I)^\theta h_I\right)\right|.$$

Inserting the equation (2.3.26) for u and applying Theorem 2.3.10 gives

$$||u||_{H^p} \leq C2\sqrt{2}R_0^{1-\theta}R_1^\theta||x||_{H^1}^{1-\theta}||y||_2^\theta. \qquad \square$$

Recall that we are working with an analytic family of operators $\{J_z : z \in V\}$ such that J_{it}, and J_{1+it} have bounded extensions to H^1 and L^2 respectively, and

$$||J_{it}||_{H^1} \leq R_0 \quad\text{and}\quad ||J_{1+it}||_{H^1} \leq R_1.$$

The next corollary shows that the norms of analytic families satisfy a convexity estimate. In particular it shows that J_θ is bounded on H^p when $\theta = 2 - 2/p$. The proof below merges the three lines theorem and the decomposition Theorem 2.3.4 into one single estimate.

Corollary 2.3.12. *For $p \in (1,2)$ and $\theta = 2 - 2/p$, the operator J_θ has a bounded extension to H^p, satisfying the norm estimate*

$$||J_\theta||_{H^p} \leq C R_0^{1-\theta} R_1^\theta,$$

where $C > 0$ is independent of p.

Proof. Let $u \in H^p$ with $\int u = 0$. By Theorem 2.3.4 there exists $x \in H^1$ and $y \in L^2$, such that

$$u = \sum x(I)^{1-\theta} y(I)^\theta h_I,$$

and

$$||x||_{H^1}^{1-\theta} ||y||_2^\theta \leq 2^5 ||u||_{H^p}.$$

With Corollary 2.3.11 this gives

$$\begin{aligned} ||J_\theta u||_{H^p} &\leq C R_0^{1-\theta} R_1^\theta ||x||_{H^1}^{1-\theta} ||y||_2^\theta \\ &\leq C R_0^{1-\theta} R_1^\theta ||u||_{H^p}. \end{aligned}$$

□

Next we present a typical application of Corollary 2.3.12. We give a second proof that the norm of rearrangement operators satisfies a convexity estimate. Recall that

$$T_p : \frac{h_I}{|I|^{1/p}} \mapsto \frac{h_{\tau(I)}}{|\tau(I)|^{1/p}},$$

and that we denote $||T_p||_p = ||T_p : H^p \to H^p||$. Using Corollary 2.3.12 we prove that for $1 \leq p \leq 2$ and $\theta = 2 - 2/p$,

$$||T_p||_p \leq C ||T_1||_1^{1-\theta},$$

where $C = 2^6\sqrt{2}$. The plan is to embed the rearrangement operators $\{T_p : p \in (1,2)\}$ into an analytic family of operators as follows. For $z \in V$ define operators J_z by their action on single Haar functions,

$$J_z(h_I) = h_{\tau(I)} \left(\frac{|I|}{|\tau(I)|} \right)^{1-(z/2)}.$$

Observe that for $t \in \mathbb{R}$,

$$\Re(1 - \frac{(it)}{2}) = 1, \quad \text{and} \quad \Re(1 - \frac{(1+it)}{2}) = \frac{1}{2}.$$

Moreover as $\theta = 2 - 2/p$, we have

$$1 - \frac{\theta}{2} = \frac{1}{p}.$$

Consequently,

$$J_\theta = T_p,$$

and for every function f finitely supported over the Haar system we have

$$||J_{it}(f)||_{H^1} = ||T_1(f)||_{H^1}, \qquad \text{and} \qquad ||J_{1+it}(f)||_2 = ||T_2(f)||_2.$$

Applying Corollary 2.3.12, the maximum principle for analytic families of operators, gives

$$||J_\theta(f)||_{H^p} \leq C||T_1||_1^{1-\theta}||T_2||_2^\theta||f||_{H^p}.$$

As $||T_2||_2 = 1$ and $J_\theta = T_p$, the preceding inequality proves that $||T_p||_p \leq C||T_1||_1^{1-\theta}$.

The three lines theorem. We review quickly basic notions of potential theory necessary to understand the three lines theorem, in particular the concept of harmonic measure in its most elementary form. We let D be a simply connected domain, bounded by Jordan curves. (In the applications to follow the domain D is the vertical strip V bounded by two vertical lines.) Then for every $z \in D$ the Riemann mapping theorem provides an analytic and bijective map onto the unit disk $\varphi : D \to \mathbb{D}$ such that $\varphi(z) = 0$. For domains bounded by Jordan curves the Riemann map φ can be extended continuously to the boundary; for every $\zeta \in \partial D$ the following limit exists,

$$\varphi(\zeta) = \lim_{w \to \zeta} \varphi(w).$$

For any measurable set $E \subseteq \partial D$, we define the harmonic measure of E in D evaluated at z by the equation

$$\omega_D^z(E) = |\varphi(E)|.$$

Often we simply call $\omega_D^z(E)$ the harmonic measure of E in D. The dependence $z \mapsto \omega_D^z(E)$, defines a harmonic function in D. If, moreover, the domain D is bounded, then every harmonic function $u : D \to \mathbb{R}$ with continuous extension to the boundary can be obtained from its boundary values and harmonic measure of the domain D. Indeed, for $z \in D$,

$$u(z) = \int_{\partial D} u(w) d\omega_D^z(w).$$

In the case of analytic functions defined on bounded domains $f : D \to \mathbb{C}$, the logarithm $\log|f(z)|$ satisfies the following subharmonicity estimate: Assume that f extends continuously to the boundary. Then for $n \in \mathbb{N}$ and $z \in D$ we have

$$\max\{-n, \log|f(z)|\} \leq \int_{\partial D} \max\{-n, \log|f(w)|\} d\omega_D^z(w).$$

This is a very interesting result that incidentally distinguishes analytic functions from harmonic functions. It becomes really important when it is combined with estimates of harmonic measure which are derived from the geometry of the domain D and its boundary. We point out that in general subharmonicity estimates by their very nature imply maximum principles.

In the context of interpolation, we apply the above facts in very simple domains that are obtained by truncating the vertical strip at a certain height M. We fix $M > 0$ and define

$$\begin{aligned} V_M &= \{x + iy : x \in (0,1)\,,\ |y| < M\}, \\ O_M &= \{x + iy : x \in (0,1)\,,\ y = M\}, \\ U_M &= \{x + iy : x \in (0,1)\,,\ y = -M\}. \end{aligned}$$

The sets O_M and U_M denote the upper and lower part of the truncated vertical strip. When evaluated at $x \in (0,1)$ the harmonic measure in V_M puts very little mass on the sets $O_M \cup U_M$. It turns out that the harmonic measure of $O_M \cup U_M$ decreases exponentially as M increases; this is of crucial importance; not only for the proof of the maximum principle below but also for many other important results of potential theory. Specifically the following inequality holds,

$$\omega_{V_M}^x(O_M \cup U_M) \leq e^{-Ma}\,, \ x \in (0,1).$$

We will now show that under very mild assumptions the subharmonicity estimate for analytic functions can be extended to unbounded domains such as the vertical strip. Thus we obtain the generalization of the maximum principle that is called the three lines theorem.

Theorem 2.3.13. *Let $F(z)$ be analytic in $V = \{x + iy : x \in (0,1),\ y \in \mathbb{R}\}$. Assume that F extends continuously to the boundary of V, and that there exists a polynomial P such that*

$$\log|F(x + iy)| \leq P(y), \ x + iy \in V.$$

If $\sup_t |F(it)| \leq R_0$ and $\sup_t |F(1 + it)| \leq R_1$, then for every $\theta \in (0,1)$,

$$|F(\theta)| \leq R_0^{1-\theta} R_1^{\theta}.$$

Proof. Let $w \in \partial V$, then by hypothesis, $|F(w)| \leq \max\{R_0, R_1\}$. Hence

$$\int_{\partial V} \log^+ |F(w)| d\omega_V^z(w) < \infty. \tag{2.3.27}$$

It follows from (2.3.27) that in (2.3.28) below the integral on the left-hand side is well defined as the limit appearing on the right-hand side. We define

$$\int_{\partial V} \log|F(w)| d\omega_V^z(w) = \lim_{n\to\infty} \int_{\partial V} \max\{-n, \log|F(w)|\} d\omega_V^z(w). \tag{2.3.28}$$

The value of the limit may be finite or equal to $-\infty$. We shall show below that the assumption $\log|F(x+iy)| \le P(y)$, guarantees the following subharmonicity estimate. For $z \in V$,

$$\log|F(z)| \le \int_{\partial V} \log|F(w)| d\omega_V^z(w). \tag{2.3.29}$$

We postpone the proof of (2.3.29) and derive first one of its consequences: In $\mathbb{C}$, the boundary of V consists of the components, $\{it : t \in \mathbb{R}\}$ and $\{1+it : t \in \mathbb{R}\}$. For $\theta \in (0,1)$ we obtain with (2.3.29) that

$$\begin{aligned}
\log|F(\theta)| &\le \int_{\mathbb{R}} \log|F(it)| d\omega_V^\theta(it) + \int_{\mathbb{R}} \log|F(1+it)| d\omega_V^\theta(1+it) \\
&\le \log R_0 \int_{\mathbb{R}} d\omega_V^\theta(it) + \log R_1 \int_{\mathbb{R}} d\omega_V^\theta(1+it) \\
&= \log R_0 \int_{\mathbb{R}} d\omega_V^\theta(it) + \log R_1 \left(1 - \int_{\mathbb{R}} d\omega_V^\theta(it)\right).
\end{aligned}$$

We proceed by identifying the value of the integral $\int_{\mathbb{R}} d\omega_V^\theta(it)$. For $z \in V$, consider now $f(z) = \int_{\mathbb{R}} d\omega_V^z$. Note that f is harmonic in V. Moreover by the geometry of the vertical strip, the value of the function $f(x+iy)$ does not depend on $y \in \mathbb{R}$. Hence $f(x)$ is linear in $x \in (0,1)$. Moreover $f(0) = 1$, $f(1) = 0$, and $f(x) = 1 - x$. Summing up, we proved that

$$\int_{\mathbb{R}} d\omega_V^\theta(it) = 1 - \theta.$$

Inserting we obtain

$$\log|F(\theta)| \le (1-\theta)\log R_0 + \theta \log R_1.$$

Exponentiating gives that

$$|F(\theta)| \le R_0^{1-\theta} R_1^\theta.$$

Now we turn to the proof of the crucial subharmonicity estimate (2.3.29) for the unbounded domain V. With two approximation procedures we reduce the estimate (2.3.29) to the case of bounded subharmonic functions in bounded domains. By hypothesis,

$$\sup_{w \in O_M \cup U_M} \log|F(w)| \le \max\{P(M), P(-M)\}.$$

The upper bound of the harmonic measure of $O_M \cup U_M$ in V_M gives

$$\int_{O_M \cup U_M} \max\{-n, \log|F(w)|\} d\omega_{V_M}^{\theta}(w) \leq e^{-Ma} \max\{P(M), P(-M)\}. \quad (2.3.30)$$

Now we keep n fixed and let M tend to infinity. The estimate (2.3.30) implies that

$$\lim_{M\to\infty} \left| \int_{O_M \cup U_M} \max\{-n, \log|F(w)|\} d\omega_V^{\theta}(w) \right| = 0.$$

Consequently,

$$\lim_{M\to\infty} \int_{\partial V_M} \max\{-n, \log|F(w)|\} d\omega_{V_M}^{\theta}(w) = \int_{\partial V} \max\{-n, \log|F(w)|\} d\omega_V^{\theta}(w).$$

Invoking the subharmonicity estimate for the bounded domain V_M gives

$$\max\{-n, \log|F(z)|\} \leq \int_{\partial V_M} \max\{-n, \log|F(w)|\} d\omega_{V_M}^{z}(w),$$

and letting M tend to infinity gives

$$\max\{-n, \log|F(z)|\} \leq \int_{\partial V} \max\{-n, \log|F(w)|\} d\omega_V^{\theta}(w).$$

The assumed upper bound of $\log|F(w)|$ on ∂V allows us to take the limit $n \to \infty$, which gives that

$$\log|F(\theta)| \leq \int_{\partial V} \log|F(w)| d\omega_V^{\theta}(w). \qquad \square$$

2.4 Notes

The book by C. Goffman and G. Pedrick [76] is an introduction to basic functional analysis treating also some of the advanced topics like complemented subspaces, Schauder bases and the biorthogonal systems of Haar, Rademacher and Walsh. Our sources for concepts in Banach space theory are the books by J. Lindenstrauss and L. Tzafriri [130] and P. Wojtaszczyk [213]. The proof of Proposition 2.1.1 is taken from B. Maurey [144].

The book of A. M. Garsia [75] is the standard reference to martingale H^1 spaces. Theorem 2.1.2 is due to B. Maurey [144].

The importance of atomic H^1 spaces is established in the classic paper [56] by R. Coifman and G. Weiss. Recent developments, in particular the T(1) theorem for operators in spaces of homogeneous type, are presented by M. Christ [50]. Theorem 2.1.3 is from [155].

Three-valued martingale difference sequences are basic tools in the construction of complemented subspaces in $L^p (1 < p < \infty)$. See J. Bourgain, H. Rosenthal and G. Schechtman [29], J. Bourgain [16] and W. B. Johnson, B. Maurey, G. Schechtman and L. Tzafriri [93]. The proof of Proposition 2.1.4 uses the methods introduced by B. Maurey in [139] and in Section 4 of [144]. Theorem 2.1.5 asserting that martingale H^1 spaces generated by sequences of finite σ algebras have an unconditional basis is due to B. Maurey [140]. Theorem 2.1.6, providing the isomorphic classificaton of the spaces $X[\mathcal{E}]$, appears in [151].

In [158] the Rosenthal space and the independent sum of BMO_n are shown to be isomorphic to a complemented subspace of BMO. Proposition 2.1.7, Theorem 2.1.8 and Proposition 2.1.10 are from [158].

The space $\mathcal{S}^\infty$ appears as canonical endpoint to a class of spaces studied by D. Kleper and G. Schechtman in [117]. Theorem 2.1.11, showing that $\mathcal{S}^\infty$ is isomorphic to a complemented subspace of BMO, is proven in [157].

The decomposition method was introduced by A. Pełczinski in [169]. It is reproduced in many books, for instance in [213] and [130]. Significant finite dimensional refinements of the decomposition method were obtained by G. Bennett, L.E. Dor, V. Goodman, W.B. Johnson, and C.N. Newman [9], W.B. Johnson, B. Maurey, G. Schechtman and L. Tzafriri [93] and by W.B. Johnson and G. Schechtman [96]. The proof of Wojtaszczyk's Theorem 2.2.3 is taken from Chapter II.E of [213]. The representation of the sequence $\{H^1(\ell^2_n)\}$ as copies of uniformly complemented subspaces of H^1 appears in [154].

M. Schmuckenschläger informed me about the elementary proof of the Riesz convexity theorem by which we introduced the section on interpolation. He pointed out that it is a widely known folklore construction. Real-variables methods for the characterization of the complex interpolation spaces between H^1 and L^p $(1 < p < \infty)$ are due to C. Fefferman and E. M. Stein. In particular Corollary 2.3.12, the interpolation theorem for analytic families of operators appears in [64]. The stopping time argument in Theorem 2.3.3 is due to S. Janson and P. W. Jones [90]. The proofs for Theorem 2.3.1, Proposition 2.3.2, Theorem 2.3.4, Theorem 2.3.7 and Corollary 2.3.8 use ingredients extracted from Section 8 of [70] by M. Frazier and B. Jawerth. The survey paper [99] by P. W. Jones contains information about history, methods and results on interpolation problems, H^1 and H^∞. Theorem 2.3.4 shows that the Calderón product of the Banach lattices H^1 and H^2 coincides with H^p. The Calderón product is introduced by A. Calderón in [38]. See the monograph [119] by S. Krein, J. Petunin and E. M. Semenov for further information on the Calderón product.

The norm for H^1 appearing in Theorem 2.3.5 was invented by G. Pisier in [175] and [176]. In Chapter 3 of [57] by M. Cwikel, P. Nilsson and G. Schechtman, the norm devised by G. Pisier is shown to express a uniqueness result for the endpoints of complex interpolation scales of Banach lattices. Specializing the uniqueness result to the context of H^1, gives Theorem 2.3.5. The proof on page 154 for the left-hand side estimate in (2.3.16) emerged in conversations with V. Pillwein. The duality proof of Theorem 2.3.5 is obtained by specialization from G. Pisier [176] and M. Cwikel, P. Nilsson and G. Schechtman [57]. The content of Chapter 3 in [57] has circulated in preprint form for some time. Early references to

[57] and the uniqueness result for complex interpolation scales include the papers by N. Kalton [105] and [106]. See also the related work by P. Nilsson [164]. I don't know who is to be credited for the proof of Theorem 2.3.6 presented in the text. The original proof, however, of Theorem 2.3.6, is in [156] and will be discussed in Chapter 3 below.

For the three lines theorem, analytic families of operators and related results see the books by E. M. Stein and G. Weiss [195], C. Bennett and R. Sharpley [8] or [119] by S. Krein, J. Petunin and E. M. Semenov.

Chapter 3

Combinatorics of Colored Dyadic Intervals

In this chapter we establish the combinatorial connections between the Carleson packing condition and the boundedness of orthogonal projections in BMO. We introduce the Gamlen–Gaudet construction and prove the uniform approximation property for BMO in a special case. Rearrangement operators of the Haar basis that preserve BMO and L^p are characterized.

3.1 The Carleson packing condition

In the unit interval $[0, 1]$, the dyadic subintervals are those of the form

$$\left[\frac{k-1}{2^n}, \frac{k}{2^n}\right[,$$

where $n \in \mathbb{N}$ and $1 \leq k \leq 2^n$. We let $\mathcal{D}$ denote the collection of all dyadic intervals. Note that $\mathcal{D}$ is a nested family of sets. It satisfies the following property.

$$\text{If } I, J \in \mathcal{D} \text{ are not disjoint, then either } I \subseteq J \text{ or } J \subseteq I.$$

The maximal dyadic intervals of any collection $\mathcal{C} \subseteq \mathcal{D}$ are pairwise disjoint. To denote the maximal dyadic intervals of $\mathcal{C}$ we use the notation $G_0(\mathcal{C})$, and also $\max \mathcal{C}$. The collection $G_0(\mathcal{C})$, is called the founding generation of $\mathcal{C}$. Note that $G_0(\mathcal{C})$ covers the same set as $\mathcal{C}$. Later generations are defined inductively. Let $p \in \mathbb{N}$, and assume that $G_0(\mathcal{C}), \ldots, G_{p-1}(\mathcal{C})$ are already defined, then we put $\mathcal{H} = \mathcal{C} \setminus G_0(\mathcal{C}) \cup \cdots \cup G_{p-1}(\mathcal{C})$ and define

$$G_p(\mathcal{C}) = G_0(\mathcal{H}).$$

Following standard notation we let $J \cap \mathcal{C} = \{I \in \mathcal{C} : I \subseteq J\}$. A localized version of the above construction is obtained as follows. We let $J \in \mathcal{D}$ and for $p \in \mathbb{N}$ we denote

$$G_p(J, \mathcal{C}) = G_p(J \cap \mathcal{C}).$$

Note that for $J \in \mathcal{C}$ we have $G_0(J, \mathcal{C}) = \{J\}$, and $G_1(J, \mathcal{C})$ consists of the maximal dyadic intervals of $\{I \in \mathcal{C} : I \subset J\}$.

Throughout this book we denote by $\mathcal{C}^*$ the point-set which is covered by the collection $\mathcal{C}$, that is, we put $\mathcal{C}^* = \bigcup_{I \in \mathcal{C}} I$. Often we measure the size of $\mathcal{C}$ by considering $\limsup \mathcal{C}$, the subset of $[0, 1]$ which is covered by infinitely many intervals of $\mathcal{C}$. Note that $G_{n-1}(\mathcal{C})^*$ is the subset of $[0, 1[$ which is covered by at least n intervals in $\mathcal{C}$. Hence,

$$\limsup \mathcal{C} = \bigcap_{n=0}^{\infty} G_n(\mathcal{C})^*.$$

The next proposition provides a link between the generations $G_n(I, \mathcal{C})$, defined above, and the measure of the set $\limsup \mathcal{C}$.

Proposition 3.1.1. *If $\mathcal{C}$ is a collection of dyadic intervals such that $|\limsup \mathcal{C}| = 0$, then for any dyadic interval I there exists $N = N(I)$ such that*

$$|G_N(I, \mathcal{C})^*| \le \frac{1}{4}|I|.$$

Proof. Suppose that the proposition fails. Then there exists $I \in \mathcal{C}$ such that $|G_n(I, \mathcal{C})^*| > |I|/4$, for every $n \in \mathbb{N}$. Note that each point in $G_n(I, \mathcal{C})^*$ is contained in more than n intervals of $\mathcal{C}$. Moreover, observe that

$$G_{n+1}(I, \mathcal{C})^* \subseteq G_n(I, \mathcal{C})^*$$

for any $n \in \mathbb{N}$. This implies that

$$\left| \bigcap_{n \in \mathbb{N}} G_n(I, \mathcal{C})^* \right| \ge \frac{1}{4}|I|,$$

and hence $|\limsup \mathcal{C}| \ge |I|/4$. This contradicts the assumption that $\limsup \mathcal{C}$ has vanishing Lebesgue measure. □

The Carleson constant and generations of dyadic intervals

A quantitative and scale invariant measure for the size of $\mathcal{C}$ is the Carleson packing condition. Recall that a collection $\mathcal{C}$ of dyadic intervals satisfies the Carleson packing condition if, the following expression is finite,

$$[\![\mathcal{C}]\!] = \sup_{I \in \mathcal{C}} \frac{1}{|I|} \sum_{J \in I \cap \mathcal{C}} |J|.$$

The number $[\![\mathcal{C}]\!]$ is then called the Carleson constant of the collection $\mathcal{C}$. If there exists $M > 0$ such that $[\![\mathcal{C}]\!] \leq M$, then we say that $\mathcal{C}$ satisfies the M-Carleson condition.

Observe that $|\limsup \mathcal{C}| > 0$ implies that $[\![\mathcal{C}]\!] = \infty$, and conversely $[\![\mathcal{C}]\!] < \infty$ implies $|\limsup \mathcal{C}| = 0$. The next proposition is similar in spirit to the previous one. Note however, that the qualitative consideration in the proof of Proposition 3.1.1 is replaced by a simple estimate.

Proposition 3.1.2. *Let $\mathcal{C}$ be a collection of dyadic intervals which satisfies the M-Carleson condition. Let A be the largest integer which is less than $4M + 1$. Then for every $I \in \mathcal{C}$,*

$$|G_A(I, \mathcal{C})^*| \leq \frac{1}{4}|I|.$$

Proof. Suppose that the proposition fails. Then there exists $I \in \mathcal{C}$ such that $|G_A(I, \mathcal{C})^*| > |I|/4$. Clearly, the following inclusions hold between generations, $G_s(I, \mathcal{C})^* \supseteq G_A(I, \mathcal{C})^*$, for $s \leq A$. Consequently the following lower bound holds true for any $s \leq A$.

$$\sum_{J \in G_s(I, \mathcal{C})} |J| \geq |G_A(I, \mathcal{C})^*| > \frac{1}{4}|I|.$$

Summing over $s \leq A$ gives a lower bound for the Carleson constant of $\mathcal{C}$ as follows,

$$\begin{aligned} [\![\mathcal{C}]\!] &\geq \frac{1}{|I|} \sum_{s=0}^{A} \sum_{J \in G_s(I, \mathcal{C})} |J| \\ &\geq \frac{1}{|I|} \sum_{s=0}^{A} |G_{4M}(I, \mathcal{C})^*| \\ &> M. \end{aligned}$$

This contradicts our assumption that $\mathcal{C}$ satisfies the M-Carleson condition. □

We will now discuss two basic techniques to decompose collections of dyadic intervals. We begin by showing that the Carleson constant $[\![\mathcal{C}]\!]$ satisfies a property resembling the positive homogeneity of norms: Each collection $\mathcal{C}$ of dyadic intervals can be decomposed into $4[\![\mathcal{C}]\!] + 1$ subcollections $\mathcal{C}_i$ such that $[\![\mathcal{C}_i]\!] \leq 4/3$. We obtain this by considering arithmetic progressions of the generations of $\mathcal{C}$.

Lemma 3.1.3. *Suppose that $\mathcal{C}$ satisfies the M-Carleson condition. Let A be the largest integer less than $4M + 1$. Then $\mathcal{C}$ can be decomposed into $\mathcal{C}_0, \ldots, \mathcal{C}_{A-1}$ such that each $\mathcal{C}_i$ satisfies the $4/3$-Carleson condition.*

Proof. First we split $\mathcal{C}$ into its generations $G_0(\mathcal{C}), G_1(\mathcal{C}), \ldots, G_n(\mathcal{C}), \ldots$. Then for $k = 0, 1, \ldots, A - 1$ we put

$$\mathcal{C}_k = \bigcup_{j=0}^{\infty} G_{jA+k}(\mathcal{C}).$$

To obtain the claimed upper bound for the Carleson constant of $\mathcal{C}_k$, we choose a dyadic interval $I \in \mathcal{C}_k$. It follows from the previous Proposition 3.1.2 that

$$|G_A(I, \mathcal{C})^*| \leq \frac{|I|}{4}.$$

Note that the first generation of I in $\mathcal{C}_k$, coincides with generation A of I in $\mathcal{C}$, so

$$G_1(I, \mathcal{C}_k) = G_A(I, \mathcal{C}).$$

Applying this observation repeatedly gives a geometrically decreasing estimate for the point-set covered by the generations $G_j(I, \mathcal{C}_k)$. Indeed we obtain that

$$|G_j(I, \mathcal{C}_k)^*| \leq 4^{-j}|I|.$$

Summing over $j \geq 0$ we arrive at the upper estimate

$$\begin{aligned} \sum_{J \in \mathcal{C}_k \cap I} |J| &= |I| + \sum_{j \geq 1} |G_j(I, \mathcal{C}_k)^*| \\ &\leq |I| \left(1 + \sum_{j \geq 1} 4^{-j}\right). \end{aligned}$$

This gives $[\![\mathcal{C}_k]\!] \leq 4/3$. □

Next we discuss the so-called condensation lemma. This is a widely used result asserting that if the Carleson constant of a collection $\mathcal{C}$ is very large, then somewhere inside this collection the Carleson constant is as large as possible, thus condensation has occurred and it can be measured with the estimate of Lemma 3.1.4. Thereby we obtain a natural converse to Proposition 3.1.2.

Lemma 3.1.4. *Let $\epsilon > 0$, $n \in \mathbb{N}$, and let $\mathcal{C}$ be a collection of dyadic intervals such that*

$$\sup_{I \in \mathcal{C}} \frac{1}{|I|} \sum_{J \in I \cap \mathcal{C}} |J| > \frac{n}{\epsilon}.$$

Then there exists $K \in \mathcal{C}$ such that

$$|G_n(K, \mathcal{C})^*| > (1 - \epsilon)|K|.$$

We point out that the conclusion $|G_n(K, \mathcal{C})^*| > (1 - \epsilon)|K|$ implies, of course, that for each of the generations $1 \leq s \leq n$ the same measure estimate holds, namely

$$|G_s(K, \mathcal{C})^*| > (1 - \epsilon)|K|. \tag{3.1.1}$$

Thus the condensation lemma asserts that everywhere between the interval K and the n-th generation of K the density of the collection $\mathcal{C}$ is larger than $1 - \epsilon$.

Proof of Lemma 3.1.4. Let $\epsilon > 0$ and $n \in \mathbb{N}$ be given, then by assumption we have $[\![\mathcal{C}]\!] > n/\epsilon$. Now we suppose that the lemma fails, that is,

$$|G_n(J, \mathcal{C})^*| \leq (1 - \epsilon)|J|$$

for any $J \in \mathcal{C}$. Next we will use this condition to obtain an upper estimate for the Carleson constant of $\mathcal{C}$. We fix $I \in \mathcal{C}$. Under the assumption that $|G_n(J, \mathcal{C})^*| \leq (1 - \epsilon)|J|$, for $J \in \mathcal{C}$ we obtain by induction that

$$|G_{nm+k}(I, \mathcal{C})^*| \leq (1 - \epsilon)^m |I| \tag{3.1.2}$$

for any $m \in \mathbb{N}$ and any $0 \leq k \leq n - 1$. Summing the estimates (3.1.2) gives an upper bound for the Carleson constant of $\mathcal{C}$ that contradicts the hypothesis. Indeed we have

$$\begin{aligned} \sum_{J \in \mathcal{C} \cap I} |J| &= \sum_{k=0}^{n-1} \sum_{m=0}^{\infty} |G_{mn+k}(I, \mathcal{C})^*| \\ &\leq \sum_{k=0}^{n-1} \sum_{m=0}^{\infty} (1 - \epsilon)^m |I| \\ &= \frac{n}{\epsilon} |I|. \end{aligned}$$

As $I \in \mathcal{C}$ was arbitrary we have thus obtained the following upper bound for the Carleson constant of $\mathcal{C}$,

$$[\![\mathcal{C}]\!] \leq n/\epsilon.$$

This contradicts our assumption that $[\![\mathcal{C}]\!] > n/\epsilon$. □

The Proposition 3.1.2 and Lemma 3.1.3 are especially useful in combination with Proposition 3.1.5, below. Estimates similar to those of Proposition 3.1.5 yielded the atomic decomposition of H^1 (Theorem 1.2.4) and also the closely related decomposition theorems in Section 2.3.

Recall that for $f \in L^p$ the Haar support of f is the following collection of dyadic intervals $\{I : \langle f, h_I \rangle \neq 0\}$.

Proposition 3.1.5. *Let $1 \leq p \leq 2$, and let $\{f_i\}$ be a sequence in H^p. Let F_i denote the support of the square function $S(f_i)$. Assume that any dyadic interval J in the Haar support of f_i satisfies*

$$\sum_{j=i+1}^{\infty} |J \cap F_j| \leq \frac{|J|}{2}. \tag{3.1.3}$$

Then for a linear combination

$$h = \sum a_i f_i,$$

the following estimates hold true,

$$||S(h)||_p^p \leq \sum |a_i|^p ||S(f_i)||_p^p \leq 2||S(h)||_p^p. \tag{3.1.4}$$

Moreover if $\{f_i\}$ is a sequence in BMO, then

$$||h||_{BMO} \leq 16 \sup_i |a_i| ||f_i||_{BMO}. \tag{3.1.5}$$

Proof. We consider the H^p case first, and start proving the left-hand inequality. Note that assumption (3.1.3) implies that the sequence $\{f_i\}$ is disjointly supported over the Haar system. Hence

$$S(h) = (\sum a_i^2 S^2(f_i))^{1/2}.$$

As $1 \leq p \leq 2$, we apply Hölder's inequality to obtain

$$(\sum a_i^2 S^2(f_i))^{p/2} \leq \sum |a_i|^p S^p(f_i). \tag{3.1.6}$$

Integrating the pointwise inequality (3.1.6) gives

$$||S(h)||_p^p \leq \sum |a_i|^p ||S(f_i)||_p^p.$$

Thus we verified the left-hand side of (3.1.4).

Now we turn to the right-hand side estimate. Recall that we denote the support set of $S(f_i)$ by F_i. Next we let

$$D_i = F_i \setminus \bigcup_{j=i+1}^{\infty} F_j.$$

By definition the difference sets D_i are pairwise disjoint. Our hypothesis (3.1.3) gives the following lower bound $|I \cap D_i| \geq 1/2|I|$, for every interval I in the Haar support of f_i. Consequently,

$$\int S(f_i)^p \leq 2 \int S(f_i)^p 1_{D_i}.$$

Summing up we obtained the lower bound

$$\begin{aligned} \int S(h)^p &\geq \int (\sum |a_i|^2 S^2(f_i) 1_{D_i})^{p/2} \\ &= \sum |a_i|^p \int S(f_i)^p 1_{D_i} \\ &\geq \frac{1}{2} \sum |a_i|^p \int S(f_i)^p. \end{aligned}$$

The BMO case follows from the above estimate with $p = 1$, and the H^1 – BMO duality theorem. By Fefferman's inequality and Theorem 1.2.3 there exists $g \in H^1$ with

$$\|g\|_{H^1} \le 2\sqrt{2},$$

such that

$$\langle h, g \rangle = \|h\|_{\mathrm{BMO}}.$$

Let $\mathcal{F}_i$ be the Haar support of the function f_i. Using the collections $\mathcal{F}_i$ we split $g \in H^1$ as follows,

$$g_i = \sum_{I \in \mathcal{F}_i} \langle g, h_I \rangle \frac{h_I}{|I|}.$$

Then of course $\langle f_i, g \rangle = \langle f_i, g_i \rangle$. The sequence (g_i) satisfies the assumption (3.1.3). Hence (3.1.4) with $p = 1$ holds for the sequence (g_i). Thus we obtain that

$$\sum_i \|g_i\|_{H^1} \le 2\|g\|_{H^1} \le 4\sqrt{2}.$$

Summing up, we estimate

$$\begin{aligned} \|h\|_{\mathrm{BMO}} &= \langle h, g \rangle \\ &= \sum_i a_i \langle f_i, g_i \rangle \\ &\le 2\sqrt{2} \sup_i |a_i| \, \|f_i\|_{\mathrm{BMO}} \sum_i \|g_i\|_{H^1} \\ &\le 16 \sup_i |a_i| \, \|f_i\|_{\mathrm{BMO}}. \end{aligned}$$

Thus (3.1.5) is verified. □

Remarks. 1. One obtains (3.1.5) also directly, without using duality. (The direct proof sketched below yields also better constants than the duality argument.) The starting observation is that (3.1.3) implies that $\mathcal{F} = \bigcup_i \mathcal{F}_i$. satisfies the 2-Carleson condition. Having made this observation it is then straightforward to estimate $\|h\|_{\mathrm{BMO}}$ so that (3.1.5) holds true.

2. Assume that the hypothesis (3.1.3) is replaced by the weaker condition

$$\sum_{j=i+2}^{\infty} |J \cap F_j| \le \frac{|J|}{2}. \tag{3.1.7}$$

Then by considering separately the sequences $\{a_{2i} f_{2i}\}$ and $\{a_{2i+1} f_{2i+1}\}$ we obtain from (3.1.4) that

$$\sum |a_i|^p ||S(f_i)||_p^p \le 4||S(h)||_p^p. \tag{3.1.8}$$

Large Carleson constants and the Gamlen–Gaudet construction

Following is an application of the condensation Lemma 3.1.4. It is selected to illustrate the relationship between large Carleson constants, the Gamlen–Gaudet construction and Jones's compatibility condition (J). Let $\mathcal{C}_1, \dots, \mathcal{C}_N$ be disjoint collections of dyadic intervals. Let C_i denote the point-set covered by the collection $\mathcal{C}_i$, that is,

$$C_i = \bigcup_{I \in \mathcal{C}_i} I.$$

Recall from Section 1.5 that disjoint collections of dyadic intervals $\mathcal{C}_1, \dots, \mathcal{C}_N$ are said to satisfy Jones' compatibility condition (J) if the following conditions (J 1)–(J 4) are satisfied.

(J 1) Each $\mathcal{C}_i$ is a collection of pairwise disjoint dyadic intervals.

(J 2) There exists a constant B, so that for each $J \in \mathcal{C}_j$ and each $i \geq j$ the following estimate holds,

$$\sum_{I \in \mathcal{C}_i, I \subseteq J} \frac{|I|}{|J|} \leq B \frac{|C_j \cap C_i|}{|C_j|}.$$

(J 3) The collections $\{\mathcal{C}_i : 1 \leq i \leq N\}$ are ordered in the following sense: If there exists $I \in \mathcal{C}_i$ and $J \in \mathcal{C}_j$, such that $I \subseteq J$, then $i \geq j$. Conversely, if $i \geq j$, if $I \in \mathcal{C}_i$ and $J \in \mathcal{C}_j$, such that $I \cap J \neq \emptyset$, then $I \subseteq J$.

(J 4) If $j \leq i$ and if $C_j \cap C_i \neq \emptyset$, then $C_i \subseteq C_j$. In particular the sets $\{C_i : 1 \leq i \leq N\}$ are nested; if $C_j \cap C_i \neq \emptyset$, then either $C_i \subseteq C_j$ or $C_j \subseteq C_i$.

Fix a collection of dyadic intervals $\mathcal{K}$ and assume that the Carleson constant of $\mathcal{K}$ is bounded below by a large constant. Our aim is to find functions $\{k_I : |I| \geq 2^{-n}\}$ for which the Haar support is contained in $\mathcal{K}$ and which behave like the first 2^{n+1} Haar functions. This can be most easily achieved through the construction of Gamlen and Gaudet which we present now in its simplest form.

Proposition 3.1.6. *Let $n \in \mathbb{N}$. Let $\mathcal{K}$ be a collection of dyadic intervals such that*

$$\sup_{I \in \mathcal{K}} \sum_{J \in \mathcal{K}, J \subseteq I} \frac{|J|}{|I|} \geq n4^n.$$

Then there exists $K \in \mathcal{K}$ and a family of pairwise disjoint collections of dyadic intervals $\{\mathcal{C}_I : |I| \geq 2^{-n}\}$ so that the following properties hold.

1. *Each of the collections $\mathcal{C}_I$, is contained in $\{J \in \mathcal{K} : J \subseteq K\}$.*
2. *The family $\{\mathcal{C}_I : |I| \geq 2^{-n}\}$ satisfies the compatibility condition (J).*
3. *Let $k_I = \sum_{J \in \mathcal{C}_I} h_J$. Then relative to the interval K the joint distribution of the functions $\{S^2(k_I) : |I| \geq 2^{-n}\}$ is equivalent to the joint distribution of the Haar functions $\{1_I : |I| \geq 2^{-n}\}$. The constant of equivalence is bounded by 2.*

4. *In BMO,* $\{k_I : |I| \ge 2^{-n}\}$ *is equivalent to the Haar functions* $\{h_I : |I| \ge 2^{-n}\}$ *so that for any choice of scalars* $\{x_I : |I| \ge 2^{-n}\}$,

$$\frac{1}{2}\|\sum x_I h_I\|_{BMO} \le \|\sum x_I k_I\|_{BMO} \le 2\|\sum x_I h_I\|_{BMO}.$$

Proof. Let $\mathcal{K}$ be a collection of dyadic intervals such that $[\![\mathcal{K}]\!] > n4^n$. Applying the condensation Lemma 3.1.4 we obtain $K \in \mathcal{K}$ such that

$$|G_n^*(K,\mathcal{K})| > (1-4^{-n})|K|. \tag{3.1.9}$$

Note that the estimate (3.1.9) implies that the generations prior to $G_n^*(K,\mathcal{K})$ are covering a set of measure $> (1-4^{-n})|K|$. We have

$$|G_k^*(K,\mathcal{K})| > (1-4^{-n})|K|, \qquad \text{for } k \le n\ . \tag{3.1.10}$$

Now we describe the inductive construction of the families $\{\mathcal{C}_I : |I| \ge 2^{-n}\}$. We start by putting

$$\mathcal{C}_{[0,1)} = \{K\}.$$

Let $I_1 = [0, \frac{1}{2})$, and $I_2 = [\frac{1}{2}, 1)$. In the next step we define $\mathcal{C}_{I_1}$ and $\mathcal{C}_{I_2}$. Recall that $\mathcal{C}_{[0,1)}$ contains only one interval, namely K. Let K_1 be the left half of K and K_2 be the right half of K. Then define

$$\mathcal{C}_{I_1} = \{L \in G_1(K,\mathcal{K}) : L \subseteq K_1\} \quad \text{and} \quad \mathcal{C}_{I_2} = \{L \in G_1(K,\mathcal{K}) : L \subseteq K_2\}.$$

Let $k \le n$. Assume that the collections $\mathcal{C}_I$ have been constructed for all dyadic intervals of length $\ge 2^{-k+1}$. Then we fix a dyadic interval I of length $|I| = 2^{-k+1}$. Let I_1 be the left half of I, and let I_2 be the right half of I. Next pick any interval $J \in \mathcal{C}_I$. Splitting J in the same way we let J_1 be the left half of J, and J_2 be the right half of J. Then we form the auxiliary collections,

$$\mathcal{C}_{I_1}(J) = \{L \in G_k(K,\mathcal{K}) : L \subseteq J_1\} \quad \text{and} \quad \mathcal{C}_{I_2}(J) = \{L \in G_k(K,\mathcal{K}) : L \subseteq J_2\}.$$

Completing the induction step we take the union over all intervals $J \in \mathcal{C}_I$ thus forming,

$$\mathcal{C}_{I_1} = \bigcup_{J\in\mathcal{C}_I} \mathcal{C}_{I_1}(J) \quad \text{and} \quad \mathcal{C}_{I_2} = \bigcup_{J\in\mathcal{C}_I} \mathcal{C}_{I_2}(J).$$

Having completed the Gamlen–Gaudet construction of the collections $\{\mathcal{C}_I : |I| \ge 2^{-n}\}$ we analyze their properties. We let $C_I = \bigcup_{L\in\mathcal{C}_I} L$. Thus $\{C_I : |I| \ge 2^{-n}\}$ is a nested collection of sets satisfying the same intersection pattern as the collection of dyadic intervals. That is,

$$C_I \subseteq C_J \quad \text{iff} \quad I \subseteq J.$$

Observe that by (3.1.10) the following condition holds true. If $L \in \mathcal{C}_J$ and $I \subseteq J$, then

$$\frac{|I|}{2|J|} \leq \frac{|L \cap C_I|}{|L|} \leq \frac{2|I|}{|J|}. \tag{3.1.11}$$

By (3.1.11) the collection $\{\mathcal{C}_I : |I| \geq 2^{-n}\}$ satisfies Jones's compatiblity condition (J).

Next we analyze the block bases $k_I = \sum_{L \in \mathcal{C}_I} h_L$ and the joint distribution of $\{S^2(k_I)\}$. Given coefficients $\{a_I : |I| \geq 2^{-n}\}$ we form

$$g = \sum a_I h_I \quad \text{and} \quad f = \sum a_I k_I.$$

Since $\{\mathcal{C}_I : |I| \geq 2^{-n}\}\}$ are disjoint collections consisting of pairwise disjoint dyadic intervals we obtain $S(k_I) = 1_{C_I}$ and

$$S(f) = \left(\sum a_I^2 1_{C_I}\right)^{1/2}.$$

Hence it follows from (3.1.11) that for $t > 0$,

$$\frac{1}{2}|\{S(g) > t\}| \leq \frac{|\{S(f) > t\}|}{|K|} \leq 2|\{S(g) > t\}|. \tag{3.1.12}$$

The BMO part follows by restriction and re-normalization from (3.1.12). □

Consequences. Next we draw some consequences of proposition 3.1.6. Using the functions $\{k_I : |I| \geq 2^{-n}\}$ obtained by the Gamlen–Gaudet construction we define the orthogonal projection

$$P(f) = \sum_{\{I : |I| \geq 2^{-n}\}} \langle f, \frac{k_I}{\|k_I\|_2} \rangle \frac{k_I}{\|k_I\|_2}.$$

By Theorem 1.5.9 the projection P is a bounded operator on BMO, with norm estimates depending only on the fact that the collections $\{\mathcal{C}_I : |I| \geq 2^{-n}\}$ satisfy the compatibility condition (J). Hence $\|P\|_{\mathrm{BMO}} \leq C$, where C is independent of n. The assertions of Proposition 3.1.6 show that the linear extension of the map

$$E : h_I \to k_I, \quad \text{where} \quad |I| \geq 2^{-n},$$

defines an isomorphism from BMO_n onto its range, so that

$$\|E\| \cdot \|E^{-1}\| \leq 4.$$

Thus the range of P in BMO is well isomorphic to BMO_n.

The H^1 – BMO duality implies the analogous boundedness of P in H^1, and $\|P\|_{H^1} \leq C$. Moreover also in H^1 the map

$$E : h_I \to k_I, \quad \text{where} \quad |I| \geq 2^{-n},$$

defines an isomorphism from H^1_n onto its range.

Small Carleson constants and Jones's compatibility condition

Next we give an application of Proposition 3.1.2. Recall the following convenient notation. For a collection of pairwise disjoint dyadic intervals $\mathcal{C}$ we denote

$$Q(\mathcal{C}) = \bigcup_{K \in \mathcal{C}} \{J \in \mathcal{D} : J \subseteq K\}.$$

Thus $Q(\mathcal{C})$ is the collection of intervals which are contained in one of the intervals of $\mathcal{C}$. For a single dyadic interval I we write

$$Q(I) = \{J \in \mathcal{D} : J \subseteq I\}.$$

Next recall the notion of a block of intervals which was introduced in Chapter 1 following the atomic decomposition of H^1. Given a collection $\mathcal{K}$ of dyadic intervals, and a subcollection $\mathcal{L} \subseteq \mathcal{K}$. We say that $\mathcal{L}$ is a block of intervals in $\mathcal{K}$ if the following conditions are satisfied.

1. $\mathcal{L}$ contains exactly one maximal interval, say I_0.
2. For every $J \in \mathcal{L}$ and $K \in \mathcal{K}$ the following implication holds true.

$$\text{If } J \subseteq K \subseteq I_0, \text{ then } K \in \mathcal{L}.$$

Let $\mathcal{L} \subseteq \mathcal{K}$ be a block of intervals in $\mathcal{K}$. Fix now one interval, say $I \in \mathcal{L}$. Then the collection $\mathcal{L} \setminus \{I\}$ need not be a block of intervals. However if we remove $Q(I)$ from the block $\mathcal{L}$, then the result is again a block. Expressed formally, we have the following implication.

$$\text{If } \mathcal{L} \text{ is a block in } \mathcal{K}, \text{ then also } \mathcal{L} \setminus Q(I) \text{ is a block in } \mathcal{K}. \tag{3.1.13}$$

We continue by giving an illustration of Proposition 3.1.2. Let $\mathcal{K}$ be a collection of dyadic intervals $\mathcal{K}$ which satisfies the Carleson packing condition. Assume that $\mathcal{K}$ has just one maximal interval, call it I_0. Suppose we wish to decompose $\mathcal{K}$ into families $\mathcal{C}_0, \ldots, \mathcal{C}_M$ satisfying Jones's compatibility condition (J). Can one do that with a good upper bound for M? Not in general, but the following is true:

Proposition 3.1.7. *Let $\mathcal{K}$ be a collection of dyadic intervals $\mathcal{K}$, such that* $\max\{\mathcal{K}\} = \{I_0\}$. *Assume that $\mathcal{K}$ satisfies the Carleson packing condition,*

$$[\![\mathcal{K}]\!] = \sup_{I \in \mathcal{K}} \sum_{J \in \mathcal{K}, J \subseteq I} \frac{|J|}{|I|} < \infty.$$

Then there exists $A \leq 4[\![\mathcal{K}]\!]$, so that $\mathcal{K}$ contains disjoint subcollections $\mathcal{C}_0, \ldots, \mathcal{C}_A$ satisfying Jones's compatibility condition, and so that the following holds.

1. *The union*

$$\mathcal{L} = \mathcal{C}_0 \cup \cdots \cup \mathcal{C}_A$$

is a block of intervals such that

$$|\{\mathcal{K} \setminus \mathcal{L}\}^*| \leq \frac{1}{4}|I_0|.$$

2. *Each of the* $\mathcal{C}_i$ *satisfies*

$$\sum_{I \in \mathcal{C}_i} |I| \geq \frac{1}{4}|I_0|.$$

Proof. We construct $\mathcal{L}$ by a stopping time argument, where the size of the generations $G_m(I_0, \mathcal{K})$ determines the stopping rule. Denote $\mathcal{C}_m = G_m(I_0, \mathcal{K})$ and let C_m be the point-set covered by the collection $\mathcal{C}_m$. Define $A \in \mathbb{N}$ by the following rule: Let A be the largest integer for which

$$|C_A| > \frac{1}{4}|I_0|. \tag{3.1.14}$$

Proposition 3.1.2 implies the following upper bound holds for A,

$$A \leq 4[\![\mathcal{K}]\!]. \tag{3.1.15}$$

Moreover the stopping time definition of A implies that

$$|C_{A+1}| \leq \frac{1}{4}|I_0|, \tag{3.1.16}$$

since otherwise we would have chosen $A+1$ and not A. Define now

$$\mathcal{L} = \bigcup_{j=0}^{A} \mathcal{C}_j.$$

It is easy to see that $\mathcal{L}$ is a block of dyadic intervals. By (3.1.14) the linearly ordered collections $\{\mathcal{C}_0, \ldots, \mathcal{C}_A\}$ satisfy Jones's compatibility condition. By (3.1.16) the block $\mathcal{L}$ is large in the following sense,

$$|\{\mathcal{K} \setminus \mathcal{L}\}^*| \leq \frac{1}{4}|I_0|, \tag{3.1.17}$$

and by (3.1.15) we needed less than $4[\![\mathcal{K}]\!]$ generations to achieve (3.1.17). In that sense $\{\mathcal{C}_0, \ldots, \mathcal{C}_A\}$, is an efficient decomposition of $\mathcal{L}$. □

3.2 Orthogonal projections and colored intervals

A Banach space X satisfies the uniform approximation property if there exists $\lambda > 0$ so that for every $n \in \mathbb{N}$ there exists $f(n) \in \mathbb{N}$ so that for every $x_1, \dots, x_n \in X$ there exists a linear operator $T : X \to X$ satisfying

$$\begin{aligned} ||T|| &\leq \lambda, \\ Tx_i &= x_i, \\ \operatorname{rank} T &\leq f(n). \end{aligned}$$

In Chapter 5 we prove the deep result by P. W. Jones that BMO satisfies the uniform bounded approximation property. As a preparation we treat in Theorem 3.2.3 the special case when

$$x_i = \sum \{h_I \, : \, I \in \mathcal{B}_i\}.$$

To be able to do this we need to establish the combinatorial connection between the Carleson constant of collections of dyadic intervals and Jones's compatibility condition (J). The chief difficulties are addressed in Theorem 3.2.1.

We view Theorem 3.2.1 as a multicolored version of Proposition 3.1.7. Formally, colors are attached to intervals in a collection $\mathcal{K}$ as follows. Let $L \in \mathbb{N}$. Let $\mathcal{B}_1, \dots, \mathcal{B}_L$ be disjoint collections of dyadic intervals such that

$$\mathcal{K} = \mathcal{B}_1 \cup \cdots \cup \mathcal{B}_L. \tag{3.2.1}$$

Then the color of $J \in \mathcal{K}$ is defined to be $j \in \{1, \dots, L\}$ if $J \in \mathcal{B}_j$. Accordingly, on $\mathcal{K}$ we define the function

$$\text{color} : \mathcal{K} \to \{1, \dots, L\} \tag{3.2.2}$$

by the relation

$$\text{color}(J) = j \quad \text{iff} \quad J \in \mathcal{B}_j. \tag{3.2.3}$$

Let $\mathcal{C} \subseteq \mathcal{K}$. We say that $\mathcal{C}$ is monochromatic if

$$\text{color}(J) = \text{color}(I) \quad \text{for all} \quad I, J \in \mathcal{C}. \tag{3.2.4}$$

Jones's compatibility condition and colored intervals

Next we obtain a decomposition of a colored collection into monochromatic subcollections satisfying Jones's compatibility condition. The following theorem plays an important role in proving a special case of the uniform approximation property for BMO (Theorem 3.2.3 below). It also constitutes a crucial component in the proof of the general case.

Theorem 3.2.1. *Let $L \in \mathbb{N}$. Let $\mathcal{K}$ be a collection of dyadic intervals with one maximal interval, I_0; each $I \in \mathcal{K}$ has attached to it one of L different colors. Assume that $\mathcal{K}$ satisfies the Carleson packing condition,*

$$\sup_{I\in\mathcal{K}} \sum_{J\in\mathcal{K},J\subseteq I} \frac{|J|}{|I|} \leq K.$$

Then $\mathcal{K}$ contains disjoint subcollections $\mathcal{C}_1, \cdots, \mathcal{C}_M$ satisfying Jones's compatibility condition and also the following properties:

1. *The union*
$$\mathcal{L} = \mathcal{C}_1 \cup \cdots \cup \mathcal{C}_M$$
is a block of intervals such that
$$|\{\mathcal{K} \setminus \mathcal{L}\}^*| \leq \frac{1}{2}|I_0|.$$
2. *Each of the collections $\mathcal{C}_i$ satisfies*
$$\sum_{I\in\mathcal{C}_i} |I| \geq \delta|I_0|.$$
3. *Each of the collections $\mathcal{C}_i$ is monochromatic (with respect to the colors attached to the intervals of $\mathcal{K}$).*
4. *There exist upper and lower bounds $M \leq M(K, L)$, and $\delta \geq \delta(K, L)$.*

Proof. The proof is divided into four separate parts. The first part contains a summary and a description of the main steps in the argument.

Part 1. Preparation and overview. For the generations of $\mathcal{K}$ we use the abbreviation $G_m = G_m(I_0, \mathcal{K})$. Define $s \in \mathbb{N}$ to be the largest integer for which

$$|G_s^*| > \frac{1}{4}|I_0|.$$

Since $\mathcal{K}$ satisfies the Carleson packing condition with constant $\leq K$, Proposition 3.1.2 implies that the following upper bound holds for s,

$$s \leq 4K. \tag{3.2.5}$$

In the course of the proof we will define the block $\mathcal{L}$ to be a subcollection of $\bigcup_{m=0}^{s} G_m(I_0, \mathcal{K})$. The construction starts by considering *first* the intervals in G_s and later the intervals in G_{s-1} until the generation G_1 is reached.

Recall that we have attached one of L colors to the intervals in $\mathcal{K}$. Put $L_s = L$. For $k \leq L_s$ define the following subcollection of G_s,

$$\mathcal{O}_{(s,k)} = \{I \in G_s : \text{color}(I) = k\}.$$

Now let $1 \le m < s$. In parts 2 and 3 of the proof we find $L_m \in \mathbb{N}$, and monochromatic, disjoint subcollections of G_m

$$\mathcal{O}_{(m,n)} \quad \text{where} \quad n \le L_m,$$

so that the following conditions hold.

1. There exists $L_0 = L_0([\![\mathcal{K}]\!], L)$, depending only on the Carleson constant $[\![\mathcal{K}]\!]$ and the number of colors L, so that
$$L_m \le L_0.$$

2. The union
$$\mathcal{L} = \{I_0\} \cup \bigcup_{m=1}^{s} \bigcup_{n=1}^{L_m} \mathcal{O}_{(m,n)}$$
forms a block of dyadic intervals in $\mathcal{K}$ so that
$$|\{\mathcal{K} \setminus \mathcal{L}\}^*| \le \frac{1}{2}|I_0|.$$

3. Each of the collections $\mathcal{O}_{(m,n)}, n \le L_m$ is monochromatic with respect to the colors in $\mathcal{K}$.

4. For $1 \le m < s$ the collections $\{\mathcal{O}_{(m,n)} : n \le L_m\}$ and $\{\mathcal{O}_{(m+1,k)} : k \le L_{m+1}\}$ are related by the following (extremely restrictive) homogeneity condition: For $n \le L_m$ and $k \le L_{m+1}$ there exists $A = A(m,n,k)$ so that for each $I \in \mathcal{O}_{(m,n)}$,
$$\left(2^{-\frac{1}{s}}\right) A \le \sum_{J \in I \cap \mathcal{O}_{(m+1,k)}} \frac{|J|}{|I|} \le \left(2^{\frac{1}{s}}\right) A. \tag{3.2.6}$$

The hard part of the proof consists of constructing the block $\mathcal{L}$ and the collections $\mathcal{O}_{(m,n)}$ where $n \le L_m$ and $1 \le m \le s$. This is done in parts 2 and 3 below. In part 4 we further split the monochromatic collections $\{\mathcal{O}_{(m,n)} : n \le L_m\}$, into $\{\mathcal{F}_{(m,k)} : k \le N_m\}$, in such a way that $\{\mathcal{F}^*_{(m,k)} : k \le N_m, m \le s\}$, is a nested collection of sets. (This is done by a straightforward procedure.) After that we relabel $\{\mathcal{F}_{(m,k)}\}$, as $\{\mathcal{C}_j\}$ transforming lexicographic order into linear order. We will easily see that $\{\mathcal{C}_j\}$ satisfies the compatibility conditions $(J\,1), (J\,3)$, and $(J\,4)$. Exploiting (3.2.6) we will verify that also $(J\,2)$ holds true.

Part 2. The construction of the block $\mathcal{L}$. Step s. Recall that L is the number of colors in $\mathcal{K}$. Put $L_s = L$. At step s we are concerned with the intervals in G_s. Each interval $I \in G_s$ has attached to it one of L_s colors. Then for $k \le L_s$ we define

$$\mathcal{O}_{(s,k)} = \{I \in G_s : \text{ color}(I) = k\}.$$

The combinatorial idea of the proof (as described below) is encoded in the following construction which successively selects in G_m ($m \le s$) certain subcollections $\mathcal{O}_{(m,n)}$. The first appearance of $\mathcal{O}_{(m,n)}$ is at step m of the construction. At each of the following steps (that is at the steps $m-1, m-2, \ldots, 1$) it may become necessary to modify the already existing collection $\mathcal{O}_{(m,n)}$, by deleting some of its intervals. The following rule is applied throughout: If at some point we decide to discard an interval I from the collection $\mathcal{O}_{(m,n)}$, then simultaneously we delete from $\mathcal{O}_{(m+j,n')}$ the collections

$$I \cap \mathcal{O}_{(m+j,n')}, \quad \text{for} \quad n' \le L_{m+j} \quad \text{and} \quad 1 \le j \le s-m.$$

This rule gives that the collection $\mathcal{L}$ defined by (3.2.15) is a block of intervals. We will make sure that the cumulative effect of updating $\mathcal{O}_{(m,n)}$ remains small compared to $|I_0|$.

<u>Step $s-1$.</u> At step $s-1$ we define $L_{s-1} \in \mathbb{N}$ and subcollections of G_{s-1} denoted $\mathcal{O}_{(s-1,n)}$, where $n \le L_{s-1}$. We start by modifying the already existing collections $\{\mathcal{O}_{(s,k)} : k \le L_s\}$. Fix $I \in G_{s-1}$. For each $k \le L_s$ we now determine the density of $\mathcal{O}_{(s,k)}$ in I. Let $O_{(s,k)}$ be the point-set covered by $\mathcal{O}_{(s,k)}$. If

$$\frac{|I \cap O_{(s,k)}|}{|I|} \le \frac{1}{8sL_s}, \tag{3.2.7}$$

then we define the collection $I \cap \mathcal{E}_{(s,k)}$ by putting

$$I \cap \mathcal{E}_{(s,k)} := I \cap \mathcal{O}_{(s,k)}.$$

Applying this rule to each of the collections $\{\mathcal{O}_{(s,k)}, k \le L_s\}$ defines

$$I \cap \mathcal{E}_s = \bigcup_{k=1}^{L_s} I \cap \mathcal{E}_{(s,k)}.$$

Having created the collections $I \cap \mathcal{E}_s$ for each $I \in G_{s-1}$, we take the union and form

$$\mathcal{E}_s = \bigcup_{I \in G_{s-1}} I \cap \mathcal{E}_s.$$

Next we modify $\mathcal{O}_{(s,k)}$ by renaming

$$\mathcal{O}_{(s,k)} := \mathcal{O}_{(s,k)} \setminus \mathcal{E}_s. \tag{3.2.8}$$

Let $I \in G_{s-1}$, and let $k \le L_s$. <u>After</u> the modifications defined by (3.2.8) the following implication holds.

$$\text{If} \quad I \cap O_{(s,k)} \ne \emptyset, \quad \text{then} \quad \frac{|I \cap O_{(s,k)}|}{|I|} > \frac{1}{8sL_s}.$$

Consequently for each $k \leq L_s$ and each $I \in G_{s-1}$, there exists an integer $n(k)$ with

$$n(k) \in \{1, \ldots, s \log_2(8sL_s)\} \cup \{\infty\}$$

and such that

$$2^{-\frac{n(k)+1}{s}} \leq \frac{|I \cap O_{(s,k)}|}{|I|} \leq 2^{-\frac{n(k)}{s}}.$$

In summary, for a fixed $I \in G_{s-1}$ and for each $k \leq L_s$ we determined $n(k) \in \mathbb{N}$ satisfying $n(k) \leq s \log(8sL_s)$ or $n(k) = \infty$ so that the vector

$$\alpha = (n(1), \ldots, n(L))$$

records the densities of $\mathcal{O}_{(s,k)}$ in I.

Now we change the point of view (thereby testing our understanding of the above) . We fix a color $k' \leq L$. Then we collect all intervals $I \in G_{s-1}$, for which the color is k' and which produce the same vector of indices when the densities of $\mathcal{O}_{(s,k)}$ in I are computed. Precisely, given $k' \leq L$ and

$$\alpha = (n(1), \ldots, n(L))$$

where $n(k) \leq s \log(8sL)$, or $n(k) = \infty$, we define $\mathcal{O}_{(s-1,k',\alpha)}$ to be the collection of all dyadic intervals $I \in G_{s-1}$, which have the color k' attached to them and for which the following two-sided estimates hold,

$$2^{-\frac{n(k)+1}{s}} < \frac{|I \cap O_{(s,k)}|}{|I|} \leq 2^{-\frac{n(k)}{s}} \qquad \text{for } k \leq L. \tag{3.2.9}$$

In this way we are producing at most L_{s-1} different collections of the form $\mathcal{O}_{(s-1,k',\alpha)}$; these are pairwise disjoint collections of dyadic intervals in G_{s-1}, hence they are covering disjoint subsets of the unit interval. This completes step $s-1$ of the construction. We now relabel the collections $\{\mathcal{O}_{(s-1,k,\alpha)}\}$ in the form

$$\mathcal{O}_{(s-1,n)}, \text{ where } n \leq L_{s-1}.$$

Before we proceed to step $s-2$ we change our point of view once more, and ask ourselves the following question: What follows when two different intervals I, I' belong to the same collection $\mathcal{O}_{(s-1,n)}$? It follows that first of all the intervals I, I' have the same color. Second, for each $k \leq L_s$ the density of $\mathcal{O}_{(s,k)}$ in I and in I' are almost the same; the relative error is of size $2^{1/s}$. That is,

$$\frac{|I' \cap O_{(s,k)}|}{|I'|} < \left(2^{\frac{1}{s}}\right) \frac{|I \cap O_{(s,k)}|}{|I|}, \quad \text{where } k \leq L_s. \tag{3.2.10}$$

The same estimate holds by symmetry with I, I' interchanged.

Step $s-2$. Now we describe step $s-2$ of the proof. By a mere change of notation the construction presented for $s-2$ applies to the general step $m \leq s-2$.

We start again by modifying the already existing families $\{\mathcal{O}_{(s-1,n)} : n \le L_{s-1}\}$ and $\{\mathcal{O}_{(s,k)} : k \le L_s\}$. Recall our convention to let $O_{(s-1,n)}$ be the point-set covered by $\mathcal{O}_{(s-1,n)}$. Fix $I \in G_{s-2}$. We apply the following rule to each of the collections $\mathcal{O}_{(s-1,n)}$, where $n \le L_{s-1}\}$. If $n \le L_{s-1}$ satisfies

$$\frac{|I \cap O_{(s-1,n)}|}{|I|} \le \frac{1}{8sL_{s-1}}, \tag{3.2.11}$$

then define

$$I \cap \mathcal{E}_{(s-1,n)} = \{I \cap \mathcal{O}_{(s-1,n)}\} \cup \bigcup_{k=1}^{L_s} \{J \in \mathcal{O}_{(s,k)} : J \subseteq I \cap O_{(s-1,n)}\}.$$

Next we take the union over $n \le L_{s-1}$ and $I \in G_{s-2}$. This gives

$$\mathcal{E}_{s-1} = \bigcup_{I \in G_{s-2}} \bigcup_{n=1}^{L_{s-1}} I \cap \mathcal{E}_{s-1,n}.$$

Having done that we delete the intervals of $\mathcal{E}_{s-1}$ from the already existing collections $\{\mathcal{O}_{(s-1,n)}\}$ and $\{\mathcal{O}_{(s,k)}\}$. For $n \le L_{s-1}$ and $k \le L_s$, we define

$$\begin{aligned} \mathcal{O}_{(s-1,n)} &:= \mathcal{O}_{(s-1,n)} \setminus \mathcal{E}_{s-1}, \\ \mathcal{O}_{(s,k)} &:= \mathcal{O}_{(s,k)} \setminus \mathcal{E}_{s-1}. \end{aligned} \tag{3.2.12}$$

Fix $I \in G_{s-2}$, and let $n \le L_{s-1}$. <u>After</u> the modifications defined by (3.2.12) the following implication holds.

$$\text{If} \quad I \cap O_{(s-1,n)} \ne \emptyset, \quad \text{then} \quad \frac{|I \cap O_{(s-1,n)}|}{|I|} \ge \frac{1}{8sL_{s-1}}. \tag{3.2.13}$$

It follows from (3.2.13) that for each $n \le L_{s-1}$ there exists $m(n)$ so that

$$m(n) \in \{1, \ldots, s\log_2(8sL_{s-1})\} \cup \{\infty\}$$

and

$$2^{-\frac{m(n)+1}{s}} < \frac{|I \cap O_{(s-1,n)}|}{|I|} \le 2^{-\frac{m(n)}{s}}.$$

Hence up to a small relative error, the value of $m(n)$ determines the density of $\mathcal{O}_{(s-1,n)}$ in I. Summing up, for $I \in G_{s-2}$ the vector

$$\alpha = (m(1), \ldots, m(L))$$

stores the densities of $\mathcal{O}_{(s-1,n)}$, $n \le L_{s-1}$ in I.

Now we change our point of view. We fix a color $k' \le L$ and collect all those intervals $I \in G_{s-2}$ for which the color is k' and which produce the same vector of indices when the densities of $\mathcal{O}_{(s-1,n)}$ in I are computed. Precisely, we are given given $n \leq L_{s-1}$, and

$$m(n) \in \{1, \ldots, s \log_2(8sL_{s-1})\} \cup \{\infty\},$$

and hence

$$\alpha = (m(1), \ldots, m(L_{s-1})).$$

Then define $\mathcal{O}_{(s-2,k',\alpha)}$ to be the collection of all intervals $I \in G_{s-2}$ for which the color is k' and for which the following inequalities hold true,

$$2^{-\frac{m(n)+1}{s}} < \frac{|I \cap O_{(s-1,n)}|}{|I|} \leq 2^{-\frac{m(n)}{s}}, \quad \text{where } n \leq L_{s-1}. \tag{3.2.14}$$

In this way we define at most $L_{s-2} = L(s, L_{s-1})$ different subcollections of G_{s-2}; they are pairwise disjoint, hence they are covering pairwise disjoint subsets of the unit interval. To simplify notation we relabel the collections $\mathcal{O}_{(s-2,k',\alpha)}$ as follows,

$$\mathcal{O}_{(s-2,n)} \text{ where } 1 \leq n \leq L_{s-2}.$$

We continue inductively, until we reach step 1. In this way we define the collections

$$\begin{aligned} &\mathcal{O}_{(m,n)}, \quad 1 \leq m \leq s, \quad \text{and} \quad n \leq L_m, \\ &\mathcal{E}_m, \quad 1 \leq m \leq s. \end{aligned}$$

We summarize the important properties of these collections. First, we define

$$\mathcal{L} = \left\{ \bigcup_{m=0}^{s} G_m \right\} \setminus \left\{ \bigcup_{m=1}^{s} \mathcal{E}_m \right\}. \tag{3.2.15}$$

It is easy to see that the collection $\mathcal{L}$ is a block of dyadic intervals in $\mathcal{K}$. We show next that $\mathcal{L}$ is large enough so that $|\{\mathcal{K} \setminus \mathcal{L}\}^*| \leq 1/2|I_0|$. Note that the point-set E_m covered by $\mathcal{E}_m$ is small in measure. By (3.2.11) it satisfies

$$|E_m| \leq \frac{1}{8s}|I_0|.$$

Taking the sum over $1 \leq m \leq s$ gives the upper bound

$$\begin{aligned} |\{\mathcal{K} \setminus \mathcal{L}\}^*| &\leq \frac{1}{4}|I_0| + \sum_{m=1}^{s} |E_m| \\ &\leq \frac{1}{2}|I_0|. \end{aligned}$$

Part 3. Homogeneity Properties of the collections $\mathcal{O}_{(m,n)}$. Now we analyze more closely the properties of $\mathcal{O}_{(m,n)}$, and prove that (3.2.6) holds true. Let $O_{(m,n)}$ be the point-set covered by $\mathcal{O}_{(m,n)}$.

In (3.2.6) we claim that there holds a very strong homogeneity condition between the collections $\mathcal{O}_{(m,n)}$ and $\mathcal{O}_{(m+1,k)}$, taken from consecutive levels. We fix $k \le L_{m+1}$, and define

$$A = \frac{|O_{(m,n)} \cap O_{(m+1,k)}|}{|O_{(m,n)}|}.$$

We prove now that for every $I \in \mathcal{O}_{(m,n)}$, the following estimates hold,

$$\left(2^{-\frac{1}{s}}\right) A \le \frac{|I \cap O_{(m+1,k)}|}{|I|} \le \left(2^{\frac{1}{s}}\right) A. \tag{3.2.16}$$

To verify (3.2.16) we fix two different intervals $I, I' \in \mathcal{O}_{(m,n)}$. Note that in (3.2.14) we established that the density of $\mathcal{O}_{(m+1,k)}$ in I and in I' is the same, up to a relative error of $2^{1/s}$. Precisely, (3.2.14) implies

$$\frac{|I \cap O_{(m+1,k)}|}{|I|} < \left(2^{\frac{1}{s}}\right) \frac{|I' \cap O_{(m+1,k)}|}{|I'|}. \tag{3.2.17}$$

By symmetry, the same estimate holds with the roles of I and I' interchanged. Multiply the inequalities (3.2.17) with $|I'|$, then take the sum over all $I' \in \mathcal{O}_{(m,n)}$. This gives the estimates

$$|O_{(m,n)}| \frac{|I \cap O_{(m+1,k)}|}{|I|} < \left(2^{\frac{1}{s}}\right) B, \tag{3.2.18}$$

where we used the abbreviation

$$B = |O_{(m,n)} \cap O_{(m+1,k)}|.$$

By symmetry we also obtain the corresponding lower bound,

$$|O_{(m,n)}| \frac{|I \cap O_{(m+1,k)}|}{|I|} \ge \left(2^{-\frac{1}{s}}\right) B. \tag{3.2.19}$$

Dividing the estimates (3.2.18) and (3.2.19) by $|O_{(m,n)}|$ gives the claimed homogeneity condition (3.2.16). It is important at this stage of the proof that we have constants like $\left(2^{\frac{1}{s}}\right)$ in (3.2.16), which are getting closer to 1 as the number of stages s is increasing. (This correlation makes it possible to transfer the homogeneity condition (3.2.16) across s different levels .) Next we apply (3.2.16) inductively. Fix $m \le s$. Let $s \ge l \ge m$, choose indices

$$k(m),\ k(m+1),\ \ldots,\ k(l),$$

and define the intersection

$$O = O_{(m,k(m))} \cap \cdots \cap O_{(l,k(l))}.$$

Fix $I, I' \in \mathcal{O}_{(m,k(m))}$. A repeated application of (3.2.16) shows that, up to a relative error of size $\left(2^{\frac{l-m}{s}}\right)$, the density of O is the same in I, and I'. That is

$$\frac{|I \cap O|}{|I|} \leq \left(2^{\frac{l-m}{s}}\right) \frac{|I' \cap O|}{|I'|}. \tag{3.2.20}$$

By symmetry, the estimate (3.2.20) holds with I and I' interchanged.

Part 4. The construction of the nested family $\{F_{(m,k)}\}$. We construct the family $\{F_{(m,k)}\}$ by splitting the collections $\{O_{(m,n)}\}$ in a canonical way as follows:

Recall first that for each fixed m the sets $\{O_{(m,n)} : n \leq L_m\}$ are pairwise disjoint. Let $\mathcal{A}_1$ denote the collection $\{O_{(1,n)} : n \leq L_1\}$. By induction we define $\mathcal{A}_m$ to be the atoms of the σ-algebra generated by $\mathcal{A}_{m-1}$ and $\{O_{(m,n)} : n \leq L_m\}$. Let $N(m)$ be the cardinality of $\mathcal{A}_m$, and let

$$F_{(m,1)}, \ldots, F_{(m,N(m))}$$

be the list of sets contained in $\mathcal{A}_m$.

The family $\{F_{(m,k)} : m \leq s,\ k \leq N(m)\}$ forms a nested family of measurable sets. There exists $\delta = \delta(K, L)$ and $M = M(K, L)$ such that

$$|F_{(m,k)}| \geq \delta |I_0| \tag{3.2.21}$$

and

$$\sum_{m=1}^{s} N(m) \leq M. \tag{3.2.22}$$

Observe that for each set $F_{(m,k)}$ there exists a unique $O_{(m,n)}$ such that

$$F_{(m,k)} \subseteq O_{(m,n)}.$$

For $m \leq s$ and $k \leq N(m)$, we define the following collections of dyadic intervals,

$$\mathcal{F}_{(m,k)} = \{J \in \mathcal{O}_{(m,n)} : J \subseteq F_{(m,k)}\}.$$

The remainder of the proof is devoted to showing that (up to relabeling) $\{\mathcal{F}_{(m,k)}\}$ is a decomposition of $\mathcal{L}$ and satisfies Jones 's compatibility condition. To see that $\{\mathcal{F}_{(m,k)}\}$ is a decomposition of $\mathcal{L}$ it suffices to rewrite $\mathcal{L}$ using the collections $\mathcal{O}_{(m,n)}$ as

$$\mathcal{L} = \{I_0\} \cup \bigcup_{m=1}^{s} \bigcup_{n=1}^{L_m} \mathcal{O}_{(m,n)}.$$

Next we pass the homogeneity condition (3.2.16) on to the families $\{\mathcal{F}_{m,k}\}$. Choose $l \geq m$, and

$$F_{(m,n)} \supseteq F_{(l,p)}.$$

Then by definition of the sets $F_{(l,p)}$ and $F_{(m,n)}$, there exist indices

$$k(m),\ k(m+1),\ \dots,\ k(l),$$

(uniquely determined) such that with $O = O_{(m,k(m))} \cap \cdots \cap O_{(l,k(l))}$ we have

$$F_{(l,p)} = F_{(m,n)} \cap O. \tag{3.2.23}$$

Now we fix two different intervals $I, I' \in \mathcal{F}_{(m,n)}$. The identity (3.2.23) provides us with an induced representation for $I \cap F_{(l,p)}$ and $I' \cap F_{(l,p)}$. Indeed, (3.2.23) implies that

$$I \cap F_{(l,p)} = I \cap O \quad \text{and} \quad I' \cap F_{(l,p)} = I' \cap O. \tag{3.2.24}$$

We compare the measure of $I \cap F_{(l,p)}$ to that of $I' \cap F_{(l,p)}$ using the right-hand side of (3.2.24). Hence by (3.2.20) we obtain that

$$\frac{|I \cap F_{(l,p)}|}{|I|} \leq \left(2^{\frac{l-m}{s}}\right) \frac{|I' \cap F_{(l,p)}|}{|I'|}. \tag{3.2.25}$$

Next we multiply (3.2.25) on both sides by $|I| \cdot |I'|$ and the result is then summed over all $I' \in \mathcal{F}_{(m,n)}$. This gives

$$|F_{(m,n)}| \cdot |I \cap F_{(l,p)}| \leq \left(2^{\frac{l-m}{s}}\right) |I| \cdot |F_{(m,n)} \cap F_{(l,p)}|.$$

Dividing by $|I| \cdot |F_{(m,n)}|$ shows that for each $I \in \mathcal{F}_{(m,n)}$ the homogeneity condition holds,

$$\frac{|I \cap F_{(l,p)}|}{|I|} \leq \left(2^{\frac{l-m}{s}}\right) \frac{|F_{(m,n)} \cap F_{(l,p)}|}{|F_{(m,n)}|}. \tag{3.2.26}$$

Finally we point out that the symmetry between I and I' in (3.2.25) implies also the converse inequality to (3.2.26).

Part 5. Conclusion. It remains to define the families $\{\mathcal{C}_i\}$ satisfying the compatibility condition. Recall that in (3.2.22) we defined $M = \sum_{m=1}^{s} N(m)$. Take a bijection between the pairs

$$(m,n) \quad \text{where} \quad 1 \leq m \leq s, \quad \text{and} \quad 1 \leq n \leq N(m),$$

and the set

$$\{1, \dots, M\}$$

mapping the lexicographic order onto the linear order. Using this bijection we relabel the families $\{\mathcal{F}_{(m,n)} : 1 \leq m \leq s,\ 1 \leq n \leq N(m)\}$ as $\{\mathcal{C}_i : i \leq M\}$. Clearly $\{\mathcal{C}_j\}$ satisfies the compatibility conditions $(J\,1)$, $(J\,3)$, and $(J\,4)$. By (3.2.25) it also satisfies $(J\,2)$. By (3.2.21) and (3.2.22) the assertions 2) and 3) of Theorem 3.2.1 are verified. □

We continue with an extension of Theorem 3.2.1 that provides a profound understanding of the relation between the Carleson packing condition and Jones's compatibility condition. What follows is a deep result; its depth stems from the combinatorial power of Theorem 3.2.1.

Theorem 3.2.2. *Let $\mathcal{K}$ be a collection of dyadic intervals, where each $I \in \mathcal{K}$ has attached to it one of L different colors. Assume that $\mathcal{K}$ satisfies the Carleson packing condition,*

$$\sup_{I\in\mathcal{K}} \sum_{J\in\mathcal{K}, J\subseteq I} \frac{|J|}{|I|} \leq K.$$

Then $\mathcal{K}$ can be decomposed into disjoint blocks of dyadic intervals $\{\mathcal{L}(I) : I \in \Omega\}$, so that the following conditions hold.

1. *Each $I \in \Omega$ is the unique maximal interval in $\mathcal{L}(I)$. The index set Ω satisfies the 2-Carleson condition.*
2. *For $I \in \Omega$ there exists $M = M(I)$ and disjoint collections $\mathcal{C}_1(I), \ldots, \mathcal{C}_M(I)$ satisfying Jones's compatibility condition, such that*

$$\mathcal{L}(I) = \mathcal{C}_1(I) \cup \cdots \cup \mathcal{C}_M(I).$$

3. *Each of the $\mathcal{C}_i(I)$ is monochromatic with respect to the colors of $\mathcal{K}$.*
4. *There exists $\delta = \delta(K, M)$ such that each of the collections $\mathcal{C}_i(I)$ satisfies $|\mathcal{C}_i(I)^*| \geq \delta|I|$.*
5. *There exists an upper bound, $\sup_{I\in\Omega} M(I) \leq M(K, L)$.*

Proof. The proof consists of iteratively applying Theorem 3.2.1. Many proofs in this book will follow the pattern that we are going to establish now. Let $\mathcal{G}_0$ denote the maximal intervals of $\mathcal{K}$. Fix $I \in \mathcal{G}_0$. Let

$$\mathcal{K}(I) = \{J \in \mathcal{K} : J \subseteq I\}.$$

Then I is the unique maximal interval in $\mathcal{K}(I)$. Applying Theorem 3.2.1 to $\mathcal{K}(I)$, gives a large block of dyadic intervals $\mathcal{L}(I) \subseteq \mathcal{K}(I)$, satisfying the conditions 1)–3) listed in Theorem 3.2.1. In particular $\mathcal{L}(I)$ satisfies

$$|\{\mathcal{K}(I) \setminus \mathcal{L}(I)\}^*| \leq \frac{1}{2}|I|. \tag{3.2.27}$$

Define $\mathcal{G}_1(I)$ to be the maximal intervals in $\mathcal{K}(I) \setminus \mathcal{L}(I)$. Note that (3.2.27) implies that

$$|\mathcal{G}_1(I)^*| \leq \frac{1}{2}|I|.$$

Then take the union using the intervals of $\mathcal{G}_0$ as index set, so

$$\mathcal{G}_1 = \bigcup_{I\in\mathcal{G}_0} \mathcal{G}_1(I).$$

Suppose now that we have already constructed $\mathcal{G}_0, \ldots, \mathcal{G}_{n-1}$. Then $\mathcal{G}_n$ is defined as follows. (The argument below is an exact replica of the one defining $\mathcal{G}_1$.) Fix $I \in \mathcal{G}_{n-1}$. Let

$$\mathcal{K}(I) = \{J \in \mathcal{K} : J \subseteq I\}.$$

I is the unique maximal interval in $\mathcal{K}(I)$. Apply Theorem 3.2.1 to $\mathcal{K}(I)$. This gives a large block of dyadic intervals $\mathcal{L}(I) \subseteq \mathcal{K}(I)$, satisfying the conclusions 1) –3) of Theorem 3.2.1. In particular

$$|\{\mathcal{K}(I) \setminus \mathcal{L}(I)\}^*| \leq \frac{1}{2}|I|.$$

Next define $\mathcal{G}_n(I)$ to be the maximal intervals in $\mathcal{K}(I) \setminus \mathcal{L}(I)$. Note that

$$|\mathcal{G}_n(I)^*| \leq \frac{1}{2}|I|. \tag{3.2.28}$$

Then we take the union using the intervals of $\mathcal{G}_{n-1}$ as index set and define

$$\mathcal{G}_n = \bigcup_{I \in \mathcal{G}_{n-1}} \mathcal{G}_n(I).$$

We thus completed the inductive definition of the families $\{\mathcal{G}_n\}$. Each of them consists of pairwise disjoint dyadic intervals. Next we put

$$\Omega = \bigcup_{n=0}^{\infty} \mathcal{G}_n.$$

We verify that Ω satisfies the 2-Carleson condition. This follows from (3.2.28) and the observation that the n-th generation of Ω just coincides with the collection $\mathcal{G}_n$, constructed above. Indeed for each fixed $n \in \mathbb{N}$ and $I \in \mathcal{G}_n$ we have by (3.2.28) that

$$|G_1(I, \Omega)^*| \leq \frac{1}{2}|I|.$$

Iterating gives an estimate for the k-th generation of I in Ω,

$$|G_k(I, \Omega)^*| \leq 2^{-k}|I|. \tag{3.2.29}$$

With (3.2.29) we estimate the Carleson constant of Ω as follows. Fix $I \in \Omega$, then

$$\begin{aligned}\sum_{J \in I \cap \Omega} |J| &\leq \sum_{k=0}^{\infty} |G_k(I, \Omega)^*| \\ &\leq \sum_{k=0}^{\infty} 2^{-k}|I| \\ &\leq 2|I|.\end{aligned}$$

Next observe that for every $I \in \mathcal{G}_{n-1}$, we have the representation

$$\mathcal{L}(I) = \mathcal{K}(I) \setminus \bigcup_{J \in \mathcal{G}_n(I)} \mathcal{K}(J),$$

which holds by definition of $\mathcal{K}(J), J \in \mathcal{G}_n$. Consequently $\{\mathcal{L}(I) : I \in \Omega\}$, is a decomposition of $\mathcal{K}$. It remains to recall that $\mathcal{L}(I)$ satisfies the conclusion of Theorem 3.2.1. Hence there exists M (with M depending on I) of monochromatic collections $\mathcal{C}_1(I), \ldots, \mathcal{C}_M(I)$ satisfying Jones's compatibility condition, and

$$\mathcal{L}(I) = \mathcal{C}_1(I) \cup \cdots \cup \mathcal{C}_M(I).$$

Each of the $\mathcal{C}_i(I)$ satisfies $|\mathcal{C}_i(I)^*| \geq \delta|I|$. Moreover, there exist upper and lower bounds $M \leq M(K, L)$, and $\delta \geq \delta(K, L)$. □

The first step towards the uniform approximation property

Now we summarize and crystallize the combinatorial results of the present section and merge them with norm estimates we obtained for orthogonal projections in Section 1.5. Our next theorem is the first step — a giant one — towards the uniform approximation property of BMO.

Let $\mathcal{B}_1, \ldots, \mathcal{B}_L$ be disjoint collections of dyadic intervals. For $i \leq L$ let

$$b_i = \sum \{h_J : J \in \mathcal{B}_i\}.$$

Theorem 3.2.3. *If each of the $\mathcal{B}_i$ satisfies the Carleson packing condition, and*

$$\sup_{I \in \mathcal{B}_i} \sum_{J \in I \cap \mathcal{B}_i} \frac{|J|}{|I|} \leq M,$$

then there exists an orthogonal projection $P : BMO \to BMO$, satisfying

$$\begin{aligned} &||P||_{BMO} \leq 4\sqrt{2}, \\ &P(b_i) = b_i \text{ for } \ i \leq L, \\ &\operatorname{rank}(P) \leq f(M, L). \end{aligned}$$

Proof. Theorem 3.2.2 allows us to define the desired projection P. Once this is done we will verify that P reproduces b_i, that its rank is bounded as claimed and that its norm is bounded independent of L. At this point we use the pigeon hole principle and the glueing process defined in Section 1.5.

The definition of the projection P. Define $\mathcal{K} = \mathcal{B}_1 \cup \cdots \cup \mathcal{B}_L$, then by hypothesis the Carleson constant of $\mathcal{K}$ is bounded by $M_0 = L \cdot M$. We color the intervals of $I \in \mathcal{K}$ using the following rule: If $I \in \mathcal{B}_i$, then by definition the color of I equals $i \in \{1, \ldots, L\}$.

Next apply Theorem 3.2.2 to the colored collection of dyadic intervals $\mathcal{K}$. This gives a decomposition of $\mathcal{K}$ into blocks of intervals $\{\mathcal{L}(I) : I \in \Omega\}$, so that the following holds:

1. Each $I \in \Omega$ is the unique maximal interval in $\mathcal{L}(I)$. The index set Ω satisfies the 2-Carleson condition.
2. For $I \in \Omega$ there exists $M(I)$ and disjoint collections $\mathcal{C}_1(I), \dots, \mathcal{C}_{M(I)}(I)$ satisfying Jones's compatibility condition, such that
$$\mathcal{L}(I) = \mathcal{C}_1(I) \cup \cdots \cup \mathcal{C}_{M(I)}(I).$$
3. For $i \leq M(I)$, the collection $\mathcal{C}_i(I)$ is monochromatic with respect to the colors of $\mathcal{K}$.
4. There exists $\delta_0 = \delta(M_0)$ such that each of the collections $\mathcal{C}_i(I)$ satisfies $|\mathcal{C}_i(I)^*| \geq \delta_0 |I|$.
5. There exists an upper bound, $\sup_{I \in \Omega} M(I) \leq M(M_0, L)$.

The conclusion of Theorem 3.2.2 allows us to apply the pigeon hole principle to the index set Ω. By counting, there exists $K = K(\delta_0, M_0)$ and a decomposition of Ω, into
$$\Omega = \mathcal{A}_1 \cup \cdots \cup \mathcal{A}_K,$$
so that for each fixed collection $\mathcal{A}_k$, $k \leq K$, the following three conditions hold. First, there exists $M(k)$ so that for each $J \in \mathcal{A}_k$,
$$M(J) = M(k).$$
Second, if $I, J \in \mathcal{A}_k$ and $m, n \leq M(k)$, then
$$\mathcal{C}_m^*(I) \subseteq \mathcal{C}_n^*(I) \quad \text{if} \quad \mathcal{C}_m^*(J) \subseteq \mathcal{C}_n^*(J),$$
and
$$\frac{|\mathcal{C}_m^*(I)|}{|\mathcal{C}_n^*(I)|} \leq 2\frac{|\mathcal{C}_m^*(J)|}{|\mathcal{C}_n^*(J)|}.$$
The third condition states that if $m \leq M(k)$, and $I, J \in \mathcal{A}_k$, then the colors of the monochromatic collections $\mathcal{C}_m(I)$ and $\mathcal{C}_m(J)$ coincide.

Now for $k \leq K$ and $m \leq M(k)$ define
$$\mathcal{C}_{k,m} = \bigcup \{\mathcal{C}_m(J) : J \in \mathcal{A}_k\}, \tag{3.2.30}$$
to form
$$b_{k,m} = \sum \{h_I : I \in \mathcal{C}_{k,m}\}.$$
Let P_k be defined by
$$P_k(f) = \sum_{m=1}^{M(k)} \langle f, \frac{b_{k,m}}{\|b_{k,m}\|_2} \rangle \frac{b_{k,m}}{\|b_{k,m}\|_2}, \tag{3.2.31}$$

and let P be the orthogonal projection onto $\operatorname{span}\{b_{k,m}\}$, given by

$$P = \sum_{k=1}^{K} P_k.$$

P reproduces b_i. Note that each of the collections $\mathcal{C}_{k,m}$, where $k \leq M(k)$ and $k \leq K$ is a monochromatic sub-collection of $\mathcal{K}$. They form a decomposition of $\mathcal{K}$ that refines the initial decomposition $\{\mathcal{B}_i\}$ where $i \leq L$. Indeed stating that $\mathcal{C}_{k,m}$ is monochromatic implies that there exists $i \leq L$ so that $\mathcal{C}_{k,m} \subseteq \mathcal{B}_i$. Thus it is easy to see that

$$P(b_i) = b_i.$$

The rank estimate for P. By (3.2.31) the rank of P_k is bounded by $M(k)$. Hence the rank of P is bounded by

$$f(L, M) = \sum_{k=1}^{K} M(k).$$

The norm estimate for P. Now we turn to estimating the operator norm of P on BMO. We fix $J_0 \in \mathcal{K}$. There exists $I_0 \in \Omega$, so that $J_0 \in \mathcal{L}(I_0)$. Moreover there exists $k_0 \leq M$ so that $I_0 \in \mathcal{A}_{k_0}$. It follows that $J_0 \cap \mathcal{K}$ can be rewritten as

$$\begin{aligned} J_0 \cap \mathcal{K} &= J_0 \cap \mathcal{L}(I_0) \cup \bigcup \{\mathcal{L}(I) : I \in \Omega, I \subset I_0\} \\ &= J_0 \cap \mathcal{L}(I_0) \cup \bigcup_{k=1}^{K} \bigcup \{\mathcal{L}(I) : I \in \mathcal{A}_k, I \subset I_0\}. \end{aligned} \tag{3.2.32}$$

If $I \in \mathcal{A}_k$, then the following identity holds for the square function localized to the block $\mathcal{L}(I)$,

$$S^2(P(f)|\mathcal{L}(I)) = S^2(P_k(f)|\mathcal{L}(I)). \tag{3.2.33}$$

Next we expand the square function $S^2(P(f)|J_0 \cap \mathcal{K})$ using the decomposition (3.2.32) in combination with (3.2.33). This gives

$$\begin{aligned} S^2(P(f)|J_0 \cap \mathcal{K}) &= S^2(P_{k_0}(f)|J_0 \cap \mathcal{L}(I_0)) \\ &\quad + \sum_{k=1}^{K} \sum_{\substack{I \in \mathcal{A}_k \\ I \subset I_0}} S^2(P_k(f)|\mathcal{L}(I)). \end{aligned} \tag{3.2.34}$$

Next we apply Theorem 1.5.11 to the averaging projection P_k; for $I_0 \in \mathcal{A}_{k_0}$, and $I \in \mathcal{A}_k$, it asserts that the following estimates hold,

$$\int S^2(P_{k_0}(f)|J_0 \cap \mathcal{L}(I_0)) \leq 16B\|f\|_{\mathrm{BMO}}^2 |J_0| \tag{3.2.35}$$

and

$$\int S^2(P_k(f)|\mathcal{L}(I)) \leq 16B\|f\|^2_{\mathrm{BMO}}|I|. \tag{3.2.36}$$

Integrate the equation (3.2.34) and insert the estimates (3.2.35) and (3.2.36) obtained by Theorem 1.5.11. As $\Omega = \mathcal{A}_1 \cup \cdots \cup \mathcal{A}_K$, we obtain that

$$\int S^2(P(f)|J_0 \cap \mathcal{K}) \leq 16B\|f\|^2_{\mathrm{BMO}}|J_0| \cdot [\![\Omega]\!] . \tag{3.2.37}$$

Finally recall that Ω satisfies the 2-Carleson condition, thus (3.2.37) implies the operator bound $\|P\|_{\mathrm{BMO}} \leq 4\sqrt{2}$. □

3.3 Rearrangement operators

In this section we will study operators induced by rearrangements acting on the Haar system. Here we will determine when these operators are bounded and when not. Several equivalent conditions for their boundedness will be obtained. In BMO we study

$$T : h_I \mapsto h_{\tau(I)},$$

where h_I is the L^∞-normalized Haar function, and where τ is an injective map defined on dyadic intervals. In the case of L^p one considers rearrangements of the L^p normalized Haar system. Rearrangements of the L^p normalized Haar system are given by

$$T_p : \frac{h_I}{|I|^{1/p}} \mapsto \frac{h_{\tau(I)}}{|\tau(I)|^{1/p}},$$

where $1 < p < \infty$. The operator T_p coincides with the operator

$$h_I \mapsto h_{\tau(I)} \left(\frac{|\tau(I)|}{|I|}\right)^{-1/p} .$$

The family of operators $\{T_p : 2 \leq p < \infty\}$ is embedded in the *analytic* family of operators

$$R_z : h_I \mapsto h_{\tau(I)} \left(\frac{|\tau(I)|}{|I|}\right)^{(z-1)/2} .$$

The correspondence is given by the equation $T_p = R_{1-2/p}$. Therefore we will use the BMO result and the interpolation Corollary 2.3.12 valid for analytic families of operators to deduce L^p results. By the duality, Corollary 2.3.12 applies here since for rearrangements the transposed operator coincides with the inverse operator.

Theorem 3.3.1 below contains a combinatorial and geometric description of rearrangements that induce bounded operators. For example, boundedness is guaranteed as soon as the rearrangement preserves the Carleson condition. Which means that there exists $M > 0$ such that for every collection of dyadic intervals $\mathcal{C}$ we have

$$[\![\tau(\mathcal{C})]\!] \leq M[\![\mathcal{C}]\!].$$

On the other hand the boundedness of the operators imposes a geometric condition on the underlying rearrangement τ. The condition we derive says that, up to an exceptional set with bounded Carleson constant, τ maps blocks of intervals into blocks of intervals; and on each block the rearrangement acts homogeneously. Most of the work in this chapter will focus on identifying and analyzing this condition (its precise definition is on page 207).

The following theorem summarizes the relation between the boundedness of rearrangement operators on BMO, respectively L^p, the preservation of the Carleson condition, and geometric homogeneity conditions on the underlying rearrangement (called below Property $\mathcal{P}$).

Theorem 3.3.1. *For a rearrangement τ the following are equivalent:*

1. *The operator $T : BMO \to BMO$ is bounded.*
2. *The rearrangement τ preserves the Carleson condition.*
3. *There exists $2 < p < \infty$ such that $T_p : L^p \to L^p$ is a bounded operator.*
4. *For every $2 < p < \infty$, $T_p : L^p \to L^p$ is a bounded operator.*
5. *The rearrangement τ satisfies Property $\mathcal{P}$.*

The implications of Theorem 3.3.1 are formulated and proved one by one in Proposition 3.3.2, Theorem 3.3.4, Theorem 3.3.5, Proposition 3.3.6, and Theorem 3.3.7. These results form the content of the next subsection. Analogous results hold for rearrangement operators on L^p $(1 < p < 2)$ and on H^1. To see this, it suffices to observe that transposing a rearrangement operator gives again a rearrangement operator. The transpose is defined by the inverse of τ.

E. M. Semenov [190] initiated the systematic investigation of operators rearranging the Haar system. Semenov's main result asserts that the linear extension of the map

$$h_I \to h_{\tau(I)}, \quad |\tau(I)| = |I|,$$

is bounded in L^p $(1 < p < 2)$ if the following condition on the underlying rearrangement is satisfied,

$$\sup_{\mathcal{C} \subset \mathcal{D}} \frac{|\tau(\mathcal{C})^*|}{|\mathcal{C}^*|} < \infty. \tag{3.3.1}$$

While working directly with the space L^p, E. M. Semenov uncovered his criterion and its scale invariant nature. This is quite remarkable since there aren't any obvious links between the integral norm of L^p and scale invariant properties. Later F. Schipp [188] observed the connection between (3.3.1) and BMO.

General rearrangements are studied in [156] where a priori restrictions, such as $|\tau(I)| = |I|$, have been removed from the assumption of Semenov's theorems.

Rearrangement operators on BMO

The analysis of *general* rearrangement operators on BMO begins with a combinatorial result asserting that the operator is bounded if the underlying rearrangement preserves the Carleson constant. More precisely we can read off the norm of the operator from the ratio

$$\sup_{\mathcal{C}\subset\mathcal{D},\text{ finite}} \frac{[\![\tau(\mathcal{C})]\!]}{[\![\mathcal{C}]\!]},$$

where the supremum is taken over all finite collections of dyadic intervals. Equivalently boundedness of rearrangement operators can be determined by their response to the class of simple functions

$$x = \sum_{I\in\mathcal{C}} h_I. \tag{3.3.2}$$

This result will be followed by an example of an orthogonal projection of rank equal to n which maps the simple functions (3.3.2) into $[\![\mathcal{C}]\!] \times$ {the unit ball of BMO}, while the operator norm of the projection exceeds $3\sqrt{n}/4$. Here we encounter, for the first time, projections onto the range of order-inversing embeddings. These operators originate with J. Bourgain's solution of the dimension conjecture for H^1 spaces, a detailed account of which is presented in Section 5.2.

Proposition 3.3.2. *Suppose that there exists $M > 0$ such that for every collection of dyadic intervals $\mathcal{C}$,*

$$[\![\tau(\mathcal{C})]\!] \le M[\![\mathcal{C}]\!].$$

Then the linear extension of the map $T : h_I \mapsto h_{\tau(I)}$ defines a bounded operator on BMO satisfying

$$||T||_{BMO} \le (3M)^{1/2}.$$

Conversely, if T is bounded on BMO, then τ preserves the Carleson condition.

Proof. Let $x = \sum x_I h_I$ be a finite linear combination of Haar functions such that $||x||_{\mathrm{BMO}} \le 1$. Using the hypothesis that τ preserves the Carleson condition we will prove that $||Tx||_{\mathrm{BMO}} \le (3M)^{1/2}$. It suffices to consider x for which the Haar coefficients x_I are rational numbers. In that case there exist $K \in \mathbb{N}$ and $k_I \in \mathbb{N}$ such that

$$x_I^2 = k_I K^{-1} \text{ for } I \in \mathcal{D}. \tag{3.3.3}$$

Note that $x_I^2 \le ||x||_{\mathrm{BMO}}^2 \le 1$, implies that $k_I \le K$. Now we fix $\epsilon > 0$ and choose a dyadic interval J such that

$$(1-\epsilon)||Tx||_{\mathrm{BMO}}^2 \le \frac{1}{|J|} \sum_{\tau(I)\subseteq J} x_I^2|\tau(I)|. \tag{3.3.4}$$

The idea of the proof is to define collections of dyadic intervals $\mathcal{E}_1, \dots, \mathcal{E}_K$, with the following properties: Each of the $\mathcal{E}_i$ satisfies the 3-Carleson condition,

$\tau(\mathcal{E}_i)^*$ is contained in J, and the following crucial convexity relation holds true,

$$\frac{1}{K}\sum_{\tau(I)\subseteq J} k_I|\tau(I)| = \frac{1}{K}\sum_{i=1}^{K}\sum_{I\in\mathcal{E}_i}|\tau(I)|. \tag{3.3.5}$$

Now we will carry out the construction of the collections $\mathcal{E}_i$. Let $n \in \mathbb{N}$. Let v_n be the vector whose entries are dyadic intervals I of length 2^{-n} such that $\tau(I) \subseteq J$. Moreover in the vector v_n we let I appear consecutively k_I times, where we defined k_I in (3.3.3). Now we define the rule which determines when a dyadic interval belongs to the collection $\mathcal{E}_i$. Fix $I \in \mathcal{D}$ and suppose that $|I| = 2^{-n}$. Choose $i \le K$. Then we define I to be in $\mathcal{E}_i$ iff I occupies the p-th position in v_n, and $p = i \bmod K$. Note that since $k_I \le K$, the entries of the vectors v_n are bijectively distributed among the collections $\mathcal{E}_1, \dots, \mathcal{E}_K$, by the above rule. It is also easy to see from that definition that the convexity relation (3.3.5) is satisfied.

The decisive property of our definition is however that simultaneously with (3.3.5) each of the collections $\mathcal{E}_i$ satisfies the 3-Carleson condition. To verify this claim we fix $\mathcal{E}_i$. Then we choose $n \in \mathbb{N}$ and $I_0 \in \mathcal{D}$. Let $A_{n,i}$ be the cardinality of the collection $\{I \subseteq I_0 : I \in \mathcal{E}_i, |I| = 2^{-n}\}$. Taking another look at the defining properties of $\mathcal{E}_i$ lets us observe the estimate

$$A_{n,i} \le 1 + \frac{1}{K}\sum_{I\subseteq I_0, |I|=2^{-n}} k_I.$$

Having made these observations we verify the claim that $\mathcal{E}_i$ satisfies the 3-Carleson condition.

$$\begin{aligned}\sum_{I\in\mathcal{E}_i\cap I_0}|I| &= \sum_{n=-\log_2|I_0|}^{\infty} 2^{-n}A_{n,i}\\ &\le 2|I_0| + \frac{1}{K}\sum_{I\subseteq I_0} k_I|I|\\ &\le 3|I_0|.\end{aligned}$$

So, $[\![\mathcal{E}_i]\!] \le 3$. Now we finish the proof by showing that $||Tx|| \le (3M)^{1/2}$. Recall that in (3.3.4) the interval J was chosen so that

$$(1-\epsilon)||Tx||^2 \le \frac{1}{|J|}\sum_{\tau(I)\subseteq J} x_I^2|\tau(I)|. \tag{3.3.6}$$

Recall also that by (3.3.3) the Haar coefficients of x satisfy $x_I^2 = k_I K^{-1}$. With the convexity relation (3.3.5) we may rewrite the sum in the right-hand side of (3.3.6) as follows,

$$\sum_{\tau(I)\subseteq J} x_I^2|\tau(I)| = \frac{1}{K}\sum_{i=1}^{K}\sum_{I\in\mathcal{E}_i}|\tau(I)|. \tag{3.3.7}$$

Next we give an obvious upper bound for the inner sum appearing in (3.3.7) above,

$$\sum_{I\in\mathcal{E}_i} |\tau(I)| \leq [\![\tau(\mathcal{E}_i)]\!]|\tau(\mathcal{E}_i)^*|. \tag{3.3.8}$$

We showed already that the collection $\mathcal{E}_i$ satisfies the 3-Carleson condition, and by hypothesis the rearrangement τ preserves the Carleson condition. So we have $[\![\tau(\mathcal{E}_i)]\!] \leq 3M$. Next recall that, by construction, $|\tau(\mathcal{E}_i)^*| \leq |J|$. Inserting into the right-hand side of (3.3.8) we showed that the right-hand side of (3.3.6) admits the upper bound

$$\frac{1}{|J|}\sum_{\tau(I)\subseteq J} x_I^2|\tau(I)| \leq 3M.$$

This proves that $||Tx||_{\mathrm{BMO}} \leq (3M)^{1/2}$.

The converse follows from the observation that for special functions of the form $x = \sum_{I\in\mathcal{C}} h_I$, the norm in BMO satisfies

$$||x||_{\mathrm{BMO}} = [\![\mathcal{C}]\!]^{1/2},$$

and that $Tx = \sum_{I\in\mathcal{C}} h_{\tau(I)}$. Hence $||Tx||_{\mathrm{BMO}} = [\![\tau(\mathcal{C})]\!]^{1/2}$. □

Theorem 3.3.2 raises the question if for general operators on BMO the norm can be tested on special functions of the form

$$x = \sum_{I\in\mathcal{C}} h_I. \tag{3.3.9}$$

Equivalently we ask if in some renorming of BMO the simple functions (3.3.9) are the extreme points of the unit ball. That this is not so can be seen by looking at the class of orthogonal projections discussed in Section 1.5. Recall that pairwise disjoint collections $\{\mathcal{S}_j\}$ determine an orthogonal projection in the following way.

$$Pf = \sum_{j=1}^{n}\langle f, k_j\rangle k_j,$$

where $k_j = b_j/||b_j||_2$ and where

$$b_j = \sum\{h_I : I \in \mathcal{S}_j\}.$$

Below we will select pairwise disjoint collections $\{\mathcal{S}_j : 1 \leq j \leq n\}$, such that

$$||P||_{\mathrm{BMO}} \geq \frac{3}{4}\sqrt{n},$$

and so that for functions of the form $x = \sum_{I\in\mathcal{C}} h_I$, the following upper bound holds true,

$$||Px||_{\mathrm{BMO}} \leq [\![\mathcal{C}]\!].$$

To begin with, we fix $n \in \mathbb{N}$. Then define

$$I_j = [0, 4^{-j}[, \text{ where } 1 \leq j \leq n. \tag{3.3.10}$$

We attach frequencies to the intervals I_j as follows. We start with the smallest interval I_n. By our choice $|I_n| = 4^{-n}$. For $1 \leq k \leq n$, we define the collection $\mathcal{B}_{n,k}$ to consist of all dyadic intervals with length $= 2^{-k}4^{-n}$, which are contained in I_n. Thus, we form

$$\mathcal{B}_{n,k} = \{J \subseteq I_n : |J| = 2^{-k}4^{-n}\}.$$

To the remaining intervals $I_j, 1 \leq j \leq n-1$ we attach frequencies in a similar fashion. We define

$$\mathcal{B}_{j,k} = \{J \subseteq I_j : |J| = 2^{-k}4^{-n(n-j+1)}\}.$$

Next we fix j and take the union over $k \leq n$, thus defining

$$\mathcal{S}_j = \mathcal{B}_{j,1} \cup \cdots \cup \mathcal{B}_{j,n}.$$

Note that the collections $\{\mathcal{S}_j : 1 \leq j \leq n\}$ are pairwise disjoint and that $\mathcal{S}_j$ is an n-fold cover of the interval I_j. Moreover we observe the following crucial relation among the collections $\{\mathcal{S}_j : 1 \leq j \leq n\}$.

$$\text{If } K \in \mathcal{S}_j \text{ and } L \in \mathcal{S}_{j+1}, \text{ then } |K| \leq \frac{1}{2}|L|. \tag{3.3.11}$$

The collections $\{\mathcal{S}_j : 1 \leq n\}$, are thus inversely ordered in the following sense: The larger I_j is, the shorter are the intervals in the collection $\mathcal{S}_j$. Now we form the block bases

$$b_j = \sum\{h_I : I \in \mathcal{S}_j\}.$$

It is straightforward to evaluate the square function of b_j, We get $S(b_j) = n^{1/2}1_{I_j}$. Hence for the norm of b_j in the spaces H^1, L^2 and BMO, we obtain that

$$||b_j||_{H^1} = n^{1/2}|I_j|, \quad ||b_j||_2 = n^{1/2}|I_j|^{1/2} \quad \text{and} \quad ||b_j||_{\mathrm{BMO}} = n^{1/2}.$$

Let $k_j = b_j/||b_j||_2$ and define the orthogonal projection

$$Pf = \sum_{j=1}^{n} \langle f, k_j \rangle k_j.$$

Proposition 3.3.3. *The norm of P on BMO is larger than $3\sqrt{n}/4$. For a collection of dyadic intervals $\mathcal{C}$, the following estimate holds when $x = \sum_{J \in \mathcal{C}} h_J$,*

$$||Px||_{BMO} \leq [\![\mathcal{C}]\!].$$

Proof. First we show the lower bound on the norm of P on BMO. We consider the collections of dyadic intervals

$$\mathcal{L}_j = \{J \in \mathcal{S}_j : J \cap I_{j+1} = \emptyset\}. \tag{3.3.12}$$

Observe that $\mathcal{L}_j^* = I_j \setminus I_{j+1}$. Now we define,

$$f_j = n^{-1/2} \sum \{h_I : I \in \mathcal{L}_j\}.$$

Note that $||f_j||_{\mathrm{BMO}} \leq 1$, since the square functions $S(f_j)$ are pointwise ≤ 1. Moreover the support of $S(f_j)$ equals $I_j \setminus I_{j+1}$. Hence the sum

$$f = \sum_{j=1}^{n} f_j$$

satisfies the upper bound $||f||_{\mathrm{BMO}} \leq 1$. Now we give a lower estimate for $||Pf||_{\mathrm{BMO}}$. Observe that, by (3.3.12), we have the identities

$$\langle f, k_j \rangle = \frac{|I_j \setminus I_{j+1}|}{|I_j|^{1/2}} = \frac{3}{4}|I_j|^{1/2},$$

and therefore,

$$Pf = \frac{3}{4} \sum_{j=1}^{n} k_j |I_j|^{1/2}.$$

Next we prove that $||Pf||_{\mathrm{BMO}} \geq 3\sqrt{n}/4$. To this end we expand Pf in its Haar series. This gives

$$Pf = \sum \alpha_I h_I,$$

where $\alpha_I = 3n^{-1/2}/4$, for $I \in \bigcup \mathcal{S}_j$, and $\alpha_I = 0$ otherwise. Next we fix an interval $J \in \mathcal{S}_n$. By the condition of order inversion that is by (3.3.11) it follows that

$$\begin{aligned} \frac{16}{9} \sum_{I \subseteq J} \alpha_I^2 |I| &= \sum_{j=1}^{n} ||k_j||_2^2 |J| \\ &= n|J|. \end{aligned} \tag{3.3.13}$$

Clearly the estimate (3.3.13) implies that $||Pf||_{\mathrm{BMO}} \geq 3\sqrt{n}/4$. Recall that the norm of f in BMO is bounded by 1. Hence the operator norm satisfies the lower bound $||P||_{\mathrm{BMO}} \geq 3\sqrt{n}/4$.

Next we prove an upper bound for $||Px||_{\mathrm{BMO}}$, when $x = \sum_{J \in \mathcal{C}} h_J$. It hinges on the following estimate for the coefficients $\langle x, k_j \rangle$. As $||b_j||_2 = n^{1/2}|I_j|^{1/2}$, and $k_j = b_j/||b_j||_2$ obtain,

$$\begin{aligned} \langle x, k_j \rangle &= n^{-1/2}|I_j|^{-1/2} \sum_{J \in \mathcal{C} \cap \mathcal{S}_j} |J| \\ &\leq [\![\mathcal{C}]\!] n^{-1/2} |I_j|^{1/2}. \end{aligned} \tag{3.3.14}$$

Next we observe that $k_j|I_j|^{1/2} = b_j n^{-1/2}$, and that $\|b_j\|_{\mathrm{BMO}} = n^{1/2}$. Recall also that the Haar support of $k_j|I_j|^{1/2}$ is pairwise disjoint. Thus

$$||\sum_{j=1}^{n} k_j|I_j|^{1/2}||_{\mathrm{BMO}} \leq n^{1/2}. \tag{3.3.15}$$

Now we combine the upper bounds in (3.3.15) and (3.3.14). This causes the factors $n^{-1/2}$ in (3.3.14) and $n^{1/2}$ in (3.3.15) to cancel. We find,

$$\begin{aligned} ||\sum_{j=1}^{n} \langle x, k_j \rangle k_j||_{\mathrm{BMO}} &\leq [\![\mathcal{C}]\!] n^{-1/2} || \sum_{j=1}^{n} k_j |I_j|^{1/2} ||_{\mathrm{BMO}} \\ &\leq [\![\mathcal{C}]\!]. \end{aligned}$$

Thus $||Px||_{\mathrm{BMO}} \leq [\![\mathcal{C}]\!]$, as claimed. □

Now we continue the analysis of rearrangement operators by merging the conclusion of Proposition 3.3.2 with the results on complex interpolation. This gives a criterion for the boundedness of rearrangement operators on L^p $(2 < p < \infty)$. Recall that a rearrangement τ is said to preserve the Carleson condition, if there exists $M > 0$ so that $[\![\tau\mathcal{C}]\!] \leq M[\![\mathcal{C}]\!]$, for every collection of dyadic intervals $\mathcal{C}$.

Theorem 3.3.4. *Suppose that τ is a rearrangement that preserves the Carleson condition. Then for $2 \leq p < \infty$,*

$$T_p : \frac{h_I}{|I|^{1/p}} \mapsto \frac{h_{\tau(I)}}{|\tau(I)|^{1/p}},$$

extends to a bounded operator on L^p.

Proof. The operator T_p coincides with the operator

$$h_I \mapsto h_{\tau(I)} \left(\frac{|\tau(I)|}{|I|}\right)^{-1/p}.$$

The rearrangement operators $\{T_p : 2 \leq p < \infty\}$ are embedded in the *analytic* family of operators

$$R_z : h_I \mapsto h_{\tau(I)} \left(\frac{|\tau(I)|}{|I|}\right)^{(z-1)/2},$$

since $T_p = R_{1-2/p}$. By the BMO result, Proposition 3.3.2, and the interpolation Corollary 2.3.12 for analytic families of operators, the conclusion of Theorem 3.3.4 holds. The duality Theorem 1.2.8 and Corollary 2.3.12 apply here since for rearrangements the transposed operator coincides with the inverse operator. □

We have proven so far that if τ preserves the Carleson condition, then permuting the L^p-normalized Haar system by τ leads to a bounded operator on L^p whenever $2 < p < \infty$. We now turn to the converse of the interpolation result in Theorem 3.3.4. We wish to prove that if a rearrangement of the L^p normalized Haar system is bounded in L^p, then the same rearrangement of the L^∞ normalized Haar system is bounded in BMO. Thus we intend to establish an extrapolation result from L^p to BMO. We saw already a proof of this as a corollary to G. Pisier's renorming of H^1. Different in spirit is the proof given below.

Theorem 3.3.5. *If there exists $p > 2$ such that the linear extension of*

$$T_p : \frac{h_I}{|I|^{1/p}} \mapsto \frac{h_{\tau(I)}}{|\tau(I)|^{1/p}}$$

defines a bounded operator on L^p, then τ preserves the Carleson condition, and $T_\infty : h_I \mapsto h_{\tau(I)}$ extends linearly to a bounded operator on BMO satisfying

$$||T_\infty||_{BMO}^{1-2/p} \leq C||T_p||_p.$$

Proof. We assume that τ fails to preserve the Carleson constant. We will show that this assumption implies that the operator T_p maps the unit vector basis of l^p_N into the unit vector basis of l^2_N, for any $N \in \mathbb{N}$. However for $p > 2$, and N large enough, this contradicts the hypothesis that T_p is bounded on L^p.

Now we fix $N \in \mathbb{N}$. We assume that there exists a collection of dyadic intervals $\mathcal{C}$ such that

$$[\![\tau(\mathcal{C})]\!] > 16N[\![\mathcal{C}]\!]. \tag{3.3.16}$$

In a first step we show how (3.3.16) implies that there exists a collection of dyadic intervals $\mathcal{A}$, such that

$$[\![\mathcal{A}]\!] \leq 2, \quad \text{and} \quad [\![\tau(\mathcal{A})]\!] \geq 2N. \tag{3.3.17}$$

Assuming that $[\![\mathcal{C}]\!]$ is an integer, we let $A = 4[\![\mathcal{C}]\!]$. By the positive homogeneity of Carleson constants, as expressed in Lemma 3.1.3, there exist pairwise disjoint collections of dyadic intervals $\mathcal{C}_1, \dots, \mathcal{C}_A$, such that

$$\mathcal{C} = \mathcal{C}_1 \cup \dots \cup \mathcal{C}_A,$$

and

$$[\![\mathcal{C}_j]\!] \leq 2, \quad \text{for} \quad j \leq A.$$

Next we let $M \geq 1$ be such that

$$[\![\tau(\mathcal{C}_j)]\!] \leq M[\![\mathcal{C}_j]\!],$$

for $j \le A$. We claim that this is possible only if $M > 2N$. To see that M has to be larger that $2N$ we split $\tau(\mathcal{C})$ and estimate as follows,

$$\begin{aligned} [\![\tau(\mathcal{C})]\!] &\le \sum_{j=1}^{A} [\![\tau(\mathcal{C}_j)]\!] \\ &\le M \sum_{j=1}^{A} [\![\mathcal{C}_j]\!] \\ &\le 8M[\![\mathcal{C}]\!]. \end{aligned} \tag{3.3.18}$$

On the other hand we assume that (3.3.16) holds, that is

$$[\![\tau(\mathcal{C})]\!] > 16N[\![\mathcal{C}]\!].$$

Comparing the inequalities (3.3.16) and (3.3.18) gives that

$$M > 2N.$$

Thus we showed the following statement: If τ fails to preserve the Carleson condition, then for any $N \in \mathbb{N}$, there exists a collection of dyadic intervals $\mathcal{A}$ for which,

$$[\![\mathcal{A}]\!] \le 2, \quad \text{and} \quad [\![\tau(\mathcal{A})]\!] \ge 2N.$$

Now we apply the condensation lemma to the collection $\tau(\mathcal{A})$. By Lemma 3.1.4, the condensation lemma, there exists $K \in \tau(\mathcal{A})$, such that

$$|G_N(K, \tau(\mathcal{A}))^*| > \frac{1}{2}|K|. \tag{3.3.19}$$

Combining the estimate (3.3.19) with the fact that $\mathcal{A}$ satisfies the 2-Carleson condition, we will next construct $f \in L^p$, with Haar support in $\mathcal{A}$, such that

$$||S(T_p f)||_p \ge ||S(f)||_p \frac{1}{32} N^{1/2-1/p}.$$

Clearly for $p > 2$ and N large enough, this estimate contradicts the boundedness of $||T_p||_p$. For $i \le N$ we define

$$\mathcal{E}_i = G_i(K, \tau(\mathcal{A})) \quad \text{and} \quad \mathcal{F}_i = \tau^{-1}(\mathcal{E}_i). \tag{3.3.20}$$

For $i \le N$, and $J \in \mathcal{F}_i$ we define coefficients as

$$b_J = |\tau(J)|^{1/p}. \tag{3.3.21}$$

Then let

$$f = \sum_{i=1}^{N} \sum_{J \in \mathcal{F}_i} b_J \frac{h_J}{|J|^{1/p}}.$$

Applying the rearrangement operator T_p to f gives

$$T_p f = \sum_{i=1}^{N} \sum_{J \in \mathcal{F}_i} b_J \frac{h_{\tau(J)}}{|\tau(J)|^{1/p}}.$$

We continue with an upper bound for $||S(f)||_p$. Note that the Haar support of f is the collection $\bigcup_{i=1}^{N} \mathcal{F}_i$ which is contained in $\mathcal{A}$. Hence as $[\![\mathcal{A}]\!] \le 2$, we apply Proposition 3.1.2 (the decomposition lemma) and Proposition 3.1.5 to obtain

$$||S(f)||_p \le 16 \left(\sum_{i=1}^{N} \sum_{J \in \mathcal{F}_i} |b_J|^p \right)^{1/p}. \tag{3.3.22}$$

Next recall that the coefficients are defined by $|b_J|^p = |\tau(J)|$. Hence for each i the inner sum in (3.3.22) is bounded as follows,

$$\begin{aligned} \sum_{J \in \mathcal{F}_i} |b_J|^p &= \sum_{J \in \mathcal{F}_i} |\tau(J)| \\ &\le |K|. \end{aligned}$$

Inserting this bound in the inequality (3.3.22) gives an upper bound for the norm of f as

$$\begin{aligned} ||S(f)||_p &\le 16 (\sum_{i=1}^{N} 1)^{1/p} |K|^{1/p} \\ &= 16 N^{1/p} |K|^{1/p}. \end{aligned} \tag{3.3.23}$$

Next we give a lower estimate for $||S(T_p f)||_p$. Here we will exploit (3.3.19) which followed from the condensation lemma. By (3.3.21) we obtain the following identities for $S(T_p f)$,

$$\begin{aligned} S^2(T_p f) &= \sum_{i=1}^{N} \sum_{J \in \mathcal{F}_i} |b_J|^2 |h_{\tau(J)}|^2 |\tau(J)|^{-2/p} \\ &= \sum_{i=1}^{N} \sum_{I \in \mathcal{E}_i} |h_I|^2. \end{aligned} \tag{3.3.24}$$

By (3.3.20) we have $\mathcal{E}_i = G_i(K, \tau(\mathcal{A}))$. Hence with (3.3.24) and (3.3.19) we obtain the lower bound

$$\begin{aligned} ||S(T_p f)||_p &\ge \frac{1}{4} ||(\sum_{i=1}^{N} 1_K)^{1/2}||_p \\ &= \frac{1}{4} (\sum_{i=1}^{N} 1)^{1/2} |K|^{1/p} \\ &= \frac{1}{4} N^{1/2} |K|^{1/p}. \end{aligned} \tag{3.3.25}$$

Clearly for N big enough and $p > 2$, the estimates (3.3.23) and (3.3.25) contradict the boundedness of T_p. Thus we showed that if a rearrangement τ does not preserve the the Carleson condition, then the induced rearrangement operator T_p is unbounded on L^p.

Let us now review the preceding argument to obtain extrapolation formulas for the norm, that is

$$||T_\infty||_{\mathrm{BMO}}^{1-2/p} \leq C||T_p||_p.$$

Let $N \in \mathbb{N}$ be a large integer. Combine Proposition 3.3.2 and (3.3.17) to see that

$$||T_\infty||_{\mathrm{BMO}} \geq CN^{1/2}$$

implies the existence of a collection $\mathcal{A}$ with $[\![\mathcal{A}]\!] \leq 2$ and $[\![\tau(\mathcal{A})]\!] \geq N$. Using the collection $\mathcal{A}$ we defined $f \in L^p$ so that

$$||S(T_p f)||_p \geq ||S(f)||_p \frac{1}{32} N^{1/2-1/p}.$$

Consequently the hypothesis that $||T_\infty||_{\mathrm{BMO}} \geq CN^{1/2}$ implies that

$$||T_p||_p \geq N^{1/2-1/p}.$$

As N was arbitrary we have

$$||T_p||_p \geq c||T_\infty||_{\mathrm{BMO}}^{2(1/2-1/p)}. \qquad \square$$

Rearrangements with Property $\mathcal{P}$

Now we introduce Property $\mathcal{P}$. We prove that a rearrangement preserves the Carleson packing condition if and only if it satisfies Property $\mathcal{P}$.

We say that a rearrangement τ satisfies the Property $\mathcal{P}$ if there exists $M > 0$ so that for every $\mathcal{B} \subset \mathcal{D}$ and $J \in \mathcal{D}$, the collection $\tau(\mathcal{B}) \cap J$ can be decomposed as a disjoint union

$$\bigcup_{i=1}^{\infty} \tau(\mathcal{L}_i) \cup \bigcup_{i=1}^{\infty} \mathcal{E}_i,$$

so that:

$$\bigcup_{i=1}^{\infty} \mathcal{E}_i \quad \text{satisfies the } M\text{-Carleson condition.} \tag{P 1}$$

$$\text{For every } I \in \mathcal{L}_i, \quad \frac{|\tau(I)|}{|I|} \leq M \frac{|\tau(\mathcal{L}_i)^*| + |\mathcal{E}_i^*|}{|\mathcal{L}_i^*|}. \tag{P 2}$$

$$\sum_{i=1}^{\infty} |\tau(\mathcal{L}_i)^*| \leq M|J| \sup_i [\![\tau^{-1}(\max \tau(\mathcal{L}_i))]\!]. \tag{P 3}$$

We start with the simple implication that a rearrangement preserves the Carleson packing condition provided that it satisfies Property $\mathcal{P}$.

Proposition 3.3.6. *If τ satisfies Property $\mathcal{P}$, then τ preserves Carleson' s condition.*

Proof. We let $\mathcal{B}$ be a collection of dyadic intervals satisfying Carleson's packing condition, and we assume that τ satisfies Property $\mathcal{P}$. We will show that then $\tau(\mathcal{B})$ also satisfies Carleson's packing condition. To this end we fix an arbitrary dyadic interval J, and we seek good estimates for the sum

$$S = \sum_{\tau(I)\in\tau(\mathcal{B})\cap J} |\tau(I)|.$$

Now, by Property $\mathcal{P}$ we may write down the decomposition

$$\tau(\mathcal{B}) \cap J = \bigcup_{i=1}^{\infty} \tau(\mathcal{L}_i) \cup \bigcup_{i=1}^{\infty} \mathcal{E}_i,$$

so that the conditions (P 1)–(P 3) are satisfied. Note that the collection $\tau(\mathcal{B})\cap J$ is the index set of the sum S. Thus the splitting of $\tau(\mathcal{B})\cap J$ induces a corresponding splitting of S as

$$S = S_1 + S_2,$$

where

$$S_1 = \sum_i \sum_{I\in\mathcal{L}_i} |\tau(I)|$$

and

$$S_2 = \sum_i \sum_{K\in\mathcal{E}_i} |K|.$$

We first focus on estimating S_1. We begin by rewriting the homogeneity condition (P 2) as

$$|\tau(I)| \leq M \frac{|\tau(\mathcal{L}_i)^*| + |\mathcal{E}_i^*|}{|\mathcal{L}_i^*|} |I|.$$

Inserting this estimate in the sum defining S_1, we obtain the upper bound

$$S_1 \leq M \sum_{i=1}^{\infty} \frac{|\tau(\mathcal{L}_i)^*| + |\mathcal{E}_i^*|}{|\mathcal{L}_i^*|} \sum_{I\in\mathcal{L}_i} |I|. \tag{3.3.26}$$

Next we will further analyze the inner sums of the form $\sum_{I\in\mathcal{L}_i} |I|$ appearing on the right-hand side of the estimate (3.3.26) for S_1. Note that the obvious inclusion $\mathcal{L}_i \subseteq \mathcal{B}$ gives the upper estimate

$$\sum_{I\in\mathcal{L}_i} |I| \leq [\![\mathcal{B}]\!] |\mathcal{L}_i^*|. \tag{3.3.27}$$

Next we insert the bounds (3.3.27) into the estimate (3.3.26). This gives that

$$S_1 \leq M[\![\mathcal{B}]\!] \sum_{i=1}^{\infty} |\tau(\mathcal{L}_i)^*| + M[\![\mathcal{B}]\!] \sum_{i=1}^{\infty} |\mathcal{E}_i^*|. \tag{3.3.28}$$

We continue by noting the following obvious inclusions $\tau^{-1}(\max \tau(\mathcal{L}_i)) \subseteq \mathcal{L}_i \subseteq \mathcal{B}$. This gives the upper bounds

$$[\![\tau^{-1}(\max \tau(\mathcal{L}_i))]\!] \le [\![\mathcal{B}]\!],$$

for any i. We invoke condition (P 3) to obtain,

$$\sum_{i=1}^{\infty} |\tau(\mathcal{L}_i)^*| \le M|J|[\![\mathcal{B}]\!]. \tag{3.3.29}$$

Next we observe that the sets $\mathcal{E}_i^*$ are contained in J. Thus for $\mathcal{E} = \bigcup \mathcal{E}_i$ we have by condition (P 1),

$$\begin{aligned}\sum_{i=1}^{\infty} |\mathcal{E}_i^*| &\le [\![\mathcal{E}]\!]|J| \\ &\le M|J|.\end{aligned} \tag{3.3.30}$$

Inserting the estimates (3.3.29) and (3.3.30) into (3.3.28), we obtain the upper bound for S_1,

$$\begin{aligned}S_1 &\le M[\![\mathcal{B}]\!] \sum_{i=1}^{\infty} |\tau(\mathcal{L}_i)^*| + |\mathcal{E}_i^*| \\ &\le M^2[\![\mathcal{B}]\!]([\![\mathcal{B}]\!]+1)|J|.\end{aligned}$$

Now we turn to the estimate for S_2. Note, that with $\mathcal{E} = \bigcup \mathcal{E}_i$ we have,

$$\begin{aligned}S_2 &\le [\![\mathcal{E}]\!]|J| \\ &\le M|J|.\end{aligned}$$

Now we combine the estimates for S_1 and S_2. We observe that $[\![\mathcal{B}]\!] \ge 1$ whenever $\mathcal{B}$ is not empty, hence we can combine the estimates for S_1 and S_2 to obtain

$$S_1 + S_2 \le 3M^2[\![\mathcal{B}]\!]^2|J|.$$

As $J \in \mathcal{D}$ was an arbitrary dyadic interval, we obtained that

$$[\![\tau(\mathcal{B})]\!] \le 3M^2[\![\mathcal{B}]\!]^2. \tag{3.3.31}$$

To complete the proof it remains to remove the quadratic term $[\![\mathcal{B}]\!]^2$ from the estimate (3.3.31) and replace it by the correct linear term $8[\![\mathcal{B}]\!]$.

To do so we use Lemma 3.1.3 asserting that the Carleson constant measures the size of collections like a positively homogeneous gauge function. We assume that $[\![\mathcal{B}]\!]$ is an integer and put $A = 4[\![\mathcal{B}]\!]$. By Lemma 3.1.3 we can decompose $\mathcal{B}$ into $\mathcal{B}_1, \dots, \mathcal{B}_A$ such that $[\![\mathcal{B}_i]\!] \le 2$, for $i \le A$. Now recall that in the first part of the proof we showed the estimates $[\![\tau(\mathcal{B}_i)]\!] \le 3M^2[\![\mathcal{B}_i]\!]^2$. Taking the sum we obtain finally that

$$\begin{aligned}[\![\tau(\mathcal{B})]\!] &\le \sum_{i=1}^{A} [\![\tau(\mathcal{B}_i)]\!] \\ &\le 48M^2[\![\mathcal{B}]\!].\end{aligned}$$

□

Now we will continue with a detailed analysis of rearrangements which preserve the Carleson condition. We show that for a rearrangement τ, Property $\mathcal{P}$ is equivalent to preserving the Carleson condition.

Theorem 3.3.7. *Let τ be a rearrangement of dyadic intervals. Suppose that there exists $M \geq 1$ such that*

$$[\![\tau(\mathcal{E})]\!] \leq M[\![\mathcal{E}]\!],$$

for any collection of dyadic intervals $\mathcal{E}$. Let $\mathcal{B}$ be a fixed collection of dyadic intervals, and let J be a fixed interval. Then, the collection $\tau(\mathcal{B}) \cap J$ can be decomposed as

$$\bigcup_{i=1}^{\infty} \tau(\mathcal{L}_i) \cup \bigcup_{i=1}^{\infty} \mathcal{E}_i,$$

so that the following conditions hold.

$$\bigcup \mathcal{E}_i \text{ satisfies the } 2M\text{-Carleson condition.} \tag{3.3.32}$$

$$\frac{|\tau(I)|}{|I|} \leq 2M \frac{|\tau(\mathcal{L}_i)^*| + |\mathcal{E}_i^*|}{|\mathcal{L}_i^*|}, \qquad \text{for } I \in \mathcal{L}_i. \tag{3.3.33}$$

$$\sum_{i=1}^{\infty} |\tau(\mathcal{L}_i)^*| \leq |J| 2M \sup_i [\![\tau^{-1}(\max \tau(\mathcal{L}_i))]\!]. \tag{3.3.34}$$

The natural approach to prove this decomposition of $\tau(\mathcal{B}) \cap J$ is by iterating the following stopping time rule: "Select the largest block of intervals $\mathcal{L}$ on which τ acts homogeneously." After a moments reflection we observe however that our rule formulates two competing (if not conflicting) conditions. Hence in the actual implementation of this idea we are confronted with combinatorial difficulties which we have to address first. We isolate the combinatorial aspects of the proof in Lemma 3.3.8. We recall first the following notation introduced earlier. For a collection of pairwise disjoint dyadic intervals $\mathcal{C}$ we let

$$Q(\mathcal{C}) = \bigcup_{K \in \mathcal{C}} \{J \in \mathcal{D} : J \subseteq K\}.$$

This is the collection of intervals which are contained in (at least) one of the intervals of $\mathcal{C}$. For a single dyadic interval I we let

$$Q(I) = \{J \in \mathcal{D} : J \subseteq I\}.$$

Lemma 3.3.8. *Let τ be a rearrangement of dyadic intervals. Suppose that there exists $M \geq 1$ such that*

$$[\![\tau(\mathcal{E})]\!] \leq M[\![\mathcal{E}]\!],$$

for any collection of dyadic intervals $\mathcal{E}$. Let $\mathcal{B}$ be a fixed collection of dyadic intervals, and let I_0 be a fixed interval. Then there exists a collection of pairwise disjoint dyadic intervals $\mathcal{C} \subseteq \mathcal{B} \cap I_0$ so that the following conditions hold:

1.
$$\sum_{L\in\mathcal{C}} |L| \le \frac{|I_0|}{2}.$$

2. *Let*
$$\mathcal{L} = \mathcal{B} \cap [Q(I_0) \setminus Q(\mathcal{C})],$$
then for $I \in \mathcal{L}$,
$$\frac{|\tau(I)|}{|I|} \le 2M \frac{|\tau(\mathcal{L})^*| + |\tau(\mathcal{C})^*|}{|I_0|}.$$

Proof. We will first construct an auxiliary collection $\mathcal{K}$ of colored dyadic intervals. In each step of the construction we modify $\mathcal{K}$ by either changing the color of a particular $I \in \mathcal{K}$, or by adding a new interval to $\mathcal{K}$. We write $\mathcal{K} = \mathcal{K} \cup \{L\}$ when we decide to add the dyadic interval L to the already existing collection $\mathcal{K}$. We shall now define two construction rules and a stopping rule. Following these rules we will eventually construct $\mathcal{K}$ and $\mathcal{C}$.

Rule 1. Suppose there exists a green interval $I \in \mathcal{K}$ such that at least one of the intervals in $G_1(I, \mathcal{B})$ is not contained in $\mathcal{K}$. Call it I_1. Now we define the color of I_1 as follows:

If
$$\frac{|\tau(I_1)|}{|I_1|} \le 2M \frac{|\tau(\mathcal{K} \cup \{I_1\})^*|}{|I_0|},$$
then we define I_1 to be a green interval and put $\mathcal{K} = \mathcal{K} \cup \{I_1\}$.

If
$$\frac{|\tau(I_1)|}{|I_1|} > 2M \frac{|\tau(\mathcal{K} \cup \{I_1\})^*|}{|I_0|},$$
then we define I_1 to be a red interval and put $\mathcal{K} = \mathcal{K} \cup \{I_1\}$.

Rule 2. Suppose $I \in \mathcal{K}$ is a red interval.

If
$$\frac{|\tau(I)|}{|I|} \le 2M \frac{|\tau(\mathcal{K})^*|}{|I_0|},$$
then we redefine the color of I from red to green. (Otherwise we don't change the color of I.)

Rule 3. Suppose that for each red interval $I \in \mathcal{K}$ we have
$$\frac{|\tau(I)|}{|I|} > 2M \frac{|\tau(\mathcal{K})^*|}{|I_0|},$$
then we define $\mathcal{C}$ to be the collection of red intervals in $\mathcal{K}$.

Next we decree how these rules are to be applied.

We start by defining I_0 to be a green interval and by putting $\mathcal{K} := \{I_0\}$. Then we apply Rule 1 until for every green interval $I \in \mathcal{K}$, the following condition holds true

$$G_1(I, \mathcal{B}) \subseteq \mathcal{K}.$$

Clearly in this case Rule 1 can't be applied any longer.

We now apply Rule 2 to the red intervals of $\mathcal{K}$.

If a new green interval is created by applying Rule 2, then we apply Rule 1, as above, and thereafter Rule 2 again, etc.

If we don't create new green intervals by applying Rule 2 to the red intervals of $\mathcal{K}$, then the stopping criterion of Rule 3 holds true. Then finally we define $\mathcal{C}$ to be the red intervals in $\mathcal{K}$.

Let $\mathcal{C} \subset \mathcal{K} \subseteq \mathcal{B} \cap I_0$ be the collections constructed by applying our three rules in the way described above. Now we will verify that $\mathcal{C}$ is a collection of pairwise disjoint intervals satisfying

$$\sum_{L \in \mathcal{C}} |L| \leq \frac{|I_0|}{2}.$$

We begin by checking that $\mathcal{C}$ is in fact a collection of pairwise disjoint dyadic intervals: Recall that during the construction, each interval in $\mathcal{K}$ has been colored. Note that only subintervals of green intervals can be placed into $\mathcal{K}$ by Rule 1. Hence, if $I \in \mathcal{K}$ is red and K is strictly contained in I, then $K \notin \mathcal{K}$. Therefore any two red intervals $L, K \in \mathcal{K}$ are necessarily disjoint. Hence, $\mathcal{C}$ is a collection of pairwise disjoint intervals.

Next we show that $\mathcal{C}$ covers only a small fraction of I_0. First we note that by summing over intervals for which the stopping condition of Rule 3 is satisfied, we obtain the upper bound

$$\frac{1}{|I_0|} \sum_{I \in \mathcal{C}} |I| \leq \frac{1}{2M|\tau(\mathcal{K})^*|} \sum_{I \in \mathcal{C}} |\tau(I)|.$$

Second we recall that τ preserves the Carleson condition. Hence the collection $\tau(\mathcal{C})$ satisfies the M-Carleson condition, since $\mathcal{C}$ satisfies the 1-Carleson condition. Therefore using that $\tau(\mathcal{K}) \supseteq \tau(\mathcal{C})$, we obtain the estimate

$$\frac{1}{|\tau(\mathcal{K})^*|} \sum_{I \in \mathcal{C}} |\tau(I)| \leq \frac{1}{|\tau(\mathcal{C})^*|} \sum_{I \in \mathcal{C}} |\tau(I)| \leq M.$$

Combining the last two estimates we obtain that

$$\sum_{I \in \mathcal{C}} |I| \leq \frac{|I_0|}{2}.$$

Next we verify that the collection of green intervals of $\mathcal{K}$ coincides with the collection $\mathcal{L} = \mathcal{B} \cap [Q(I_0) \setminus Q(\mathcal{C})]$. Recall that $K \notin \mathcal{K}$ when K is strictly contained in $I \in \mathcal{C}$. Recall also that $I_0 \in \mathcal{K}$ and if $K \in \mathcal{K}$, then every interval L satisfying $L \in \mathcal{B}$ and $I_0 \supseteq L \supseteq K$ is also an element of $\mathcal{K}$. Summing up we showed that

$$\mathcal{B} \cap [Q(I_0) \setminus Q(\mathcal{C})] = \mathcal{K} \setminus \mathcal{C},$$

or equivalently that $\mathcal{L}$ consists of the green intervals of $\mathcal{K}$. Recall that $I \in \mathcal{K}$ is a green interval when

$$\frac{|\tau(I)|}{|I|} \leq 2M \frac{|\tau(\mathcal{K})^*|}{|I_0|}.$$

Clearly, $|\tau(\mathcal{K})^*| \leq |\tau(\mathcal{L})^*| + |\tau(\mathcal{C})^*|$. Therefore we arrived at the following estimate holding for intervals $I \in \mathcal{L}$,

$$\frac{|\tau(I)|}{|I|} \leq 2M \frac{|\tau(\mathcal{L})^*| + |\tau(\mathcal{C})^*|}{|I_0|}.$$

This completes the proof of the lemma. □

Repeatedly applying Lemma 3.3.8 gives the proof of Theorem 3.3.7.

Proof of Theorem 3.3.7. Fix $J \in \mathcal{D}$. Let $\mathcal{G}_0$ be the maximal intervals of $\tau^{-1}\{\tau(\mathcal{B}) \cap J\}$. Clearly $\mathcal{G}_0$ is a collection of pairwise disjoint intervals. Now starting with $\mathcal{G}_0$ we construct a collection $\mathcal{G}_1$ as follows. We fix $I \in \mathcal{G}_0$, and apply Lemma 3.3.8 to the collection $\mathcal{B} \cap I$. This gives a collection of pairwise disjoint dyadic intervals $\mathcal{C}_I$, and $\mathcal{L}_I = \mathcal{B} \cap [Q(I) \setminus Q(\mathcal{C}_I)]$, which satisfy the conditions

$$\sum_{L \in \mathcal{C}_I} |L| \leq \frac{|I|}{2}$$

and

$$\frac{|\tau(K)|}{|K|} \leq 2M \frac{|\tau(\mathcal{L}_I)^*| + |\tau(\mathcal{C}_I)^*|}{|I|},$$

for $K \in \mathcal{L}_I$.

Now we let $G_1(\mathcal{C}_I, \mathcal{B})$ be the collection of maximal intervals of $\mathcal{B}$ which lie underneath $\mathcal{C}_I$. That is we put

$$G_1(\mathcal{C}_I, \mathcal{B}) = \bigcup_{\{L \in \mathcal{C}_I\}} G_1(L, \mathcal{B}).$$

Clearly, we defined $G_1(\mathcal{C}_I, \mathcal{B})$ to be a collection of pairwise disjoint intervals. Hence also the following is a collection of pairwise disjoint intervals,

$$\mathcal{G}_1 = \bigcup_I G_1(\mathcal{C}_I, \mathcal{B}),$$

here I ranges over $\mathcal{G}_0$. This completes the first step of the proof.

Now we repeat the first step and construct $\mathcal{G}_2$ from $\mathcal{G}_1$. We choose now $K \in \mathcal{G}_1$, and we let $\mathcal{C}_K$, respectively $\mathcal{L}_K$, satisfy the conclusions of Lemma 3.3.8. Now we define

$$G_1(\mathcal{C}_K, \mathcal{B}) = \bigcup_{\{L \in \mathcal{C}_K\}} G_1(L, \mathcal{B}),$$

and

$$\mathcal{G}_2 = \bigcup_K G_1(\mathcal{C}_K, \mathcal{B}),$$

where K ranges over $\mathcal{G}_1$. Note that $\mathcal{G}_2$ is a collection of pairwise disjoint dyadic intervals, since $\mathcal{G}_1$ was. Continuing now inductively we construct the sequence $\{\mathcal{G}_k\}$, and simultaneously, we obtain the decomposition

$$\tau^{-1}\{\tau(\mathcal{B}) \cap J\} = \bigcup \mathcal{L}_I \cup \bigcup \mathcal{C}_I,$$

where I ranges over $\mathcal{G} = \bigcup_{k=0}^{\infty} \mathcal{G}_k$. Note that our recursive definitions give automatically the identity

$$\bigcup_{I \in \mathcal{G}} G_1(\mathcal{C}_I, \mathcal{B}) = \bigcup_{k=1}^{\infty} \mathcal{G}_k.$$

We will now verify that $\mathcal{G}$ satisfies the 2-Carleson condition. Recall that $\mathcal{G}_k$ is a collection of disjoint intervals and that $\mathcal{G}_{k+1} = \bigcup G_1(\mathcal{C}_I, \mathcal{B})$ where I ranges over $\mathcal{G}_k$. Hence,

$$\mathcal{G}_{k+1} \subset Q(\mathcal{G}_k). \tag{3.3.35}$$

Note that by Lemma 3.3.8, for $I \in \mathcal{G}_k$, the following estimate holds,

$$\sum_{K \in \mathcal{G}_{k+1} \cap I} |K| \leq \frac{|I|}{2}.$$

Hence, by induction, for $I \in \mathcal{G}_k$,

$$\sum_{K \in \mathcal{G}_{k+l} \cap I} |K| \leq \frac{|I|}{2^l}. \tag{3.3.36}$$

From these properties we can easily evaluate the Carleson constant of $\mathcal{G}$. Indeed, by (3.3.36) and summing a geometric series we obtain that $[\![\mathcal{G}]\!] \leq 2$. We will show next that the collections $\{\mathcal{L}_I : I \in \mathcal{G}\}$ satisfy the estimate

$$\frac{1}{|J|} \sum_{\{I \in \mathcal{G}\}} |\tau(\mathcal{L}_I)^*| \leq 2M \sup_{\{I \in \mathcal{G}\}} [\![\tau^{-1}(\max \tau(\mathcal{L}_I))]\!].$$

To see why this holds we will examine the inductive construction of the collections $\{\mathcal{L}_I : I \in \mathcal{G}\}$. Recall that $\mathcal{L}_I \subseteq Q(I) \setminus Q(\mathcal{C}_I)$. Hence

$$\bigcup_{I \in \mathcal{G}_k} \mathcal{L}_I \subseteq Q(\mathcal{G}_k) \setminus Q(\mathcal{G}_{k+1}). \tag{3.3.37}$$

Recall also that $\mathcal{L}_I^* \cap \mathcal{L}_K^* = \emptyset$, when $I, K \in \mathcal{G}_k$ and clearly the following inclusion holds true,

$$\tau^{-1}(\max \tau(\mathcal{L}_I)) \subseteq \mathcal{L}_I.$$

Now let

$$\mathcal{V}_k = \bigcup \tau^{-1}(\max \tau(\mathcal{L}_I)),$$

where the union is taken over $I \in \mathcal{G}_k$. By (3.3.35) – (3.3.37), the hypothesis of Proposition 3.1.5 is satisfied. Its BMO assertion gives

$$[\![\bigcup \mathcal{V}_k]\!] \leq 2 \max [\![\mathcal{V}_k]\!].$$

Consequently we obtain that

$$\left[\!\!\left[\bigcup_{I \in \mathcal{G}} \tau^{-1}(\max \tau(\mathcal{L}_I))\right]\!\!\right] \leq 2 \sup_{I \in \mathcal{G}} [\![\tau^{-1}(\max \tau(\mathcal{L}_I))]\!].$$

With these observation we now obtain the estimates

$$\begin{aligned} \frac{1}{|J|} \sum_{I \in \mathcal{G}} |\tau(\mathcal{L}_I)^*| &\leq \left[\!\!\left[\bigcup \max \tau(\mathcal{L}_I)\right]\!\!\right] \\ &\leq M \left[\!\!\left[\bigcup \tau^{-1}(\max \tau(\mathcal{L}_I))\right]\!\!\right] \\ &\leq 2M \sup_I [\![\tau^{-1}(\max \tau(\mathcal{L}_I))]\!]. \end{aligned}$$

We now let $\mathcal{E}_I = \tau(\mathcal{C}_I)$. We prove next that the union

$$\mathcal{E} = \bigcup_{I \in \mathcal{G}} \mathcal{E}_I$$

satisfies the $2M$-Carleson condition. To do so we only need to combine the fact that $\bigcup \mathcal{C}_I$ is a subcollection of $\mathcal{G}$, and the fact $\mathcal{G}$ satisfies the 2-Carleson condition. Indeed, as τ preserves the Carleson condition, we have the following string of estimates.

$$\begin{aligned} [\![\mathcal{E}]\!] &= [\![\tau(\mathcal{G})]\!] \\ &\leq M[\![\mathcal{G}]\!] \\ &\leq 2M. \end{aligned}$$

Finally we relabel the collections $\{\mathcal{C}_I, \mathcal{L}_I : I \in \mathcal{G}\}$ as $\{\mathcal{C}_i, \mathcal{L}_i : i \in \mathbf{N}\}$. We also let $\mathcal{E}_i = \tau(\mathcal{C}_i)$, where the actual enumeration is not at all important. We have thus verified that the decomposition

$$\tau(\mathcal{B}) \cap J = \bigcup_{i=1}^{\infty} \tau(\mathcal{L}_i) \cup \bigcup_{i=1}^{\infty} \mathcal{E}_i,$$

satisfies the conclusions of Theorem 3.3.7. □

Norm estimates for U_m and T_m

In Section 1.4 and Section 1.5 we presented Figiel's work on biorthogonal systems and integral operators. We used there (without proof) that on L^p, $1 < p < \infty$ the rearrangement operator T_m and the shift operator U_m have norm bounded by $C_p \log_2(4+|m|)$. In this subsection we verify these estimates and thereby complete the proof of Theorem 1.5.3. Let $m \in \mathbb{Z}$, and define

$$\mathcal{R}_m = \{I \in \mathcal{D} : I + m|I| \subseteq [0,1]\}.$$

In Figiel's work the analysis of integral operators is based on the rearrangement respectively splitting operators defined by the equations

$$U_m(h_I) = 1_{I+m|I|} - 1_I \quad \text{and} \quad T_m(h_I) = h_{I+m|I|},$$

for $I \in \mathcal{R}_m$. The logarithmic dependence on m in the estimates for $\|T_m\|$ and $\|U_m\|$ was crucial in applying T.Figiel's integral representation to the kernels of Calderon–Zygmund operators. The norm estimate for T_m as stated in Theorem 3.3.9 follows directly from E. M. Semenov's criterion (3.3.1) on boundedness of rearrangement operators. We present an independent proof below for which we use just Burkholder's Theorem 1.4.1 and E. M. Stein's martingale inequality (Theorem 1.4.2). Thus we show that

$$\int_0^1 \|\sum x_I T_m(h_I)\|_E^p \le C_p^p \log_2(4+|m|)^p \int_0^1 \|\sum x_I h_I\|_E^p, \tag{3.3.38}$$

for $x_I \in E$ whenever E is a Banach space with the UMD property. This conclusion can't be drawn from the methods of the previous subsections, since they are strictly(!) limited to the scalar case. Estimates analogous to (3.3.38) hold with T_m replaced by U_m.

Theorem 3.3.9. *On L^p $(1 < p < \infty)$ the following norm estimates hold true,*

$$\|U_m\|_{L^p} \le C_p \log_2(4+|m|) \quad \textit{and} \quad \|T_m\|_{L^p} \le C_p \log_2(4+|m|).$$

We prepare the proof by formulating a general criterion on $\tau : \mathcal{A} \mapsto \mathcal{D}$ ensuring that the linear extension of

$$U_\tau(h_I) = 1_{\tau(I)} - 1_I, \quad I \in \mathcal{A},$$

respectively $T_\tau(h_I) = h_{\tau(I)}$, defines a bounded operator on L^p. Let $\mathcal{A}$ be a collection of dyadic intervals. We say that an injective map $\tau : \mathcal{A} \mapsto \mathcal{D}$ satisfies Figiel's compatibility condition (F) if the following three conditions hold:

$$|\tau(J)| = |J| \text{ for } J \in \mathcal{A}. \tag{F 1}$$

$$\text{If } I \in \mathcal{A}, \text{ then } \tau(I) \notin \mathcal{A}. \tag{F 2}$$

$$\mathcal{C} = \{I, \tau(I), I \cup \tau(I) : I \in \mathcal{A}\} \text{ is a nested collection of sets.} \tag{F 3}$$

Lemma 3.3.10. *Let $\tau : \mathcal{A} \mapsto \mathcal{D}$ satisfy Figiel's compatibility condition (F). On L^p $(1 < p < \infty)$ the following norm estimates hold,*

$$\|U_\tau\|_p \leq B_p \quad \text{and} \quad \|T_\tau\|_p \leq K_p,$$

where $B_p = C2(p^* - 1)^3$ and $K_p = C2(p^* - 1)^3$.

Proof. Since, by hypothesis, the family

$$\mathcal{C} = \{I, \tau(I), I \cup \tau(I) : I \in \mathcal{A}\}$$

forms a nested collection of sets, we may split it canonically into generations. Let $G_n(\mathcal{C})$ denote the n-th generation of $\mathcal{C}$. We define first a sequence of σ-algebras using the generations of the nested collection $\mathcal{C}$ as follows. For $n \in \mathbb{N}_0$ let $\mathcal{F}_n$ denote the σ-algebra generated by $G_0(\mathcal{C}) \cup \cdots \cup G_n(\mathcal{C})$.

Observe that for every $I \in \mathcal{A}$ there exists a uniquely determined $n \in \mathbb{N}$ so that the union $I \cup \tau(I)$ is an atom for the σ-algebra $\mathcal{F}_n$. Thus we define

$$\mathcal{K}_n = \{I \in \mathcal{A} : I \cup \tau(I) \text{ is an atom in } \mathcal{F}_n\}.$$

Hence for every $I \in \mathcal{K}_n$, we obtain the identity

$$2\mathbb{E}(1_I \,|\mathcal{F}_n) = 1_I + 1_{\tau(I)}. \tag{3.3.39}$$

Next we prove the estimate for $\|U_\tau\|_p$. Let $\{a_I\}$ be a sequence of scalars. Let $\{r_I\}$ be an enumeration of the Rademacher sequence. The assumption that $\mathcal{C} = \{I, \tau(I), I \cup \tau(I) : I \in \mathcal{A}\}$ is a nested collection implies that the functions $\{1_{\tau(I)} - 1_I : I \in \mathcal{A}\}$ can be relabelled to become a martingale difference sequence. Since martingale differences are unconditional in L^p, by Burkholder's Theorem 1.4.1, we get

$$\left\|\sum_{I \in \mathcal{A}} a_I(1_{\tau(I)} - 1_I)\right\|_{L^p}^p \leq (p^* - 1)^p \int_0^1 \left\|\sum_{I \in \mathcal{A}} a_I r_I(t)(1_{\tau(I)} - 1_I)\right\|_{L^p}^p dt. \tag{3.3.40}$$

Recall (3.3.39) that $2\mathbb{E}(1_I \,|\mathcal{F}_n) = 1_I + 1_{\tau(I)}$. Hence

$$1_{\tau(I)} - 1_I = 2\mathbb{E}(1_I \,|\mathcal{F}_n) - 2 \cdot 1_I. \tag{3.3.41}$$

The purpose of rewriting $1_{\tau(I)} - 1_I$ in (3.3.41) is that the conditional expectation of 1_I appears on the right-hand side. Hence by Bourgain's version of Stein's inequality the right-hand side of (3.3.40) is bounded by

$$C_p^p \int_0^1 \left\|\sum_{I \in \mathcal{A}} a_I r_I(t) 1_I\right\|_{L^p}^p dt, \tag{3.3.42}$$

where $C_p = 4(p^* - 1)^2$. Combining (3.3.40)–(3.3.42) and applying again Burkholder's inequality we obtain that $\|U_\tau\|_p \leq B_p$.

Next we turn to estimating $||T_\tau||_p$. Applying the same method of proof we relate $T_\tau(h_I)$ to a conditional expectation of 1_I. We start with the Burkholder inequality (Theorem 1.1.5).

$$\left\|\sum_{I\in\mathcal{A}} a_I h_{\tau(I)}\right\|_{L^p}^p \le (p^*-1)^p \int_0^1 \left\|\sum_{I\in\mathcal{A}} a_I r_I(t) 1_{\tau(I)}\right\|_{L^p}^p dt. \tag{3.3.43}$$

By equation (3.3.39), the integral on the right-hand side of (3.3.43) is bounded by

$$2^p \int_0^1 \left\|\sum_{n=0}^{\infty}\sum_{I\in\mathcal{K}_n} a_I r_I(t)\mathbb{E}(1_I|\mathcal{F}_n)\right\|_{L^p}^p dt + \int_0^1 \left\|\sum_{I\in\mathcal{A}} a_I r_I(t) 1_I\right\|_{L^p}^p dt. \tag{3.3.44}$$

Next, Bourgain's version of Stein's inequality (Theorem 1.4.2) asserts that

$$\int_0^1 \left\|\sum_{n=0}^{\infty}\sum_{I\in\mathcal{K}_n} a_I r_I(t)\mathbb{E}(1_I|\mathcal{F}_n)\right\|_{L^p}^p dt \le (p^*-1)^p \int_0^1 \left\|\sum_{I\in\mathcal{A}} a_I r_I(t) 1_I\right\|_{L^p}^p dt. \tag{3.3.45}$$

Summing up, the estimates (3.3.43)–(3.3.45) imply that $||T_\tau||_p \le K_p$. □

By combinatorial considerations we reduce the norm estimates of Theorem 3.3.9 to those of the above lemma. This is done by decomposing $\mathcal{R}_m$ into roughly $\log(4+m)$ families of dyadic intervals, in such a way that on each sub-family Figiel's compatibility condition is satisfied for rearrangements closely related to the original one given by $\tau(I) = I + m|I|$.

Proof of Theorem 3.3.9. Let $m \in \mathbb{Z}$. We first treat the case where $|m| \ge 2$. By symmetry it suffices to consider only $m \ge 2$. Below we give a separate and simple argument for the case $|m| = 1$. Let

$$\mathcal{R}_m = \{I \in \mathcal{D} : I + m|I| \subseteq [0,1[\}.$$

We study in detail the action of the rearrangement $\tau(I) = I + m|I|$, for $I \in \mathcal{R}_m$.

For $n \in \mathbb{N}$ let $\mathcal{L}_n = \{I \in \mathcal{R}_m : |I| = 2^{-n}\}$. Next split $\mathcal{L}_n$ into four collections $\mathcal{K}_n^{(1)}, \dots, \mathcal{K}_n^{(4)}$ so that for $i \in \{1, \dots, 4\}$, the following conditions hold. Let $K_n^{(i)}$ be the point-set covered by the collection $\mathcal{K}_n^{(i)}$.

$$I \in \mathcal{K}_n^{(i)} \quad \text{implies} \quad \operatorname{dist}(\tau(I), K_n^{(i)}) \ge |I|,$$

and

$$I, J \in \mathcal{K}_n^{(i)} \quad \text{implies} \quad \operatorname{dist}(I \cup \tau(I), J \cup \tau(J)) \ge |I|.$$

Since τ is of the simple form $\tau(I) = I + m|I|$, it is straightforward to obtain the splitting of $\mathcal{L}_n$ into sub-collections with the above property. Define $k \in \mathbb{N}$ by the relation $k - 1 \le \log_2(m) \le k$. Next we fix $0 \le j \le k$, and take the union

$$\mathcal{E}_j^{(i)} = \bigcup_{n=1}^{\infty} \mathcal{K}_{n(k+1)+j}^{(i)}.$$

Thus we decomposed $\mathcal{R}_m$ into $C\log(4+m)$ families of the form $\mathcal{E}_j^{(i)}$. We drop the subscript j and superscript (i) and put

$$\mathcal{E} = \mathcal{E}_j^{(i)}.$$

We aim to show that the linear extension of the maps

$$U_m(h_I) = 1_{I+m|I|} - 1_I \quad \text{and} \quad T_m(h_I) = h_{I+m|I|} \quad \text{for} \quad I \in \mathcal{E},$$

are defining bounded operators on L^p with norm independent of m. Now we examine closely the collection $\mathcal{E}$. Let $I \in \mathcal{E}$, then $|I| \le 2^{-k}$. By $I^{(k)}$ we denote the dyadic interval satisfying $I \subset I^{(k)}$ and $|I^{(k)}| = 2^k|I|$. Depending on the position of I in the unit interval the following cases may hold for $\tau(I) = I + m|I|$:

1. $\tau(I) \subseteq I^{(k)}$.
2. $\tau(I) \not\subseteq I^{(k)}$. If this holds, then the following three, mutually exclusive subcases may occur:
   (a) Neither $\tau(I)$ nor I are touching the right endpoint of $I^{(k)}$. In that case $\tau(I)$ and I are called good intervals.
   (b) I touches the right endpoint of $I^{(k)}$. In that case we call I a bad interval.
   (c) $\tau(I)$ touches the right endpoint of $I^{(k)}$. In that case we call $\tau(I)$ a bad interval.

Note that for $I \in \mathcal{E}$ precisely one of the above conditions (1), (2.a), (2.b) or (2.c) are satisfied. Accordingly the collection $\mathcal{E}$ splits into the following four disjoint subcollections.

$$\begin{aligned}
\mathcal{A} &= \{I \in \mathcal{E} : \tau(I) \subseteq I^{(k)}\} \\
\mathcal{G} &= \{I \in \mathcal{E} : \tau(I) \not\subseteq I^{(k)}, \text{ and } \tau(I) \text{ and } I \text{ are good intervals.}\} \\
\mathcal{B}_1 &= \{I \in \mathcal{E} : \tau(I) \not\subseteq I^{(k)}, I \text{ is a bad interval.}\} \\
\mathcal{B}_2 &= \{I \in \mathcal{E} : \tau(I) \not\subseteq I^{(k)}, \tau(I) \text{ is a bad interval.}\}
\end{aligned}$$

Case 1. Observe that by our definition of $\mathcal{E}$ and $\mathcal{A}$ every $I \in \mathcal{A}$ satisfies Figiel's compatibility condition (F). Hence, by Lemma 3.3.10, the linear extension of the maps

$$U_m(h_I) = 1_{I+m|I|} - 1_I \quad \text{and} \quad T_m(h_I) = h_{I+m|I|} \quad \text{for} \quad I \in \mathcal{A},$$

define bounded operators on L^p with norm independent of m.

Case 2a. Next we turn to the analysis of the collection $\mathcal{G}$. We will split $\mathcal{G}$ into $\mathcal{G}_1$ and $\mathcal{G}_2$ so that the restriction of $I \mapsto I + m|I|$ to $\mathcal{G}_i$, $i \in \{1,2\}$, satisfies Figiel's compatibility condition. With the unconditionality of the Haar system and Lemma 3.3.10 we thus obtain that the linear extension of

$$U_m(h_I) = 1_{I+m|I|} - 1_I \quad \text{and} \quad T_m(h_I) = h_{I+m|I|} \quad \text{for} \quad I \in \mathcal{G}_i,$$

define bounded operators on L^p.

Now we turn to defining the splitting $\mathcal{G} = \mathcal{G}_1 \cup \mathcal{G}_2$. Fix $K \in \mathcal{G} \cup \tau(\mathcal{G})$ and $J \in \mathcal{G}$ with $|J| < |K|$. Denote by $\text{co}[J, \tau(J)]$ the convex hull of the intervals J and $\tau(J)$. We let $\mathcal{L}(K)$ be those intervals $J \in \mathcal{G}$ for which $\text{co}[J, \tau(J)]$ intersects both K and the complement of K. Thus

$$\mathcal{L}(K) = \{J \in \mathcal{G} : |J| < |K|, \text{co}[J, \tau(J)] \cap K \neq \emptyset, \text{ and } \text{co}[J, \tau(J)] \cap ([0,1] \setminus K) \neq \emptyset\}.$$

Finally we let

$$\mathcal{C}(K) = \{J, \tau(J) : J \in \mathcal{L}(K)\}.$$

Due to the defining properties of $\mathcal{G}$ the collection $\mathcal{C}(K)$ is a collection of *pairwise disjoint* dyadic intervals. Moreover it satisfies the following remarkable properties.

(a) For each $J \in \mathcal{G} \cup \tau(\mathcal{G})$ there exists at most one $K \in \mathcal{G} \cup \tau(\mathcal{G})$ so that

$$J \in \mathcal{C}(K).$$

Thus each $J \in \mathcal{C}(K)$ determines uniquely K and hence $\mathcal{C}(K)$.

(b) If $L \in \mathcal{G} \cup \tau(\mathcal{G})$ and $L \notin \mathcal{C}(K)$, then

$$L \cup \tau(L) \subseteq K, \text{ or } L \cup \tau(L) \subseteq [0,1] \setminus K. \tag{3.3.46}$$

Exploiting the above properties of $\mathcal{C}(K)$ allows us to define a two-coloring of $\mathcal{G} \cup \tau(\mathcal{G})$ so that the following conditions hold.

(a) For each $K \in \mathcal{G} \cup \tau(\mathcal{G})$ the collection $\mathcal{C}(K)$ is monochromatic.

(b) If the color of $K \in \mathcal{G} \cup \tau(\mathcal{G})$ is already determined, then each $J \in \mathcal{C}(K)$ satisfies

$$\text{color}(K) \neq \text{color}(J).$$

(c) For each $K \in \mathcal{G}$,

$$\text{color}(K) = \text{color}(\tau(K)).$$

Assume that we used the colors black and white to describe the two-coloring of $\mathcal{G}$. Then we define

$$\mathcal{G}_1 = \{K \in \mathcal{G} : \text{color}(K) = \text{white } \} \text{ and } \mathcal{G}_2 = \{K \in \mathcal{G} : \text{color}(K) = \text{black } \}.$$

Thus we obtained a splitting $\mathcal{G} = \mathcal{G}_1 \cup \mathcal{G}_2$ so that $\mathcal{G}_1$ and $\mathcal{G}_2$ satisfy Figiel's compatibility condition. By Lemma 3.3.10 for $i \in \{1,2\}$, the linear extension of

$$h_I \mapsto 1_{I+m|I|} - 1_I \quad \text{for} \quad I \in \mathcal{G}_i,$$

and

$$h_I \mapsto h_{I+m|I|} \quad \text{for} \quad I \in \mathcal{G}_i,$$

gives bounded operators on L^p.

Case 2b. Next we turn to the analysis of the collection $\mathcal{B}_1$. By reduction to the previous case we will show that the linear extension of the map

$$U_m(h_I) = 1_{I+m|I|} - 1_I \quad \text{for} \quad I \in \mathcal{B}_1,$$

defines a bounded operator on L^p.

Let $I \in \mathcal{B}_1$. Recall that then, by definition, the right endpoint of I coincides with the right endpoint of $I^{(k)}$, and that the interval $\tau(I)$ is well inside of $\tau(I)^{(k)}$, satisfying

$$\operatorname{dist}(\tau(I), [0,1] \setminus \tau(I)^{(k)}) \geq |I|.$$

Next we let J be the dyadic predecessor of I. Thus J is a dyadic interval such that $J \supseteq I$ and $|J| = 2|I|$. Let J_1 be the left half of J and let J_2 be the right half of J. Our assumption that the right endpoint of I coincides with the right endpoint of $I^{(k)}$, implies that the interval I is the right half of its dyadic predecessor. Thus

$$I = J_2.$$

Next we define an injective map σ by interchanging J_2 and J_1. Hence,

$$\sigma(I) = \sigma(J_2) = J_1 \quad \text{and} \quad \sigma(J_1) = J_2 = I.$$

Observe that $\sigma : \mathcal{B}_1 \mapsto \mathcal{D}$ satisfies Figiel's compatibility condition (F). Hence by Lemma 3.3.10 the linear extension of the map

$$S(h_I) = h_{\sigma(I)} \quad \text{for} \quad I \in \mathcal{B}_1$$

defines a bounded operator in L^p. Let $\mathcal{R} = \sigma(\mathcal{B}_1)$, and define $\rho = \tau \circ \sigma$. It is easy to see that for $K \in \mathcal{R}$ the intervals $\rho(K)$ and K are good intervals with respect to the map $\rho = \tau \circ \sigma$. Consequently, by case 2a, the linear extension of

$$R(h_K) = 1_{\rho(K)} - 1_K \quad \text{for} \quad K \in \mathcal{R}$$

is bounded on L^p. Hence the composition RS is a well-defined and bounded linear operator on L^p. As $\sigma(\sigma(I)) = I$ the composition RS coincides with the linear extension of the map

$$U_m(h_I) = 1_{I+m|I|} - 1_I \quad \text{for} \quad I \in \mathcal{B}_1.$$

A moment's reflection yields that simultaneously we proved that the same holds for the map

$$T_m(h_I) = h_{I+m|I|} \quad \text{for} \quad I \in \mathcal{B}_1.$$

Case 2c. It remains to analyze the contribution of the collection $\mathcal{B}_2$. Again we proceed by reduction to case 2a. Recall that $I \in \mathcal{B}_2$ if $\tau(I) \not\subseteq I^{(k)}$, so that $\tau(I)$ touches the right endpoint of $I^{(k)}$. Precisely, the right endpoint of $I^{(k)}$ coincides with the left endpoint of $\tau(I)$. Define a new rearrangement by shifting the interval $\tau(I)$ to the right by an amount of $|I|$. Thus for $I \in \mathcal{B}_2$ we let

$$\tau_1(I) = \tau(I) + |I|.$$

Note that for $I \in \mathcal{B}_2$ the union of $\tau(I)$ and $\tau_1(I)$ is a dyadic interval. Define $P(\tau(I))$ to be the dyadic predecessor of $\tau(I)$. Note that for $I \in \mathcal{B}_2$ we have that

$$P(\tau(I)) = \tau(I) \cup \tau_1(I).$$

Observe that $\tau_1(I)$ is well inside of $\tau(I)^{(k)}$. That is,

$$\operatorname{dist}(\tau_1(I), [0,1] \setminus \tau(I)^{(k)}) \geq |I|.$$

On the other hand for $I \in \mathcal{B}_2$ we have that, by definition,

$$\operatorname{dist}(I, [0,1] \setminus I^{(k)}) \geq |I|.$$

Thus for $I \in \mathcal{B}_2$ the intervals $\tau_1(I)$ and I are good intervals with respect to the rearrangement τ_1. By the estimate of case 2a, the linear extension of the shift

$$h_I \mapsto 1_{\tau_1(I)} - 1_I \quad \text{for} \quad I \in \mathcal{B}_2$$

extends to a bounded operator on $L^p, 1 < p < \infty$. For the same reason the rearrangement operator defined by

$$h_I \mapsto h_{\tau_1(I)} \quad \text{for} \quad I \in \mathcal{B}_2$$

extends boundedly to $L^p, 1 < p < \infty$. Notice that the map $h_{\tau_1(I)} \mapsto h_{P(\tau(I))}$ for $I \in \mathcal{B}_2$ extends to a bounded linear operator on L^p. Next observe that for $I \in \mathcal{B}_2$ the following identity holds,

$$1_{\tau(I)} = 1_{\tau_1(I)} + h_{P(\tau(I))},$$

and consequently

$$1_{\tau(I)} - 1_I = \{1_{\tau_1(I)} - 1_I\} + h_{P(\tau(I))}. \tag{3.3.47}$$

We observed that the two terms on the right-hand side of (3.3.47) represent maps that extend to a bounded linear operator on $L^p, 1 < p < \infty$. Thus we showed that

$$U_m(h_I) = 1_{\tau(I)} - 1_I \quad \text{for} \quad I \in \mathcal{B}_2$$

defines a bounded linear operator on $L^p, 1 < p < \infty$.

The case $|m| = 1$. We now analyze T_m and U_m when $|m| = 1$. We present also the estimates for the rearrangement operators since we are still interested in having a proof for vector-valued coefficients taken from a UMD space. The scalar case is of course covered by E. M. Semenov's criterion (3.3.1). By symmetry it suffices to consider $m = 1$. Let

$$\mathcal{R}_1 = \{I \in \mathcal{D} : I + |I| \subseteq [0,1]\}.$$

For $I \in \mathcal{R}_1$ let $I^{(1)}$ denote the dyadic predecessor of I. Next split $\mathcal{R}_1$ as follows,

$$\mathcal{A} = \{I \in \mathcal{R}_1 : I + |I| \subseteq I^{(1)}\} \quad \text{and} \quad \mathcal{B} = \mathcal{R}_1 \setminus \mathcal{A}.$$

Observe that the rearrangement

$$I \mapsto I + |I| \quad \text{for} \quad I \in \mathcal{A}$$

satisfies Figiel's compatibility condition, hence the linear extension of the maps

$$T_1(h_I) = h_{I+|I|} \quad \text{and} \quad U_1(h_I) = 1_{I+|I|} - 1_I \quad \text{for} \quad I \in \mathcal{A}$$

define bounded operators on L^p $(1 < p < \infty)$.

Now we consider the collection $\mathcal{B} = \mathcal{R}_1 \setminus \mathcal{A}$. Define the map σ by shifting the intervals of $\mathcal{B}$ to the left by one unit, thus

$$\sigma(I) = I - |I|.$$

Note that $\sigma : \mathcal{B} \to \mathcal{D}$ satisfies Figiel's compatibility condition, hence the linear extension of

$$S(h_I) = h_{\sigma(I)} \quad \text{for} \quad I \in \mathcal{B}$$

defines a bounded operator on L^p. Moreover the following identity holds,

$$T_1(h_I) = T_2(S(h_I)) \quad \text{for} \quad I \in \mathcal{B}.$$

Consequently T_1 extends to a bounded operator since T_2 is bounded and S extends. Finally we analyze U_1 on $\mathcal{B}$. The starting point is the identity

$$1_{I+|I|} = 1_{I+2|I|} + h_{I^{(1)}+|I^{(1)}|} \quad \text{for} \quad I \in \mathcal{B}, \tag{3.3.48}$$

where $I^{(1)}$ denotes the dyadic predecessor of I. Recall that $U_1(h_I) = 1_{I+|I|} - 1_I$ and $U_2(h_I) = 1_{I+2|I|} - 1_I$. Next subtract on both sides of (3.3.48) the function 1_I. Then rewrite (3.3.48) as

$$U_1(h_I) = U_2(h_I) + h_{I^{(1)}+|I^{(1)}|} \quad \text{for} \quad I \in \mathcal{B}. \tag{3.3.49}$$

Finally $h_{I^{(1)}+|I^{(1)}|} = T_1(h_{I^{(1)}})$, hence

$$U_1(h_I) = U_2(h_I) + T_1(h_{I^{(1)}}) \quad \text{for} \quad I \in \mathcal{B}. \tag{3.3.50}$$

The right-hand side of (3.3.50) consists of maps that extend to bounded linear operators on L^p, hence the same is true for the left-hand side. □

Rearrangement operators on $\mathbf{L}^1$

We close this section with a discussion of rearrangement operators acting in L^1. We begin with an example showing that boundedness in L^1 requires a much more restrictive condition on the underlying permutation than boundedness in $L^p, 1 < p < \infty$. Let $I_n = [1/2, 1/2 + 2^{-n})$, $n \in \mathbb{N}$. We define a rearrangement τ first on the even numbered intervals $\{I_{2n} : n \in \mathbb{N}\}$,

$$\tau(I_{2n}) = [\frac{1}{2}, \frac{1}{2} - 2^{-2n}).$$

Thus τ reflects the even numbered intervals at the point 1/2. Next we define that τ leaves invariant the dyadic intervals $J \notin \{I_{2n} : n \in \mathbb{N}\}$, that is

$$\tau(J) = J, \quad J \notin \{I_{2n} : n \in \mathbb{N}\}.$$

Now we show that τ defines an unbounded operator on L^1. Let $M \in \mathbb{N}$ and put

$$f = \sum_{n=1}^{M} h_{I_n} |I_n|^{-1}.$$

We have $||f||_1 \leq 2$. On the other hand a moment's reflection shows that

$$||\sum_{n=1}^{M} h_{\tau(I_n)} |I_n|^{-1}||_1 \geq M/2.$$

Thus the induced rearrangement operator T_τ is unbounded on L^1. By contrast, Theorem 3.3.4 shows clearly that $||T_\tau||_p \leq C_p$, when $1 < p < \infty$.

It would be interesting to analyze the L^1 case more closely and to describe the bounded rearrangement operators on L^1. As an introduction to this open problem we discuss now the work of N. Kislovskaya and V. Osipov [114] who solved a special case. We consider rearrangements τ which satisfy the additional hypothesis that

$$|\tau(I)| = |I|.$$

Their result shows that there are just two possibilities for how a rearrangement satisfying $|\tau(I)| = |I|$ can induce a bounded operator on L^1. Roughly speaking they obtain that either τ is a reflection or τ is a translation.

In the analysis given below these cases are represented by green, respectively red, colors attached to the intervals in $\mathcal{D}$. Black intervals also appear and they indicate that the rearrangement had done something very bad there. With the theorem of N. Kislovskaya and V. Osipov [114] we may read off the correct upper and lower estimates for $||T_\tau||_{L^1}$, from the number of non-empty generations of black intervals obtained by the following coloring process. Before we define the coloring algorithm in detail we will summarize its essential properties: Let $\mathcal{B}$ denote the collection of black intervals obtained, and let

$$\mathcal{B}_n = G_n(\mathcal{B})$$

be the the n-th generation of $\mathcal{B}$. Let $Q(\mathcal{B}_n) = \bigcup_{K\in\mathcal{B}_n}\{J \in \mathcal{D} : J \subseteq K\}$. Then define

$$\mathcal{G}_n = Q(\mathcal{B}_n) \setminus Q(\mathcal{B}_{n+1}).$$

Note that the maximal intervals of $\mathcal{G}_n$ is just the collection $\mathcal{B}_n$. Removing $\mathcal{B}_n$ from $\mathcal{G}_n$ we form the collection $\mathcal{L}_n$, that is we put

$$\mathcal{L}_n = \mathcal{G}_n \setminus \mathcal{B}_n.$$

The coloring will be such that each of the collections $\mathcal{L}_n$ is monochromatic: Either all intervals in $\mathcal{L}_n$ are red or all intervals in $\mathcal{L}_n$ are green. Next we point out that by the definition (given below) of the coloring algorithm, the restriction of the rearrangement operator to $\mathcal{L}_n$ is a bounded operator.

We start by calling the unit interval $[0,1]$ a black interval. Now suppose that $J \in \mathcal{D}$ has been colored. Black, red or green are the colors possible for J. We let J_1 be the left half of J, and J_2 be the right half of J. We assume that J_1 or J_2 have not yet been colored. First we formulate a simple rule which attaches the color black to the interval J_i : If $\tau(J_i)$ is not contained in $\tau(J)$ then J_i is a black interval. We call this rule simple, since the color of J_i is determined independently of the color of J.

The following rules define the color of J_1, J_2 depending on the color of J. There are three possible colors for J, hence there are three set of rules. Note that by the simple rule we only have to consider the cases when $\tau(J_i) \subseteq \tau(J)$. (Since otherwise the color of J_i is already determined to be black.) The following rules determine the color of J_i when $\tau(J_i) \subseteq \tau(J)$. The color depends on the position of $\tau(J_i)$ inside $\tau(J)$. The only possibilities are that $\tau(J_i)$ is the right half of $\tau(J)$, or that $\tau(J_i)$ is the left half of $\tau(J)$.

Case 1. J is a black interval: In this case the colors of J_1, J_2 can be green or red.

We start with the color of J_1, using the following set of rules.

1. If $\tau(J_1)$ is the left half of $\tau(J)$, then J_1 is a green interval.
2. If $\tau(J_1)$ is the right half of $\tau(J)$, then J_1 is a red interval.

Now we define the set of rules which determine the color of J_2.

1. If $\tau(J_2)$ is the left half of $\tau(J)$, then J_2 is a red interval.
2. If $\tau(J_2)$ is the right half of $\tau(J)$, then J_2 is a green interval.

Case 2. J is a green interval: In this case the colors of J_1, J_2 can be green or black.

We start with the color of J_1, using the following set of rules.

1. If $\tau(J_1)$ is the left half of $\tau(J)$, then J_1 is a green interval.

2. If $\tau(J_1)$ is the right half of $\tau(J)$, then J_1 is a black interval.

Now we define the set of rules which determine the color of J_2.

1. If $\tau(J_2)$ is the left half of $\tau(J)$, then J_2 is a black interval.
2. If $\tau(J_2)$ is the right half of $\tau(J)$, then J_2 is a green interval.

Case 3. J is a red interval: In this case the colors of J_1, J_2 can be red or black. We start with the color of J_1, using the following set of rules.

1. If $\tau(J_1)$ is the left half of $\tau(J)$, then J_1 is a black interval.
2. If $\tau(J_1)$ is the right half of $\tau(J)$, then J_1 is a red interval.

Now we define the set of rules which determine the color of J_2.

1. If $\tau(J_2)$ is the left half of $\tau(J)$, then J_2 is a red interval.
2. If $\tau(J_2)$ is the right half of $\tau(J)$, then J_2 is a black interval.

Having defined the color of each dyadic interval, we let $\mathcal{B}$ be the collection of all black intervals, and we let

$$\mathcal{B}_n = G_n(\mathcal{B})$$

be the generations of $\mathcal{B}$. Now we can decide whether the operator $T_\tau : L^1 \to L^1$ is bounded or not.

Theorem 3.3.11. *Let $\tau : \mathcal{D} \to \mathcal{D}$ be a rearrangement satisfying $|\tau(I)| = |I|$. The rearrangement operator $T_\tau : L^1 \to L^1$ is bounded if the number of non-empty generations of $\mathcal{B}$ is finite. Moreover*

$$\frac{N-1}{2} \leq ||T_\tau|| \leq 2N,$$

where N is the smallest number such that

$$G_N(\mathcal{B}) = \emptyset.$$

Proof. Let $n \leq N$. Fix $J \in \mathcal{B}_n$. Then define a block of intervals,

$$\mathcal{G}(J) = Q(J) \setminus Q(\mathcal{B}_{n+1} \cap J).$$

Next observe that once we remove the black interval J from the block $\mathcal{G}(J)$, then the entire remaining collection is monochromatic. In other words the intervals $\mathcal{G}(J) \setminus J$ are either all red or all green. If they are all green, the collection $\tau(\mathcal{G}(J))$ is again a block of intervals and more importantly, on $\mathcal{G}(J)$, the rearrangement τ acts by translation, and there exists $a \in \mathbb{R}$, such that

$$\tau(I) = a + I,$$

for $I \in \mathcal{G}(J)$. Now consider the case when the intervals in $\mathcal{G}(J) \setminus J$ are all red. In that case τ acts by first translating, then reflecting and then translating again. In either case we have the identity

$$||\sum_{\mathcal{G}(J)} h_{\tau(I)} a_I||_{L^1} = ||\sum_{\mathcal{G}(J)} h_I a_I||_{L^1}.$$

Next, as $\mathcal{G}(J)$ is a block of intervals, we obtain by conditional expectation,

$$||\sum_{J \in \mathcal{B}_n} \sum_{\mathcal{G}(J)} h_I a_I||_{L^1} \leq 2||f||_{L^1}.$$

Now we show the boundedness of T_τ. Let $f \in L^1$ with Haar expansion $f = \sum a_I h_I$. Then we estimate as follows,

$$\begin{aligned} ||T_\tau(f)||_{L^1} &\leq \sum_{n=1}^N \sum_{J \in \mathcal{B}_n} ||\sum_{\mathcal{G}(J)} h_{\tau(I)} a_I||_{L^1} \\ &= \sum_{n=1}^N \sum_{J \in \mathcal{B}_n} ||\sum_{\mathcal{G}(J)} h_I a_I||_{L^1} \\ &= \sum_{n=1}^N || \sum_{J \in \mathcal{B}_n} \sum_{\mathcal{G}(J)} h_I a_I||_{L^1} \\ &\leq 2N||f||_{L^1}. \end{aligned}$$

Next we show that $||T_\tau||_{L^1} \geq (N-1)/4$. Here we will find $f \in L^1$, with $||f||_1 = 2$ such that $||T_\tau(f)||_{L^1} \geq (N-1)/2$. Fix $J \in \mathcal{B}_{N-1}$. By assumption $\mathcal{B}_{N-1}$ is non-empty. Next we let $\mathcal{U}(J)$ be the collection of dyadic intervals which contain the interval J, that is

$$\mathcal{U}(J) = \{K \in \mathcal{D} : K \supseteq J\}.$$

We let $\{I_n\}$ be an enumeration of the intervals in $\mathcal{U}(J)$ such that $|I_n| = 2^{-n}$. Define

$$f = \sum h_{I_n} 2^n s_n,$$

where $s_n \in \{-1, 1\}$ is given as follows. If I_n is the left half of I_{n-1}, then $s_n = 1$, otherwise $s_n = -1$. With this choice of s_n there is cancellation in the sum defining f and $||f||_{L^1} \leq 2$. Now we analyze the action of τ on $\mathcal{U}(J)$. First note that for any $1 \leq k \leq N-1$, the intersection $\mathcal{U}(J) \cap \mathcal{B}_k$ consists of exactly one interval. Next we write

$$J_k = \mathcal{U}(J) \cap \mathcal{B}_k.$$

Then we define $\mathcal{U}_k$ to be the collection of dyadic intervals I satisfying, $J_k \subset I \subseteq J_{k+1}$. Observe that $\mathcal{U}_k \setminus \{J_{k+1}\}$ is again monochromatic, and, incidentally, the collections $\{\mathcal{U}_k\}$ define the largest monochromatic decomposition of $\mathcal{U}(J)$. Now let

$$f_k = \sum_{I \in \mathcal{U}_k} \langle f, h_I \rangle \frac{h_I}{|I|}.$$

Then $||T_\tau(f_k)||_{L^1} = ||f_k||_{L^1} \geq 1$, and $f = \sum f_k$. And moreover we have the following crucial lower bound,

$$||T_\tau(f)||_{L^1} \geq \frac{1}{2} \sum_{k=1}^{N-1} ||T_\tau(f_k)||_{L^1}.$$

(The best way to see this lower bound is to reconsider the example which marked the beginning of our discussion on rearrangement operators in L^1.) Summing up our observations we obtain the following lower bound for $||T_\tau(f)||_{L^1}$.

$$\begin{aligned} ||T_\tau(f)||_{L^1} &\geq \frac{1}{2} \sum_{k=1}^{N-1} ||T_\tau(f_k)||_{L^1} \\ &\geq \frac{1}{2} \sum_{k=1}^{N-1} ||f_k||_{L^1} \\ &\geq \frac{1}{2}(N-1). \end{aligned}$$

□

3.4 Notes

Sources for Section 3.1 are L. Carleson and J. Garnett [46] and J. Garnett and P.W. Jones [74]. See also Chapter VII in the book by J Garnett [72]. Proposition 3.1.6 is from [149].

Section 3.2 is based entirely on the work of P.W. Jones in [102].

The systematic investigation of rearrangement operators started with E. M. Semenov [190] who treated rearrangements satisfying $|\tau(I)| = |I|$. Further extensions were obtained by E. M. Semenov and B. Stöckert [191]. These results are presented in detail in the monograph [165] by I. Novikov and E. M. Semenov. In [188] F. Schipp obtained E. M. Semenov's theorem from the corresponding BMO result and interpolation. In Section 3.3 the treatment of general rearrangements in L^p and BMO follows [156]. The proof of Proposition 3.3.2 is extracted from P.W. Jones's work [98] on Carleson measures and the Fefferman–Stein decomposition of BMO($\mathbb{R}$).

The norm estimates for the operators U_m and T_m in Theorem 3.3.9 are due to T. Figiel [65]. The use of Stein's martingale inequality in the proof of Theorem 3.3.9 replaces two algebraic identities of T. Figiel's original proof presented in Section 6 of [65].

Rearrangements in L^1 are treated by N. Kislovkaya and V. Osipov in [114]. In particular Theorem 3.3.11 is taken from [114]. Our discussion follows their approach.

Chapter 4

Martingale H^1 Spaces

On the probability space $(\Omega, \mathcal{F}, |\cdot|)$ we are given an increasing sequence of σ-algebras $\mathcal{F}_0, \ldots, \mathcal{F}_n, \ldots$ which generate the σ-algebra $\mathcal{F}$. Then the martingale space, $H^1[(\mathcal{F}_n)]$, consists of integrable $f : \Omega \to \mathbb{R}$, with integrable square function, that is,

$$\|f\|_{H^1[(\mathcal{F}_n)]} = \int \left(\mathbb{E}(f|\mathcal{F}_0)^2 + \sum_n |\mathbb{E}(f|\mathcal{F}_n) - \mathbb{E}(f|\mathcal{F}_{n-1})|^2 \right)^{1/2} < \infty.$$

In this chapter we will be treating the case when each $\mathcal{F}_n$ is generated by a finite collection of pairwise disjoint sets $\mathcal{A}_n$. Under this hypothesis the resulting family of Banach spaces $H^1[(\mathcal{F}_n)]$ will be classified. We show that there are exactly three different isomorphic representatives for this family.

In Section 4.1 we present important work of B. Maurey who established the isomorphism between the martingale $H^1[(\mathcal{F}_n)]$ spaces and the class of $X[\mathcal{E}]$ spaces, consisting of three-valued martingale differences. The method presented there is quite explicit. In Section 4.2 we analyze the spaces $X[\mathcal{E}]$ and show that their isomorphic representation is given by the Banach spaces $H^1, (\sum H^1_n)_1$ or ℓ^1.

4.1 Maurey's isomorphism

Maurey's Isomorphism provides the equivalence of the martingale spaces $H^1[(\mathcal{F}_n)]$, to the spaces $X[\mathcal{E}]$. We recall the definition of $X[\mathcal{E}]$ as given in Chapter 2. We say that $\mathcal{E}$ is a nested collection of measurable sets if for $A, B \in \mathcal{E}$,

$$A \cap B \neq \emptyset \text{ implies } A \subseteq B \text{ or } B \subseteq A.$$

Let $\{r_A : A \in \mathcal{E}\}$ be an enumeration of the independent Rademacher functions, and form the three-valued basis functions

$$d_A(w, t) = 1_A(w) r_A(t), \tag{4.1.1}$$

where $A \in \mathcal{E}$. Then $X[\mathcal{E}]$ is the space of functions $f = \sum a_A d_A$ for which

$$||f||_{X[\mathcal{E}]} = \int \left(\sum a_A^2 1_A\right)^{1/2} < \infty.$$

The aim of this section is to prove the following theorem of B. Maurey [144].

Theorem 4.1.1. *Let $\{\mathcal{F}_n : n \in \mathbb{N}\}$ be a sequence of increasing σ-algebras, where each $\mathcal{F}_n$ is generated by finitely many atoms. Then there exists a nested collection of measurable sets $\mathcal{E}$ such that $H^1[(\mathcal{F}_n)]$ and $X[\mathcal{E}]$ are isomorphic Banach spaces. Moreover the nested collection $\mathcal{E}$ satisfies the following condition.*

If $A, B \in \mathcal{E}$ and if A is strictly contained in B, then $|A| \leq |B|/2$.

As a corollary to Maurey's theorem we obtain that $H^1[(\mathcal{F}_n)]$ has an unconditional basis. The isomorphism between $H^1[(\mathcal{F}_n)]$ and $X[\mathcal{E}]$, will be explicitly constructed below. It will provide us with a detailed pictorial description of the unconditional basis in $H^1[(\mathcal{F}_n)]$.

The proof of Maurey's theorem blends a hard combinatorial construction and analytic estimates based on martingale inequalities. Combinatorial problems arise when we construct a local isomorphism that acts on the martingale differences between $\mathcal{F}_n$ and $\mathcal{F}_{n-1}$, for a fixed $n \in \mathbb{N}$. Addressing these problems forms the first step in the construction of Maurey's isomorphism. In the second step we will sum the local operators, and then we use martingale transform techniques to show that the resulting series defines an isomorphism between $H^1[(\mathcal{F}_n)]$ and $X[\mathcal{E}]$.

Operations on martingale differences

We begin with the construction of the local isomorphism. Clearly if we restrict the attention to a fixed atom of $\mathcal{F}_{n-1}$, then the martingale differences at that level appear as $\mathcal{F}_n$ measurable functions with mean zero. Let $\mathcal{A}_{n-1}$ denote the atoms of $\mathcal{F}_{n-1}$. Let $A \in \mathcal{A}_{n-1}$, and let $L^\circ(A)$ denote the finite dimensional space of all functions $f : A \to \mathbb{R}$ which are measurable with respect to $\mathcal{F}_n \cap A$, and for which $\mathbb{E}(f) = 0$. Let $A^* \subset A$ be an atom in $\mathcal{A}_n$, satisfying

$$|A^*| = \max_{B \in \mathcal{A}_n\ B \subseteq A} |B|.$$

Then $L^*(A)$ denotes the space of functions $f : A \to \mathbb{R}$ which are $\mathcal{F}_n \cap A$ measurable, and which vanish on A^*. Note that $L^\circ(A)$ and $L^*(A)$ are finite dimensional and of the same algebraic dimension. Below we need a linear bijection $T_A : L^*(A) \mapsto L^\circ(A)$, which is an isomorphism for both the L^1-norm and the L^2-norm (with constants independent of the dimensions). The existence of such an isomorphism is the content of the next proposition.

Proposition 4.1.2. *There exists a linear bijection between $T_A : L^*(A) \to L^\circ(A)$ which satisfies*

$$\frac{1}{16}||f||_1 \leq ||T_A f||_1 \leq 6||f||_1$$

and

$$\frac{1}{16}||f||_2 \le ||T_A f||_2 \le 6||f||_2,$$

for $f \in L^*(A)$.

Proof. Ones first guess for such an operator might be

$$Tf = f - \mathbb{E}(f)\frac{1_{A^*}}{|A^*|}.$$

T maps $L^*(A)$ bijectively onto $L^\circ(A)$ and satisfies the correct L^1 estimates. However the L^2 bounds for T depend on the ratio $|A|/|A^*|$, and hence for the case where $|A^*| \ll |A|$ the operator T is just not the right one. A very careful adjustment is required. We proceed by defining the operators T_A. The first step consists in defining an auxiliary shift operator D_A and a carefully chosen sigma algebra $\mathcal{C}_A$.

Defining the shift operator. Assume that $|A^*| \le |A|/4$. Then there exist pairwise disjoint sets $F_1, \dots, F_m$ in $\mathcal{F}_n \cap A$ such that $F_1 \cup \dots \cup F_m = A \setminus A^*$ and so that

$$\begin{aligned} |A|/2 &\le |F_1| \le |A|, \\ |F_i|/2 &\le |F_{i+1}| \le |F_i|, \quad \text{for } i \le m-1, \\ |F_m|/4 &\le |A^*| \le 2|F_m|. \end{aligned}$$

Next put $F_{m+1} = A^*$ and let $\mathcal{C}_A$ be the sub σ-algebra of $\mathcal{F}_n \cap A$ generated by $F_1, \dots, F_{m+1}$. For a function g, measurable with respect to $\mathcal{C}_A$, we denote its value on F_i by $g(F_i)$. Next we define the L^1 normalized shift operator D_A as follows.

$$\begin{aligned} D_A(g)(t) &= 0 \quad \text{for } t \in F_1, \\ D_A g(t) &= g(F_{i-1})\frac{|F_{i-1}|}{|F_i|} \qquad \text{for } t \in F_i \quad \text{and} \quad 2 \le i \le m+1. \end{aligned}$$

Note that D_A maps $\mathcal{C}_A$ measurable functions into $\mathcal{C}_A$ measurable functions.

Defining T_A. For $f \in L^*(A)$, we define $T_A(f)$ depending on the ratio $|A^*|/|A|$. If $|A^*|/|A| \ge \frac{1}{4}$, then

$$T_A f = f - \mathbb{E}(f)\frac{1_{A^*}}{|A^*|}.$$

If $|A^*|/|A| \le \frac{1}{4}$, define

$$T_A f = f - \mathbb{E}(f|\mathcal{C}_A) + D_A \mathbb{E}(f|\mathcal{C}_A) - \mathbb{E}(f)\frac{1_{F_1}}{|F_1|}.$$

By definition, the operator T_A is a linear bijection from $L^*(A)$ onto $L^\circ(A)$. To prove the norm estimates we distinguish between the cases $|A^*|/|A| \ge 1/4$ and $|A^*|/|A| \le 1/4$.

Norm estimates for T_A. In the case when $|A^*|/|A| \geq 1/4$ we only need to observe that $f \in L^*(A)$ is supported on $A \setminus A^*$ and that $\mathbb{E}(f)1_{A^*}/|A^*|$ is supported on A^*.

Now we turn to the case $|A^*|/|A| \leq 1/4$. We analyze the operator T_A in two steps. First we consider the main part of T_A, namely

$$Uf = f - \mathbb{E}(f|\mathcal{C}_A) + D_A(\mathbb{E}(f|\mathcal{C}_A)).$$

Note that $\mathbb{E}(Uf|\mathcal{C}_A) = D_A(\mathbb{E}(f|\mathcal{C}_A))$ and that $||\mathbb{E}(f|\mathcal{C}_A)||_p \leq 2||D_A(\mathbb{E}(f|\mathcal{C}_A))||_p$, for $1 \leq p \leq 2$. Then we write

$$f = Uf + \mathbb{E}(f|\mathcal{C}_A) - D_A(\mathbb{E}(f|\mathcal{C}_A)).$$

The above observations give that

$$\begin{aligned} ||f||_p &\leq ||Uf||_p + 3||D_A(\mathbb{E}(f|\mathcal{C}_A))||_p \\ &\leq ||Uf||_p + 3||\mathbb{E}(Uf|\mathcal{C}_A)||_p \\ &\leq 4||Uf||_p. \end{aligned} \tag{4.1.2}$$

On the other hand we have, by the triangle inequality,

$$\begin{aligned} ||Uf||_p &\leq 2||f||_p + ||D_A\mathbb{E}(f|\mathcal{C}_A)||_p \\ &\leq 4||f||_p. \end{aligned} \tag{4.1.3}$$

Now we study the interaction between Uf and $\mathbb{E}(f)1_{F_1}/|F_1|$. We observe that by construction the mean value of Uf over the set F_1 vanishes, that is $\int_{F_1} Uf = 0$. We will now give a lower bound for the L^p norm of $T_Af = Uf - \mathbb{E}(f)1_{F_1}/|F_1|$. Fix $p \in [1,2]$ and choose $h \in L^q$ with $||h||_q = 1$, such that $||Uf||_p = \mathbb{E}(Uf \cdot h)$. Next we let

$$g = h - \mathbb{E}(h1_{F_1})\frac{1_{F_1}}{|F_1|},$$

then $||g||_q \leq 4||h||_q$, $\mathbb{E}(Uf \cdot g) = \mathbb{E}(Uf \cdot h)$ and $\mathbb{E}(g1_{F_1}) = 0$. It follows that

$$\begin{aligned} ||T_Af||_p &\geq \frac{1}{4}\mathbb{E}(T_A(f) \cdot g) \\ &= \frac{1}{4}(\mathbb{E}(Uf \cdot g) + \mathbb{E}(g \cdot 1_{F_1})\mathbb{E}(f)/|F_1|) \\ &= \frac{1}{4}\mathbb{E}(Uf \cdot h). \end{aligned}$$

Thus $||T_A(f)||_p \geq (1/4)||U(f)||_p$, and by estimates (4.1.2) for the operator U we have

$$||T_Af||_p \geq ||Uf||_p/4 \geq ||f||_p/16.$$

On the other hand by (4.1.3) the following upper bound holds for the L_p norm of T_A,

$$\begin{aligned} ||T_Af||_p &\leq ||Uf||_p + 2||f||_p \\ &\leq 6||f||_p. \end{aligned}$$

□

Martingale transform techniques

Next we define a nested collection of measurable sets $\mathcal{E}$, and construct the isomorphism between $X[\mathcal{E}]$ and $H^1[(\mathcal{F}_n)]$. Let us fix $A \in \mathcal{A}_{n-1}$ and let A^* denote an atom in $\mathcal{A}_n$ such that $A^* \subset A$ and each $B \in \mathcal{A}_n$, with $B \subset A$ satisfies $|B| \leq |A^*|$. Clearly such an $A^* \in \mathcal{A}_n$ exists for each $A \in \mathcal{A}_{n-1}$, but in general it is not unique. We let

$$\mathcal{E}(A) = \{B \subset A : B \in \mathcal{A}_n\} \setminus A^*, \tag{4.1.4}$$

and define the nested collection of sets

$$\mathcal{E} = \bigcup_{n=1}^{\infty} \bigcup_{A \in \mathcal{A}_{n-1}} \mathcal{E}(A).$$

Note that the nested collection $\mathcal{E}$ satisfies the following property which becomes crucial later, when we determine the isomorphic type of $X[\mathcal{E}]$.

$$\text{If } A, B \in \mathcal{E} \text{ and } A \subset B, \text{ then } |A| \leq |B|/2.$$

Now we construct the isomorphism between $X[\mathcal{E}]$ and $H^1[(\mathcal{F}_n)]$. Let T be defined by its action on the unconditional basis $\{d_B : B \in \mathcal{E}\}$ as follows: For $n \in \mathbb{N}$, $A \in \mathcal{A}_{n-1}$ and $B \in \mathcal{E}(A)$, put

$$T(d_B) = T_A(1_B).$$

The next theorem specifies the isomorphism whose existence is claimed in Theorem 4.1.1.

Theorem 4.1.3. *The isomorphism between $X[\mathcal{E}]$ and $H^1[(\mathcal{F}_n)]$ is given by the operator*

$$Tf = \sum_{n \in \mathbb{N}} \sum_{A \in \mathcal{A}_{n-1}} \sum_{B \in \mathcal{E}(A)} a_B T_A(1_B),$$

where

$$f = \sum_{n \in \mathbb{N}} \sum_{A \in \mathcal{A}_{n-1}} \sum_{B \in \mathcal{E}(A)} a_B d_B.$$

Proof. The proof of Maurey's theorem is long and hard. We therefore break it up into six connected components of manageable size. We begin by isolating special properties of T.

Local L^1 estimates. We fix $n \in \mathbb{N}$ and consider the n-th level in the basis expansion of $f \in X[\mathcal{E}]$. For $A \in \mathcal{A}_{n-1}$ let $\mathcal{E}(A)$ be given by (4.1.4). Then put

$$\delta_n = \sum_{A \in \mathcal{A}_{n-1}} \sum_{B \in \mathcal{E}(A)} a_B d_B.$$

On δ_n the operator T acts as follows,

$$T\delta_n = \sum_{A\in\mathcal{A}_{n-1}} T_A\left(\sum_{B\in\mathcal{E}(A)} a_B d_B\right).$$

Observe that the family of support sets

$$\left\{\operatorname{supp} T_A\left(\sum_{B\in\mathcal{E}(A)} a_B d_B\right) : A\in\mathcal{A}_{n-1}\right\}$$

is a collection of pairwise disjoint, $\mathcal{F}_n$-measurable sets. This holds since the support of $T_A(d_B)$ is contained in A, for each $B\in\mathcal{E}(A)$. Hence we obtain the following identity for the L^1 norm of $T\delta_n$,

$$||T\delta_n||_1 = \sum_{A\in\mathcal{A}_{n-1}} \left\|T_A\left(\sum_{B\in\mathcal{E}(A)} a_B d_B\right)\right\|_1.$$

By Proposition 4.1.2, the operator T_A is an L^1 isomorphism (from its domain onto its range), and therefore

$$\frac{1}{A_1}||T\delta_n||_1 \le \sum_{A\in\mathcal{A}_{n-1}} \left\|\sum_{B\in\mathcal{E}(A)} a_B d_B\right\|_1 \le A_1||T\delta_n||_1.$$

Here $A_1 = 16$. It remains to observe that the collection $\bigcup_{A\in\mathcal{A}_{n-1}}\bigcup_{B\in\mathcal{E}(A)}\{B\}$ is a collection of pairwise disjoint elements in $\mathcal{E}$ and as $\operatorname{supp} d_B = B\times[0,1]$ we have

$$||\delta_n||_1 = \sum_{A\in\mathcal{A}_{n-1}} \left\|\sum_{B\in\mathcal{E}(A)} a_B d_B\right\|_1.$$

This proves the following local L^1 estimates for T:

$$\frac{||\delta_n||_1}{A_1} \le ||T\delta_n||_1 \le A_1||\delta_n||_1, \tag{4.1.5}$$

with $A_1 = 16$.

Global L^2-estimates. Let $n\in\mathbb{N}$. Define

$$\mathcal{E}_n = \bigcup_{A\in\mathcal{A}_{n-1}} \mathcal{E}(A) \quad \text{where} \quad \mathcal{E}(A) = \{B\subset A : B\in\mathcal{A}_n\}\setminus A^*. \tag{4.1.6}$$

We consider as before,

$$\delta_n = \sum_{B\in\mathcal{E}_n} a_B d_B,$$

for $n \in \mathbb{N}$. Clearly the sequence $\{\delta_n\}$ is biorthogonal in the Hilbert space $L^2(\Omega \times [0,1])$. Observe that the operator T maps $\{\delta_n\}$ to a biorthogonal sequence in $L^2(\Omega)$. Indeed, by Proposition 4.1.2, $T\delta_n$ is $\mathcal{F}_n$ measurable and $\mathbb{E}(T\delta_n|\mathcal{F}_{n-1}) = 0$, hence $\{T\delta_n\}$ is a martingale difference sequence, hence biorthogonal. Again by Proposition 4.1.2, the operator T satisfies local L^2 estimates,

$$\frac{||\delta_n||_2}{A_2} \leq ||T\delta_n||_2 \leq A_2||\delta_n||_2. \tag{4.1.7}$$

We can take $A_2 = 16$ again. Next we combine the local L^2-estimates (4.1.7) and biorthogonality, to obtain global L^2-estimates as follows.

$$\begin{aligned}\left\|\sum_{n=1}^{\infty} \delta_n\right\|_2^2 &= \sum_{n=1}^{\infty} ||\delta_n||_2^2 \\ &\leq A_2^2 \sum_{n=1}^{\infty} ||T\delta_n||_2^2 \\ &= A_2^2 \left\|T\left(\sum \delta_n\right)\right\|_2^2.\end{aligned}$$

Also the converse holds. With the same proof we obtain,

$$\left\|T\left(\sum_{n=1}^{\infty} \delta_n\right)\right\|_2 \leq A_2 \left\|\sum_{n=1}^{\infty} \delta_n\right\|_2.$$

T acts like a martingale transform. Let $g : \Omega \to \mathbb{R}$ be $\mathcal{F}_{n-1}$ measurable, and let $B \in \mathcal{E}(A)$ for $A \in \mathcal{A}_{n-1}$. Then by construction of T we have the crucial pointwise identity

$$g(w)d_B(w,t) = T^{-1}(gTd_B)(w,t). \tag{4.1.8}$$

An elementary inequality. We continue by proving a nice, elementary inequality. Let $\{a_n\}$ be positive real numbers and $s_m = \sum_{n=1}^{m} a_n$, then the following inequality holds,

$$s_m^{1/2} \geq \frac{1}{2} \sum_{n=1}^{m} s_n^{-1/2} a_n, \quad m \in \mathbb{N}. \tag{4.1.9}$$

For the proof we put $s_0 = 0$ and estimate using integrals.

$$\begin{aligned}2s_m^{1/2} &= \int_0^{s_m} t^{-1/2}\, dt \\ &= \sum_{n=1}^{m} \int_{s_{n-1}}^{s_n} t^{-1/2}\, dt \\ &\geq \sum_{n=1}^{m} s_n^{-1/2}\{s_n - s_{n-1}\}.\end{aligned}$$

After the preliminary observations we will show next that T is bounded. Later we prove that T^{-1} is bounded. Thus we will establish that

$$||T|| \cdot ||T^{-1}|| < \infty.$$

T **is bounded.** Recall $T : X[\mathcal{E}] \to H^1[(\mathcal{F}_n)]$ acts on the space of three-valued martingale differences. We let $f \in X[\mathcal{E}]$ and expand,

$$f = \sum \delta_n,$$

where

$$\delta_n = \sum_{B \in \mathcal{E}_n} a_B d_B$$

and where $\mathcal{E}_n$ is given by (4.1.6). We perform now a canonical decomposition of δ_n by splitting the index set $\mathcal{E}_n$ as follows. For $B \in \mathcal{E}_n$ we say that B belongs to $\mathcal{C}_n$ iff on the support of d_B, the following holds,

$$4 \sum_{k=1}^{n-1} \delta_k^2 \le a_B^2 d_B^2. \tag{4.1.10}$$

Let

$$\alpha_n = \sum_{B \in \mathcal{C}_n} a_B d_B,$$

and split δ_n as follows,

$$\delta_n = \alpha_n + \beta_n.$$

Next put

$$h = \sum \alpha_n,$$

and decompose $f = g + h$. We continue giving separate estimates for $Th = \sum T\alpha_n$ and $Tg = \sum T\beta_n$. We start with estimates for the norm of Th in $H^1[(\mathcal{F}_n)]$. Let

$$h_n = \sum_{k=1}^{n} \alpha_k.$$

For h_n the associated square function is given as

$$\mathbb{S}(h_n) = \left(\sum_{k=1}^{n} \alpha_k^2 \right)^{1/2}.$$

By (4.1.10) we have the crucial estimates

$$\mathbb{S}(h_n) \ge \frac{1}{2}|\alpha_n| + \mathbb{S}(h_{n-1}).$$

Summing the telescoping series gives that

$$\begin{aligned}\sum_{n=1}^{\infty}|\alpha_n| &\le 2\mathbb{S}(h)-2\mathbb{S}(h_0)\\ &\le 2\mathbb{S}(f).\end{aligned} \tag{4.1.11}$$

The local L^1 estimates for T in combination with (4.1.11) yield upper bounds for Th. Indeed, as $T\alpha_n$ is $\mathcal{F}_n$ measurable and $\mathbb{E}(T\alpha_n|\mathcal{F}_{n-1})=0$,

$$||Th||_{H^1[(\mathcal{F}_n)]}=\int(\sum(T\alpha_n)^2)^{1/2}, \tag{4.1.12}$$

and

$$\begin{aligned}\int(\sum(T\alpha_n)^2)^{1/2} &\le \int\sum|T\alpha_n|\\ &\le A_1\int\sum|\alpha_n|.\end{aligned} \tag{4.1.13}$$

By (4.1.12), (4.1.13) and (4.1.11) we obtain

$$\|Th\|_{H^1[(\mathcal{F}_n)]}\le 2A_1\|f\|_{X[\mathcal{E}]}. \tag{4.1.14}$$

Next we will estimate Tg where

$$g=\sum_{k=1}^{\infty}\beta_k.$$

Here we will use global L^2 estimates for T together with martingale transform techniques. Let

$$f_k=\sum_{\ell=1}^{k}\delta_\ell,\quad\text{and}\quad \mathbb{S}(f_k)=\left(\sum_{\ell=1}^{k}\delta_\ell^2\right)^{1/2}.$$

We begin with the following pointwise estimate for square functions, which uses just that $\mathbb{S}(f_{k-1})\le\mathbb{S}(f)$,

$$\sum_{k=1}^{\infty}(T\beta_k)^2\le\mathbb{S}(f)\sum_{k=1}^{\infty}(T\beta_k)^2\mathbb{S}(f_{k-1})^{-1}.$$

As $||Tg||_{H^1[(\mathcal{F}_n)]}=\int(\sum_{k=1}^{\infty}(T\beta_k)^2)^{1/2}$ we obtain the following preliminary estimate using the inequality of Cauchy-Schwarz,

$$||Tg||_{H^1[(\mathcal{F}_n)]}\le\left(\int\sum_{k=1}^{\infty}(T\beta_k)^2\mathbb{S}(f_{k-1})^{-1}\right)^{1/2}\left(\int\mathbb{S}(f)\right)^{1/2}.$$

We will now show the following bound for the first factor on the right-hand side,

$$\int \sum_{k=1}^{\infty} (T\beta_k)^2 \mathbb{S}(f_{k-1})^{-1} \leq 2A_2^2 \int \mathbb{S}(f).$$

By construction $\mathbb{S}(f_{k-1})$ is measurable with respect to $\mathcal{F}_{k-1}$. By (4.1.8), T acts like a martingale transform and we have the identity

$$(T\beta_k)\mathbb{S}(f_{k-1})^{-1/2} = T(\beta_k \mathbb{S}(f_{k-1})^{-1/2}).$$

Global L^2 bounds for T give then the estimate

$$\int \sum_{k=1}^{\infty} (T\beta_k)^2 \mathbb{S}(f_{k-1})^{-1} \leq A_2^2 \int \sum \beta_k^2 \mathbb{S}(f_{k-1})^{-1}. \tag{4.1.15}$$

Note that by the negation of (4.1.10) β_k was defined to satisfy the pointwise estimate

$$(\sum_{\ell=1}^{k} \beta_\ell^2)^{1/2} \leq 3\mathbb{S}(f_{k-1}).$$

Inserting this in the right-hand side of (4.1.15) gives

$$\int \sum_{k=1}^{\infty} (T\beta_k)^2 \mathbb{S}(f_{k-1})^{-1} \leq 3A_2^2 \int \sum_{k=1}^{\infty} \beta_k^2 (\sum_{\ell=1}^{k} \beta_\ell^2)^{-1/2}. \tag{4.1.16}$$

Applying the elementary inequality (4.1.9) gives

$$\sum_{k=1}^{\infty} \beta_k^2 (\sum_{\ell=1}^{k} \beta_\ell^2)^{-1/2} \leq 2(\sum_{k=1}^{\infty} \beta_k^2)^{1/2}. \tag{4.1.17}$$

Inserting (4.1.17) into (4.1.16) we finally obtain

$$\begin{aligned} \int \sum_{k=1}^{\infty} (T\beta_k)^2 \mathbb{S}(f_{k-1})^{-1} &\leq 6A_2^2 \int (\sum_{k=1}^{\infty} \beta_k^2)^{1/2} \\ &\leq 6A_2^2 \int \mathbb{S}(f). \end{aligned}$$

We thus have proved the estimate

$$||Tg||_{H^1[(\mathcal{F}_n)]} \leq \sqrt{6}A_2 ||f||_{X[\mathcal{E}]}. \tag{4.1.18}$$

To complete the proof that T is bounded it remains to combine the above estimates (4.1.14) and (4.1.18) obtained for Th and Tg, and to use that $f = g + h$. Thus

$$||Tf||_{H^1[(\mathcal{F}_n)]} \leq (2A_1 + \sqrt{6}A_2)||f||_{X[\mathcal{E}]}.$$

T^{-1} **is bounded.** Recall that $T^{-1}: H^1[(\mathcal{F}_n)] \to X[\mathcal{E}]$ acts on the space of $(\mathcal{F}_n)$ martingales. We fix $f \in H^1[(\mathcal{F}_n)]$ with $\int f = 0$, and expand it into a sum of martingale differences, so that

$$f = \sum \delta_k, \quad \text{where} \quad \delta_k = \mathbb{E}(f|\mathcal{F}_k) - \mathbb{E}(f|\mathcal{F}_{k-1}).$$

For $f_n = \sum_{k=1}^n \delta_k$ we denote by $S(f_n)$ the square function of f_n with respect to the filtration $(\mathcal{F}_n)$. Thus

$$S(f_n) = \left(\sum_{k=1}^n \delta_k^2\right)^{1/2}.$$

We now define a decomposition procedure with many similarities to the one used in the previous paragraphs. There are differences however, that stem from the fact that δ_k^2 is not, in general, measurable with respect to $\mathcal{F}_{k-1}$. We let A be an atom for $\mathcal{F}_n$. Then we put $A \in \mathcal{C}_n$ iff

$$2S(f_{n-1})(\omega) \le |\delta_n(\omega)|, \quad \text{for} \quad \omega \in A. \tag{4.1.19}$$

Note that on both sides of the equation (4.1.19) the functions are actually constant when restricted to A. We define

$$\mathcal{B}_n = \mathcal{A}_n \setminus \mathcal{C}_n. \tag{4.1.20}$$

By the negation of (4.1.19), for $B \in \mathcal{B}_n$, and $\omega \in B$, the following estimate holds,

$$S(f_n)(\omega) \le 3S(f_{n-1})(\omega). \tag{4.1.21}$$

Let

$$\alpha_n = \sum_{A \in \mathcal{C}_n} \delta_n 1_A - \mathbb{E}(\delta_n 1_A|\mathcal{F}_{n-1}),$$

and decompose $\delta_n = \alpha_n + \beta_n$. Notice that α_n and β_n are martingale difference sequences with respect to the filtration $(\mathcal{F}_n)$. We put

$$h = \sum_{n=1}^\infty \alpha_n,$$

and decompose $f = g + h$. Now we give separate estimates for $T^{-1}h = \sum T^{-1}\alpha_n$, and $T^{-1}g = \sum T^{-1}\beta_n$. We begin with $T^{-1}h$: By (4.1.19), for $A \in \mathcal{C}_n$ the pointwise estimate holds,

$$S(f_n) \ge \frac{1}{2}|\delta_n|1_A + S(f_{n-1}).$$

Hence, using monotonicity of conditional expectations we obtain the pointwise estimate

$$|\alpha_n| \le 2S(f_n) - 2S(f_{n-1}) + 2\mathbb{E}(S(f_n) - S(f_{n-1})|\mathcal{F}_{n-1}).$$

Summing and integrating gives

$$\int \sum_{n=1}^{\infty} |\alpha_n| \leq 4 \int S(f).$$

By construction the operator T^{-1} satisfies

$$||T^{-1}h||_{X[\mathcal{E}]} = \int \left(\sum (T^{-1}\alpha_n)^2\right)^{1/2}. \tag{4.1.22}$$

Then the obvious estimate

$$\left(\sum (T^{-1}\alpha_n)^2\right)^{1/2} \leq \sum |T^{-1}\alpha_n|,$$

together with local L^1 estimates for T^{-1}, (4.1.5), gives

$$\begin{aligned} ||T^{-1}h||_{X[\mathcal{E}]} &\leq \sum ||T^{-1}\alpha_n||_1 \\ &\leq A_1 \int \sum |\alpha_n| \\ &\leq 4A_1 \int S(f). \end{aligned} \tag{4.1.23}$$

Next we prove estimates for the $X[\mathcal{E}]$ norm of $T^{-1}g$, where

$$g = \sum \beta_n.$$

Notice that by the definition of T^{-1},

$$||T^{-1}g||_{X[\mathcal{E}]} = \int \left(\sum (T^{-1}\beta_n)^2\right)^{1/2}.$$

By martingale transform techniques and L^2 estimates we give an upper bound for $||T^{-1}g||_{X[\mathcal{E}]}$. We begin with the pointwise estimate

$$\sum (T^{-1}\beta_n)^2 \leq S(f) \sum (T^{-1}\beta_n)^2 S(f_{n-1})^{-1}.$$

The Cauchy–Schwarz inequality gives then

$$||T^{-1}g||_{X[\mathcal{E}]} \leq \left(\int S(f)\right)^{1/2} \left(\int \sum (T^{-1}\beta_n)^2 S(f_{n-1})^{-1}\right)^{1/2}.$$

By the martingale transform property (4.1.8), we have the identity

$$(T^{-1}\beta_n) S(f_{n-1})^{-1/2} = T^{-1}(\beta_n S(f_{n-1})^{-1/2}).$$

Hence, with the L^2 estimates (4.1.7) we obtain,

$$\int \sum (T^{-1}\beta_n)^2 S(f_{n-1})^{-1} \le A_2^2 \int \sum \beta_n^2 S(f_{n-1})^{-1}. \tag{4.1.24}$$

Next we claim that

$$\int \beta_n^2 S(f_{n-1})^{-1} \le 3 \int \delta_n^2 S(f_n)^{-1}. \tag{4.1.25}$$

We postpone proving the estimate (4.1.25) and continue as follows: Use inequality (4.1.9) with $a_n = \delta_n^2$ and $s_n = S^2(f_n)$. This gives the pointwise estimate

$$\sum \delta_n^2 S(f_n)^{-1} \le 2S(f). \tag{4.1.26}$$

Now combine (4.1.24)–(4.1.26) to obtain

$$\begin{aligned} \int \sum (T^{-1}\beta_n)^2 S(f_{n-1})^{-1} &\le A_2^2 \sum \int \beta_n^2 S(f_{n-1})^{-1} \\ &\le 3A_2^2 \sum \int \delta_n^2 S(f_n)^{-1} \\ &\le 6A_2^2 \int S(f). \end{aligned}$$

We now give the proof for the claim (4.1.25). Recall that $\mathcal{A}_n$ is the collection of atoms in $\mathcal{F}_n$, and that $\mathcal{B}_n$ is the subcollection defined by (4.1.20). Next we observe that β_n has the representation

$$\beta_n = \sum_{B \in \mathcal{B}_n} \delta_n 1_B - \mathbb{E}(\delta_n 1_B | \mathcal{F}_{n-1}). \tag{4.1.27}$$

To see that (4.1.27) holds we write

$$\sum_{A \in \mathcal{C}_n} \mathbb{E}(\delta_n 1_A | \mathcal{F}_{n-1}) + \sum_{B \in \mathcal{B}_n} \mathbb{E}(\delta_n 1_B | \mathcal{F}_{n-1}) = \mathbb{E}(\delta_n | \mathcal{F}_{n-1}),$$

hence, as $\mathbb{E}(\delta_n | \mathcal{F}_{n-1}) = 0$,

$$\delta_n = \alpha_n + \sum_{B \in \mathcal{B}_n} \delta_n 1_B - \mathbb{E}(\delta_n 1_B | \mathcal{F}_{n-1}). \tag{4.1.28}$$

Comparing (4.1.28) with $\delta_n = \alpha_n + \beta_n$ gives (4.1.27). For $A \in \mathcal{A}_{n-1}$ the L^2 estimate holds,

$$\begin{aligned} \int_A \Bigg[\sum_{B \in \mathcal{B}_n, B \subseteq A} (\delta_n 1_B - \mathbb{E}(\delta_n 1_B | \mathcal{F}_{n-1})) \Bigg]^2 &\le \int_A \Bigg[\sum_{B \in \mathcal{B}_n, B \subseteq A} \delta_n 1_B \Bigg]^2 \\ &= \int_A \sum_{B \in \mathcal{B}_n, B \subseteq A} \delta_n^2 1_B. \end{aligned} \tag{4.1.29}$$

We square the identity (4.1.27) and multiply both sides with $S(f_{n-1})^{-1}$, then integrate. Exploiting that $S(f_{n-1})^{-1}$ is measurable with respect to $\mathcal{F}_{n-1}$, we obtain from (4.1.29) that

$$\int \beta_n^2 S(f_{n-1})^{-1} \leq \sum_{B\in\mathcal{B}_n} \int_B S(f_{n-1})^{-1}\, \delta_n^2. \tag{4.1.30}$$

Next recall (4.1.21) asserting that $S(f_n)(\omega) \leq 3S(f_{n-1})(\omega)$ for $B \in \mathcal{B}_n$ and $\omega \in B$. Inserting (4.1.21) into the right-hand side integral of (4.1.30) gives that

$$\begin{aligned}\sum_{B\in\mathcal{B}_n} \int_B S(f_{n-1})^{-1}\delta_n^2 &\leq 3 \sum_{B\in\mathcal{B}_n} \int_B S(f_n)^{-1}\delta_n^2 \\ &\leq 3 \int S(f_n)^{-1}\delta_n^2.\end{aligned} \tag{4.1.31}$$

Combining (4.1.30) and (4.1.31) finishes the proof of the claim (4.1.25).

Summing up we showed the following estimates for $f \in H^1[(\mathcal{F}_n)]$,

$$\begin{aligned}||T^{-1}(f)||_{X[\mathcal{E}]} &\leq ||T^{-1}(h)||_{X[\mathcal{E}]} + ||T^{-1}(g)||_{X[\mathcal{E}]} \\ &\leq (4A_1 + \sqrt{6}A_2)||f||_{H^1[(\mathcal{F}_n)]}.\end{aligned}$$

□

4.2 Isomorphic classification

In this section we prove the classification theorem for martingale H^1 spaces.

We determine the isomorphic representatives of the spaces generated by three-valued nested martingale differences which appear through B. Maurey's isomorphism. Furthermore we classify the family of $\mathcal{S}^\infty$ spaces obtained as the weighted intersection between BMO and ℓ^∞.

For the spaces $H^1[(\mathcal{F}_n)]$ B. Maurey identified an intrinsic criterion which implies the existence of an isomorphism between the martingale H^1 space and dyadic H^1 : Let

$$A_n^\epsilon = \bigcup\{A \in \mathcal{F}_n : |A| \leq \epsilon\} \quad \text{and} \quad A_\infty = \bigcap_{\epsilon>0} \bigcup_{n=1}^{\infty} A_n^\epsilon.$$

It turns out that $|A_\infty| > 0$, precisely when $H^1[(\mathcal{F}_n)]$ and H^1 are isomorphic Banach spaces. See Section 4 of B. Maurey [144].

Classification of martingale H^1 spaces

Maurey's isomorphism reduces the classification for $H^1[(\mathcal{F}_n)]$ spaces to that of the class $X[\mathcal{E}]$. The nested collections $\mathcal{E}$ arising through Maurey's isomorphism satisfy the following condition:

$$\text{If } A, B \in \mathcal{E}, \text{and if } A \text{ is strictly contained in } B, \text{ then } |A| \leq |B|/2. \tag{4.2.1}$$

Condition (4.2.1) makes $\mathcal{E}$ very similar to a subcollection of dyadic intervals. This in turn is the reason that the classification of $X[\mathcal{E}]$ is remarkably simple when $\mathcal{E}$ satisfies (4.2.1).

We will prove that under the hypothesis (4.2.1) the isomorphic types of the class $X[\mathcal{E}]$ are given by the spaces H^1, $(\sum H_n^1)_1$ or ℓ^1. Each representative is characterized by simple geometric parameters derived from $\mathcal{E}$. These are $\limsup \mathcal{E}$, the set which is contained in infinitely many elements of $\mathcal{E}$, and the Carleson constant,

$$[\![\mathcal{E}]\!] = \sup_{A\in\mathcal{E}} \frac{1}{|A|} \sum_{B\subseteq A, B\in\mathcal{E}} |B|.$$

The following theorem presents this in detail.

Theorem 4.2.1. *Let $\mathcal{E}$ be a nested collection of measurable sets such that for $A, B \in \mathcal{E}$,*

$$A \subset B \text{ implies } |A| \le |B|/2.$$

Then $X[\mathcal{E}]$ is isomorphic to one of the spaces H^1, $(\sum H_n^1)_1$ or ℓ^1. Each of these cases is determined as follows.

(a) *If $|\limsup \mathcal{E}| > 0$, then $X[\mathcal{E}]$ is isomorphic to H^1.*

(b) *If $|\limsup \mathcal{E}| = 0$ and $[\![\mathcal{E}]\!] = \infty$, then $X[\mathcal{E}]$ is isomorphic to $(\sum H_n^1)_1$.*

(c) *If $[\![\mathcal{E}]\!] < \infty$, then $X[\mathcal{E}]$ is isomorphic to ℓ^1.*

This result owes its existence to Pełczyński's decomposition method. Currently explicit isomorphisms are unknown. The nested collection $\mathcal{E}$ consists of measurable sets in the probability space $(\Omega, \mathcal{F}, |\,.\,|)$. Recall that $\mathcal{E}$ is canonically decomposed into a sequence of generations. For $A \in \mathcal{E}$ put $G_0(A, \mathcal{E}) = \{A\}$ and let $G_1(A, \mathcal{E})$ denote the maximal elements of $\mathcal{E}$ which are strictly contained in A. By induction for $p \geq 2$ we define

$$G_p(A, \mathcal{E}) = \bigcup_{B\in G_{p-1}(A,\mathcal{E})} G_1(B, \mathcal{E}).$$

For convenience of notation we write $G_p(\mathcal{E}) = G_p(\Omega, \mathcal{E})$. Note that $\limsup \mathcal{E}$ and the generations $G_p(\mathcal{E})$ are related by the identity

$$\limsup \mathcal{E} = \bigcap_{p=0}^{\infty} G_p^*(\mathcal{E}),$$

where we put $G_p^*(\mathcal{E}) = \bigcup_{A\in G_p(\mathcal{E})} A$. It is easy to see that the following version of Proposition 3.1.1 holds: If $|\limsup \mathcal{E}| = 0$, then for every $A \in \mathcal{E}$ there exists $p = p(A)$ so that

$$|G_p^*(A, \mathcal{E})| \le \frac{1}{4}|A|.$$

Similarly the proof of the condensation Lemma 3.1.4 shows that if

$$[\![\mathcal{E}]\!] > \frac{n}{\epsilon},$$

then there exists $B \in \mathcal{E}$ so that

$$|G_n^*(B, \mathcal{E})| \geq (1 - \epsilon)|B|.$$

The first step in the proof of Theorem 4.2.1 determines when $X[\mathcal{E}]$ is isomorphic to H^1. We move to the front an elementary remark that is often helpful in evaluating the dyadic square function and its integral. Let

$$f = \sum_{I \in \mathcal{D}_N} a_I h_I,$$

where $\mathcal{D}_N = \{I \in \mathcal{D} : |I| \geq 2^{-N}\}$. Let $J \in \mathcal{D}_N$, with $|J| = 2^{-N}$, and let $x \in J$. Then

$$\sum_{I \in \mathcal{D}_N} a_I^2 1_I(x) = \sum_{I \supseteq J} a_I^2.$$

Note that the dyadic square function $S(f)$ is constant over each dyadic interval J satisfying $|J| = 2^{-N}$. Hence

$$\int_J S(f)(x)dx = |J| \left(\sum_{I \supseteq J} a_I^2 \right)^{1/2}, \quad \text{when} \quad |J| = 2^{-N}.$$

Taking the sum over all dyadic intervals J with $|J| = 2^{-N}$, we get

$$\int_0^1 S(f)(x)dx = \sum_{\{J : |J| = 2^{-N}\}} |J| \left(\sum_{I \supseteq J} a_I^2 \right)^{1/2}. \tag{4.2.2}$$

It is easy to extend the identity (4.2.2) from the case where f is a finite linear combination of Haar functions to the case where $f \in H^1$. Indeed, by the monotone convergence theorem we pass to the limit $N \to \infty$ and obtain the H^1 norm of f. Equation (4.2.2) is merely a rewriting of the definitions involved; analogous identities hold of course for the norm in $X[\mathcal{E}]$ where dyadic intervals are replaced by nested collections of measurable sets.

Theorem 4.2.2. *Let $\mathcal{E}$ be a nested collection of measurable sets such that for $A, B \in \mathcal{E}$,*

$$A \subset B \text{ implies } |A| \leq |B|/2.$$

Then $|\limsup \mathcal{E}| > 0$ implies that $X[\mathcal{E}]$ is isomorphic to H^1.

Proof. In Chapter 2 we proved Proposition 2.1.4 to the effect that for any nested collection of measurable sets $\mathcal{E}$ the space $X[\mathcal{E}]$ is isomorphic to a complemented subspace of H^1. Hence by the decomposition method it suffices to prove now that if $\mathcal{E}$ satisfies (4.2.1) and if the measure of $\limsup \mathcal{E}$ is strictly positive, then $X[\mathcal{E}]$ contains a complemented copy of H^1. We claim that $\mathcal{E}$ contains pairwise disjoint collections $\mathcal{C}_I, I \in \mathcal{D}$ with the following properties: The sets $\{\mathcal{C}_I^* : I \in \mathcal{D}\}$ form a nested collection with

$$\mathcal{C}_I^* \subseteq \mathcal{C}_J^* \text{ iff } I \subseteq J; \tag{4.2.3}$$

the collections $\{\mathcal{C}_I : I \in \mathcal{D}\}$ satisfy Jones compatibility condition (J), and

$$\frac{1}{2}\epsilon\,|I| \leq |\mathcal{C}_I^*| \leq 2\epsilon\,|I|, \tag{4.2.4}$$

where $\epsilon = |\limsup \mathcal{E}|$.

We prove the claim by applying the Gamlen–Gaudet construction infinitely often. Begin by recalling that $\limsup \mathcal{E}$ can be expressed using generations of $\mathcal{E}$,

$$\limsup \mathcal{E} = \bigcap_{m=1}^{\infty} G_m(\mathcal{E})^*.$$

Hence there is $m_1 \in \mathbb{N}$ such that

$$|G_{m_1}^*(\mathcal{E}) \setminus \limsup \mathcal{E}| \leq \frac{\epsilon}{4},$$

where $\epsilon = |\limsup \mathcal{E}|$. We define our first collection to be

$$\mathcal{C}_{[0,1]} = G_{m_1}(\mathcal{E}).$$

Suppose now that we have already defined the collections $\mathcal{C}_I$ for all intervals of length $\geq 2^{-n}$.

We now define the next level of collections using the Gamlen–Gaudet construction. Fix a dyadic interval I such that $|I| = 2^{-n}$ and consider $A \in \mathcal{C}_I$. Using condition (4.2.1) we obtain a splitting of $G_{n+2}(A, \mathcal{E})$ into two collections $\mathcal{P}_A$ and $\mathcal{M}_A$ such that the measures $|\mathcal{P}_A^*|$ and $|\mathcal{M}_A^*|$ are almost equal. Precisely their difference is small relative to $|A|$ as follows,

$$\big|\,|\mathcal{P}_A^*| - |\mathcal{M}_A^*|\,\big| \leq |A| 2^{-n-2}.$$

Now let I_1 be the left half of I and let I_2 be the right half of I. Then we define

$$\mathcal{C}_{I_1} = \bigcup_{A \in \mathcal{C}_I} \mathcal{M}_A \quad \text{and} \quad \mathcal{C}_{I_2} = \bigcup_{A \in \mathcal{C}_I} \mathcal{P}_A.$$

In this way we have defined now the collections $\mathcal{C}_J$ for all dyadic intervals of length $\geq 2^{-n-1}$. This completes the induction step of the construction. We define next block bases

$$b_I = \sum \{d_A : A \in \mathcal{C}_I\}, \quad \text{for } I \in \mathcal{D}.$$

The collections $\{\mathcal{C}_J\}$ being obtained through the Gamlen–Gaudet construction satisfy the Jones compatibility condition (J). By Theorem 1.5.9 the orthogonal projection onto $\operatorname{span}\{b_I : I \in \mathcal{D}\}$, which is

$$P(f) = \sum \langle f, \frac{b_I}{||b_I||_2} \rangle \frac{b_I}{||b_I||_2},$$

is a bounded operator on $X[\mathcal{E}]$.

It remains to prove that the closed linear span of $\{b_I : I \in \mathcal{D}\}$ is isomorphic to H^1. We do this by showing that in $X[\mathcal{E}]$ the block basis $\{b_I\}$ is equivalent to the Haar basis in H^1. To this end we fix a finite linear combination $\sum a_I b_I$ in $X[\mathcal{E}]$, and put $C_I = \mathcal{C}_I^*$. Then choose $N \in \mathbb{N}$ large enough so that each I with $a_I \neq 0$, satisfies $|I| \geq 2^{-N}$. Then using (4.2.3) and (4.2.4) we estimate as follows:

$$\begin{aligned} \left\| \sum a_I b_I \right\|_{X[\mathcal{E}]} &= \int (\sum a_I^2 1_{C_I})^{1/2} \\ &\geq \sum_{\{J:|J|=2^{-N}\}} (\sum_{I \supseteq J} a_I^2)^{1/2} |C_J| \\ &\geq \frac{1}{2} |C_{[0,1]}| \left\| \sum a_I h_I \right\|_{H^1} . \end{aligned} \tag{4.2.5}$$

On the other hand (4.2.3) and (4.2.4) give the upper estimate

$$\begin{aligned} \left\| \sum a_I b_I \right\|_{X[\mathcal{E}]} &\leq 2 \sum_{\{J:|J|=2^{-N}\}} (\sum_{I \supseteq J} a_I^2)^{1/2} |C_J| \\ &\leq 4 |C_{[0,1]}| \left\| \sum a_I h_I \right\|_{H^1} . \end{aligned} \tag{4.2.6}$$

Combining (4.2.5) and (4.2.6) we showed that the closure of $\operatorname{span}\{b_I : I \in \mathcal{D}\}$ is isomorphic to H^1. Summing up, the closure of $\operatorname{span}\{b_I : I \in \mathcal{D}\}$ is a complemented subspace of $X[\mathcal{E}]$, which is isomorphic to H^1. □

We derive now consequences of the fact that

$$|\limsup \mathcal{E}| = 0.$$

Proposition 4.2.3. *Let $\mathcal{E}$ be a nested family of sets for which $|\limsup \mathcal{E}| = 0$. Then $\mathcal{E}$ can be decomposed into a sequence of finite collections $\mathcal{E}_i$ such that*

$$\sum_{j=i+2}^{\infty} |\mathcal{E}_j^* \cap J| \leq \frac{|J|}{4}. \tag{4.2.7}$$

Consequently

$$X[\mathcal{E}] \sim \left(\sum X[\mathcal{E}_i] \right)_1,$$

and

$$2 \sup_i [\![\mathcal{E}_i]\!] \geq [\![\mathcal{E}]\!].$$

Proof. The proof starts by defining the decomposition of $\mathcal{E}$. We rename $\mathcal{F}_0 = \mathcal{E}$, and choose a maximal interval $I \in \mathcal{F}_0$. We have as before,

$$\limsup \mathcal{E} = \bigcap_{m=1}^{\infty} G_m(\mathcal{E})^*.$$

By hypothesis $|\limsup \mathcal{E}| = 0$. So the generations $G_m(\mathcal{E})$ are shrinking down to a set of measure zero. Hence we may pick $p = p(I)$ such that

$$|G_p(I, \mathcal{F}_0)^*| \leq \frac{1}{8}|I|. \tag{4.2.8}$$

Then choose a block of finitely many intervals, call it $\mathcal{E}(I)$, satisfying

$$\mathcal{E}(I) \subseteq G_0(I, \mathcal{F}_0) \cup \cdots \cup G_p(I, \mathcal{F}_0), \tag{4.2.9}$$

so that the remaining collection defined by

$$\mathcal{F}_1(I) = \mathcal{F}_0 \cap I \setminus \mathcal{E}(I) \tag{4.2.10}$$

is small compared to I and satisfies

$$|\mathcal{F}_1(I)^*| \leq \frac{1}{4}|I|. \tag{4.2.11}$$

Having defined $\mathcal{E}(I)$ and $\mathcal{F}_1(I)$ for every maximal element of $\mathcal{F}_0$ we form

$$\mathcal{F}_1 = \bigcup \{\mathcal{F}_1(I) : I \text{ is maximal in } \mathcal{F}_0\}.$$

Note that

$$\mathcal{F}_0 \setminus \mathcal{F}_1 = \bigcup \{\mathcal{E}(I) : I \text{ is maximal in } \mathcal{F}_0\}.$$

Next we consider $\mathcal{F}_1$ and apply the above procedure to the maximal elements of $\mathcal{F}_1$, thus forming $\mathcal{E}(I)$ for I maximal in $\mathcal{F}_1$. Proceeding by induction we obtain a decreasing chain of collections

$$\mathcal{F}_0 \supseteq \mathcal{F}_1 \supseteq \cdots \supseteq \mathcal{F}_k \ldots,$$

where

$$\mathcal{F}_k \setminus \mathcal{F}_{k+1} = \bigcup \{\mathcal{E}(I) : I \text{ is maximal in } \mathcal{F}_k\}.$$

The blocks $\mathcal{E}(I)$ consist of finitely many intervals and satisfy conditions analogous to (4.2.8)–(4.2.11) with $\mathcal{F}_0$ replaced by $\mathcal{F}_k$. We relabel the families $\{\mathcal{E}(I)\}$ as $\{\mathcal{E}_i : i \in \mathbb{N}\}$ such that

$$J \in \mathcal{E}_j, \quad K \in \mathcal{E}_k \text{ and } J \subseteq K \text{ implies } j \geq k. \tag{4.2.12}$$

With property (4.2.11) and (4.2.12) it is easy to see that for $I \in \mathcal{E}_i$,

$$\sum_{j=i+2}^{\infty} |\mathcal{E}_j^* \cap I| \leq \frac{1}{4}|I|. \tag{4.2.13}$$

The estimate (4.2.13) in combination with (4.2.12) implies that

$$2\sup[\![\mathcal{E}_j]\!] \geq [\![\mathcal{E}]\!]. \tag{4.2.14}$$

Note that in the case when $[\![\mathcal{E}]\!] = \infty$, the estimate $2\sup[\![\mathcal{E}_j]\!] \geq [\![\mathcal{E}]\!]$ implies that $\sup[\![\mathcal{E}_j]\!] = \infty$. It is in this case that we use the estimate (4.2.14) below.

We show now that $X[\mathcal{E}]$ is isomorphic to the ℓ^1 direct sum of the spaces $X[\mathcal{E}_i]$. Let $f_i \in X[\mathcal{E}_i]$. With the unconditional basis $\{d_A : A \in \mathcal{E}_i\}$ we expand

$$f_i = \sum \{a_A d_A : A \in \mathcal{E}_i\}.$$

It suffices to show that

$$8\|\sum f_i\|_{X[\mathcal{E}]} \geq \sum \|f_i\|_{X[\mathcal{E}_i]}.$$

We consider separately the sum over odd and even indices. By unconditionality we get

$$2\|\sum f_i\|_{X[\mathcal{E}]} \geq \|\sum f_{2i}\|_{X[\mathcal{E}]} + \|\sum f_{2i+1}\|_{X[\mathcal{E}]}.$$

Without loss of generality we treat only the even terms. We let $F_i = \operatorname{supp} S(f_{2i})$. Hence by (4.2.13), for $A \in \mathcal{E}_{2i}$,

$$\sum_{j=i+1}^{\infty} |A \cap F_j| \leq \frac{1}{4}|A|. \tag{4.2.15}$$

Now write

$$D_i = F_i \setminus \bigcup_{j=i+1}^{\infty} F_j.$$

Since the sets $\{D_i\}$ are disjoint we obtain, using (4.2.15),

$$\int S(f_{2i})1_{D_i} \geq \frac{1}{2}\int S(f_{2i}).$$

Now we finish as follows,

$$\begin{aligned}\int S(\sum f_{2i}) &\geq \sum \int S(f_{2i})1_{D_i} \\ &\geq \frac{1}{2}\sum \int S(f_{2i}).\end{aligned}$$

We estimate the odd numbered terms in an analogous way. Thus, $X[\mathcal{E}]$ is isomorphic to $(\sum X[\mathcal{E}_i])_1$. □

A combination of Proposition 4.2.3, the Gamlen–Gaudet construction and the condensation lemma leads to the identification of $X[\mathcal{E}]$ when $|\limsup \mathcal{E}| = 0$ and $[\![\mathcal{E}]\!] = \infty$.

Theorem 4.2.4. *Let $\mathcal{E}$ be a nested collection of measurable sets such that for $A, B \in \mathcal{E}$,*

$$A \subset B \text{ implies } |A| \leq \frac{1}{2}|B|.$$

Then $|\limsup \mathcal{E}| = 0$ and $[\![\mathcal{E}]\!] = \infty$ implies that

$$X[\mathcal{E}] \sim \left(\sum H_n^1\right)_1.$$

Proof. With Proposition 4.2.3, $\mathcal{E}$ can be decomposed into a sequence of finite nested collections $\mathcal{E}_i$ so that

$$X[\mathcal{E}] \sim \left(\sum X[\mathcal{E}_i]\right)_1,$$

and simultaneously

$$\sup_i [\![\mathcal{E}_i]\!] = \infty. \tag{4.2.16}$$

As $\mathcal{E}_i$ is finite, Proposition 2.1.4 implies that for i there exists $n = n(i)$ so that $X[\mathcal{E}_i]$ is well isomorphic to a well complemented subspace of H_n^1. Consequently the space $(\sum X[\mathcal{E}_i])_1$ is isomorphic to a complemented subspace of $\left(\sum H_n^1\right)_1$.

Next we show that conversely, $\left(\sum H_n^1\right)_1$ is isomorphic to a complemented subspace of $(\sum X[\mathcal{E}_i])_1$. Specifically we will show that for $n \in \mathbb{N}$ there exists $i \in \mathbb{N}$ such that

$$\begin{array}{ccc} H_n^1 & \xrightarrow{Id} & H_n^1 \\ E \searrow & & \nearrow Q \\ & X[\mathcal{E}_i] & \end{array}$$

where $||E|| \, ||Q|| \leq 16$ (by the decomposition principle of A. Pełczyński this proves the theorem). Fix $n \in \mathbb{N}$, then by (4.2.16) there exists $i = i(n)$ such that

$$[\![\mathcal{E}_i]\!] \geq 8^n.$$

We apply the condensation Lemma 3.1.4 and find $A \in \mathcal{E}_i$ such that

$$|G_n(A, \mathcal{E}_i)^*| \geq |A|(1 - 4^{-n}).$$

Now starting with $A \in \mathcal{E}_i$ we repeat the basic step in the Gamlen–Gaudet construction n-times. This provides us with collections $\mathcal{C}_I \subseteq \mathcal{E}_i$ where $I \in \mathcal{D}_n$, such that

$$\mathcal{C}_{[0,1]} = \{A\};$$

the sets $\{\mathcal{C}_I^* : I \in \mathcal{D}_n\}$ form a nested collection, and

$$\mathcal{C}_I^* \subseteq \mathcal{C}_J^* \text{ iff } I \subseteq J;$$

the collections $\{\mathcal{C}_I : I \in \mathcal{D}_n\}$ satisfy Jones compatibility condition (J) and

$$\frac{1}{2}|A|\,|I| \le |\mathcal{C}_I^*| \le |A|\,|I|.$$

Now for $I \in \mathcal{D}_n$, we form

$$b_I = \sum \{d_B : B \in \mathcal{C}_I\}.$$

We saw in the proof of Theorem 4.2.2 that the block basis $\{|A|^{-1}b_I\}$ is equivalent to the Haar basis of H^1_n. More precisely we showed that

$$\frac{1}{2}\left\|\sum a_I h_I\right\|_{H^1_n} \le \frac{1}{|A|}\left\|\sum a_I b_I\right\|_{X[\mathcal{E}_i]} \le 2\left\|\sum a_I h_I\right\|_{H^1_n}, \tag{4.2.17}$$

for any choice of $a_I \in \mathbb{R}$, with $I \in \mathcal{D}_n$. We define the operator $E : H^1_n \to X[\mathcal{E}_i]$ to be the linear extension of the map

$$E : h_I \to |A|^{-1}b_I.$$

The right-hand side of (4.2.17) shows that the operator E has norm ≤ 2. Next we define $Q : X[\mathcal{E}_i] \to H^1_n$ by the equation

$$Qf = \sum_{I \in \mathcal{D}_n} \langle f, \frac{b_I}{\|b_I\|_2}\rangle \frac{h_I}{\|h_I\|_2}.$$

Clearly Q inverts the action of E so that

$$\mathrm{Id} = QE.$$

Furthermore, Jones's compatibility condition (J) implies that the operator Q has norm ≤ 8. Summing up we proved that $X[\mathcal{E}]$ contains a complemented subspace isomorphic to $(\sum H^1_n)_1$. $\square$

For completeness we record again the proposition which identifies the isomorphic type of $X[\mathcal{E}]$ when $\mathcal{E}$ satisfies a Carleson condition. We stated and proved it already in Chapter 2 (see (2.1.10) and (2.1.11)).

Proposition 4.2.5. *Let $\mathcal{E}$ be a nested family of sets satisfying $[\![\mathcal{E}]\!] < \infty$. Then*

$$X[\mathcal{E}] \sim \ell^1,$$

or $X[\mathcal{E}]$ is finite dimensional.

Proof of Theorem 4.2.1. We obtain part (a) of Theorem 4.2.1 from Theorem 4.2.2. Part (b) follows from Theorem 4.2.4 and part (c) from Proposition 4.2.5.

Next note that for each nested collection $\mathcal{E}$ one of the following statements holds true, $|\limsup \mathcal{E}| > 0$, $|\limsup \mathcal{E}| = 0$ and $[\![\mathcal{E}]\!] = \infty$, or $[\![\mathcal{E}]\!] < \infty$. Since these conditions on $\mathcal{E}$ are mutually exclusive it follows that the implications in part (a), (b) and (c) are actually equivalences. Thus the following three implications hold true. If $X[\mathcal{E}]$ is isomorphic to H^1, then $|\limsup \mathcal{E}| > 0$, if $X[\mathcal{E}]$ is isomorphic to $(\sum H^1_n)_1$, then $|\limsup \mathcal{E}| = 0$ and $[\![\mathcal{E}]\!] = \infty$, and finally if $X[\mathcal{E}]$ is isomorphic to ℓ^1, then $[\![\mathcal{E}]\!] < \infty$. □

Next we obtain the classification of the dual spaces to $X[\mathcal{E}]$. Clearly, we are going to apply Theorem 4.2.1. Given the nested collection $\mathcal{E}$ the following system of three-valued martingale differences form an unconditional basis for $X[\mathcal{E}]$,

$$d_A(\omega, t) = 1_A(\omega) r_A(t) \quad A \in \mathcal{E}.$$

Recall that in Section 2.1 we observed that the dual to $X[\mathcal{E}]$ can be identified with the space $Y[\mathcal{E}]$ consisting of those functions $h = \sum_{A \in \mathcal{E}} b_A d_A$ for which

$$\|h\|_{Y[\mathcal{E}]} = \sup_{A \in \mathcal{E}} \left(\frac{1}{|A|} \sum_{B \in \mathcal{E} \cap A} b_B^2 |B| \right)^{1/2} < \infty.$$

The next theorem classifies the Banach spaces $Y[\mathcal{E}]$. Its interest comes from the fact that there exist just two different spaces in the class $Y[\mathcal{E}]$, and that the isomorphic type of $Y[\mathcal{E}]$ is determined by the value of the Carleson constant,

$$[\![\mathcal{E}]\!] = \sup_{A \in \mathcal{E}} \frac{1}{|A|} \sum_{B \in \mathcal{E} \cap A} |B|.$$

Theorem 4.2.6. *Let $\mathcal{E}$ be a nested collection of measurable sets satisfying the following condition.*

If $A, B \in \mathcal{E}$, and if A is strictly contained in B, then $|A| \le |B|/2$.

Then the space $Y[\mathcal{E}]$ is isomorphic to ℓ^∞ or BMO.

(a) *If $[\![\mathcal{E}]\!] < \infty$, then $Y[\mathcal{E}]$ is isomorphic to ℓ^∞.*

(b) *If $[\![\mathcal{E}]\!] = \infty$, then $Y[\mathcal{E}]$ is isomorphic to BMO.*

Proof. The result follows from the classification Theorem 4.2.1 and Wojtaszczyk's result that BMO is isomorphic to $(\sum \mathrm{BMO}_n)_\infty$. Indeed, if $[\![\mathcal{E}]\!] < \infty$, then $X[\mathcal{E}] \sim \ell^1$. Hence $Y[\mathcal{E}] \sim \ell^\infty$. On the other hand if $[\![\mathcal{E}]\!] = \infty$, then $X[\mathcal{E}] \sim (\sum H^1_n)_1$ or $X[\mathcal{E}] \sim H^1$. Hence $Y[\mathcal{E}] \sim (\sum \mathrm{BMO}_n)_\infty$ or $Y[\mathcal{E}] \sim \mathrm{BMO}$. By Theorem 2.2.3 the space $(\sum \mathrm{BMO}_n)_\infty$ is isomorphic to BMO. □

Let $\mathcal{C}$ be a collection of dyadic intervals. Then clearly $X[\mathcal{C}]$ is isometric to the closure of $\operatorname{span}\{h_I : I \in \mathcal{C}\}$ in H^1. Similarly the space $Y[\mathcal{C}]$ is isometric to the weak *-closure of $\operatorname{span}\{h_I : I \in \mathcal{C}\}$ in BMO. Thus Theorem 4.2.1 classifies the subspaces in H^1 which are generated by subsequences of the Haar system. The classification for the weak *-closed hull of $\{h_I : I \in \mathcal{C}\}$ in BMO is given by Theorem 4.2.6.

Classification of weighted intersections

Next we turn to the spaces $\mathcal{S}^\infty$ defined in Section 2.1. Let (x_I) be a sequence such that $0 \leq x_I \leq 1$. Then the space $\mathcal{S}^\infty$ consists of all sequences (a_I) for which

$$\|(a_I)\|_{\mathcal{S}^\infty} = \|\sum x_I a_I h_I\|_{\mathrm{BMO}} + \sup |a_I| < \infty.$$

Theorem 2.1.11 asserts that the space $\mathcal{S}^\infty$ is isomorphic to a complemented subspace of BMO so that the norm of embedding and projection are independent of the weight (x_I). In the course of proving that $\mathcal{S}^\infty$ is isomorphic to a complemented subspace of BMO we exhibited a family of BMO functions $\{g_I\}$ disjointly supported over the Haar system so that

$$g = \sum a_I g_I, \quad \text{with} \quad a_I \in \mathbb{R},$$

satisfies

$$c\|g\|_{\mathrm{BMO}} \leq \|\sum a_I x_I h_I\|_{\mathrm{BMO}} + \sup |a_I| \leq C\|g\|_{\mathrm{BMO}}. \tag{4.2.18}$$

Thus we verified that $\{g_I\}$ is equivalent to the unit vector basis of $\mathcal{S}^\infty$. Moreover in Section 2.1 we showed that the orthogonal projection

$$P(f) = \sum_I \langle f, \frac{g_I}{||g_I||_2}\rangle \frac{g_I}{||g_I||_2}$$

is bounded in BMO.

Now we continue with the analysis of $\mathcal{S}^\infty$ and we show that this space is isomorphic to either ℓ^∞ or to BMO.

Theorem 4.2.7. *For every choice of scalars $x_I \in \mathbb{R}$ with $0 \leq x_I \leq 1$ the resulting space $\mathcal{S}^\infty$ is isomorphic to BMO or to ℓ^∞. If $\sum x_I h_I \in$ BMO, then $\mathcal{S}^\infty$ is isomorphic to ℓ^∞, and if $\sum x_I h_I \notin$ BMO, then $\mathcal{S}^\infty$ is isomorphic to BMO.*

Proof. Suppose now that $f = \sum x_I h_I \in \mathrm{BMO}$. Let $\{a_I\}$ be a given sequence of scalars. Then we have the upper bound

$$\begin{aligned} ||\sum a_I x_I h_I||_{\mathrm{BMO}} &\leq \sup |a_I| \cdot ||\sum x_I h_I||_{\mathrm{BMO}} \\ &\leq \sup |a_I| \cdot ||f||_{\mathrm{BMO}}. \end{aligned}$$

Inserting this estimate in the equation defining the norm of $\mathcal{S}^\infty$ shows that $\mathcal{S}^\infty$ is isomorphic to ℓ^∞ when $f \in \mathrm{BMO}$. Next we turn to the converse case when $f \notin \mathrm{BMO}$. We begin by performing a stopping time argument on the function $f = \sum x_I h_I$, or rather we perform the stopping time argument on the set of coefficients defining $f = \sum x_I h_I$. For the unit interval $[0,1]$ we define $\mathcal{H}_{[0,1]}$ to be the largest block of dyadic intervals satisfying $[0,1] \in \mathcal{H}_{[0,1]}$ and

$$\sum_{L \in \mathcal{H}_{[0,1]}} x_L^2 h_L^2 \leq 2.$$

The maximality condition in the definition of $\mathcal{H}_{[0,1]}$, and the fact that $|x_I| \leq 1$ imply the following lower estimate. If $J \in Q([0,1]) \setminus \mathcal{H}_{[0,1]}$, then

$$\sum_{L \in \mathcal{H}_{[0,1]}} x_L^2 h_L^2 \geq 1 \text{ on the interval } J.$$

Next let K be a maximal interval in the collection $Q([0,1]) \setminus \mathcal{H}_{[0,1]}$, Then define $\mathcal{H}_K$ to be the maximal block of dyadic intervals contained in $Q(K)$ such that

$$\sum_{L \in \mathcal{H}_K} x_L^2 h_L^2 \leq 2.$$

Note that if $J \in Q(K) \setminus \mathcal{H}_K$, then

$$\sum_{L \in \mathcal{H}_K} x_L^2 h_L^2 \geq 1 \text{ on the interval } J.$$

This process defines a decomposition of the dyadic intervals into a family of blocks $\{\mathcal{H}_K : K \in \mathcal{E}\}$. In this way we also obtain a decomposition of the function f into pieces $f_K = \sum_{L \in \mathcal{H}_K} x_L h_L$. This completes the stopping time decomposition of $f = \sum x_I h_I$

Now we turn to the last part of the proof examining consequences of the fact that $f \notin \mathrm{BMO}$. The proof follows a pattern we are quite familiar with by now. If $f \notin \mathrm{BMO}$ the index set obtained from the stopping time decomposition does not satisfy the Carleson packing condition. Applying to this index set the condensation Lemma 3.1.4 and then performing the Gamlen–Gaudet construction sufficiently often we find well complemented copies of BMO_n in $\mathcal{S}^\infty$. Thus we find $(\sum \mathrm{BMO}_n)_\infty$ as a complemented subspace of $\mathcal{S}^\infty$. With the decomposition principle we conclude that $\mathcal{S}^\infty$ and $(\sum \mathrm{BMO}_n)_\infty$ are isomorphic spaces. Following are the details.

We first observe that if $f \notin \mathrm{BMO}$, then the index $\mathcal{E}$ satisfies

$$\sup_I \frac{1}{|I|} \sum_{J \in \mathcal{E}, J \subseteq I} |J| = \infty.$$

The condensation Lemma 3.1.4 implies that for any $n \in \mathbb{N}$, there exists a dyadic interval $A \in \mathcal{E}$ such that

$$|G_n^*(A, \mathcal{E})| \geq |A|(1 - 8^{-n}).$$

Now let $\mathcal{X}_{[0,1]} = \{A\}$ be the collection containing just the interval A. Let J_1 be the left half of $[0,1]$, and let J_2 be the right half of $[0,1]$. We also let A_1 be the left half of A, and let A_2 be the right half of A. Then define $\mathcal{X}_{J_i}$ to be the collection of maximal intervals in $\mathcal{E}$, which are contained in A_i. (This was the first step of the Gamlen–Gaudet construction.) Next let K be any interval in $\mathcal{X}_{J_i}$. Let K_1 be the left half of K, and let K_2 be the right half of K. Then let $\mathcal{X}(K_j)$ be the collection of maximal intervals in $\mathcal{E}$, which are contained in K_j. We let J_{i1} be the left half of J_i, and we let J_{i2} be the right half of J_i. We will now take the union and put

$$\mathcal{X}_{J_{ij}} = \bigcup_{K \in \mathcal{X}_{J_i}} \mathcal{X}(K_j).$$

This was the second step of the Gamlen–Gaudet construction. We repeat the basic step of the Gamlen–Gaudet construction n times. With the assertion of the condensation lemma, that is with

$$|G_n^*(A, \mathcal{E})| \geq |A|(1 - 8^{-n}),$$

we obtain collections of pairwise disjoint dyadic intervals $\mathcal{X}_J \subseteq \mathcal{E}$, for $|J| \geq 2^{-n}$, satisfying the following properties.

1. For $I \subseteq J$ let X_I be the point-set covered by the collection $\mathcal{X}_I$. Then for every $K \in \mathcal{X}_J$,
$$(1/2)|K| \cdot |I| \leq |J| \cdot |X_I \cap K| \leq 2|K| \cdot |I|.$$
2. If $X_I \cap X_J \neq \emptyset$, then either $X_I \subseteq X_J$, or $X_J \subseteq X_I$.
3. If $X_J \subseteq X_I$, then $J \subseteq I$.
4. $|A| \cdot |J|/2 \leq |X_J| \leq 2|A| \cdot |J|$.

Let $\{g_I\}$ be the BMO functions satisfying (4.2.18). That is, $\{g_I\}$ is equivalent to the unit vector basis of S^∞. Their construction is obtained in the course of proving Theorem 2.1.11. Now define

$$G_J = \sum_{K \in \mathcal{X}_J} \sum_{L \in \mathcal{H}_K} g_L,$$

and also

$$H_J = \sum_{K \in \mathcal{X}_J} \sum_{L \in \mathcal{H}_K} x_L h_L.$$

We use $\{e_L\}$ indexed by dyadic intervals to denote the unit vector basis of ℓ^∞. Then put

$$F_J = \sum_{K \in \mathcal{X}_J} \sum_{L \in \mathcal{H}_K} e_L.$$

In the proof of the following statements the systems $\{H_J\}$, and $\{F_J\}$ play the role of auxiliary tools: First, we claim that in BMO the system $\{G_J : |J| \geq 2^{-n}\}$ is equivalent to the Haar basis $\{h_J : |J| \geq 2^{-n}\}$. And second, the orthogonal projection Q defined by

$$Q(y) = \sum_{\{I:|I| \geq 2^{-n}\}} \langle y, \frac{G_I}{||G_I||_2} \rangle \frac{G_I}{||G_I||_2}$$

satisfies

$$||Q(y)||_{\mathrm{BMO}} \leq 4||y||_{\mathrm{BMO}},$$

for $y \in \operatorname{span}\{g_J\}$.

We begin by showing the first claim doing the calculations with the system $\{H_J\}$. We fix J and let J_1 be the left half of J and let J_2 be the right half of J. Note that the square function $S^2(H_J) = \sum_{K \in \mathcal{X}_J} \sum_{L \in \mathcal{H}_K} x_L^2 h_L^2$ satisfies the pointwise upper estimate $S^2(H_J) \leq 2$, and also the lower estimates

$$S^2(H_J) \geq 1 \text{ on the set } X_{J_1} \cup X_{J_1}.$$

From this it follows easily that

$$||\sum a_J H_J||_{\mathrm{BMO}} \sim ||\sum a_J h_J||_{\mathrm{BMO}}. \tag{4.2.19}$$

Below we verify that

$$||\sum a_J F_J||_{\ell^\infty} = \sup |a_J|. \tag{4.2.20}$$

To see this fix two dyadic intervals $J \neq J'$. Then the corresponding index sets given by $\mathcal{I}_J = \bigcup_{K \in \mathcal{X}_J} \mathcal{H}_K$ and $\mathcal{I}_{J'} = \bigcup_{K \in \mathcal{X}_{J'}} \mathcal{H}_K$, are disjoint collections of intervals. Thus we proved that (4.2.20) holds true. By (4.2.18) in combination with (4.2.19) and (4.2.20) it follows that

$$\begin{aligned} ||\sum a_J G_J||_{\mathrm{BMO}} &\sim ||\sum a_J H_J||_{\mathrm{BMO}} + \sup |a_I| \\ &\sim ||\sum a_J h_J||_{\mathrm{BMO}}. \end{aligned}$$

Thus we proved the claim that in BMO the system $\{G_J : |J| \geq 2^{-n}\}$ is equivalent to the Haar basis $\{h_J : |J| \geq 2^{-n}\}$.

Now we give the BMO estimates showing the boundedness of the orthogonal projection Q defined on $\operatorname{span}\{g_I\}$. We let

$$y = \sum a_I g_I.$$

Then

$$\langle y, G_J \rangle = \sum_{K \in \mathcal{X}_J} \sum_{L \in \mathcal{H}_K} a_L x_L^2 |L|,$$

and

$$||G_J||_2^2 = \sum_{K \in \mathcal{X}_J} \sum_{L \in \mathcal{H}_K} x_L^2 |L|.$$

Combining the above expressions with Hölder's inequality gives

$$\frac{\langle y, G_J \rangle^2}{||G_J||_2^2} \leq \sum_{K \in \mathcal{X}_J} \sum_{L \in \mathcal{H}_K} a_L^2 x_L^2 |L|.$$

Recall that $\mathcal{X}_J$ consists of pairwise disjoint intervals, hence,

$$|X_J| = \sum_{K \in \mathcal{X}_J} |K|.$$

For $K \in \mathcal{X}_J$, let $f_K = \sum_{L \in \mathcal{H}_K} x_L h_L$. Then the square function, $S^2(f_K) = \sum_{L \in \mathcal{H}_K} x_L^2 h_L^2$, satisfies

$$\sum_{L \in \mathcal{H}_K} x_L^2 |L| = \int_K S^2(f_K).$$

Next by the stopping time construction and the condensation lemma we obtain,

$$\frac{|K|}{2} \leq \int_K S^2(f_K) \leq 2|K|.$$

Taking the sum over $K \in \mathcal{X}_J$ gives

$$\frac{|X_J|}{2} \leq \sum_{K \in \mathcal{X}_J} \int_K S^2(f_K) \leq 2|X_J|.$$

Now observe that for J fixed, and for I strictly contained in J , and $K \in \mathcal{X}_J$, we have the identities,

$$\frac{1}{|K|} \int_K G_I^2 = \frac{||G_I||_2^2}{|X_J|} = \frac{|X_I|}{|X_J|}.$$

With the information collected so far we now derive estimates for the norm of $Q(y)$ in BMO. We fix J and $K \in \mathcal{X}_J$, then we compute obtaining global estimates,

$$\begin{aligned}
\frac{1}{|K|} \int_K |Q(y) - \frac{1}{|K|} \int_K Q(y)|^2 &\leq \frac{\langle y, G_J \rangle^2}{||G_J||_2^4} + \sum_{I \subset J} \langle y, \frac{G_I}{||G_I||_2} \rangle^2 \frac{1}{|K|} \int_K \frac{G_I^2}{||G_I||_2^2} \\
&\leq \sup |a_I|^2 + \frac{1}{|B_J|} \sum_{I \subset J} \sum_{L \in \mathcal{X}_I} a_L^2 x_L^2 |L| \\
&\leq \sup |a_I|^2 + || \sum a_L x_L h_L ||_{\mathrm{BMO}}^2 \\
&\leq || \sum a_L g_L ||_{\mathrm{BMO}}^2.
\end{aligned}$$

Finally we point out that by the stopping time process we have the following local estimates. For any interval $L \in \mathcal{H}_K$ with $K \in \mathcal{X}_I$,

$$\frac{1}{|L|}\int_L |G_I - \frac{1}{|L|}\int_L G_I|^2 \leq C.$$

Combining the local estimate and the global estimates above shows that

$$||Q(y)||_{\mathrm{BMO}} \leq C||y||_{\mathrm{BMO}}.$$

So far we showed that for each $n \in \mathbb{N}$ there is a well complemented copy of BMO_n in $\mathrm{span}\{g_I\}$. We obtained these copies using only finite linear combinations of Haar functions. Therefore we actually showed that the infinite direct sum $(\sum \mathrm{BMO}_n)_\infty$ is isomorphic to a complemented subspace of the weak* closure of $\mathrm{span}\{g_I\}$ in BMO. Hence by Theorem 2.2.3 (asserting that $(\sum \mathrm{BMO}_n)_\infty \sim \mathrm{BMO}$) we showed that BMO is isomorphic to a complemented subspace of $\mathcal{S}^\infty$ provided that $\sum x_I h_I \notin \mathrm{BMO}$. Since, by Theorem 2.1.11 the space $\mathcal{S}^\infty$ is isomorphic to a complemented subspace of BMO, Pełczyński's decomposition method implies that the weak* closure of $\mathrm{span}\{g_I\}$ in BMO and BMO are isomorphic when $\sum x_I h_I \notin \mathrm{BMO}$. □

4.3 More on subsystems of the Haar system

In this section we study the subspaces in $L^p([0,1])$, $1 < p < \infty$, which are spanned by subsequences of the Haar system. We present the theorem of Gamlen and Gaudet to the effect that the closed linear span of every infinite sequence of Haar functions is isomorphic to ℓ^p or L^p. We also prove a finite-dimensional version of the Gamlen–Gaudet theorem. This is accomplished by replacing certain qualitative considerations with dimension free estimates. The section ends with the discussion of several related open problems.

The theorem of Gamlen and Gaudet

We first prove the classical theorem of Gamlen and Gaudet that classifies the spaces generated by a subsequence of the Haar basis in L^p. Later we establish a quantitative refinement of the Gamlen and Gaudet theorem exploiting the properties of Maurey's isomorphism.

Let $\mathcal{C}$ be a collection of dyadic intervals. Let $1 < p < \infty$, and let $X^p[\mathcal{C}]$ denote the closure of $\mathrm{span}\{h_I : I \in \mathcal{C}\}$ in L^p. On $X^p[\mathcal{C}]$ we put the norm induced by the square function characterization of $L^p, 1 < p < \infty$. Thus for $f \in X^p[\mathcal{C}]$ with Haar expansion $f = \sum_{I \in \mathcal{C}} a_I h_I$ we define

$$\|f\|_{X^p[\mathcal{C}]} = \|S(f)\|_{L^p} \quad \text{where} \quad S(f)^2(t) = \sum_{I \in \mathcal{C}} a_I^2 h_I^2(t).$$

The following theorem of Gamlen and Gaudet asserts that the value of $|\limsup \mathcal{C}|$ determines the isomorphic type of $X^p[\mathcal{C}]$.

Theorem 4.3.1. *Let $1 < p < \infty$, and let $\mathcal{C}$ be an infinite collection of dyadic intervals.*

(a) *If $|\limsup \mathcal{C}| = 0$, then $X^p[\mathcal{C}]$ is isomorphic to ℓ^p.*

(b) *If $|\limsup \mathcal{C}| > 0$, then $X^p[\mathcal{C}]$ is isomorphic to $L^p([0,1])$.*

Proof. The proof follows from Theorem 4.2.2 and Proposition 4.2.3 by interpolation.

We start with the observation that $X^p[\mathcal{C}]$ is a complemented subspace of $L^p, 1 < p < \infty$. If $|\limsup \mathcal{C}| > 0$, then by Theorem 4.2.2 there exist sub-collections in $\mathcal{C}$, denoted $\{\mathcal{C}_I : I \in \mathcal{D}\}$ satisfying the compatibility condition (J) together with (4.2.3) and (4.2.4). By Theorem 1.5.9 and interpolation between H^1 and L^2 the following orthogonal projection is bounded in $X^p[\mathcal{C}]$,

$$P(f) = \sum_{I \in \mathcal{D}} \langle f, \frac{b_I}{\|b_I\|_2} \rangle \frac{b_I}{\|b_I\|_2},$$

where $b_I = \sum\{h_J : J \in \mathcal{C}_I\}$. Furthermore it is easy to observe that the L^p version of (4.2.5) and (4.2.6) hold true. Let $C_{[0,1]} = \bigcup_{J \in \mathcal{C}_{[0,1]}} J$. Then

$$\frac{1}{2}\|\sum a_I h_I\|_{X^p[\mathcal{D}]} \leq |C_{[0,1]}|^{-1/p}\|\sum a_I b_I\|_{X^p[\mathcal{C}]} \leq 2\|\sum a_I h_I\|_{X^p[\mathcal{D}]}. \tag{4.3.1}$$

By Burkholder's theorem the identity operator is an isomorphism between L^p and $X^p[\mathcal{D}]$. In summary if $|\limsup \mathcal{C}| > 0$, then $X^p[\mathcal{C}]$ contains a complemented copy of $L^p, 1 < p < \infty$. The decomposition principle of A. Pełczyński implies that $X^p[\mathcal{C}]$ is isomorphic to L^p.

Next we turn to the case when $|\limsup \mathcal{C}| = 0$. We will show that then $X^p[\mathcal{C}]$ is isomorphic to a complemented subspace of $\ell^p, 1 < p < \infty$. If $|\limsup \mathcal{C}| = 0$, then by Proposition 4.2.3 there exists a decomposition of $\mathcal{C}$ into *finite* collections $\{\mathcal{C}_i : i \in \mathbb{N}\}$ so that for each $i \in \mathbb{N}$ and $J \in \mathcal{C}_i$ we have

$$\sum_{j=i+2}^{\infty} |J \cap C_j| \leq \frac{|J|}{4}, \tag{4.3.2}$$

where $C_j = \bigcup_{K \in \mathcal{C}_j} K$. Recall that the estimate (4.3.2) appears in the hypothesis of Proposition 3.1.5. Applying it to

$$f = \sum f_i, \quad \text{where} \quad f_i \in X^p[\mathcal{C}_i],$$

gives that

$$\frac{1}{2}\|f\|_{X^p[\mathcal{C}]} \leq \left(\sum \|f_i\|^p_{X^p[\mathcal{C}_i]}\right)^{1/p} \leq 2\|f\|_{X^p[\mathcal{C}]}.$$

Thus $X^p[\mathcal{C}] \sim (\sum X^p[\mathcal{C}_i])_p$.

For $n \in \mathbb{N}$ let $\mathcal{D}_n = \{I \in \mathcal{D} : |I| \geq 2^{-n}\}$. Clearly, since $\mathcal{C}_i$ is finite there exists $n = n(i)$ so that $\mathcal{C}_i \subseteq \mathcal{D}_n$. By Burkholder's theorem for $1 < p < \infty$ the identity operator provides an isomorphism between $X^p[\mathcal{D}_n]$ and

$$L^p_{2^{n+1}-1} = (\text{span}\{h_I : I \in \mathcal{D}_n\}, \|\cdot\|_{L^p}).$$

Next observe that $L^p_{2^{n+1}-1}$ is 2-isomorphic to $\ell^p_{2^{n+1}-1}$. Thus $(\sum X^p[\mathcal{C}_i])_p$ is isomorphic to a complemented subspace of $(\sum \ell^p_n)_p \sim \ell^p$.

In summary we showed that if $|\limsup \mathcal{C}| = 0$, then $X^p[\mathcal{C}]$ is isomorphic to a complemented subspace of ℓ^p. It remains to cite the classic theorem of A. Pełczyński that every infinite-dimensional complemented subspace of ℓ^p is isomorphic to $\ell^p, 1 < p < \infty$ (see [169] or [130] Volume I, Theorem 2.a.3). □

The theorem of Gamlen and Gaudet as well as Theorem 4.2.1 are infinite dimensional in nature. They don't contain specific information when $\mathcal{C}$ is a finite collection of intervals. By contrast the following local version of the Gamlen–Gaudet theorem shows that for every finite collection $\mathcal{C}$ of dyadic intervals there exists an isomorphism T acting between $X^p[\mathcal{C}]$ and ℓ^p_N where $N =$ cardinality of $\mathcal{C}$, so that $\|T\| \cdot \|T^{-1}\|$ is bounded independent of N. Specifically we obtain

$$\|T\| \cdot \|T^{-1}\| \leq C_p.$$

The operator T by which we prove the local Gamlen–Gaudet theorem is just Maurey's isomorphism.

Theorem 4.3.2. *Let $\mathcal{C}$ be a collection of dyadic intervals and let $1 \leq p < \infty$. Then the closed linear span of $\{h_I : I \in \mathcal{C}\}$ in L^p is C_p-isomorphic to ℓ^p, L^p or ℓ^p_n where n is the cardinality of $\mathcal{C}$.*

Proof. We let $\mathcal{F}_n$ be the algebra of sets, generated by the collection $\{I \in \mathcal{C} : |I| \geq 2^{-n}\}$ and let $\mathcal{F}_0 = \{[0,1], \emptyset\}$. Let A be an atom for $\mathcal{F}_{n-1}$. Then the atoms of $\mathcal{F}_n$ contained in A are the dyadic intervals $\mathcal{B} = \{I \in \mathcal{C} : I \subseteq A, |I| = 2^{-n}\}$ together with $A \setminus \mathcal{B}^*$. Now we assume that our collection $\mathcal{C}$ was such that $A \setminus \mathcal{B}^* \neq \emptyset$, then clearly $|A \setminus \mathcal{B}^*| \geq 2^{-n}$. Note that on the other hand each atom $I \in \mathcal{B}$ satisfies $|I| = 2^{-n}$.

Therefore the largest atom $A^* \in \mathcal{F}_n$ contained in A can be chosen to be $A \setminus \mathcal{B}^*$, provided that $A \setminus \mathcal{B}^* \neq \emptyset$. Now we consider the collection $\mathcal{E}(A)$, formed by deleting A^* from the collection of all atoms in $\mathcal{F}_n$ that are contained in A, and we let

$$\mathcal{E} = \bigcup \mathcal{E}(A),$$

where the union is taken over all atoms in $\mathcal{F}_n$, $n \in \mathbb{N}$. Note that by construction $\mathcal{E}$ coincides with the collection $\mathcal{C}$ we started with.

We will use the space $X^p[\mathcal{C}]$ to mediate between the closure of span$\{h_I : I \in \mathcal{C}\}$ and $L^p(\mathcal{F}_\infty)$. For $I \in \mathcal{C}$ with $|I| = 2^{-n}$ we form

$$b_I = 1_I \otimes r_n.$$

By Burkholder's theorem,

$$\left\|\sum_{I\in\mathcal{C}} a_I b_I\right\|^p_{L^p([0,1]\times[0,1])} \sim \int\left(\sum_{I\in\mathcal{C}} a_I^2 1_I\right)^{p/2}.$$

Hence $X^p[\mathcal{C}]$ is isomorphic to $\overline{\operatorname{span}}\{h_I : I \in \mathcal{C}\}$, and the isomorphism is given by the linear extension of the relation

$$b_I \mapsto h_I.$$

On the other hand Maurey's operator

$$T : X^p[\mathcal{C}] \mapsto H^p[(\mathcal{F}_n)]$$

provides the isomorphism between $X^p[\mathcal{C}]$ and $H^p[(\mathcal{F}_n)]$, which is isomorphic to $L^p([0,1],\mathcal{F}_\infty)$ by Burkholder's theorem.

Finally we remark that $L^p([0,1],\mathcal{F}_\infty)$ is isometric to ℓ^p_n when $\mathcal{F}_\infty$ is generated by n atoms, and $L^p([0,1],\mathcal{F}_\infty)$ is isometric to ℓ^p when $\mathcal{F}_\infty$ is generated by an infinite sequence of atoms. Finally $L^p([0,1],\mathcal{F}_\infty)$ is isomorphic to L^p when $\mathcal{F}_\infty$ is non-atomic.

To remove the restrictive assumption

$$A \setminus \mathcal{B}^* \neq \emptyset$$

from $\mathcal{C}$, we remove from $\mathcal{C}$ one interval of length 2^{-n} when $\mathcal{B}^* = A$ for an atom A in $\mathcal{F}_{n-1}$. In this way we split $\mathcal{C}$ into $\mathcal{C}_1$ and $\mathcal{C}_2$, each of which satisfies the additional assumption, and the theorem follows from the unconditionality of the Haar basis. □

Related open problems

The space $\mathcal{S}^\infty$, consisting of sequences (a_I) for which

$$\|(a_I)\|_{\mathcal{S}^\infty} = \|\sum x_I a_I h_I\|_{\mathrm{BMO}} + \sup|a_I| < \infty,$$

appears as a natural endpoint of the scale $\mathcal{S}^p, 2 \le p < \infty$ for which the norm is given by

$$\|(a_I)\|_{\mathcal{S}^p} = \left\|\sum x_I^{1-2/p} a_I \frac{h_I}{|I|^{1/p}}\right\|_{L^p} + \left(\sum |a_I|^p\right)^{1/p} < \infty. \tag{4.3.3}$$

The result of D. Kleper and G. Schechtman in [117] asserts that $\mathcal{S}^p$ is isomorphic to a complemented subspace of L^p. By the decomposition principle of A. Pełczyński this is the first and crucial step towards the isomorphic classification of $\mathcal{S}^p$ in [157]: The space $\mathcal{S}^p$ is isomorphic to L^p or to ℓ^p. Open, however, is the corresponding finite-dimensional problem of classifying the family $\mathcal{S}^p_n$, $2 < p < \infty$, $n \in \mathbb{N}$, where

$$\mathcal{S}^p_n = \{(a_I)_{I\in\mathcal{D}_n} : \|(a_I)\|_{\mathcal{S}^p} < \infty\},$$

and where the norm on $\mathcal{S}^p_n$ is given by (4.3.3).

Problem 4.3.3. *Determine an isomorphism* $T : \mathcal{S}_n^p \to \ell^p_{2^{n+1}-1}$ *so that*

$$\|T\| \cdot \|T^{-1}\| < C_p.$$

In the special case where $x_I \in \{0, 1\}$, *the desired operator* T *is given by Maurey's isomorphism (this is the content of Theorem* 4.3.2 *and its proof).*

Next we discuss the product version of the Gamlen–Gaudet problem. Let $1 < p, r < \infty$. Then $L^p(L^r)$ is defined to be the space of measurable functions on the product $[0, 1] \times [0, 1]$ so that $f \in L^p(L^r)$ if

$$\|f\|_{L^p(L^r)} = \left(\int_0^1 \left(\int_0^1 |f(s,t)|^r dt\right)^{p/r} ds\right)^{1/p} < \infty.$$

The product form of the Haar system

$$h_{I\times J}(s, t) = h_I(s)h_J(t), \quad I, J \in \mathcal{D},$$

forms an unconditional basic sequence in $L^p(L^r)$. For $f \in L^p(L^r)$, with Haar expansion

$$f = \sum_{I\times J\in\mathcal{D}\times\mathcal{D}} a_{I\times J} h_{I\times J},$$

the norm in $L^p(L^r)$ is equivalent to the square-function integral

$$\left(\int_0^1 \left(\int_0^1 \Big(\sum_{I\times J\in\mathcal{D}\times\mathcal{D}} a^2_{I\times J} h^2_{I\times J}\Big)^{r/2} dt\right)^{p/r} ds\right)^{1/p}.$$

Hence for any collection of dyadic rectangles $\mathcal{R} \subseteq \mathcal{D} \times \mathcal{D}$ the closure in $L^p(L^r)$ of

$$\operatorname{span}\{h_{I\times J} \,:\, I \times J \in \mathcal{R}\}$$

is complemented in $L^p(L^r)$.

M. Capon in [42] established a subtle, combinatorial criterion for $\{h_{I\times J} : I \times J \in \mathcal{R}\}$ to generate a subspace of $L^p(L^r)$ which is isomorphic to $L^p(L^r)$. Following is the product version of the Gamlen–Gaudet theorem due to M. Capon [42].

Theorem 4.3.4. *Let* $\mathcal{R} \subseteq \mathcal{D} \times \mathcal{D}$ *be a collection of dyadic rectangles. For* $I \in \mathcal{D}$ *and* $t \in [0, 1]$ *define the family of dyadic intervals*

$$\mathcal{R}_I = \{J \in \mathcal{D} : I \times J \in \mathcal{R}\},$$

and put

$$R_t = \limsup\{I \in \mathcal{D} : t \in \limsup \mathcal{R}_I\}.$$

If

$$|\{t \in [0,1] : |R_t| > \frac{1}{2}\}| > 0, \tag{4.3.4}$$

then the closure in $L^p(L^r)$ *of*

$$\text{span}\,\{h_{I\times J}\ :\ I\times J \in \mathcal{R}\}$$

is isomorphic to $L^p(L^r)$.

Observe that for $\mathcal{R} \subseteq \mathcal{D}\times\mathcal{D}$ or for $\mathcal{R}' = \mathcal{R}\backslash\mathcal{D}\times\mathcal{D}$ the criterion (4.3.4) is satisfied. The insight into the structure of the product Haar system obtained by M. Capon in [42] should be helpful in solving the following problem.

Problem 4.3.5. *Determine the isomorphic type of the closure of*

$$\text{span}\,\{h_{I\times J}\ :\ I\times J \in \mathcal{R}\}$$

in $L^p(L^r)$.

We close this section with the discussion of a problem posed by P. Wojtaszczyk: Decide which subsequences of the Haar system are permutatively equivalent to the whole system. That is, given a collection of dyadic intervals $\mathcal{B}$, when is it possible to find a bijective map $\tau : \mathcal{D} \to \mathcal{B}$ such that the induced rearrangement operator $T_1 : H^1 \to H^1$, is bounded and invertible on its range ?

This problem links the results on rearrangement operators in Chapter 2 to the classification theorems of the present chapter. By definition, for $p < 1$ the space H^p consists of the sequences $\{a_I\}$ for which

$$\|\{a_I\}\|^p_{H^p} = \int (\sum a_I^2 h_I^2)^{p/2} < \infty.$$

Defined that way $H^p, p < 1$, is a quasi-normed space. The importance of Wojtaszczyk's problem becomes apparent only in the context of H^p when $p < 1$. Indeed for $H^p, p < 1$, P. Wojtaszczyk [215] proved the following results which show fundamental and unexpected differences between $H^p, p < 1$ and H^1.

1. Any unconditional, normalized basis in a complemented subspace of $H^p, p < 1$ is equivalent to a sub-collection of the Haar basis in $H^p, p < 1$.
2. If a subsystem $\{h_I : I \in \mathcal{B}\}$ of the Haar basis spans a subspace isomorphic to $H^p, p < 1$, then there exists a bijective mapping $\tau : \mathcal{D} \to \mathcal{B}$, such that the associated permutation operator is bounded and invertible on its range. In other words the subsystem $\{h_I : I \in \mathcal{B}\}$ is permutatively equivalent to the whole dyadic system.
3. Let two subsystems of the Haar basis $\{h_I : I \in \mathcal{B}_1\}$ and $\{h_I : I \in \mathcal{B}_2\}$ be spanning subspaces of $H^p, p < 1$, that are isomorphic. Assume that these spaces are isomorphic to their own square. Then there exists a bijective mapping $\tau : \mathcal{B}_1 \to \mathcal{B}_2$, which extends to an isomorphism from the closure of $\text{span}\{h_I : I \in \mathcal{B}_1\}$ onto the closure of $\text{span}\{h_I : I \in \mathcal{B}_2\}$.

4. There exist uncountably many collections $\mathcal{B}_\alpha$, such that the subsystems of the Haar basis $\{h_I : I \in \mathcal{B}_\alpha\}$ are pairwise non-isomorphic subspaces of $H^p, p < 1$.

Note that in particular the last statement is in strong distinction to the H^1 result of Theorem 4.2.1. Indeed, there we classified the subspaces generated by subcollections of the Haar basis, and we showed that in H^1 there are exactly three isomorphic types. In H^p with $p > 1$, there are exactly two isomorphic types by the Gamlen–Gaudet Theorem 4.3.1.

Nevertheless the answer to Wojtaszczyk's conjecture should be the same in all Hardy spaces, hence we discuss it in the context of H^1. Let $\mathcal{B}$ be a collection of dyadic intervals, and let

$$\mathcal{B}_k = \mathcal{B} \cap \{I \in \mathcal{D} : |I| = 2^{-k}\}.$$

We say that $\mathcal{B}$ satisfies Property W if there exists $\delta > 0$ and $s \in \mathbb{N}$ so that for any $k \in \mathbb{N}$, there exists $n \in [k, k+s]$ such that

$$|\mathcal{B}_n^*| > \delta.$$

The conjecture of P. Wojtaszczyk states that $\mathcal{B}$ satisfies Property W if (and only if) there exists a bijective map $\tau : \mathcal{D} \to \mathcal{B}$ such that the induced rearrangement operator $T_1 : H^1 \to H^1$, is bounded and invertible on its range. Support for this conjecture comes from two results of Wojtaszczyk. First, the conjecture is true for special cases, namely when $\mathcal{B}$ is defined by the rule

$$I \in \mathcal{B} \text{ if } |I| = 2^{-k_n},$$

where k_n is a given increasing sequence of natural numbers. For these examples it is easy to see that $\mathcal{B}$ satisfies Property W if

$$\sup_n (k_{n+1} - k_n) < \infty.$$

Second, one implication of Wojtaszczyk 's conjecture holds in general. There we have the following theorem [215].

Theorem 4.3.6. *Let $\mathcal{B}$ be a collection of dyadic intervals. Suppose that there exists a bijective map $\tau : \mathcal{D} \to \mathcal{B}$ such that the induced rearrangement operator $T_1 : H^1 \to H^1$, is bounded and invertible on its range. Then, the collection $\mathcal{B}$ satisfies Property W.*

Proof. Let $\tau : \mathcal{D} \to \mathcal{B}$ be a rearrangement and suppose that the induced rearrangement operator extends to an isomorphism onto its range. Now assume that the theorem is false, and that $\mathcal{B}$ does not satisfy Property W. Then for $\delta = 1/n$ and $s = n$ there exists k_n, such that $\lim k_n = \infty$, and for all $m \in [k_n, k_n + n]$ we have

$$|\mathcal{B}_m^*| \leq \frac{1}{n}.$$

Note that by the injectivity of τ there are at most 2^{k_n+1} dyadic intervals I for which $|\tau(I)| \leq 2^{-k_n}$. Let $K_n = [k_n + n/3, k_n + 2n/3]$, and let $m \in K_n$. Now put

$$\mathcal{C}_m = \{I \in \mathcal{D} : |I| = 2^{-m} \text{ and } |\tau(I)| \leq 2^{-k_n}\}.$$

By the injectivity of τ each of the collections $\mathcal{C}_m$ covers almost all of the unit interval, more precisely, for $m \in K_n$, $|\mathcal{C}_m^*| \geq (1 - 2^{-n/3})$. Now put

$$x_m = \sum_{I \in \mathcal{C}_m} h_I.$$

Then we have that $||x_m||_{H^1} \geq 1 - 2^{-n/3}$, and also

$$||\sum_n \sum_{m \in K_n} a_m x_m||_{H^1} \leq C(\sum_n \sum_{m \in K_n} a_m^2)^{1/2}.$$

The boundedness of the rearrangement operator T_1 gives that

$$||\sum_n \sum_{m \in K_n} a_m T_1 x_m||_{H^1} \leq C||T_1||(\sum_n \sum_{m \in K_n} a_m^2)^{1/2}.$$

Note that the Haar support of the function $T_1 x_m$ is a pairwise disjoint collection of dyadic intervals. Summing up the first set of observations, we showed that there exists $\eta > 0$, (not depending on n) such that for $m \in K_n$,

$$\sum_{I \in \mathcal{C}_m} |\tau(I)| \geq \eta.$$

Hence summing over $m \in K_n$ gives

$$\sum_{m \in K_n} \sum_{I \in \mathcal{C}_m} |\tau(I)| \geq \frac{n}{3}\eta.$$

Now we prove that this lower estimate contradicts the assumption that Property W does not hold. To see this we consider the index set of the above sum, and split it as follows. First we form

$$\mathcal{A}_0 = \{\tau(I) : I \in \mathcal{C}_m, m \in K_n\} \cap \{J : |J| > 2^{-k_n}\}.$$

Note that by definition of $\mathcal{C}_m$ this collection is actually empty. Then for $k \in [k_n, k_n + n]$, we form

$$\mathcal{A}_k = \{\tau(I) : I \in \mathcal{C}_m, m \in K_n\} \cap \{J : |J| = 2^{-k}\}.$$

Note that $\mathcal{A}_k \subseteq \mathcal{B}_k$, and that by assumption for each $k \in [k_n, k_n + n]$,

$$|\mathcal{B}_k^*| \leq \frac{1}{n}.$$

And finally recall that for $m \in K_n$ the cardinality of $\mathcal{C}_m$ is bounded by $2^{k_n+2n/3}$. Combining these observations into a single estimate we have that

$$\sum_{m\in K_n}\sum_{I\in\mathcal{C}_m}|\tau(I)| \le n \times \frac{1}{n} + \sum_{m\in K_n}\sum_{I\in\mathcal{C}_m} 2^{-k_n-n}$$
$$\le 1 + n \times 2^{-k_n-n} \times 2^{k_n+2n/3+1}.$$

Note that for n large enough the last upper bound is ≤ 2, and this contradicts the previous lower estimate . $\square$

4.4 Notes

Maurey's isomorphism represents one of several fundamental techniques introduced by B. Maurey in [140], [142], [143], and [144]. The classification of martingale H^1 spaces was obtained as a consequence of Maurey's isomorphism in [150]. In the text we follow the presentation of [159] and [151]. The subspaces of H^1 that are spanned by subsequences of the Haar basis are classified in [149].

The family $\mathcal{S}^\infty$ is classified in [157]. Very interesting applications of three-valued martingale difference sequences to related classification problems are contained in the books of W. Johnson, B. Maurey, G. Schechtman and L. Tzafriri [93], in J. Bourgain [16] and in J. Bourgain, H. Rosenthal and G. Schechtman [29].

Theorem 4.3.1 is a classic result of J. Gamlen and R. Gaudet [71]. Theorem 4.3.2, the finite-dimensional version of the Gamlen–Gaudet theorem, is in [152]; the proof in the text follows that in [159].

Problem 4.3.3 emerged with the analysis of the peculiar rearrangement considered in [160]. A systematic study of the Haar system in vector-valued L^p spaces is contained in the important series of papers by M. Capon [41], [40],[42] and [43]. The conjecture of P. Wojtaszczyk is formulated in [215]. Its discussion in the text is taken from P. Wojtaszczyk's paper [215]. For its motivation see also [109] and [107].

Chapter 5

Isomorphic Invariants for H^1

In this chapter we give a fairly complete discussion of the known isomorphic invariants of H^1. We study copies of ℓ^2 and prove two dichotomies for complemented subspaces: First, if a complemented subspace of H^1 does not contain a copy of ℓ^2, then it is isomorphic to a complemented subspace of $(\sum H^1_n)_1$. Second, if BMO $\sim X \oplus Y$, then one of the spaces X or Y is isomorphic to BMO. We prove that the vector-valued space $H^1(\ell^2)$, is not isomorphic to H^1. The chapter concludes with the proof that BMO satisfies the uniform approximation property.

5.1 Complemented copies of Hilbert spaces

In this section we study subspaces of H^1 containing copies of Hilbert spaces. We prove the theorem of S. Kwapien and A. Pełczyński, establish Johnson's factorization for H^1 and J. Bourgain's description of uniformly complemented copies of ℓ^2_n in H^1.

Existence, abundance

Subspaces of a Banach space which are isomorphic to ℓ^2 are the so-called Hilbertian subspaces. We start with the basic result of S. Kwapien and A. Pełczyński, that every Hilbertian subspace of H^1 contains a complemented copy of ℓ^2 (where the complementation is in H^1). This holds in spite of the fact that there are plenty of Hilbertian subspaces of H^1 which are not complemented. For example the Walsh functions of multiplicity 2 span a non-complemented copy of ℓ^2.

We prepare the theorem of S. Kwapien and A. Pełczyński, by showing first a general upper ℓ^2 estimate in BMO and by duality a corresponding lower estimate for H^1. Following is a simple but useful observation.

Proposition 5.1.1. *Let $\{x_\ell\}$ be a sequence in H^1. Let $\{y_\ell\}$ be a sequence in BMO. Assume that the sequences $\{x_\ell\}$ and $\{y_\ell\}$ are disjointly supported over the Haar* *system. Let $\{a_\ell\}$ and $\{b_\ell\}$ be sequences in $\mathbb{R}$. Then the following estimates hold,*

$$\|\sum_{\ell=1}^{\infty} b_\ell y_\ell\|_{BMO} \leq \left(\sum_{\ell=1}^{\infty} b_\ell^2 \|y_\ell\|_{BMO}^2\right)^{1/2} \tag{5.1.1}$$

and

$$\left(\sum_{\ell=1}^{\infty} a_\ell^2 \|x_\ell\|_{H^1}^2\right)^{1/2} \leq 2\sqrt{2}\|\sum_{\ell=1}^{\infty} a_\ell x_\ell\|_{H^1}. \tag{5.1.2}$$

Proof. We recall the definition of the localized square function. Fix $g \in$ BMO with Haar expansion $g = \sum z_I h_I$. Fix a dyadic interval I, then write $Q(I) = \{J \in \mathcal{D} : J \subseteq I\}$ and

$$S^2(g|Q(I)) = \sum_{J \in Q(I)} z_J^2 h_J^2.$$

Now fix a sequence $\{y_\ell\}$ in BMO that is disjointly supported over the Haar system. Let $\{b_\ell\}$ be a sequence in $\mathbb{R}$, and assume that only finitely many of its entries are non-vanishing. Then define

$$f = \sum_{\ell=1}^{\infty} b_\ell y_\ell.$$

Fix a dyadic interval I. Since the sequence $\{y_\ell\}$ is disjointly supported over the Haar system we obtain the identity

$$S^2(f|Q(I)) = \sum_{\ell=1}^{\infty} b_\ell^2 S^2(y_\ell|Q(I)).$$

The definition of the BMO norm gives $\int S^2(y_\ell|Q(I)) \leq |I| \cdot \|y_\ell\|_{\mathrm{BMO}}^2$. Thus

$$\int S^2(f|Q(I)) \leq |I| \sum_{\ell=1}^{\infty} b_\ell^2 \|y_\ell\|_{\mathrm{BMO}}^2.$$

Dividing by $|I|$ and taking the supremum over all dyadic intervals I gives the estimate (5.1.1).

We derive the lower H^1 estimate (5.1.2) from (5.1.1) and H^1-BMO duality. Let $\{x_\ell\}$ be a sequence in H^1, disjointly supported over the Haar system. By H^1-BMO duality, Theorem 1.2.3, there exists $w_\ell \in$ BMO, such that $\|w_\ell\|_{\mathrm{BMO}} = 1$, and $\langle x_\ell, w_\ell\rangle = \|x_\ell\|_{H^1}$. Clearly we may select w_ℓ in such a way that its Haar support coincides with that of x_ℓ. Thus the sequence $\{w_\ell\}$ is disjointly supported over the Haar system. Next we select a sequence of scalars $\{b_\ell\}$ such that $\sum b_\ell^2 = 1$ and

$$\begin{aligned}\left(\sum_{\ell=1}^{\infty} a_\ell^2 \|x_\ell\|_{H^1}^2\right)^{1/2} &= \sum_{\ell=1}^{\infty} a_\ell \langle x_\ell, w_\ell\rangle b_\ell \\ &= \left\langle \sum_{\ell=1}^{\infty} a_\ell x_\ell, \sum_{\ell=1}^{\infty} b_\ell w_\ell \right\rangle.\end{aligned} \tag{5.1.3}$$

Applying Fefferman's inequality to the scalar product appearing in the last line of (5.1.3) we obtain that

$$\left(\sum_{\ell=1}^{\infty} a_\ell^2 \|x_\ell\|_{H^1}^2\right)^{1/2} \leq 2\sqrt{2}\left\|\sum_{\ell=1}^{\infty} a_\ell x_\ell\right\|_{H^1}\left\|\sum_{\ell=1}^{\infty} b_\ell w_\ell\right\|_{\mathrm{BMO}}.$$

Finally we apply the upper ℓ^2 estimate (5.1.1) to $\{w_\ell\}$. In BMO the sequence satisfies $\|w_\ell\|_{\mathrm{BMO}} = 1$ and $\sum b_\ell^2 = 1$, this gives (5.1.2). □

Let $\{u_j\}$ be a sequence in a Banach space X with Schauder basis $\{y_n\}$. Assume that the basis constant of $\{y_n\}$ equals K and let $\{y_n^*\}$ denote the associated biorthogonal functionals defined by the equations

$$\langle y_n,\ y_m^*\rangle = \delta_{n,m}.$$

Recall that $\{u_j\}$ is said to be equivalent to the unit vector basis of ℓ^2 if there exists $c > 0$, and $C > 0$ so that

$$c\left(\sum a_j^2\right)^{1/2} \leq \|\sum a_j u_j\|_X \leq C\left(\sum a_j^2\right)^{1/2},$$

for every sequence $\{a_j\} \in \ell^2$. Below we use that for $\epsilon > 0$ there exists a sequence of increasing integers $\{N_j\}$ and a subsequence $\{u_{n_j}\}$ such that

$$x_j = \sum_{n=N_j+1}^{N_{j+1}} \langle u_{n_j}, y_n^*\rangle y_n$$

satisfies

$$\|u_{n_j} - x_j\|_X \leq \epsilon 2^{-j}.$$

Moreover, if the closure of $\operatorname{span}\{x_j\}$ is complemented in X by a projection P and if $\epsilon = \epsilon(K, \|P\|, C, c)$ is sufficiently small, then also the closure of $\operatorname{span}\{u_{n_j}\}$ is complemented in X. The same conclusion can be derived for sequences $\{u_j\}$ satisfying

$$c \leq \|u_j\|_X \leq C \quad \text{and} \quad \langle f, u_j\rangle \to 0 \quad \forall f \in X^*.$$

This result is proved using gliding hump arguments and perturbation results. It is contained in standard texts of functional analysis (for instance in [130] page 7 or [213] page 42).

Theorem 5.1.2. *Let $\{u_j\}$ be a sequence in H^1 which is equivalent to the unit vector basis on ℓ^2. Then there exists a subsequence $\{x_j\}$ of $\{u_j\}$ such that the closed linear span of $\{x_j\}$ is complemented in H^1.*

Proof. The sequence $\{u_j\}$ is equivalent to the unit vector basis on ℓ^2, hence $u_j \to 0$ weakly in H^1. By a standard perturbation argument we may therefore assume that there exists a subsequence $\{x_j\}$, and pairwise disjoint collections $\{\mathcal{B}_j\}$ consisting of finitely many dyadic intervals, such that

$$x_j \in \operatorname{span}\{h_J : J \in \mathcal{B}_j\}.$$

Next by $H^1 - \mathrm{BMO}$ duality, that is by Theorem 1.2.3, we may choose $z_j \in \mathrm{BMO}$, with $||z_j||_{\mathrm{BMO}} \leq 1$, so that

$$\langle x_j, z_j \rangle = ||x_j||_{H^1},$$

and

$$z_j \in \operatorname{span}\{h_J : J \in \mathcal{B}_j\}.$$

Now we define the projection

$$Pf = \sum_j \langle f, z_j \rangle \frac{x_j}{\langle x_j, z_j \rangle}.$$

Clearly the range of P equals $\operatorname{span}\{x_j\}$. It remains to verify that P is a bounded operator in H^1. To this end we fix $h \in \mathrm{BMO}$, and we evaluate

$$\begin{aligned}\langle Pf, h \rangle &= \sum \langle f, z_j \rangle \frac{\langle h, x_j \rangle}{\langle x_j, z_j \rangle} \\ &= \langle f, P^* h \rangle.\end{aligned}$$

We continue by obtaining an upper bound for $||P^*h||_{\mathrm{BMO}}$. Recall that we defined the sequence $\{z_j\}$ to be disjointly supported over the Haar system. Applying the upper ℓ^2 estimate of Proposition 5.1.1 gives that

$$\left\|\sum \frac{\langle h, x_j \rangle}{\langle x_j, z_j \rangle} z_j \right\|_{\mathrm{BMO}}^2 \leq \sum \frac{\langle h, x_j \rangle^2}{\langle x_j, z_j \rangle^2} ||z_j||_{\mathrm{BMO}}^2.$$

Inserting this gives

$$\begin{aligned}||P^*h||_{\mathrm{BMO}}^2 &\leq C \sum \langle h, x_j \rangle^2 \\ &= C \sup\{\sum \langle h, x_j \rangle a_j : \sum a_j^2 = 1\} \\ &= C \sup\{\langle h, \sum x_j a_j \rangle : \sum a_j^2 = 1\}.\end{aligned} \tag{5.1.4}$$

We continue giving an upper estimate for the scalar product $|\langle h, \sum x_j a_j \rangle|$ under the hypothesis that $\sum a_j^2 = 1$. Now recall that the sequence x_j is equivalent to the unit vector basis of ℓ^2. Therefore the following upper bound holds true,

$$||\sum x_j a_j||_{H^1} \leq C(\sum a_j^2)^{1/2}.$$

Finally we apply Fefferman's inequality to the scalar product appearing in the last line of (5.1.4). Thus we obtain the norm estimate of P^*,

$$\begin{aligned}||P^*h||_{\mathrm{BMO}} &\leq C||h||_{\mathrm{BMO}} ||\sum x_j a_j||_{H^1} \\ &\leq C||h||_{\mathrm{BMO}}.\end{aligned}$$

□

Johnson's factorization on H^1

In this subsection we aim at proving dichotomies for complemented subspaces of H^1. We determine when a projection on H^1 factors through $(\sum H_n^1)_1$ and when not. To this end we introduce a combinatorial method called Johnson's factorization.

Before we continue we should recall some standard language from Banach space theory. Let $T : X \mapsto Y$ be a linear operator acting between two given Banach spaces, X and Y. Let Z be a third Banach space given to us. If there exist bounded linear operators $R : X \mapsto Z$ and $S : Z \mapsto Y$ such that the following diagram commutes,

$$\begin{array}{ccc} X & \xrightarrow{T} & Y \\ R\searrow & & \nearrow S \\ & Z & \end{array}$$

then we say that the operator T factors through Z. In Theorem 4.2.2 we showed that $\mathrm{Id} : H^1 \to H^1$ factors through $X[\mathcal{E}]$ provided that $|\limsup \mathcal{E}| > 0$.

Let $\{X_n\}$ be a given sequence of finite-dimensional Banach spaces. We say that the Banach space X contains uniformly complemented copies of $\{X_n\}$ if the following condition holds true. There exist two sequences of linear operators $R_n : X_n \to X$ and $S_n : X \to X_n$, such that

$$\begin{array}{ccc} X_n & \xrightarrow{\mathrm{Id}} & X_n \\ R_n\searrow & & \nearrow S_n \\ & X & \end{array} \qquad \text{and} \qquad \sup_{n\in\mathbb{N}} \|R_n\| \cdot \|S_n\| < \infty.$$

By Theorem 4.2.4, the space $X[\mathcal{E}]$ contains uniformly complemented copies of $\{H_n^1\}$ if $[\![\mathcal{E}]\!] = \infty$.

We prove next a dichotomy for complemented subspaces in H^1.

Theorem 5.1.3. *Let X be a complemented subspace of H^1. Then either ℓ^2 embeds into X, or X is isomorphic to a complemented subspace of $(\sum H_n^1)_1$.*

The proof of Theorem 5.1.3 requires a sufficiently good understanding of operators which factor through the space $(\sum H_n^1)_1$. Below we will establish the following criterion for a given operator $P : H^1 \to X$ to admit such a factorization:

$$\text{If} \quad L^2 \xrightarrow{\mathrm{Id}} H^1 \xrightarrow{P} X \quad \text{is a compact operator,}$$

then

$$\begin{array}{ccc} H^1 & \xrightarrow{P} & X \\ R\searrow & & \nearrow S \\ & (\sum H_n^1)_1 & \end{array} \qquad \text{and} \qquad \|R\| \cdot \|S\| < \infty.$$

The proof of this criterion is based on a remarkable combinatorial construction, due to W. B. Johnson. Before we proceed we mention several open problems related to Theorem 5.1.3. Their aim is to obtain richer, and more complete, dichotomies for complemented subspaces of H^1.

Fix a subspace $X \subseteq H^1$ and assume that there exists a bounded projection $P : H^1 \to X$ whose range is X.

Problem 5.1.4. *Give an intrinsic characterization of operators $P : H^1 \to X$ which factor through $(\sum \ell^2_n)_1$. Show, specifically, that if the projection $P : H^1 \to X$ does not factor through $(\sum \ell^2_n)_1$, then X contains uniformly complemented copies of H^1_n.*

Problem 5.1.5. *Give an intrinsic characterization of operators which factor through ℓ^1. Show, specifically, if a projection $P : H^1 \to X$ does not factor through ℓ^1, then X contains uniformly complemented copies of ℓ^2_n.*

Good solutions to the above problems should allow one to settle the following conjectures concerning complemented subspaces of H^1:

Conjecture 5.1.6. *Let X be a complemented subspace of H^1. Then the following two dichotomies hold true.*

1. *Either X contains a complemented copy of $(\sum H^1_n)_1$, or X is isomorphic to a complemented subspace of $(\sum \ell^2_n)_1$.*
2. *Either X contains a complemented copy of $(\sum \ell^2_n)_1$ or X is isomorphic to ℓ^1.*

Now we turn to the proof of Theorem 5.1.3. First we describe the isomorphic invariant which controls the dichotomy expressed there. We let F_ℓ be the orthogonal projection onto the span of the Haar functions $\{h_I : |I| \leq 2^{-\ell}\}$. It is given by

$$F_\ell f = \sum_{\{I:|I|\leq 2^{-\ell}\}} \langle f, h_I\rangle \frac{h_I}{|I|}.$$

Now suppose that the composition

$$L^2 \xrightarrow{\text{Id}} H^1 \xrightarrow{P} X$$

is a compact operator. Then of course the sequence of operators

$$L^2 \xrightarrow{F_\ell} H^1 \xrightarrow{P} X,$$

converges in norm to zero, it satisfies

$$\lim_{\ell\to\infty} ||PF_\ell : L^2 \to H^1|| = 0.$$

The main step in the proof of Theorem 5.1.3 is the following result relating compactness and factorization.

Proposition 5.1.7. *Let X be a complemented subspace of H^1, and let $P : H^1 \to X$ be the projection onto X. Assume that*

$$\lim_{\ell \to \infty} ||PF_\ell : L^2 \to H^1|| = 0.$$

Then P factors through $(\sum H_n^1)_1$. Consequently X is isomorphic to a complemented subspace of $(\sum H_n^1)_1$.

We break up the proof of Proposition 5.1.7 into two separate but combinatorially intertwined steps. *Combined* they constitute Johnson's factorization. The first step gives the embedding into $(\sum H_n^1)_1$, the second one gives the projection from $(\sum H_n^1)_1$ onto a subspace isomorphic to X.

Below, the Rosenthal space plays an important role. It enters, disguised, with the following spaces that we introduced in Chapter 2, Proposition 2.1.10. Fix natural numbers $n_\ell < n_{\ell+1}$ and a non-negative M_ℓ, which plays the role of a weight below. Then define the space $T(n_\ell, n_{\ell+1}, M_\ell)$ to consist of

$$\operatorname{span}\{h_J : 2^{-n_{\ell+1}} \leq |J| \leq 2^{-n_\ell}\},$$

equipped with the norm

$$|x|_{T_\ell} = \inf\{||y||_{H^1} + ||z||_{L^2} M_\ell^{-1} : x = y + z\}. \tag{5.1.5}$$

For notational simplicity we will write

$$T_\ell = T(n_\ell, n_{\ell+1}, M_\ell),$$

when there is no confusion about the choice of $n_\ell < n_{\ell+1}$ and M_ℓ.

The following lemma is the first step in the proof of Proposition 5.1.7. In the course of its proof we make several choices for the sequences $\{n_\ell\}$ and $\{M_\ell\}$. Note that these choices bear fruit only when we prove Lemma 5.1.9.

Lemma 5.1.8. *Assume that*

$$\lim_{\ell \to \infty} ||PF_\ell : L^2 \to H^1|| = 0.$$

Then there exist increasing sequences of natural numbers, $\{n_\ell\}$ and $\{M_\ell\}$ such that

$$\Big\|\sum Px_\ell\Big\|_{H^1} \leq ||P|| \sum |x_\ell|_{T_\ell},$$

whenever $x_\ell \in T_\ell$.

Proof. First we give an inductive definition of the sequences $M_{\ell+1}$ and n_ℓ. This determines the spaces $T_\ell = T(n_\ell, n_{\ell+1}, M_\ell)$.

Put $M_1 = 1$, $n_1 = 1$, $M_2 = 2$, and $n_2 = 2$. Suppose $M_1 < \cdots < M_\ell$ and $n_1 < \cdots < n_\ell$ have been selected. Then we define

$$M_{\ell+1} = 2^{\ell+1} 2^{n_\ell}. \tag{5.1.6}$$

Next we pick $n_{\ell+1} \geq n_\ell$ so large that the following two conditions are satisfied.

$$\begin{aligned} ||PF_{n_{\ell+1}}||_{L^2,H^1} &\leq ||P||M_{\ell+1}^{-1}, \\ M_{\ell+1}^2 &\leq 2^{n_{\ell+1}}. \end{aligned} \tag{5.1.7}$$

Having determined the sequences $\{M_{\ell+1}\}$ and $\{n_\ell\}$, the spaces $T(n_\ell, n_{\ell+1}, M_\ell)$ are now defined. We abbreviate $T_\ell = T(n_\ell, n_{\ell+1}, M_\ell)$. Fix $\ell \in \mathbb{N}$ and $x_\ell \in T_\ell$. Evaluating the norm in the space T_ℓ, by (5.1.5) we may choose $y_\ell, z_\ell \in T_\ell$, such that

$$|x_\ell|_{T_\ell} = ||y_\ell||_{H^1} + ||z_\ell||_2 M_\ell^{-1}.$$

Note that we have $Pz_\ell = PF_{n_\ell} z_\ell$. Hence by our choice of M_ℓ we obtain from (5.1.7) that

$$||Pz_\ell||_{H^1} \leq M_\ell^{-1}||P|| \cdot ||z_\ell||_2. \tag{5.1.8}$$

By the triangle inequality the estimate (5.1.8) gives the upper bound

$$\begin{aligned} ||Px_\ell||_{H^1} &\leq ||Py_\ell||_{H^1} + ||Pz_\ell||_{H^1} \\ &\leq ||P||(\,||y_\ell||_{H^1} + M_\ell^{-1}||z_\ell||_2) \\ &= ||P|| \cdot |x_\ell|_{T_\ell}. \end{aligned}$$

Therefore we just apply the triangle inequality to the left-hand side expression below to obtain

$$\left\|\sum_{\ell=1}^{\infty} Px_\ell\right\|_{H^1} \leq ||P|| \sum_{\ell=1}^{\infty} |x_\ell|_{T_\ell}. \qquad \square$$

Next we are required to find upper bounds for the norm of x_ℓ in T_ℓ. An optimal estimate for $|x_\ell|_{T_\ell}$ is the result of a decomposition $x_\ell = y_\ell + z_\ell$ that minimizes the weighted sum $||y_\ell||_{H^1} + M_\ell^{-1}||z_\ell||_2$. With stopping time arguments it is quite easy to obtain close to optimal decompositions. In the previous chapters we applied this method for the first time in the course of proving the atomic decomposition and later again in the section on interpolation.

Lemma 5.1.9. *Let $x_\ell \in T(n_\ell, n_{\ell+1}, M_\ell)$, be any sequence such that $||\sum x_\ell||_{H^1} < \infty$. Then,*

$$\sum_{\ell=1}^{\infty} |x_\ell|_{T_\ell} \leq 8||\sum_{\ell=1}^{\infty} x_\ell||_{H^1}.$$

Proof. We start by giving good upper bounds for the norm of x_ℓ in T_ℓ. We intend to split x_ℓ in T_ℓ as $x_\ell = y_\ell + z_\ell$ so that the weighted sum $||y_\ell||_{H^1} + M_\ell^{-1}||z_\ell||_2$ becomes sufficiently small. Below we determine the splitting so that

$$\sum ||y_\ell||_{H^1} + M_\ell^{-1}||z_\ell||_2 \leq 8||\sum x_\ell||_{H^1}.$$

We use a stopping time argument on the square function of $S(x_\ell)$. Thus we first decompose the Haar support of x_ℓ and then split x_ℓ into $y_\ell + z_\ell$. Let N be the smallest integer so that the set

$$\Omega = \{t \in [0,1] : S(x_\ell)(t) \geq N\} \tag{5.1.9}$$

satisfies the measure estimate

$$|\Omega| \leq 2^{-n_{\ell-1}-2}. \tag{5.1.10}$$

Let $\mathcal{F}$ be the collection of dyadic intervals which are contained in Ω. Then define

$$\mathcal{C} = Q([0,1]) \setminus \mathcal{F}$$

and

$$z_\ell = \sum_{J \in \mathcal{C}} \langle x_\ell, \frac{h_J}{|J|} \rangle h_J.$$

Then put $y_\ell = x_\ell - z_\ell$. By (5.1.9) z_ℓ satisfies the pointwise estimate, $S(z_\ell) \leq N$. Hence

$$\int S(z_\ell)^2 \leq N||x_\ell||_{H^1}. \tag{5.1.11}$$

Next we recall that N was selected by a stopping time rule. Its definition implies that the measure of the set $\{S(x_\ell) > N-1\}$ is larger than $2^{-n_{\ell-1}}/4$. Hence,

$$N - 1 \leq 4 \cdot 2^{n_{\ell-1}} ||x_\ell||_{H^1}. \tag{5.1.12}$$

Inserting (5.1.12) in the estimate (5.1.11) gives an upper bound for the L^2 norm of z_ℓ.

$$\begin{aligned} \int S^2(z_\ell) &\leq 8 \cdot 2^{n_{\ell-1}} ||x_\ell||^2_{H^1} \\ &\leq 8 \cdot 2^{n_{\ell-1}} ||\sum x_j||^2_{H^1}. \end{aligned} \tag{5.1.13}$$

By (5.1.6), M_ℓ satisfies $M_\ell = 2^\ell 2^{n_{\ell-1}}$. We exploit now the seemingly unmotivated factor 2^ℓ appearing in the definition of M_ℓ. We have

$$\sum_{\ell=1}^{\infty} 2^{n_{\ell-1}} M_\ell^{-1} \leq \sum_{\ell=1}^{\infty} 2^{-\ell} \leq 1. \tag{5.1.14}$$

The calculation done in (5.1.14) and the estimate of (5.1.13) combined give that

$$\sum ||z_\ell||_2 M_\ell^{-1} \leq 2\sqrt{2} ||\sum x_\ell||_{H^1}. \tag{5.1.15}$$

Next we give estimates for the norm in H^1 of y_ℓ. We will exploit that the sequence $\{y_\ell\}$ is essentially disjointly supported. Let $K_\ell = \operatorname{supp} S(y_\ell)$. For any interval J in the Haar support of y_ℓ it follows from (5.1.10) that

$$\sum_{j=\ell+2}^{\infty} |J \cap K_j| \leq \frac{1}{2}|J|.$$

By Proposition 3.1.5 and the remark thereafter we obtain that

$$\sum \|y_\ell\|_{H^1} \le 4\|\sum y_\ell\|_{H^1}. \tag{5.1.16}$$

Recall that the decomposition $x_\ell = y_\ell + z_\ell$ is obtained by splitting the Haar support of x_ℓ, hence

$$\|\sum y_\ell\|_{H^1} \le \|\sum x_\ell\|_{H^1}. \tag{5.1.17}$$

Combining (5.1.16) and (5.1.17) with (5.1.15) we obtain finally,

$$\sum |x_\ell|_{T_\ell} \le 4\|\sum x_\ell\|_{H^1}. \qquad \square$$

Proof of Proposition 5.1.7. We let $\{n_\ell\}$ and $\{M_\ell\}$ be the sequences defined by (5.1.6) and (5.1.7). We claim that $P : H^1 \to X$ factors through $(\sum T_\ell)_1$. For $x \in H^1$, we put

$$x_\ell = (F_{n_\ell} - F_{n_{\ell-1}})x,$$

and define

$$Ex = (x_\ell)_{\ell=1}^\infty.$$

By Lemma 5.1.9 the operator $E : H^1 \to (\sum T_\ell)_1$ is bounded with norm ≤ 8. Next on $(\sum T_\ell)_1$ define Q by

$$Q((x_\ell)) = \sum Px_\ell.$$

By Lemma 5.1.8, $Q : (\sum T_\ell)_1 \to X$ is a bounded operator with norm $\le \|P\|$. Clearly the mappings E and Q are defined such that $P = QE$. Hence P factors through $(\sum T_\ell)_1$. Let us combine this with Proposition 2.1.10 asserting that the identity on $(\sum T_\ell)_1$ factors through $(\sum H^1_n)_1$. Expressed in diagrams we have that

$$\begin{array}{ccc} H^1 & \xrightarrow{P} & X \\ E\searrow & & \nearrow Q \\ & (\sum T_\ell)_1 & \end{array} \quad \text{and} \quad \begin{array}{ccc} (\sum T_\ell)_1 & \xrightarrow{\mathrm{Id}} & (\sum T_\ell)_1 \\ R\searrow & & \nearrow S \\ & (\sum H^1_n)_1 & \end{array} .$$

Combining the diagrams gives that $P : H^1 \to X$ factors through $(\sum H^1_n)_1$. Consequently X is isomorphic to a complemented subspace of $(\sum H^1_n)_1$. $\square$

Proof of Theorem 5.1.3. Let us assume that there exists $\eta > 0$ such that

$$\lim_{\ell\to\infty} \|PF_\ell\|_{L^2,H^1} > \eta.$$

Then it is easy to find an increasing sequence of integers $\{n_\ell\}$ and a sequence $\{y_\ell\}$ in L^2 such that

$$\|P(F_{n_\ell} - F_{n_{\ell+1}})y_\ell\|_{H^1} \ge \eta \quad \text{and} \quad \|(F_{n_\ell} - F_{n_{\ell+1}})y_\ell\|_{L^2} = 1. \tag{5.1.18}$$

Next we define

$$x_\ell = P(F_{n_\ell} - F_{n_{\ell+1}})y_\ell.$$

We will show now that in H^1 the sequence $\{x_\ell\}$ is equivalent to the unit vector basis of ℓ^2. Fix real numbers $\{a_\ell\}$, such that $\sum a_\ell^2 < \infty$. Recall the lower bound $||x_\ell||_{H^1} > \eta$. By passing to a subsequence we may assume that x_ℓ is disjointly supported over the Haar system. By Proposition 5.1.1 we find the lower ℓ^2 estimate in H^1,

$$\begin{aligned} 2\sqrt{2}\left\|\sum a_\ell x_\ell\right\|_{H^1} &\geq \left(\sum a_\ell^2 ||x_\ell||_{H^1}^2\right)^{1/2} \\ &\geq \eta\left(\sum a_\ell^2\right)^{1/2}. \end{aligned} \tag{5.1.19}$$

Next we observe that the converse estimate is also true. By (5.1.18),

$$\begin{aligned} \left\|\sum a_\ell x_\ell\right\|_{H^1} &\leq ||P|| \left\|\sum a_\ell(F_{n_\ell} - F_{n_{\ell+1}})y_\ell\right\|_{H^1} \\ &\leq ||P|| \left(\sum a_\ell^2\right)^{1/2}. \end{aligned} \tag{5.1.20}$$

Combining (5.1.19) and (5.1.20) shows that the sequence $\{x_\ell\}$ is equivalent to the unit vector basis of ℓ^2. In summary we showed that if $\lim_{\ell\to\infty} ||PF_\ell||_{2,1} > 0$, then X contains a subspace isomorphic to ℓ^2. And conversely by Proposition 5.1.7, if $\lim_{\ell\to\infty} ||PF_\ell||_{2,1} = 0$, then X is isomorphic to a complemented subspace of $(\sum H_n^1)_1$. Thus holds the dichotomy announced in Theorem 5.1.3. □

Intrinsic description

By Theorem 1.1.4 the Rademacher functions span complemented copies of ℓ^2 in the reflexive L^p spaces. Indeed, by Khintchine's inequality the unit vectors in ℓ^2 are equivalent to the Rademacher functions and the orthogonal projection onto their span is bounded in L^p. Also for H^1 is it true that the Rademacher functions span complemented copies of ℓ^2. This fact is a direct consequence of the definitions. Of particular interest are the converse questions which we describe as follows: Start with a complemented subspace $X \subset H^1$, and assume that X is isomorphic to ℓ^2. Let $\{f_i\}$ in X be equivalent to the unit vector basis of ℓ^2. We ask now whether isomorphic properties (like being equivalent to the unit vectors in ℓ^2) are reflected in pointwise estimates of $\{f_i\}$.

We turn to important work of J. Bourgain describing intrinsically the uniformly complemented copies of ℓ_n^2 in H^1. Loosely speaking we will prove that their unit vectors have to look (more or less) like Rademacher functions. A result like this is not only pleasing aesthetically, but also it leads us to an important isomorphic invariant for H^1. Below we will test the invariant on the space $H^1(\ell^2)$ and obtain that H^1 and $H^1(\ell^2)$ are not isomorphic Banach spaces. This is how J. Bourgain solved the important dimension conjecture formulated by B. Maurey in [144].

Fix a complemented copy of ℓ_n^2 in H^1. This is done by specifying linear operators E, P, such that

$$\begin{array}{ccc} \ell_n^2 & \xrightarrow{Id} & \ell_n^2 \\ E \searrow & & \nearrow P \\ & H^1 & \end{array}.$$

We denote the unit vectors in ℓ_n^2 by $\{e_j : j \leq n\}$. Consulting the above diagram shows that the H^1 functions Ee_j, $j \in \{1, \dots, n\}$ satisfy the following upper and lower ℓ^2 estimates.

$$\frac{1}{||P||}\left(\sum_{j=1}^{n} \alpha_j^2\right)^{1/2} \leq \left\|\sum_{j=1}^{n} \alpha_j E e_j\right\|_{H^1} \leq ||E|| \left(\sum_{j=1}^{n} \alpha_j^2\right)^{1/2}. \tag{5.1.21}$$

We restate the above estimates in a more symmetric form. Notice (5.1.21) implies that the map

$$T : e_j \mapsto (||P|| \cdot ||E||)^{1/2} E e_j$$

extends linearly to an isomorphism from ℓ_n^2 onto its range in H^1 so that for $x \in \ell_n^2$,

$$\left(\frac{||E||}{||P||}\right)^{1/2} \|x\|_{\ell_n^2} \leq \|Tx\|_{H^1} \leq \left(\frac{||P||}{||E||}\right)^{1/2} \|x\|_{\ell_n^2},$$

and EP is an idempotent map onto $T(\ell_n^2)$.

Dualizing the operators E, P gives a complemented copy of ℓ_n^2 in BMO. Let P^* denote the adjoint operator of P. The BMO functions $P^* e_j$, $j \in \{1, \dots, n\}$ are equivalent to the unit vectors of ℓ_n^2,

$$\frac{1}{||E^*||}\left(\sum_{j=1}^{n} \alpha_j^2\right)^{1/2} \leq \left\|\sum_{j=1}^{n} \alpha_j P^* e_j\right\|_{\mathrm{BMO}} \leq ||P|| \left(\sum_{j=1}^{n} \alpha_j^2\right)^{1/2}.$$

As $Id = PE$ the duality between H^1 and BMO preserves biorthogonality as follows,

$$\begin{aligned} \langle E e_j, P^* e_i \rangle &= \langle e_j, e_i \rangle \\ &= \delta_{i,j}. \end{aligned} \tag{5.1.22}$$

The next proposition clearly indicates that the ℓ_n^2 unit vectors $f_j = E e_j$ and $h_j = P^* e_j$ display the same pointwise behavior as (rescaled copies of) Rademacher functions. We will apply Proposition 5.1.10 repeatedly below. To ease the reference we rename the operators as

$$M = E \quad \text{and} \quad N = P^*.$$

Proposition 5.1.10. $\displaystyle\sum_{i=1}^{n} \sum_{j=i+1}^{n} \int S(f_j) S(h_i) \geq \frac{n(n-1)}{64^2 ||M||^4 ||N||^4}.$

The analytic backbone of this estimate is the following variant of Fefferman's inequality.

Lemma 5.1.11. *Let $h \in BMO$, $g \in L^1$ with $g \geq 0$, and let $f \in H^1$, then*

$$\int_0^1 gS(h)\,dx \geq \frac{1}{2}\int_0^1 fh\,dx - 2||S(f)-g||_{L^1}||h||_{BMO}.$$

Proof. Following is a repetition of the proof of Fefferman's inequality. First expand $h \in \mathrm{BMO}$ into its Haar series $h = \sum b_J h_J$. Let I be a fixed dyadic interval. The square function localized to $Q(I) = \{J \subseteq I\}$ is given by

$$S^2(h|Q(I))(x) = \sum_{J\in Q(I)} b_J^2 h_J^2(x).$$

If $x \in [0,1]$ satisfies $S(h)(x) < \infty$, then there exists a maximal dyadic interval $I(x)$ so that for $S(h\,|\,Q(I(x)))(x)$ the following pointwise estimate holds,

$$S(h\,|\,Q(I(x)))(x) \leq \sqrt{2}||h||_{\mathrm{BMO}},$$

Moreover it is clear that $S(h\,|\,Q(I(x)))(x) \leq S(h)(x)$. Next recall (1.2.3) asserting that

$$\int fh\,dx \leq 2\int S(f)(x)\cdot S(h\,|\,Q(I(x)))(x)\,dx.$$

Inserting g gives

$$\begin{aligned}\int fh &\leq 2\int (S(f)-g)(x)S(h\,|\,Q(I(x)))(x)\,dx + 2\int g(x)S(h\,|\,Q(I(x)))(x)\,dx\\ &\leq 4||S(f)-g||_1||h||_{\mathrm{BMO}} + 2\int g(x)S(h)\,dx.\end{aligned}$$

□

Proof of Proposition 5.1.10. Fix $r \leq n$ and $i \in \{1,\ldots,r\}$. Recall that we put $f_i = Me_i$ and $h_i = Ne_i$. Let

$$g_i = \min\left\{S(f_i), \left(\sum_{j=i+1}^{r} S^2(f_j)\right)^{1/2}\right\},$$

and apply Lemma 5.1.11 with this choice of g_i. By the biorthogonality relation (5.1.22) we have $\langle f_i, h_i\rangle = 1$. Thus we find that

$$\begin{aligned}\int g_i S(h_i) &\geq \frac{1}{2}\langle f_i, h_i\rangle - 2||S(f_i)-g_i||_1||h_i||_{\mathrm{BMO}}\\ &\geq \frac{1}{2} - 2\epsilon_i||N||,\end{aligned}$$

where we put $\epsilon_i = ||S(f_i) - g_i||_1$, and used that $||h_i||_{\mathrm{BMO}} \le ||N||$. We continue with the pointwise minorization

$$\sum_{j=i+1}^{r} S(f_j) \ge \left(\sum_{j=i+1}^{r} S^2(f_j) \right)^{1/2}$$
$$\ge g_i.$$

Integrating against $S(h_i)$ gives

$$\int \left\{ \sum_{j=i+1}^{r} S(f_j) \right\} S(h_i) \ge \int g_i S(h_i) \ge \frac{1}{2} - 2\epsilon_i ||N||. \tag{5.1.23}$$

We would like to say now that in the above estimate, $\frac{1}{2}$ is dominating while $\epsilon_i ||N||_{\mathrm{BMO}}$ is a negligible error term because ϵ_i is small. Unfortunately this need not be true for each individual ϵ_i, however in the average this statement is true. Indeed in Lemma 5.1.12 we verify that the following inequality holds,

$$\sum_{i=1}^{r} \epsilon_i \le \sqrt{2} ||M|| r^{1/2}.$$

Now continue the proof by specifying $r = 64||M||^2||N||^2$ (where we tacitly assume that r is an integer). Take the sum of the estimate (5.1.23) over $i \le r$. This gives

$$\begin{aligned} \sum_{i=1}^{r} \sum_{j=i+1}^{r} \int S(f_j) S(h_i) &\ge \frac{r}{2} - \sum_{i=1}^{r} 2\epsilon_i ||N|| \\ &\ge \frac{r}{2} - 2\sqrt{2r} ||M|| \, ||N|| \\ &\ge 1. \end{aligned} \tag{5.1.24}$$

The last inequality in (5.1.24) follows from our choice of r. Going over this argument again one notices that there is nothing special about the set $\{1, \dots, r\}$, all we used was a lower bound for its cardinality. Hence we showed that for every subset $A \subseteq \{1, \dots, n\}$ with cardinality larger than r we have

$$\sum_{i \in A} \sum_{\{j \ge i+1\} \cap A} \int S(f_j) S(h_i) \ge 1. \tag{5.1.25}$$

We continue with a simple double counting argument. For $r \le n$ we let $\mathcal{P}_r$ be the set of all subsets $A \subseteq \{1, \dots, n\}$ of cardinality r. Note that $\mathcal{P}_r$ is a set of cardinality $\binom{n}{r}$. Given $A \in \mathcal{P}_r$ we define the indicator function

$$b_A(i,j) = \begin{cases} 1 & i \geq j+1 \quad \text{and } i,j \in A, \\ 0 & \text{otherwise.} \end{cases}$$

Now for $i, j, \leq n$ we consider

$$a_{i,j} = \int S(f_j)S(h_i).$$

Note that for any given $i, j, \in \{1, \dots, n\}$ the cardinality of the set $\{A \in \mathcal{P}_r : i, j, \in A\}$ equals $\binom{n-2}{r-2}$. Changing the order of summation twice gives

$$\begin{aligned} \sum_{A \in \mathcal{P}_r} \sum_{i=1}^{n} \sum_{j=1}^{n} a_{ij} b_A(i,j) &= \sum_{i=1}^{n} \sum_{j=1}^{n} \sum_{A \in \mathcal{P}_r} a_{ij} b_A(i,j) \\ &= \sum_{i=1}^{n} \sum_{j=i+1}^{n} a_{ij} |\{A \in \mathcal{P}_r : i,j \in A\}| \qquad (5.1.26) \\ &= \sum_{i=1}^{n} \sum_{j=i+1}^{n} a_{ij} \binom{n-2}{r-2}. \end{aligned}$$

Next we recall (5.1.25) asserting that for any set A of cardinality greater than r we have

$$\sum_{i}^{n} \sum_{j}^{n} a_{ij} b_A(i,j) \geq 1.$$

We also recall that $|\mathcal{P}_r| = \binom{n}{r}$. Inserting this in identity (5.1.26) gives

$$\binom{n}{r} \leq \binom{n-2}{r-2} \sum_{i=1}^{n} \sum_{j=i+1}^{n} a_{ij},$$

or simply,

$$\sum_{i=1}^{n} \sum_{j=i+1}^{n} a_{ij} \geq \frac{n(n-1)}{r(r-1)}.$$

Recalling that r is the integer part of $64||M||^2||N||^2$, and $a_{ij} = \int S(f_j)S(h_i)$ gives the desired estimate. □

We still need to prove that most of the numbers $\epsilon_i = ||S(f_i) - g_i||_1$ are small. This is expressed in the following auxiliary estimate which finishes the proof of Proposition 5.1.10.

Lemma 5.1.12. *The error terms* $\epsilon_i = ||S(f_i) - g_i||_1$ *satisfy the upper bound*

$$\sum_{i=1}^{r} \epsilon_i \leq \sqrt{2}||M||r^{1/2}.$$

Proof. Let $1 \leq i \leq r-1$ put $A_i = \left(\sum_{j=i+1}^{r} S^2(f_j)\right)^{1/2}$, and $A_r = 0$. We start with the pointwise estimate

$$0 \leq S(f_i) - \min\left\{ S(f_i), \left(\sum_{j=i+1}^{r} S^2(f_j)\right)^{1/2} \right\} \leq A_{i-1} - A_i.$$

Evaluating the telescoping sum from $i = 1$ to $i = r$ and integrating gives

$$\sum_{i=1}^{r} \epsilon_i \leq \int_0^1 \left(\sum_{j=1}^{r} S^2(f_j)\right)^{1/2}. \tag{5.1.27}$$

Theorem 1.1.3 (the Khintchine inequality) implies that

$$\left(\sum_{j=1}^{r} S^2(f_j)\right)^{1/2} \leq \sqrt{2} \int_0^1 S\left(\sum_{j=1}^{r} r_j(s) f_j\right) ds, \tag{5.1.28}$$

where $\{r_j\}$ denotes the Rademacher system. Taking the integral of the pointwise estimate (5.1.28) we obtain that

$$\int_0^1 \left(\sum_{j=1}^{r} S^2(f_j)\right)^{1/2} \leq \sqrt{2} \int_0^1 \left\|\sum_{j=1}^{r} r_j(s) f_j\right\|_{H^1} ds. \tag{5.1.29}$$

Now we arrive at a position that allows us to exploit the hypothesis that $\{f_j\}$ is equivalent to the unit vector basis of ℓ^2. Indeed,

$$\left\|\sum_{j=1}^{r} r_j(s) f_j\right\|_{H^1} \leq ||M|| r^{1/2},$$

for $s \in [0, 1]$. It remains to insert this estimate into (5.1.29). Then Lemma 5.1.12 follows from (5.1.27). □

5.2 Complemented copies of H_n^1

We prove that BMO is primary thereby showing a further dichotomy for complemented subspaces of BMO and H^1. We obtain J. Bourgain's pointwise estimates for basis functions of uniformly complemented copies of H_n^1 in H^1.

A Banach space X is primary if for every projection $P : X \to X$ one of the spaces $P(X)$ or $(\mathrm{Id} - P)(X)$ *is isomorphic to* X. We are concerned with Banach spaces X for which

$$X \sim \left(\sum X\right)_p, \quad p = 1 \quad \text{or} \quad p = \infty. \tag{5.2.1}$$

In view of A. Pełczyński's decomposition principle (Theorem 2.2.1) a Banach space X satisfying (5.2.1) is primary, if for every projection $P : X \to X$ the range $P(X)$ or its complement $(\mathrm{Id} - P)(X)$ *contains a complemented subspace isomorphic to* X. Equivalently X is primary, if one of the operators $H = P$ or $H = \mathrm{Id} - P$ satisfies

$$\begin{array}{ccc} X & \stackrel{\mathrm{Id}}{\longrightarrow} & X \\ M \downarrow & & \uparrow N \\ X & \stackrel{H}{\longrightarrow} & X \end{array}$$

where M and N are bounded and linear.

Dichotomies

We now review Theorem 4.2.6 (on subsequences of the Haar basis) with the aim of seeing it as a result on factorization of operators. Let $\mathcal{C}$ be a collection of dyadic intervals, and let $T_{\mathcal{C}}$ be the orthogonal projection given by

$$T_{\mathcal{C}}(f) = \sum_{I \in \mathcal{C}} \langle f, h_I \rangle \frac{h_I}{|I|}.$$

Clearly one of the collections $\mathcal{C}$ or $\mathcal{D} \setminus \mathcal{C}$ has unbounded Carleson constant. Hence, by Theorem 4.2.6 the identity on BMO can be factored through $H = T_{\mathcal{C}}$ or $H = \mathrm{Id} - T_{\mathcal{C}}$, so that

$$\begin{array}{ccc} \mathrm{BMO} & \stackrel{\mathrm{Id}}{\to} & \mathrm{BMO} \\ E \downarrow & & \uparrow P \\ \mathrm{BMO} & \underset{H}{\to} & \mathrm{BMO} \end{array}$$

with bounded operators E and P. In this section we will extend this result from the special case of orthogonal projections to arbitrary operators on BMO. The following theorem shows that BMO is a primary space.

Theorem 5.2.1. *For any operator $T : BMO \to BMO$, either $H = T$ or $H = \mathrm{Id} - T$ satisfies*

$$\begin{array}{ccc} BMO & \stackrel{\mathrm{Id}}{\longrightarrow} & BMO \\ E \downarrow & & \uparrow P \\ BMO & \stackrel{H}{\longrightarrow} & BMO \end{array}$$

where E and P are bounded operators .

The proof of Theorem 5.2.1 consists of two independent components. First, by P. Wojtaszczyk's Theorem 2.2.3, we replace BMO with $(\sum \mathrm{BMO}_n)_\infty$ and we show that it suffices to prove Theorem 5.2.1 for diagonal operators of the form

$$D : (\sum \mathrm{BMO}_n)_\infty \to (\sum \mathrm{BMO}_n)_\infty$$
$$(x_n) \to (D_n x_n).$$

In the second step we prove a localized version of Theorem 5.2.1. (A localized version of a given theorem is characterized by the following two properties: First the spaces involved are replaced by their finite-dimensional analogs. Second the norm estimates appearing in the conclusion of the theorem are independent of the dimension.)

Following is a localized version of the assertion that BMO is primary.

Theorem 5.2.2. *For any $n \in \mathbb{N}$ there exists $N = N(n)$ such that for any operator $T : BMO_N \to BMO_N$, the identity on BMO_n factors through $H = T$ or $H = \mathrm{Id} - T$. That is,*

$$\begin{array}{ccc} BMO_n & \xrightarrow{\mathrm{Id}} & BMO_n \\ E\downarrow & & \uparrow P \\ BMO_N & \xrightarrow{H} & BMO_N \end{array}$$

where E, P are linear operators satisfying $||E||\,||P|| \le 16$.

The proof of Theorem 5.2.2 involves combinatorial difficulties which we isolate and address in Theorem 5.2.3 below. Its effect is that we find a well complemented copy of BMO_n on which T and $\mathrm{Id} - T$ are acting similar to a multiplier operator on the Haar basis. Thus with Theorem 5.2.3 we find a well complemented copy of BMO_n which has a basis of approximate eigenvectors for T. Moreover, the basis is well equivalent to the Haar system in BMO_n. Recall that $\mathcal{D}$ denotes the collection of all dyadic intervals. For $N \in \mathbb{N}$ we defined also that $\mathcal{D}_N = \{J \in \mathcal{D} : |J| \ge 2^{-N}\}$.

Theorem 5.2.3. *Let $n \in \mathbb{N}$. Let $\log N = 8^n$. Suppose that $H : BMO_N \mapsto BMO_N$ is a linear operator with $\|H\| = 1$. Then in $\mathcal{D}_N$ there exist collections of pairwise disjoint dyadic intervals $\mathcal{E}_1, \ldots, \mathcal{E}_{2^{n+1}-1}$ so that*

$$b_i = \sum \{h_I : I \in \mathcal{E}_i\}, \quad i \le 2^{n+1} - 1,$$

satisfy the following condition:

1. *The orthogonal projection*

$$Q(f) = \sum_{i=1}^{2^{n+1}-1} \langle f, \frac{b_i}{||b_i||_2}\rangle \frac{b_i}{||b_i||_2}$$

   is a bounded operator on BMO with norm ≤ 4.

2. *The system $b_1, \ldots, b_{2^{n+1}-1}$ is equivalent to the Haar basis of BMO_n with constants independent of n.*

3. *For each b_i,*

$$\sum_{i \ne j} |\langle Hb_j, b_i\rangle| \le 4^{-i} ||b_i||_2^2. \tag{5.2.2}$$

As we saw before the Gamlen–Gaudet construction is a very flexible tool allowing us to inductively produce a variety of complemented copies of BMO_n. In the proof below we will set up the inductive procedure so that after the completion of the last step there holds the orthogonality condition (5.2.2) We will see later that it implies the eigenvector relation

$$Hb_i = m_i b_i + \text{ a small error },$$

and that H, almost leaves invariant the linear span of $\{b_i : i \leq 2^{n+1} - 1\}$.

Proof of Theorem 5.2.3. In the first step of the construction we simply put $b_1 = h_{[0,1]}$ and $\mathcal{E}_1 = \{[0,1]\}$. Formulating the induction hypothesis we demand that at stage i of the construction we are given $\mathcal{E}_1, \ldots, \mathcal{E}_i$ and $b_1, \ldots, b_i$ such that

$$\sum_{j \leq i-1} |\langle Hb_j, b_i \rangle| + |\langle b_i, H^* b_j \rangle| \leq 4^{-i} \|b_i\|_2^2. \tag{5.2.3}$$

Let $Q(\mathcal{E}_i) = \{J \in \mathcal{D} : \text{ there exists } I \in \mathcal{E}_i \text{ and } J \subseteq I\}$. Now consider

$$\mathcal{G} = \left\{ J \in Q(\mathcal{E}_i) : \sum_{j=1}^{i} |\langle Hb_j, h_J \rangle| + |\langle h_J, H^* b_j \rangle| \leq 4^{-i-1} |J| \right\}. \tag{5.2.4}$$

We apply now Lemma 5.2.4 with the following specification of the parameters, $k = 4^{i+1}$ and $l = 8^i$. The conclusion of Lemma 5.2.4 gives $j \leq i 2^{10(i+1)}$ such that the collection of pairwise disjoint dyadic intervals

$$\mathcal{C} = \bigcup_{I \in \mathcal{E}_i} \{J \in \mathcal{G} : J \subseteq I, |J| = 2^{-j}|I|\}$$

satisfies

$$|\mathcal{C}^*| \geq (1 - 8^{-i}) |\mathcal{E}_i^*|. \tag{5.2.5}$$

Now we are ready to define the collection $\mathcal{E}_{i+1}$. We use the Gamlen–Gaudet construction. Hence we distinguish between the cases where i is odd or even, or equivalently between right and left. If i is even, then we define

$$\mathcal{E}_{i+1} = \left\{ I \in \mathcal{C} : I \subseteq \{t : b_{i/2}(t) = -1\} \right\}.$$

If i is odd, then we define

$$\mathcal{E}_{i+1} = \left\{ I \in \mathcal{C} : I \subseteq \{t : b_{(i-1)/2}(t) = +1\} \right\}.$$

In either case we put

$$b_{i+1} = \sum \{h_I : I \in \mathcal{E}_{i+1}\}.$$

The following estimate is an immediate consequence of the fact that $\mathcal{E}_{i+1} \subseteq \mathcal{G}$. In fact in trying to establish the estimate (5.2.6) we are forced to define the collection $\mathcal{G}$ as we did above.

$$\sum_{j=1}^{i} |\langle Hb_{i+1}, b_j\rangle| + |\langle b_j, H^* b_{i+1}\rangle| \leq 4^{-(i+1)} ||b_{i+1}||_2^2. \tag{5.2.6}$$

As we used the Gamlen–Gaudet construction to define $\{\mathcal{E}_j : j \leq 2^{n+1} - 1\}$ it is clear that they satisfy Jones's compatibility condition (J). Applying Theorem 1.5.9 gives the upper bound for the norm of the projection Q. By (5.2.5), the system $\{b_j : j \leq 2^{n+1} - 1\}$ is equivalent to the Haar basis of BMO_n.

Now we verify that the biorthogonality condition (5.2.2) is satisfied, that is we show that

$$\sum_{\{j : i \neq j\}} |\langle Hb_j, b_i\rangle| \leq 4^{-i} ||b_i||_2^2, \qquad \text{for every } b_i. \tag{5.2.7}$$

Recall that by (5.2.3) we obtain at once that

$$\sum_{j \leq i-1} |\langle Hb_j, b_i\rangle| \leq 4^{-i} ||b_i||_2^2.$$

Next for $j \geq i+1$ the definitions of $\mathcal{G}$ in the j-th step of the construction and (5.2.3) imply the estimates

$$\begin{aligned} |\langle Hb_j, b_i\rangle| &= |\langle b_j, H^* b_i\rangle| \\ &\leq \sum_{k=1}^{j-1} |\langle b_j, H^* b_k\rangle| \\ &\leq 4^{-j} ||b_j||_2^2 \\ &\leq 4^{-j} ||b_i||_2^2. \end{aligned} \tag{5.2.8}$$

In the last inequality we used the obvious estimate that $||b_j||_2^2 \leq ||b_i||_2^2$ for $i \leq j$. Finally we take the sum of the estimates (5.2.8) where j ranges from $i+1$ to $2^{n+1} - 1$. This gives

$$\sum_{j \geq i+1} |\langle Hb_j, b_i\rangle| \leq 4^{-i} ||b_i||_2^2. \tag{5.2.9}$$

□

Now we formulate and prove Lemma 5.2.4. Let I be a dyadic interval. Let $x \in \mathrm{BMO}$ and $y \in H^1$ such that,

$$||x||_{\mathrm{BMO}} \leq 1 \quad \text{and} \quad ||y||_{H_1} \leq |I|.$$

Fix $k \in \mathbb{N}$ and consider the collection of intervals

$$\mathcal{B} = \left\{ J \subseteq I : |\langle x, h_J\rangle| + |\langle y, h_J\rangle| \geq \frac{1}{k}|J| \right\}.$$

We claim that the collection $\mathcal{B}$ is thin in the following quantitative sense:

Lemma 5.2.4. *Let $l \in \mathbb{N}$ and let $A = 64k^2l^2 + 1$. Then there exists $j \in \{1, \ldots, A\}$ such that*

$$\mathcal{B}_j = \{J \in \mathcal{B} : |J| = 2^{-j}|I|\}$$

satisfies

$$\sum_{J \in \mathcal{B}_j} |J| \leq \frac{1}{l}|I|.$$

Proof. We assume that Lemma 5.2.4 is false and will derive a contradiction. So we are assuming that each of the collections $\mathcal{B}_j$ covers a subset of measure $\geq |I|/l$. Recall that for $J \in \mathcal{B}_j$, the coefficients satisfy the estimate

$$|\langle x, h_J\rangle| + |\langle y, h_J\rangle| \geq \frac{1}{k}|J|. \tag{5.2.10}$$

Summing the estimates (5.2.10) over all $J \in \mathcal{B}_j$, and all $j \leq A$, gives the lower bound

$$\begin{aligned}\sum_{j=1}^{A} \sum_{J \in \mathcal{B}_j} |\langle x, h_J\rangle| + |\langle y, h_J\rangle| &\geq \frac{1}{k}\sum_{j=1}^{A} \sum_{J \in \mathcal{B}_j} |J| \\ &\geq \frac{A|I|}{kl}.\end{aligned} \tag{5.2.11}$$

On the other hand we obtain an upper bound for the right-hand side of (5.2.11) by applying Fefferman's inequality. We start by observing that

$$\left\| \sum_{j=1}^{A} \sum_{J \in \mathcal{B}_j} \pm h_J \right\|_{H^1} \leq \sqrt{A}|I|,$$

and also that

$$\left\| \sum_{j=1}^{A} \sum_{J \in \mathcal{B}_j} \pm h_J \right\|_{\mathrm{BMO}} \leq \sqrt{A}.$$

Now we rewrite, and using Fefferman's inequality we obtain the estimates

$$\begin{aligned}\sum_{j=1}^{A} \sum_{J \in \mathcal{B}_j} |\langle x, h_J\rangle| + |\langle y, h_J\rangle| &= \sum_{j=1}^{A} \sum_{J \in \mathcal{B}_j} \pm\langle x, h_J\rangle \pm \langle y, h_J\rangle \\ &\leq 4\sqrt{A}|I|\,||x||_{\mathrm{BMO}} + 4\sqrt{A}||y||_{H^1} \\ &\leq 8\sqrt{A}|I|.\end{aligned} \tag{5.2.12}$$

(Recall that $||x||_{\mathrm{BMO}} \leq 1$ and $||y||_{H^1} \leq |I|$.) Comparing the lower bound (5.2.11) and the upper bound (5.2.12) gives an estimate for A. Indeed, we obtain that

$$A \leq 64k^2l^2.$$

Note however, that this estimate contradicts our previous choice of A. □

By a straightforward application of Theorem 5.2.3 we show now that BMO_n satisfies the localized version of being primary.

Proof of Theorem 5.2.2. Fix $n \in \mathbb{N}$. Then we choose $N = N(n)$ such that $\log\log N = 4n^2$. Let $T : \mathrm{BMO}_N \to \mathrm{BMO}_N$ be a given operator of norm ≤ 1. Let

$$I(n) = \{1, \ldots, 2^{n^2+1} - 1\}.$$

Apply Theorem 5.2.3 to T. This gives the system $\{b_i : i \in I(n)\}$ disjointly supported over the Haar system and equivalent to the Haar functions in BMO_{n^2}. Moreover the orthogonal projection

$$Q(f) = \sum_{i \in I(n)} \langle f, \frac{b_i}{||b_i||_2} \rangle \frac{b_i}{||b_i||_2} \tag{5.2.13}$$

is a bounded operator on BMO with norm ≤ 4, and for each b_i,

$$\sum_{i \neq j} |\langle Tb_j, b_i \rangle| \leq 4^{-i} ||b_i||_2^2.$$

Let $B_i = \operatorname{supp} b_i$. We obtained the functions b_i by the Gamlen–Gaudet construction. Hence the collection $\{B_i\}$ forms a nested family of sets. Now consider

$$\mathcal{L} = \{B_i : |\langle Tb_i, b_i \rangle| \geq ||b_i||_2^2/2\},$$

and

$$\mathcal{R} = \{B_i : |\langle (\mathrm{Id} - T)b_i, b_i \rangle| \geq ||b_i||_2^2/2\}.$$

By the triangle inequality, we have

$$\{B_i : i \in I(n)\} = \mathcal{R} \cup \mathcal{L}.$$

Hence either $\mathcal{R}$ or $\mathcal{L}$ has Carleson constant $\geq n^2/2$. Assume that this holds for $\mathcal{L}$. Thus

$$\sup_{A \in \mathcal{L}} \frac{1}{|A|} \sum_{B \in \mathcal{L},\, B \subseteq A} |B| \geq \frac{n^2}{2}. \tag{5.2.14}$$

Now we apply the condensation Lemma 3.1.4 and Gamlen–Gaudet construction to the nested collection $\mathcal{L}$. This gives subcollections of $\mathcal{L}$, say $\{\mathcal{C}_I : |I| \geq 2^{-n}\}$, indexed by dyadic intervals and satisfying the following two properties. First the system

$$f_I = \sum \{b_i : B_i \in \mathcal{C}_I\},$$

is equivalent to the Haar basis in BMO_n. Second, $\{\mathcal{C}_I : |I| \geq 2^{-n}\}$ satisfies Jones's compatibility condition (J). We let $L = \{i : B_i \in \mathcal{L}\}$, and define

$$Y = \operatorname{span}\{b_i : i \in L\},$$

equipped with the norm of BMO. Summarizing the proof up to this point, we state that there exist linear operators I and R, with norm ≤ 4, so that the following diagram commutes.

$$\begin{array}{ccc} \mathrm{BMO}_n & \stackrel{\mathrm{Id}}{\rightarrow} & \mathrm{BMO}_n \\ I\downarrow & & \uparrow R \\ Y & \stackrel{\mathrm{Id}}{\rightarrow} & Y \end{array} \tag{5.2.15}$$

Next we define the following operator on BMO,

$$Pf = \sum_{i\in L} \langle f, b_i\rangle b_i \langle Tb_i, b_i\rangle^{-1}.$$

By our definition of the index set L, and by the definition of the nested collection $\mathcal{L}$ the following estimate holds,

$$||Pf||_{\mathrm{BMO}} \leq 2||Qf||_{\mathrm{BMO}} \leq 8||f||_{\mathrm{BMO}},$$

where Q is the orthogonal projection defined in (5.2.13).

Crucial about P is this: On the subspace $Y = \mathrm{span}\{b_i : i \in L\}$, P almost inverts the action of T. Precisely we claim that for any $g \in Y$ the following estimate holds,

$$||PTg - g||_{\mathrm{BMO}} \leq \frac{1}{2}||g||_{\mathrm{BMO}}. \tag{5.2.16}$$

To prove the claim we fix $g = \sum_{i\in L} b_i c_i$, and expand PTg as

$$\begin{aligned} PTg &= \sum_{i\in L} c_i \langle Tb_i, b_i\rangle b_i \langle Tb_i, b_i\rangle^{-1} + \sum_{i\in L}\left\{\sum_{j\in L, j\neq i} c_j\langle Tb_j, b_i\rangle\right\} b_i\langle Tb_i, b_i\rangle^{-1} \\ &= g + e. \end{aligned} \tag{5.2.17}$$

With Proposition 5.2.3 we have the estimates

$$\sum_{j\in L, j\neq i} |c_j\langle Tb_j, b_i\rangle| \leq 4^{-i}||b_i||_2^2||g||_{\mathrm{BMO}}. \tag{5.2.18}$$

Recall that by our definition of the set L, the following lower bound holds,

$$|\langle Tb_i, b_i\rangle| \geq \frac{||b_i||_2^2}{2}. \tag{5.2.19}$$

Inserting (5.2.18) and (5.2.19) into (5.2.17) gives the following bound on the error term e,

$$||e||_{\mathrm{BMO}} \leq \frac{1}{2}||g||_{\mathrm{BMO}}.$$

This proves our claim in (5.2.16). Hence, by a standard perturbation argument there exists a linear operator E with norm ≤ 4 such that

$$\begin{array}{ccc} Y & \overset{\text{Id}}{\rightarrow} & Y \\ E \downarrow & & \uparrow P \\ \text{BMO}_N & \overset{T}{\rightarrow} & \text{BMO}_N \end{array} . \tag{5.2.20}$$

We complete the proof by merging the diagrams (5.2.15) and (5.2.20). □

Next we turn to proving that it suffices to show Theorem 5.2.1 for diagonal operators acting on the ℓ^∞ sum of the spaces BMO_n. By the next proposition for $N >> n$ we can always find in BMO_N a copy of BMO_n and a projection onto it that (almost) annihilates a given n-dimensional subspace in BMO_N. We prove this assertion by adapting the idea of Lemma 5.2.4.

Proposition 5.2.5. *Let $N \in \mathbb{N}$, and let $n \in \mathbb{N}$ such that*

$$4^n \leq \log \log N.$$

Let E be an n-dimensional subspace of BMO_N. Then in $\mathcal{D}_N$ there exist pairwise disjoint collections $\{\mathcal{B}_J : J \in \mathcal{D}_n\}$, so that the system

$$b_J = \sum \{h_I : I \in \mathcal{B}_J\}$$

is equivalent to the unit vector basis of BMO_n, the orthogonal projection

$$Q_n(x) = \sum_{J \in \mathcal{D}_n} \langle x, \frac{b_J}{||b_J||_2} \rangle \frac{b_J}{||b_J||_2}$$

is bounded with $||Q_n||_{BMO} \leq 4$ and so that for any $x \in E$,

$$||Q_n(x)||_{BMO} \leq 4 \frac{\log n}{n} ||x||_{BMO}.$$

Proof. Fix $n \in \mathbb{N}$. And fix an n-dimensional subspace $E \subset \text{BMO}_N$. In $\{x \in E : ||x|| = 1\}$, we choose a net $\{x_1, \ldots, x_M\}$ of width $1/n$. As $\dim E = n$ this can be done with $M = (2n)^n$.

Let $j \leq M$, then we claim that the collection

$$\mathcal{L}_j = \{J \in \mathcal{D} : n|\langle x_j, h_J \rangle| \geq |J|\}$$

satisfies the n^2-Carleson condition. Indeed, for $I \in \mathcal{L}_j$, we have the upper bound

$$\langle x_j, \frac{h_J}{|J|} \rangle|^2 \geq n^{-2} |J|.$$

Hence,

$$\begin{aligned}\sum_{J\in\mathcal{L}_j\cap I}|J| &\le \sum_{J\in\mathcal{L}_j\cap I} n^2|\langle x_j,h_J\rangle|^2|J|^{-2}|J|\\ &\le n^2\|x_j\|_{\mathrm{BMO}}^2|I|\\ &\le n^2|I|.\end{aligned}$$

Consequently the Carleson constant of the collection

$$\mathcal{G}=\mathcal{D}_N\setminus\{\mathcal{L}_1\cup\cdots\cup\mathcal{L}_M\}$$

is at least as large as $\log N - Mn^2$. We apply now the condensation Lemma 3.1.4 to $\mathcal{G}$. Then the Gamlen–Gaudet construction allows us to select in $\mathcal{G}$ collections $\{\mathcal{B}_J : |J|\ge 2^{-n}\}$ which satisfy the Jones compatibility condition (J). Moreover the system $\{b_J : |J|\ge 2^{-n}\}$ defined by

$$b_J=\sum\{h_I : I\in\mathcal{B}_J\}$$

is equivalent to the Haar basis in BMO_n. Note that we selected $\mathcal{B}_J\subseteq\mathcal{G}$, hence we ensured that for $I\in\mathcal{B}_J$ we have $|\langle x_j,h_I\rangle|\le n^{-1}|I|$. Consequently the following estimate holds for any x_j,

$$\begin{aligned}|\langle x_j,b_J\rangle| &\le \sum\{|\langle x_j,h_I\rangle| : I\in\mathcal{B}_J\}\\ &\le n^{-1}\sum\{|I| : I\in\mathcal{B}_J\}\\ &= n^{-1}\|b_J\|_2^2.\end{aligned}\tag{5.2.21}$$

Let $x\in E$ be arbitrary satisfying $\|x\|_{\mathrm{BMO}}=1$. Then there exists $x_j, j\le M$ such that $\|x-x_j\|_{\mathrm{BMO}}\le 1/n$. With Fefferman's inequality and with (5.2.21) we obtain that for any $b_J, J\in\mathcal{D}_n$, the following estimate holds,

$$\begin{aligned}|\langle x,b_J\rangle| &\le |\langle x-x_j,b_J\rangle|+|\langle x_j,b_J\rangle|\\ &\le \sqrt{2}\|b_J\|_2^2/n+\|b_J\|_2^2/n.\end{aligned}\tag{5.2.22}$$

Finally we insert the above upper bound of (5.2.22) into the definition of Q_n. This shows that Q_n almost annihilates $x\in E$, with $\|x\|_{\mathrm{BMO}}=1$.

$$\begin{aligned}\|Q_n(x)\|_{\mathrm{BMO}} &= \|\sum\{\langle x,b_J\|b_J\|_2^{-2}\rangle b_J : J\in\mathcal{D}_n\}\|_{\mathrm{BMO}}\\ &\le \frac{4}{n}\|\sum\{b_J : J\in\mathcal{D}_n\}\|_{\mathrm{BMO}}\\ &\le \frac{4}{n}\log n.\end{aligned}$$

By linearity the conclusion of Proposition 5.2.5 follows. □

Let $\{b_J : |J| \geq 2^{-n}\}$ be the system obtained by the Gamlen–Gaudet construction in the course of proving Proposition 5.2.5. By construction the linear extension of the map

$$i_n : h_J \to b_J$$

is invertible on the range of Q_n. It satisfies

$$\mathrm{Id}_{\mathrm{BMO}_n} = i_n^{-1} Q_n i_n, \tag{5.2.23}$$

and

$$\|i_n\|_{\mathrm{BMO}} \cdot \|i_n^{-1}\|_{\mathrm{BMO}} \leq 10. \tag{5.2.24}$$

Below we use the following canonical projection operators. Let K be a subset of the natural numbers. We define the projection

$$P_K : (\sum \mathrm{BMO}_n)_\infty \to (\sum \mathrm{BMO}_n)_\infty$$

as follows,

$$[P_K(x_n)]_k = \begin{cases} x_k & \text{if } k \in K, \\ 0 & \text{else.} \end{cases}$$

Thus the projection P_K acts as a $0,1$ multiplier on the entries of the vector (x_n). We say that D acting on $(\sum \mathrm{BMO}_n)_\infty$ is a diagonal operator if

$$P_{\{l\}} D P_{\{k\}} = 0, \qquad \text{for} \quad l \neq k.$$

It is convenient to write $P_k = P_{\{k\}}$ and to identify P_k with the projection onto the k-th coordinate of the vectors $(x_n) \in (\sum \mathrm{BMO}_n)_\infty$. The next theorem asserts that it suffices to prove Theorem 5.2.1 for diagonal operators on $(\sum \mathrm{BMO}_n)_\infty$.

Theorem 5.2.6. *For any operator $T : (\sum BMO_n)_\infty \to (\sum BMO_n)_\infty$ there exists a diagonal operator*

$$D : (\sum BMO_n)_\infty \to (\sum BMO_n)_\infty$$

such that

$$D = RTE \quad \text{and} \quad \mathrm{Id} - D = R(\,\mathrm{Id} - T)E,$$

where R, E are bounded linear operators on $(\sum BMO_n)_\infty$.

Proof. By induction we will determine sequences of operators $\{Q_n\}$ and $\{i_n\}$ such that

$$\begin{array}{ccc} \mathrm{BMO}_n & \xrightarrow{Id} & \mathrm{BMO}_n \\ i_n \searrow & & \nearrow i_n^{-1} Q_n \\ & \mathrm{BMO}_{N(n)} & \end{array} \quad \text{with} \quad \|Q_n\| \cdot \|i_n\| \leq 40,$$

and so that with Q and I given by (5.2.27) and (5.2.28) below, the operator $H = QTI$ is an almost diagonal operator (satisfying (5.2.31) and (5.2.32)) on the ℓ^∞ sum of BMO_n.

Let $k_1 = 1$. Then let $N(1)$ be large enough so that the conclusion of Proposition 5.2.5 holds for any 1-dimensional subspace of $\mathrm{BMO}_{N(1)}$. By a simple compactness argument we select $k_2 > k_1$ such that

$$||P_{N(1)}TP_{[k_2,\infty[}|| \le 4^{-2}||T||^{-1}.$$

At stage $n-1$ we have determined integers $k_1 < \cdots < k_{n-1}$, and $N(1) < \cdots < N(n-1)$, with $N(i) \in [k_i, \infty[$. By compactness there exists $k_n > k_{n-1}$, so that

$$||P_{N(n-1)}TP_{[k_n,\infty[}|| \le 4^{-(n-1)}||T||^{-1}. \tag{5.2.25}$$

Let $V_{n-1} = \mathrm{BMO}_1 \oplus \cdots \oplus \mathrm{BMO}_{N(n-1)}$. We view V_{n-1} as a subspace of the the ℓ^∞ sum of BMO_n. Hence T acts on V_{n-1}. Put $W_{n-1} = T(V_{n-1})$ and let w_{n-1} denote the algebraic dimension of W_{n-1}. Next choose $N(n)$ so large that

$$4^{w_{n-1}} \le \log\log N(n),$$

and also $N(n) \in [k(n), \infty)$. Applying Proposition 5.2.5 yields an orthogonal projection Q_n on $\mathrm{BMO}_{N(n)}$ so that its range is well isomorphic to BMO_n, and so that Q_n almost annihilates $Z_{n-1} = P_{N(n)}(W_{n-1})$. Specifically, for $x \in Z_{n-1}$,

$$||Q_n(x)||_{\mathrm{BMO}_{N(n)}} \le 4^{-n}||T||^{-1}||x||_{\mathrm{BMO}}. \tag{5.2.26}$$

Define now the operators Q and I acting on $(\sum \mathrm{BMO}_n)_\infty$. First fix a sequence (x_n) in $(\sum \mathrm{BMO}_n)_\infty$ and determine the sequence $I((x_n))$ by

$$[I(x_n)]_k = i_n(x_n) \quad \text{if} \quad k = N(n), \tag{5.2.27}$$

(with $[I(x_n)]_k = 0$ for $k \ne N(n)$). The projection Q is given as

$$[Q((x_n))]_k = i_k^{-1}Q_k(x_{N(k)}). \tag{5.2.28}$$

To analyze Q and I recall that the operators i_n, Q_n defined by Proposition 5.2.5 satisfy the identity

$$\mathrm{Id}_{\mathrm{BMO}_n} = i_n^{-1}Q_n i_n.$$

Hence Q inverts the action I such that $\mathrm{Id} = QI$, and the operator

$$H = QTI \tag{5.2.29}$$

satisfies

$$\mathrm{Id} - H = Q(\mathrm{Id} - T)I. \tag{5.2.30}$$

Recall that we defined the operator P_k on $(\sum \mathrm{BMO}_n)_\infty$ to be the orthogonal projection onto the k-th coordinate. The estimates (5.2.25) and (5.2.26) show that H and $\mathrm{Id} - H$ are almost orthogonal operators in the following sense.

$$\sum_{\{l:l\ne k\}} ||P_l H P_k|| \le 4^{-k}||T||^{-1}, \tag{5.2.31}$$

and

$$\sum_{\{l:l\neq k\}} ||P_l(\text{ Id} - H)P_k|| \leq 4^{-k}||T||^{-1}. \tag{5.2.32}$$

Next put

$$T_n = P_n H I_n : \text{BMO}_n \to \text{BMO}_n,$$

where I_n is the canonical embedding on BMO_n into $(\sum \text{BMO}_n)_\infty$. The operator T_n satisfies

$$\sup_n ||T_n|| \leq ||T||.$$

Finally define the diagonal operator D on $(\sum \text{BMO}_n)_\infty$ by

$$D((x_n)) = (T_n x_n).$$

By a standard perturbation argument, the estimate (5.2.31) implies that there exist operators E, R on $(\sum \text{BMO}_n)_\infty$ satisfying $\text{Id} = RE$ and

$$D = RTE \quad \text{and} \quad \text{Id} - D = R(\text{ Id} - T)E.$$

Summing up, we factored the diagonal operator D through T, and simultaneously we factored $\text{Id} - D$ through $\text{Id} - T$. □

With the above Theorem 5.2.6 we showed that it suffices to prove the factorization Theorem 5.2.1 for diagonal operators on $(\sum \text{BMO}_n)_\infty$. Furthermore we proved the localized factorization Theorem 5.2.2. Combining both gives Theorem 5.2.1.

Proof of Theorem 5.2.1. Let T be a bounded operator on $(\sum \text{BMO}_n)_\infty$. By Theorem 5.2.6 there exist bounded operators E, R on $(\sum \text{BMO}_n)_\infty$ and a diagonal operator D such that

$$D = RTE, \quad \text{and} \quad \text{Id} - D = R(\text{ Id} - T)E.$$

Let $T_n : \text{BMO}_n \to \text{BMO}_n$ be the linear maps defining the diagonal operator D, that is

$$D((x_n)) = (T_n x_n).$$

Next we apply the local factorization Theorem 5.2.2 to each of the operators T_n. This gives $E_n : \text{BMO}_n \to \text{BMO}_{N(n)}$ and $R_n : \text{BMO}_{N(n)} \to \text{BMO}_n$ with $||E_n||, ||R_n|| \leq 2$, such that for $H_n = T_n$ or $H_n = \text{ Id} - T_n$ the identity on BMO_n factors through H_n as follows,

$$\text{Id}_{\text{BMO}_n} = R_n H_n E_n.$$

We may assume that there is an infinite sequence $k(n)$, such that $H_{k(n)} = T_{k(n)}$. (Otherwise we continue with $H_{k(n)} = \text{Id} - T_{k(n)}$.) Now define operators I, P on $(\sum \text{BMO}_n)_\infty$ by

$$[I(x_n)]_m = \begin{cases} E_{k(n)}x_n & \text{if } m = k(n), \\ 0 & \text{else,} \end{cases}$$

$$[P((x_n))]_m = R_{k(m)}T_{k(m)}x_{k(m)}.$$

They satisfy

$$\begin{aligned} \text{Id}_{(\sum \text{BMO}_n)_\infty} &= PDI \\ &= PRTEI. \end{aligned}$$

Summing up, the identity on $(\sum \text{BMO}_n)_\infty$ factors boundedly through T. □

We remark in closing that the proof of Theorem 5.2.1 shows also that $(\sum H^1_n)_1$ is primary.

Intrinsic description (continued)

Next we study pointwise estimates for basis functions in uniformly complemented copies of H^1_n in H^1. Fix a complemented copy of H^1_n in H^1, by specifying operators M and P for which

$$\begin{array}{ccc} H^1_n & \xrightarrow{Id} & H^1_n \\ M \searrow & & \nearrow P \\ & H^1 & \end{array} .$$

Dualizing the operators M and P we obtain complemented copies of BMO_n in BMO,

$$\begin{array}{ccc} \text{BMO}_n & \xrightarrow{Id} & \text{BMO}_n \\ P^* \searrow & & \nearrow M^* \\ & \text{BMO} & \end{array} .$$

In H^1 the functions $\{Mh_J\}$ are $||M||\,||P||$ equivalent to the Haar basis of H^1_n, and in BMO the functions $\{P^*h_J\}$ are $||M||\,||P||$ equivalent to the Haar basis in BMO_n. Moreover H^1, BMO duality preserves the biorthogonality of Haar functions. For each $I, J \in \mathcal{D}_n$, we have that

$$\langle Mh_J, P^*h_I\rangle = \langle h_J, h_I\rangle. \tag{5.2.33}$$

For notational simplicity we write

$$N = P^*.$$

The next theorem describes the pointwise behavior of the square functions for $\{Mh_J : J \in \mathcal{D}_n\}$, and also for $\{Nh_I : I \in \mathcal{D}_n\}$. We find remarkable similarities to the original system, $\{h_J : J \in \mathcal{D}_n\}$. To formulate this precisely we use the following convention. Let $\mathcal{E}_0, \ldots, \mathcal{E}_A$ be disjoint collections of dyadic intervals, each of which consists of pairwise disjoint intervals. We call $\mathcal{E}_0, \ldots, \mathcal{E}_A$ linearly ordered if the following two conditions hold,

$$\mathcal{E}_0^* \supseteq \cdots \supseteq \mathcal{E}_A^*$$

and for $i \in \{0, \ldots, A\}$, we have that

$$I \in \mathcal{E}_i, J \in \mathcal{E}_{i+1} \text{ implies } |J| \leq \frac{1}{2}|I|.$$

The collections $\mathcal{E}_0, \ldots, \mathcal{E}_A$ are inducing functions in BMO and H^1,

$$k_i = \sum \{Nh_J : J \in \mathcal{E}_i\} \quad \text{and} \quad g_i = \sum \{Mh_J : J \in \mathcal{E}_i\} |\mathcal{E}_i^*|^{-1}.$$

The pointwise behavior of their square functions is the content of the next theorem. Its inductive proof is an iteration based on Proposition 5.1.10 that describes the unit vectors of complemented Hilbertian subspaces of H^1.

Theorem 5.2.7. *In $\mathcal{D}_n$ there exist linearly ordered collections $\mathcal{E}_0, \ldots, \mathcal{E}_A$ such that their union $\mathcal{E}_0 \cup \cdots \cup \mathcal{E}_A$ satisfies the Carleson condition with constant 2, and*

$$\int \left\{ \sum_{i=0}^{A-1} S(k_i) \right\} S(g_A) \geq \frac{A + (\ln n)^{1/2}}{32R},$$

where we put $R = 64^2 ||M||^4 ||N||^4$.

Remarks. 1. Note that $g_A \in H^1$. Hence the non-negative function $S(g_A)$ is integrable. In the conclusion of Theorem 5.2.7 the role of $S(g_A)$ is that of a weight. It compensates for possible dilations on the BMO side due to the linear map N.

2. Note that if $\mathcal{E}_0 \cup \cdots \cup \mathcal{E}_A$ satisfy the 2-Carleson condition, then the BMO functions $k_1, \ldots, k_A$ are well equivalent to the unit vectors of ℓ_A^∞.

3. Below we repeatedly use the following observation. Let (Ω, μ) be a probability space and let $F : \Omega \to \mathbb{R}^+$ be non-negative, bounded and measurable, satisfying

$$\int_\Omega F d\mu > \alpha,$$

for some $\alpha > 0$. Then

$$\mu\{F > \alpha/4\} \geq 3\alpha(4\|F\|_\infty)^{-1}. \tag{5.2.34}$$

Indeed, decompose $\Omega = \{F > \alpha/4\} \cup \{F \leq \alpha/4\}$ and estimate the integral accordingly. This gives

$$\int_\Omega F d\mu \leq \|F\|_\infty \mu\{F > \alpha/4\} + \alpha/4.$$

Recall that $\int_\Omega F d\mu > \alpha$ and subtract $\alpha/4$ to obtain

$$3\alpha \leq 4\|F\|_\infty \mu\{F > \alpha/4\}.$$

Proof of Theorem 5.2.7. The following proof proceeds by induction where the number of steps is not specified at the beginning. Instead, the induction argument contains a stopping criterion which (if satisfied) implies the validity of the Theorem 5.2.7. If on the other hand the stopping criterion is not satisfied during $C(\ln n)^{1/2}$ iterations of the induction argument, then again we will be able to conclude that Theorem 5.2.7 holds true.

Step 1. We cover the unit interval using the collections $\mathcal{G}_i = \{I : |I| = 2^{-i}\}$, for $i \in \{1, \ldots, n\}$. These collections are contained in $\mathcal{D}_n$, and generate unit vectors of ℓ^2_n in H^1 respectively BMO. Now put

$$f_i = \sum\{Mh_J : J \in \mathcal{G}_i\}, \quad h_i = \sum\{Nh_J : J \in \mathcal{G}_i\}.$$

In H^1 the functions $\{f_i\}$ are well equivalent to the unit vector basis of ℓ^2_n. The same holds in BMO for the functions $\{h_i\}$. Moreover by (5.2.33) they are linked by

$$\langle f_i, h_j \rangle = \delta_{i,j}.$$

Apply now Theorem 5.1.10 to $\{f_i\}$ and $\{h_i\}$. This gives

$$\sum_{i=1}^{n} \sum_{j=i+1}^{n} \int S(h_i)S(f_j) \geq \frac{n(n-1)}{R}, \tag{5.2.35}$$

where $R = 64^2||M||^4||N||^4$. We continue with the proof if for each $i \leq n$ and $j \geq i+1$,

$$\int S(h_i)S(f_j) \leq \frac{(\ln n)^{1/2}}{R}. \tag{5.2.36}$$

(If condition (5.2.36) does not hold, then we stop, since then there is nothing more to prove, and we put $A = 1$.) Define $\Omega = \{(i,j) : i+1 \leq j \leq n,\ i \leq n\}$ and let μ be the normalized counting measure on Ω. Next define

$$F(i,j) = \int S(h_i)S(f_j) \quad \text{where} \quad (i,j) \in \Omega.$$

By (5.2.35) and (5.2.36) the observation (5.2.34) shows that there exists a constant $K(1)$, with

$$K(1) > n(4(\ln n)^{1/2})^{-1},$$

together with $i_0 \leq n$ and a set of indices $D \subseteq \{i_0+1, \ldots, n\}$, so that the cardinality of D equals $K(1)$, and so that the following set of $K(1)$ inequalities hold true,

$$\int S(h_{i_0})S(f_j) \geq \frac{1}{4R}, \quad \forall j \in D.$$

We put

$$\mathcal{E}_0 = \mathcal{G}_{i_0} \quad \text{and} \quad k_0 = h_{i_0}.$$

Then we relabel the collections $\{\mathcal{G}_j : j \in D\}$ as

$$\mathcal{G}_{1,1}, \dots, \mathcal{G}_{1,K(1)},$$

and accordingly we rename the H^1 functions $\{f_j : j \in D\}$ as

$$f_{1,1}, \dots, f_{1,K(1)}.$$

This completes the first step of the construction.

After B steps. Next we claim that iterating the induction argument B times gives a constant $K(B)$ satisfying

$$K(B) \geq n(4(\log n)^{1/2})^{-B}$$

and collections of dyadic intervals $\mathcal{G}_{B,j}$, and $\mathcal{E}_i$ with the property that the entire string of collections,

$$\mathcal{E}_0, \dots, \mathcal{E}_{B-1}, \quad \mathcal{G}_{B,1}, \dots, \mathcal{G}_{B,K(B)},$$

is linearly ordered, and satisfies

$$|I \cap \mathcal{E}_{i+1}^*| = \frac{1}{2}|I|, \quad \text{for} \quad I \in \mathcal{E}_i \quad \text{and} \quad i \in \{0, \dots, B-2\},$$

and

$$\mathcal{G}_{B,j}^* = \mathcal{E}_{B-1}^*, \quad \text{for} \quad j \in \{0, \dots, K(B)\}.$$

Moreover for the functions induced by the collections $\mathcal{G}_{B,j}$ and $\mathcal{E}_i$, that is, for

$$f_{B,j} = 2^{B-1} \sum \{Mh_J : J \in \mathcal{G}_{B,j}\} \quad \text{and} \quad k_i = \sum \{Nh_J : J \in \mathcal{E}_i\},$$

the following set of $K(B)$ inequalities hold true,

$$\int \left\{ \sum_{i=0}^{B-1} S(k_i) \right\} S(f_{B,j}) \geq \frac{B}{4R}, \quad \forall j \leq K(B). \tag{5.2.37}$$

The induction argument. We assume the iteration has been carried out B times and we describe the next step. Start by splitting $\mathcal{G}_{B,1}$ into $\mathcal{L}$ and $\mathcal{R}$ such that

$$|I \cap \mathcal{L}^*| = |I \cap \mathcal{R}^*| = \frac{1}{2}|I|, \quad \forall I \in \mathcal{E}_{B-1}. \tag{5.2.38}$$

Then for $j \leq K(B)$ split $\mathcal{G}_{B,j}$ accordingly,

$$\mathcal{A}_j = \{J \in \mathcal{G}_{B,j} : J \subseteq \mathcal{L}^*\} \quad \text{and} \quad \mathcal{B}_j = \{J \in \mathcal{G}_{B,j} : J \subseteq \mathcal{R}^*\}.$$

With $\mathcal{A}_j, \mathcal{B}_j$ form the H^1 functions

$$a_j = 2^B \sum\{Mh_J : J \in \mathcal{A}_j\} \quad \text{and} \quad b_j = 2^B \sum\{Mh_J : J \in \mathcal{B}_j\}.$$

This choice gives

$$S(f_{B,j}) \leq \frac{1}{2}(S(a_j) + S(b_j)). \tag{5.2.39}$$

Combining the estimate (5.2.39) and the induction hypothesis (5.2.37) gives a set of indices $D \subseteq \{1, \ldots, K(B)\}$ of cardinality $\geq K(B)/2$, such that one of the following statements is true,

$$\begin{aligned} \int \left\{ \sum_{i=1}^{B-1} S(k_i) \right\} S(a_j) &\geq \frac{B}{4R}, \quad \forall j \in D, \\ \int \left\{ \sum_{i=1}^{B-1} S(k_i) \right\} S(b_j) &\geq \frac{B}{4R}, \quad \forall j \in D. \end{aligned} \tag{5.2.40}$$

We assume without loss of generality that the first set of estimates holds true. Note that, by construction, the H^1 functions $\{a_j : j \in D\}$ are equivalent to the unit vector basis of ℓ^2_n (with constant $||M||\ ||P||$). Also the BMO functions

$$h_i = \sum\{Nh_J : J \in \mathcal{A}_i\}, \quad i \in D$$

are equivalent to the unit vector basis of ℓ^2_n in BMO. The duality between H^1 and BMO preserves biorthogonality, i.e.,

$$\langle a_j, h_i \rangle = \delta_{ij}, \quad i, j \in D.$$

Hence by Theorem 5.1.10,

$$\sum_{i \in D} \sum_{\{j \geq i+1\} \cap D} \int S(h_i) S(a_j) \geq \frac{|D|(|D| - 1)}{R}, \tag{5.2.41}$$

where $|D|$ denotes the cardinality of D. We assume that for each $i \in D$ and $j \in D$ with $j \geq i + 1$,

$$\int S(h_i) S(a_j) \leq \frac{(\ln n)^{1/2}}{R}. \tag{5.2.42}$$

If the condition (5.2.42) does not hold, then we may stop, and the proof is complete. Otherwise we continue as follows. By (5.2.41) and (5.2.42) the observation (5.2.34) implies that there exists $i_0 \in D$ and a subset $E \subseteq D \cap \{i_0 + 1, \ldots, K\}$ such that

$$\int S(h_{i_0}) S(a_j) \geq \frac{1}{4R}, \quad j \in E, \tag{5.2.43}$$

and the cardinality of E, denoted by $K(B+1)$, satisfies the lower estimate

$$K(B+1) \geq n(16(\ln n)^{1/2})^{-B-1}.$$

We define

$$\mathcal{E}_B = \mathcal{A}_{i_0} \text{ and } k_B = h_{i_0},$$

and relabel $\{\mathcal{A}_j : j \in E\}$ as

$$\mathcal{G}_{B+1,1}, \ldots, \mathcal{G}_{B+1,K(B+1)}.$$

Accordingly we relabel the H^1 functions $\{a_j : j \in E\}$ as

$$f_{B+1,1}, \ldots, f_{B+1,K(B+1)}.$$

A moment's reflection shows that by (5.2.38) the collections of dyadic intervals

$$\mathcal{E}_0, \ldots, \mathcal{E}_B, \quad \mathcal{G}_{B+1,1}, \ldots, \mathcal{G}_{B+1,K(B+1)}$$

are linearly ordered. The measures covered by $\mathcal{E}_0, \ldots, \mathcal{E}_B$ decrease uniformly as follows,

$$|I \cap \mathcal{E}_{i+1}^*| = \frac{1}{2}|I|, \quad \text{for} \quad I \in \mathcal{E}_i \quad \text{and} \quad i \in \{0, \ldots, B-1\}.$$

The collections $\mathcal{G}_{B+1,j}$ cover the same set as $\mathcal{E}_{B+1}$, that is

$$\mathcal{E}_{B+1}^* = \mathcal{G}_{B+1,j}^* \quad j \in \{1, \ldots, K(B)\}.$$

Finally we add the inequalities (5.2.40) and (5.2.43) and with the notation just introduced we obtain that

$$\int \left\{ \sum_{i=0}^{B} S(k_i) \right\} S(f_{B+1,j}) \geq \frac{B+1}{4R}, \quad \forall j \leq K(B+1).$$

This completes the induction step.

Conclusion. There are two possibilities how this process will terminate. First it will terminate when at some stage of the construction we select a set of indices D for which there exist $i \in D$ and $j \in D$ with $j \geq i+1$ such that

$$\int S(h_i)S(f_j) \geq \frac{(\ln n)^{1/2}}{R}.$$

In that case the proof is complete. The second possibility is when the index set E becomes empty (that is when the estimate for the lower bound on the cardinality of E becomes < 1). This can happen only when

$$n \left[\frac{1}{16(\ln n)^{1/2}} \right]^B < 1,$$

hence $B \geq 16^{-1}(\ln n)^{1/2}$, so that by (5.2.37) the theorem is proven. □

H^1 with values in ℓ^2

Next we prove that H^1 is not isomorphic to the space of H^1 functions taking values in the Hilbert space ℓ^2. This space denoted by $H^1(\ell^2)$ consists of those measurable ℓ^2-valued functions

$$f(t) = \sum h_I(t)x_I, \quad x_I \in \ell^2,$$

for which the square function

$$Sf(t) = \left(\sum h_I^2(t)||x_I||^2\right)^{1/2}$$

is integrable. The non-isomorphism of H^1 and $H^1(\ell^2)$ is a striking result, especially in view of the obvious analogy in the definitions for H^1 and $H^1(\ell^2)$, and also in view of the multitude of analogous theorems that hold in H^1 and $H^1(\ell^2)$.

The *assumption* of the next theorem states that there exits a complemented subspace of H^1 which is isomorphic to $H^1(\ell^2)$. From this hypothesis we will derive a condition which is in contradiction to Theorem 5.2.7. Hence H^1 does not contain any complemented copy of $H^1(\ell^2)$. In particular the spaces H^1 and $H^1(\ell^2)$ are not isomorphic.

Theorem 5.2.8. *Assume that there exists a complemented subspace of H^1 which is isomorphic to $H^1(\ell^2)$. Then the following statement is true: For any $n \in \mathbb{N}$ there exist operators $M : H^1_n \to H^1$ and $Q : H^1 \to H^1_n$ satisfying the following conditions:*

1. $\mathrm{Id} = QM$, *and* $||Q||\,||M|| \le C$, *with C independent of n.*
2. *For every choice of linearly ordered collections of dyadic intervals $\mathcal{E}_0, \dots, \mathcal{E}_A$ the functions*

$$k_i = \sum\{Q^* h_J : J \in \mathcal{E}_i\},$$

and

$$f = \sum\{Mh_J : J \in \mathcal{E}_A\}|\mathcal{E}_A^*|^{-1},$$

satisfy the estimate

$$\int \left\{\sum_{i=0}^{A-1} S(k_i)\right\} S(f) \le \sqrt{A}||M|| \left\|\sum_{i=0}^{A-1} k_i\right\|_{BMO}.$$

Comment. Before we prove this theorem we will point out why its conclusion contradicts the assertions of Theorem 5.2.7. Note that the right-hand side of the above estimate contains the crucial factor $\sqrt{A}$. Next consider the collections $\mathcal{E}_0, \dots, \mathcal{E}_A$ obtained by Theorem 5.2.7. They are linearly ordered and the union $\mathcal{E}_0 \cup \cdots \cup \mathcal{E}_A$

satisfies the 2-Carleson condition. Hence $\left\|\sum_{i=0}^{A-1} k_i\right\|_{\mathrm{BMO}} \leq \sqrt{2}||Q^*||$. Now we compare the upper estimate of our present theorem,

$$\int \left\{ \sum_{i=1}^{A-1} S(k_i) \right\} S(f) \leq \sqrt{2A}\|M\| \cdot \|Q\|,$$

with the lower estimate of Theorem 5.2.7,

$$\int \left\{ \sum_{i=1}^{A-1} S(k_i) \right\} S(f) \geq \frac{A + (\log n)^{1/2}}{32R},$$

where $R = 64^2\|M\|^4 \cdot \|N\|^4$. Clearly for n large enough we have two conflicting assertions. Thus the hypothesis of Theorem 5.2.8 led to a contradiction, so that H^1 and $H^1(\ell^2)$ are not isomorphic Banach spaces.

Proof of Theorem 5.2.8. We assume that H^1 contains a complemented subspace isomorphic to $H^1(\ell^2)$. Let E, P be bounded linear operators such that

$$\begin{array}{ccc} H^1(\ell^2) & \xrightarrow{Id} & H^1(\ell^2) \\ E \searrow & & \nearrow P \\ & H^1 & \end{array}.$$

Let $\{e_k : k \in \mathbb{N}\}$ be the orthonormal unit vector basis of ℓ^2. Note that the Hilbert space-valued functions

$$\{h_I e_k : I \in \mathcal{D}, k \in \mathbb{N}\}$$

are an unconditional basis in $H^1(\ell^2)$. Now fix $I \in \mathcal{D}$. Then $h_I e_k \to 0$ weakly in $H^1(\ell^2)$ as $k \to \infty$. Consequently,

$$E(h_I e_k) \to 0 \text{ weakly in } H^1, \tag{5.2.44}$$

as $k \to \infty$. Observe that (5.2.44) implies that for $\epsilon > 0$ and $\delta > 0$ fixed there exists $m = m(\epsilon, \delta)$ and a finite collection $\mathcal{S}$ of dyadic intervals so that for

$$z = E(h_I e_m)$$

the following conditions hold.

$$\text{If } K \in \mathcal{S}, \quad \text{then} \quad |K| \leq \delta, \tag{5.2.45}$$

and

$$\|z - \sum_{K \in \mathcal{S}} \langle z, h_K \rangle \frac{h_K}{|K|}\|_{H^1} \leq \epsilon. \tag{5.2.46}$$

Next we chose $n \in \mathbb{N}$, and we start the construction of the operators M and Q which will eventually satisfy the conclusions of Theorem 5.2.8. Let $\mathcal{D}_n$ denote the collection of dyadic intervals of length$\geq 2^{-n}$. There are $2^{n+1}-1$ intervals in $\mathcal{D}_n$, we enumerate them as

$$I_1, \ldots, I_{2^{n+1}-1},$$

using lexicographic order on the intervals. Next we fix $\epsilon_k > 0$ such that

$$\sum_{k=1}^{2^{n+1}-1} \epsilon_k < (||E||\,||P||8)^{-1}. \tag{5.2.47}$$

Now we start a gliding hump process at the smallest, right-most interval $I_{2^{n+1}-1}$ and work backwards toward $I_1 = [0,1]$. We apply the remark following (5.2.44) with $\delta = 1$, $\epsilon = \epsilon_{2^{n+1}-1}$ and $I = I_{2^{n+1}-1}$. Let $\mathcal{S}_{2^{n+1}-1} = \mathcal{S}$ be the finite collection of dyadic intervals so that the conditions (5.2.45) and (5.2.46) hold true. Then put

$$\delta = \frac{1}{2}\inf\{|L| : L \in \mathcal{S}\}.$$

Apply the remark again with this choice of δ with $\epsilon = \epsilon_{2^{n+1}-2}$ and $I = I_{2^{n+1}-2}$. We continue inductively until we reach I_1. Thus we obtain a sequence (n_k) and pairwise disjoint collections of dyadic intervals $\{\mathcal{S}_k : k \leq 2^{n+1}-1\}$ so that

$$||z_k - Q_k(z_k)||_{H^1} \leq \epsilon_k ||z_k||, \tag{5.2.48}$$

where

$$z_k = E(h_{I_k} e_{n_k}),$$

and where Q_k is the orthogonal projection onto $\operatorname{span}\{h_J : J \in \mathcal{S}_k\}$ given by

$$Q_k f = \sum_{J \in \mathcal{S}_k} \langle f, h_J \rangle \frac{h_J}{|J|}.$$

Furthermore the collections $\{\mathcal{S}_k : k \leq 2^{n+1}-1\}$ are inversely ordered in the following sense.

$$K \in \mathcal{S}_k \text{ and } L \in \mathcal{S}_{k+1} \text{ implies } |K| \leq \frac{1}{2}|L|. \tag{5.2.49}$$

Note that for any sequence of scalars $\{\alpha_k\}$,

$$\left\| \sum_{k=1}^{2^{n+1}-1} \alpha_k h_{I_k} e_{n_k} \right\|_{H^1(\ell^2)} = \left\| \sum_{k=1}^{2^{n+1}-1} \alpha_k h_{I_k} \right\|_{H^1}.$$

Hence, equipped with the norm of $H^1(\ell^2)$ the space

$$X_n = \operatorname{span}\{h_{I_k} e_{n_k} : k \leq 2^{n+1}-1\}$$

is isometric to H^1_n, and the isometry is given by the linear extension of the map

$$E_1 : X_n \to H^1_n, \quad h_{I_k} e_{n_k} \to h_{I_k}.$$

We let $R_1 : H^1(\ell^2) \to X_n$ be the natural restriction operator. R_1 is given by a $\{0,1\}$ multiplier of the unconditional basis $\{h_I e_k : I \in \mathcal{D}, k \in \mathbb{N}\}$, hence $||R_1|| \leq 1$. Note that the composition

$$Q_0 = E_1 R_1 P$$

inverts the action of the embedding $M_0 : H^1_n \to H^1, h_{I_k} \to z_k$. Thus

$$\begin{array}{ccc} H^1_n & \xrightarrow{Id} & H^1_n \\ M_0 \searrow & & \nearrow Q_0 \\ & H^1 & \end{array}$$

and

$$||M_0||\,||Q_0|| \leq ||E||\,||P||.$$

This was the main step of the construction. We exploited the hypothesis to find order inverting embeddings and projections. Now we add two minor modifications to the definition of M_0 and Q_0, respectively. We start by modifying M_0. Let

$$w_k = Q_k z_k.$$

By (5.2.48) and (5.2.47) the embedding $M : H^1_n \to H^1, h_{I_k} \to w_k$, satisfies $||M|| \leq 2||M_0||$, and can be conjugated by a linear operator

$$R : H^1 \to H^1_n$$

such that $||R|| \leq 2||Q_0||$ and

$$\begin{array}{ccc} H^1_n & \xrightarrow{Id} & H^1_n \\ M \searrow & & \nearrow R \\ & H^1 & \end{array} .$$

Now we diagonalize R as follows. Put

$$Q = \sum_{k=1}^{2^{n+1}-1} T_k R Q_k,$$

where T_k is the orthogonal projection onto the one-dimensional subspace spanned by the function h_{I_k}. Then, by inspection,

$$\mathrm{Id} = QM,$$

and by unconditionality of the Haar basis we have the norm estimate $||Q|| \leq \sqrt{2}||R||$. Summarizing the effect of these modifications we observe that

$$Mh_{I_k} \in \mathrm{span}\{h_J : J \in \mathcal{S}_k\},$$

and also

$$Q^* h_{I_k} \in \operatorname{span}\{h_J : J \in \mathcal{S}_k\}.$$

We will now verify that the maps M, Q satisfy the conclusion of Theorem 5.2.8. We fix linearly ordered collections $\mathcal{E}_0, \ldots, \mathcal{E}_A$. Then we form

$$k_i = \sum\{Q^* h_J : J \in \mathcal{E}_i\} \quad \text{and} \quad f = \sum\{M h_J : J \in \mathcal{E}_A\}|\mathcal{E}_A^*|^{-1}.$$

Below we prove that $g = \sum_{i=0}^{A-1} k_i$ satisfies the surprising estimate,

$$\int S(f)S(g) \leq ||f||_{H^1} \, \|g\|_{\mathrm{BMO}} \,. \tag{5.2.50}$$

Before that we use (5.2.50) to finish the proof as follows:

$$\begin{aligned}\sum_{i=0}^{A-1} \int S(f)S(k_i) &\leq \sqrt{A} \int S(f) \left(\sum_{i=0}^{A-1} S^2(k_i) \right)^{1/2} \\ &= \sqrt{A} \int S(f) S\,(g) \\ &\leq \sqrt{A} ||f||_{H^1} \, \|g\|_{\mathrm{BMO}} \,.\end{aligned}$$

So we are left with proving the estimate (5.2.50). Now we will exploit that the collections $\mathcal{S}_k$ are inversely ordered as expressed in (5.2.49). We let $\mathcal{F}$ be the σ-algebra generated by f. This σ-algebra $\mathcal{F}$ is purely atomic. We put $g = \sum_{i=0}^{A-1} k_i$, and let

$$\phi = S\,(g)\,.$$

Now we claim that

$$||\mathbb{E}(\phi|\mathcal{F})||_\infty \leq ||g||_{\mathrm{BMO}}.$$

To prove this claim we expand g in its Haar series, thus

$$g = \sum a_J h_J.$$

Next we fix an atom B of $\mathcal{F}$, and using that $\{\mathcal{S}_k\}$ is an inversely ordered collection of dyadic intervals, we observe the pointwise equality

$$\phi^2(t) = \sum_{J \subseteq B} a_J^2 h_J^2(t), \quad \text{for} \quad t \in B. \tag{5.2.51}$$

The crucial point of equation (5.2.51) is that on the right-hand side the sum is taken over the dyadic intervals that are *contained* in that atom B (without the assumption of order inversion this does not hold). Averaging the identity (5.2.51) we obtain that

$$\mathbb{E}\left(\phi^2|\mathcal{F}\right)(t) = \frac{1}{|B|} \sum_{J \subseteq B} |J| a_J^2, \quad \text{for } t \in B. \tag{5.2.52}$$

By Hölder's inequality we estimate using (5.2.52).

$$\begin{aligned}\|\mathbb{E}(\phi|\mathcal{F})\|_\infty &\leq \left\|\mathbb{E}(\phi^2|\mathcal{F})^{1/2}\right\|_\infty \\ &\leq ||g||_{\mathrm{BMO}}.\end{aligned}$$

Finally recall that $\phi = S(\sum_{i=0}^{A-1} k_i)$, and we obtain that

$$\begin{aligned}\int S(f)\phi &= \int S(f)\mathbb{E}(\phi|\mathcal{F}) \\ &\leq \int S(f)||g||_{\mathrm{BMO}}.\end{aligned}$$

□

Next we show that $H^1(\ell^2)$ is isomorphic to a space of real-valued functions, the so-called tent space. For $0 < \alpha < 1$, and $x \in [0,1]$, we define the cone

$$\Gamma_\alpha(x) = \{(y,t) \in [-1,2] \times [0,1] : |x-y| \leq \alpha t\}.$$

Then we define the tent space T_α to be the space of all measurable functions $f : [0,1] \times [0,1] \to \mathbb{R}$ such that

$$||f||_{T_\alpha} = \int_0^1 (\int_{\Gamma_\alpha(x)} f(y,t)^2 t^{-2} dydt)^{1/2} dx < \infty.$$

The tent spaces are independent of the cone parameter α, indeed, for $\alpha \leq \beta$ there exists $C(\alpha,\beta) > 0$ such that

$$||f||_{T_\alpha} \leq C(\alpha,\beta)||f||_{T_\beta},$$

for $f \in T_\beta$.

We will show now that $H^1(\ell^2)$ is isomorphic to T_α. To begin with we discuss first convenient representations of the spaces T_α, respectively $H^1(\ell^2)$. For $I \in \mathcal{D}$ we let I_0, I_1, I_2, be three subintervals of I with length $|I_j| = |I|/3$. We let

$$T(I_j) = I_j \times [|I|, 2|I|[,$$

and we denote by S_α^j the subspace of T_α which consists of functions supported in

$$\bigcup \{T(I_j) : I \in \mathcal{D}\}.$$

It is easy to see that each of the subspaces S_α^j is isomorphic to the entire tent space T_α. It thus suffices to show that $H^1(\ell^2)$ is isomorphic to S_α^1. We continue with a representation of $H^1(\ell^2)$. For $I \in \mathcal{D}$ we say that a function f_I belongs to the Hilbert space L_I^2 iff $\operatorname{supp} f_I$ is contained in $T(I_1)$ and

$$||f_I||_2 = (\int\int_{T(I_1)} f(y,t)^2 \frac{dydt}{|I| \times |I|})^{1/2} < \infty.$$

We let $E_I : L^2_I \to \ell^2$ denote the isomorphism between L^2_I and the sequence space ℓ^2. Let Z be the space of vector-valued functions

$$F(x) = \sum h_I(x) f_I, \ f_I \in L^2_I,$$

equipped with the norm

$$||F|| = \int (\sum h_I(x)^2 ||f_I||_2^2)^{1/2}.$$

Z is isomorphic to $H^1(\ell^2)$, the isomorphism is given by

$$T(F) = \sum h_I(x) E_I(f_I).$$

Theorem 5.2.9. *a) For $f \in S^1_\alpha$, let $f_I = 1_{T(I_1)} f$ and let $a_I = E_I(f_I)$. Then the operator*

$$\begin{aligned} T : S^1_\alpha &\to H^1(\ell^2) \\ f &\to \sum h_I a_I \end{aligned}$$

is bounded and invertible.

b) For $a_I \in \ell^2$ let $f_I = E_I^{-1}(a_I)$. Then the operator

$$\begin{aligned} S : H^1(\ell^2) &\to S^1_\alpha \\ \sum h_I a_I &\to \sum f_I \end{aligned}$$

is bounded, and it is the inverse of T.

Proof. As S^1_α is independent of α we make the choice $\alpha = 2/3$. Note that for $x \in I$ we have the inclusion

$$\Gamma_\alpha(x) \supseteq T(I_1).$$

This gives the pointwise estimates,

$$\begin{aligned} \sum 1_I(x) ||a_I||_2^2 &\le 4 \sum 1_I(x) \int f_I(y,t)^2 t^{-2} dy dt \\ &\le 4 \sum \int_{\Gamma_\alpha(x)} f_I(y,t)^2 t^{-2} dy dt. \end{aligned}$$

Taking square roots, interchanging the order of the sum and the integral and finally integrating the pointwise estimate over $x \in [0,1]$, gives

$$\begin{aligned} ||\sum h_I(x) a_I||_{H^1(\ell^2)} &\le 2 \int_0^1 (\int_{\Gamma_\alpha(x)} \sum f_I(y,t)^2 t^{-2} dy dt)^{1/2} dx \\ &\le 2 ||f||_{S^1_\alpha}. \end{aligned}$$

This proves the first part of the theorem. We start the second part of the theorem by making another choice for α. We let $\alpha = 1/6$. With this choice of α we observe that for $x \notin I$ we have

$$\Gamma_\alpha(x) \cap T(I_1) = \emptyset.$$

This gives the pointwise estimate

$$\sum \int_{\Gamma_\alpha(x)} f_I(y,t)^2 t^{-2} dydt \le \sum 1_I(x) \int f_I(y,t)^2 t^{-2} dydt$$
$$\le 4 \sum 1_I(x) ||a_I||_2^2.$$

As above we obtain the boundedness of S from this pointwise estimate by taking square roots and integrating. □

5.3 The uniform approximation property of BMO

We review three of the most classical approximation properties in Banach spaces. Recall that a Banach space satisfies the approximation property if for every $\epsilon > 0$ and every compact set $K \subseteq X$ there exists a finite rank operator such that

$$||Tx - x||_X \le \epsilon ||x||_X, \text{ for } x \in K.$$

Note that the approximation property involves qualitative conditions. In that way it asks for very little, in particular the rank of the operator T and its norm are allowed to vary with K, the compact set over which T is supposed to approximate the identity operator. The bounded approximation property puts an upper bound on $||T||$, not depending on K.

A Banach space X satisfies the bounded approximation property if there exists $\lambda > 0$ so that for every $\epsilon > 0$ and every compact set $K \subseteq X$ there exists a finite rank operator satisfying $||T|| \le \lambda$, and

$$||Tx_i - x_i||_X \le \epsilon ||x_i||_X, \quad \text{for} \quad i \le n.$$

Let K be a compact subset of X and let $\epsilon > 0$. Then, by compactness, there exist $x_1, \dots, x_n \in K$ such that

$$K \subseteq \bigcup B(x_i, \epsilon).$$

Assume now that T is a finite rank operator satisfying $||T|| \le \lambda$, and

$$||Tx_i - x_i||_X \le \epsilon ||x_i||_X, \quad \text{for} \quad i \le n.$$

For $x \in K$ choose $k \le n$ such that $||x - x_k||_X < \epsilon$, and write $Tx - x = (Tx_k - x_k) + (Tx_k - Tx) + (x - x_k)$. Then the triangle inequality and the norm bound of T give

$$||Tx - x||_X \le 4\lambda\epsilon ||x||.$$

Hence in the definition of the bounded approximation property we may replace compact sets by finite sets as follows. A Banach space X satisfies the bounded approximation property if there exists $\lambda > 0$ so that for every $\epsilon > 0$ and $x_1, \dots, x_n \in X$ there exists a finite rank operator satisfying

$$\begin{aligned} ||T|| &\leq \lambda, \\ ||Tx_i - x_i||_X &\leq \epsilon ||x_i||_X, \quad \text{for} \quad i \leq n. \end{aligned}$$

The bounded approximation property allows the rank of T to vary with the choice of $x_1, \dots, x_n \in X$ even if $\epsilon > 0$ and $n \in \mathbb{N}$ remain fixed. Imposing an upper bound on the rank of T which depends on ϵ and n but not on the specific choice of $x_1, \dots, x_n \in X$ leads to the uniform approximation property. The importance of this concept was recognized when the Local Theory of Banach spaces emerged.

A Banach space X satisfies the uniform approximation property if there exists $\lambda > 0$ so that for every $\epsilon > 0$ and $n \in \mathbb{N}$ there exists $f(\epsilon, n) \in \mathbb{N}$ so that for every $x_1, \dots, x_n \in X$ there exists a linear operator $T : X \to X$ satisfying

$$\begin{aligned} ||T|| &\leq \lambda, \\ ||Tx_i - x_i|| &< \epsilon ||x_i||, \\ \text{rank} T &\leq f(\epsilon, n). \end{aligned}$$

We call $x_1, \dots, x_n \in X$ the UAP data, T is called a resolving operator, and $f(\epsilon, N)$ is called a uniformity function. Our first result establishes the uniform approximation property for the class of L^p spaces. Here the proof for L^∞ turns out to be simplest case.

We use below a technical device called the Auerbach basis. Let X be a finite-dimensional Banach space. Let n be the algebraic dimension of X. Then there exist $b_i \in X$ $(1 \leq i \leq n)$ with $\|b_i\|_X = 1$ so that for every $g \in X$ there exist uniquely determined coefficients $c_i \in \mathbb{R}$ so that

$$g = \sum_{i=i}^{n} c_i b_i \quad \text{and} \quad |c_i| \leq \|g\|_X.$$

For a proof showing the existence of such a basis see for instance [213] Lemma II.E.11.

Theorem 5.3.1. *The Banach spaces $L^p, 1 \leq p \leq \infty$, satisfy the uniform approximation property.*

Proof. Let $\epsilon > 0$. Let $x_1, \dots, x_n \in L^p$ be linearly independent. Then choose an Auerbach basis for $\text{span}\{x_1, \dots, x_n\}$, call it $b_i, i \leq n$, and define

$$f(x) = \sum_{i=1}^{n} |b_i(x)|.$$

Then clearly $|b_i(x)| \leq f(x)$, and $||f||_p \leq n$. For $i \leq n$ fixed there exists $m \leq (2/\epsilon)$, pairwise disjoint subsets of the unit interval,

$$G_{(i,1)}, \ldots, G_{(i,m)},$$

and coefficients $a_j, j \leq m$, such that the following pointwise estimate holds for every $x \in [0,1]$,

$$\left| b_i(x) - f(x) \sum_{j=1}^{m} a_j 1_{G_{(i,j)}}(x) \right| \leq \epsilon f(x).$$

Let $\mathcal{F}$ be the σ-algebra generated by the sets $\{G_{(i,j)} : i \leq n \text{ and } j \leq m(i,\epsilon)\}$. Let $\mathcal{A}$ denote the atoms of $\mathcal{F}$. Note that the cardinality of $\mathcal{A}$ is $\leq (2/\epsilon)^n$. For $i \leq n$ and $A \in \mathcal{A}$ there exist coefficients $a_A^{(i)}$, such that

$$\left| b_i(x) - \sum_{A \in \mathcal{A}} a_A^{(i)} 1_A(x) f(x) \right| \leq \epsilon f(x).$$

Raising this estimate to the power p and integrating (respectively forming the essential supremum) gives the norm approximation

$$\left\| b_i - \sum_{A \in \mathcal{A}} a_A^{(i)} 1_A f \right\|_p \leq \epsilon n.$$

(Recall that $||f||_p \leq n$.) Now fix $g \in \operatorname{span}\{x_1, \ldots, x_n\}$. Expanding g using the Auerbach basis gives

$$g = \sum_{i=1}^{n} c_i b_i,$$

where $|c_i| \leq ||g||_p$. Put

$$\alpha_A = \sum_{i=1}^{n} c_i a_A^{(i)}.$$

Then estimate the degree of approximation,

$$\begin{aligned} \left\| g - \sum_{A \in \mathcal{A}} \alpha_A 1_A f \right\|_p &\leq \left\| g - \sum_{i=1}^{n} c_i \sum_{A \in \mathcal{A}} a_A^{(i)} 1_A f \right\|_p \\ &\leq \sum_{i=1}^{n} |c_i| \left\| b_i - \sum_{A \in \mathcal{A}} a_A^{(i)} 1_A f \right\|_p \\ &\leq 2\epsilon n \sum_{i=1}^{n} |c_i| \\ &\leq 2\epsilon n^2 ||g||_p. \end{aligned}$$

We use the collection of atoms $\mathcal{A}$ to define the resolving operator

$$P(y) = \sum_{A\in\mathcal{A}} \left[\int_A y f^{p-1}\right] \frac{1_A f}{\int_A f^p}.$$

Let q be the Hölder conjugate index satisfying $(p-1)q = p$ or $p/q = p-1$. Then $(\int_A f^{(p-1)q})^{p/q} = (\int_A f^p)^{p-1}$. Hence

$$\int_A y f^{p-1} \leq \left(\int_A |y|^p\right)\left(\int_A f^p\right)^{p-1}.$$

As $\mathcal{A}$ is a collection of pairwise disjoint sets it follows that $||P||_p \leq 1$. The dimension of the range of P is the cardinality of $\mathcal{A}$, hence rank$P \leq (2/\epsilon)^n$. The following identity holds for every $A \in \mathcal{A}$,

$$P(1_A f) = 1_A f.$$

Let $g \in \text{span}\{x_1, \ldots, x_n\}$, then its distance in L^p to $\text{span}\{1_A f : A \in \mathcal{A}\}$ is bounded by $2\epsilon n^2$. Hence

$$||Pg - g||_p \leq 2\epsilon n^2 ||g||_p.$$

Thus we verified that P is a resolving operator for the UAP data $x_1, \ldots, x_n$. □

The following subsections are exclusively devoted to proving this theorem of P. W. Jones [102].

Theorem 5.3.2. *BMO satisfies the uniform approximation property.*

Before we turn to a detailed exposition we summarize the basic pattern of the argument. In Chapter 3 we showed that BMO satisfies the following special case of the uniform approximation property. Let $\{\mathcal{B}_i : i \leq N\}$ be disjoint collections of dyadic intervals. Assume that the union $\mathcal{B} = \mathcal{B}_1 \cup \cdots \cup \mathcal{B}_N$ satisfies the Carleson packing condition. Define

$$\varphi_i = \sum_{J\in\mathcal{B}_i} h_J,$$

and let

$$M = ||\varphi_1 + \cdots + \varphi_N||_{\text{BMO}}.$$

Then there exists an orthogonal projection $P : \text{BMO} \to \text{BMO}$ such that

$$\begin{aligned} ||P||_{\text{BMO}} &\leq 4, \\ P\varphi_i &= \varphi_i, \\ \text{Rank } P &\leq f(N, M). \end{aligned} \tag{5.3.1}$$

The orthogonal projection satisfying (5.3.1) was constructed in Theorem 3.2.3 of Chapter 3.

We show below that BMO satisfies the following property which is clearly weaker than the uniform approximation property: Let $\epsilon > 0$ and $N \in \mathbb{N}$, then there exists $f(N, \epsilon)$ so that the following condition holds. Given $\varphi_1, \ldots, \varphi_N \in$ BMO there exists an operator $R :$ BMO $\to$ BMO so that

$$||R|| \leq 10,$$
$$||R\varphi_j - \varphi_j|| \leq \epsilon||\varphi_j||,$$

and so that $R :$ BMO $\to$ BMO factors through ℓ^∞ as follows,

$$\begin{array}{ccc} \text{BMO} & \xrightarrow{R} & \text{BMO} \\ E \searrow & & \nearrow F \\ & \ell^\infty & \end{array} \qquad \text{with} \quad ||E|| \cdot ||F|| \leq f(N, \epsilon). \tag{5.3.2}$$

The factorization (5.3.2) will be obtained in the next subsections. In its proof we will encounter new ideas resulting in a powerful combinatorial technique.

Naturally the methods by which we obtain (5.3.1) and (5.3.2) will be important components in the proof of the uniform approximation property of BMO. We will see below that in order to obtain the complete proof that BMO has UAP we only have to merge the ideas that gave (5.3.1) and (5.3.2). It seems to be quite remarkable that proving a special case of UAP and establishing a property that is weaker than UAP suffices to show that indeed BMO satisfies the uniform approximation property.

Splitting the Haar support

Here we show that without loss of generality we may always assume that the UAP data are disjointly supported over the Haar system. Specifically we show that if one can find a resolving operator for UAP data with pairwise disjoint Haar support, then one can find a resolving operator for any UAP data. The reduction is based on the disjointification procedure introduced to show that L^p has the uniform approximation property. We apply it here to obtain functions with disjoint Haar support.

Proposition 5.3.3. *Let $\epsilon > 0$ and let $x_1, \ldots, x_n \in BMO$. Then there exists $N \leq (2n^3/\epsilon)^n$, and $\varphi_1, \ldots, \varphi_N \in BMO$ disjointly supported over the Haar system so that for $g \in \operatorname{span}\{x_1, \ldots, x_n\}$,*

$$\left\| g - \sum_{j=1}^{N} \alpha_j \varphi_j \right\|_{BMO} \leq \epsilon ||g||_{BMO},$$

where $\alpha_j \in \mathbb{R}$ are suitably chosen coefficients.

Proof. Let $b_i, i \leq n$ be an Auerbach basis for $\operatorname{span}\{x_1, \ldots, x_n\}$. Expand b_i in its Haar series,

$$b_i = \sum b_i(I) h_I.$$

For every dyadic interval I we form the coefficients

$$f(I) = \sum_{i=1}^{n} |b_i(I)|,$$

inducing the BMO function $f(x) = \sum f(I)h_I(x)$. Note that $f(I) \leq n$ and $||f||_{\mathrm{BMO}} \leq n$. For each $i \leq n$, there exist $m \leq (2/\epsilon)$, pairwise disjoint collections of dyadic intervals $\mathcal{G}_{(i,j)}, j \leq m$ and constants $a_1, \ldots, a_m$ such that for $I \in \mathcal{D}$,

$$\left| b_i(I) - f(I) \sum_{j=1}^{m} a_j G_{(i,j)}(I) \right| \leq \epsilon f(I),$$

where we denoted by $G_{(i,j)}$ the indicator function of the collection $\mathcal{G}_{(i,j)}$.

$$G_{(i,j)}(I) = \begin{cases} 1 & \text{for } I \in \mathcal{G}_{(i,j)}, \\ 0 & \text{for } I \notin \mathcal{G}_{(i,j)}. \end{cases}$$

Defined that way $G_{(i,j)}$ is a function on $\mathcal{D}$, taking the values $\{0, 1\}$. Let $\mathcal{A}$ denote the atoms of the σ-algebra generated by the functions $\{G_{(i,j)} : i \leq n, j \leq m(i, \epsilon)\}$. Observe that each $A \in \mathcal{A}$ is a collection of dyadic intervals. It is obtained by intersecting some of the collections $\{\mathcal{G}_{(i,j)} : i \leq n, j \leq m(i, \epsilon)\}$. Hence for each b_i there exist coefficients $\{a_A^{(i)} : A \in \mathcal{A}\}$, such that

$$\left| b_i(I) - \sum_{A \in \mathcal{A}} a_A^{(i)} f(I) 1_A(I) \right| \leq \epsilon f(I).$$

Next we pass from coefficients to functions. For $A \in \mathcal{A}$ we let

$$\varphi_A = \sum_{I \in \mathcal{D}} 1_A(I) f(I) h_I.$$

Then we have that

$$\left\| \sum_{I \in \mathcal{D}} b_i(I) h_I - \sum_{A \in \mathcal{A}} a_A^{(i)} \varphi_A \right\|_{\mathrm{BMO}} \leq 2\epsilon \left\| \sum_{I \in \mathcal{D}} f(I) h_I \right\|_{\mathrm{BMO}}.$$

Take $g \in \operatorname{span}\{x_1, \ldots, x_n\}$. Expand g using the Auerbach basis, then

$$g = \sum_{i=1}^{n} c_i b_i,$$

with $|c_i| \leq ||g||_{\mathrm{BMO}}$. Next fix $A \in \mathcal{A}$ and form the coefficient

$$\alpha_A = \sum_{i=1}^{n} c_i a_A^{(i)}.$$

This gives the approximation

$$\begin{aligned}\left\|g-\sum_{A\in\mathcal{A}}\alpha_A\varphi_A\right\|_{\mathrm{BMO}} &\le \left\|g-\sum_{i=1}^{n}c_i\sum_{A\in\mathcal{A}}a_A^{(i)}\varphi_A\right\|_{\mathrm{BMO}}\\ &\le \sum_{i=1}^{n}|c_i|\left\|b_i-\sum_{A\in\mathcal{A}}a_A^{(i)}\varphi_A\right\|_{\mathrm{BMO}}\\ &\le 2\epsilon n\sum_{i=1}^{n}|c_i|\\ &\le 2\epsilon n^2\|g\|_{\mathrm{BMO}}.\end{aligned}$$

Note that the cardinality of $\mathcal{A}$ is $\le (2/\epsilon)^n$. It remains to replace ϵ by $n^{-2}\epsilon$, and to relabel the functions $\varphi_A, A\in\mathcal{A}$ as

$$\varphi_1,\dots,\varphi_N,$$

with $N\le (2n^3/\epsilon)^n$. $\square$

UAP data with large Haar coefficients

We have reduced the problem to the case when the data $\varphi_1,\dots,\varphi_N\in\mathrm{BMO}$ are disjointly supported over the Haar system. We will first prove the uniform approximation property under the additional hypothesis that if a Haar coefficient does not vanish, then it satisfies the lower bound

$$|\langle\varphi_j,h_J\rangle| > \gamma|J|\cdot\|\varphi_j\|_{\mathrm{BMO}}.$$

Let $\mathcal{K}$ denote the Haar support of $\varphi_1+\cdots+\varphi_N$. The lower bound on the Haar coefficients implies an upper bound for the Carleson constant of $\mathcal{K}$, namely $[\![\mathcal{K}]\!]\le N\gamma^{-2}$. Now translate the intrinsic condition on the Haar coefficients into a statement about factorization of operators through ℓ^∞. Let R be the orthogonal projection onto $\operatorname{span}\{h_J : J\in\mathcal{K}\}$. Then $\|R\|\le 1$ and $R\varphi_j=\varphi_j$ and R factors through ℓ^∞ as follows,

$$\begin{array}{ccc}\mathrm{BMO} & \xrightarrow{R} & \mathrm{BMO}\\ E\searrow & & \nearrow F\\ & \ell^\infty & \end{array}$$

where the operators E,F are defined by

$$E\left(\sum a_I h_I\right)=(a_I),\quad\text{and}\quad F((a_I))=\sum\{a_I h_I : I\in\mathcal{K}\}.$$

They satisfy the norm estimates, $\|E\|\le 1$, and $\|F\|\le [\![\mathcal{K}]\!]^{1/2}$. Below the existence of a factorization through ℓ^∞ will indicate that the combinatorial methods of Theorem 3.2.1 are applicable. This is the value of re-expressing our hypothesis.

Theorem 5.3.4. *Let $N \in \mathbb{N}$, $\epsilon > 0$ and $1 > \gamma > 0$ be given. Let $\varphi_1, \ldots, \varphi_N \in BMO$ be disjointly supported over the Haar system. Assume that*

$$|\langle \varphi_i, h_I \rangle| > \gamma ||\varphi_i||_{BMO} |I|,$$

for each I in the Haar support of φ_i, and each $i \leq N$. Then there exists an orthogonal projection P such that,

$$\begin{aligned} ||P||_{BMO} &\leq 4, \\ ||P(\varphi_i) - \varphi_i||_{BMO} &\leq \epsilon ||\varphi_i||_{BMO} \quad \text{for} \quad i \leq N, \\ \operatorname{rank}(P) &\leq f(N, \epsilon, \gamma). \end{aligned}$$

Proof. The proof consists of a straightforward reduction to the deep combinatorial result of Theorem 3.2.3 in Chapter 3. Fix $\varphi_i \in$ BMO. Assume without loss of generality that the Haar coefficients of φ_i are non-negative and that $||\varphi_i||_{\text{BMO}} = 1$. Hence for I in the Haar support of φ_i the following estimate holds,

$$\gamma |I| < \langle \varphi_i, h_I \rangle \leq |I|.$$

Hence there exists an integer $1 \leq q \leq -\log \gamma / \log(1 + \epsilon)$ such that

$$(1 + \epsilon)^{-q} |I| < \langle \varphi_i, h_I \rangle \leq (1 + \epsilon)^{-q+1} |I|.$$

Now we change our point of view. Let $M = -\log \gamma / \log(1 + \epsilon)$, and fix $1 \leq q \leq M$. Define $\mathcal{B}_{(i,q)}$ to be the collection of dyadic intervals for which

$$(1 + \epsilon)^{-q} |I| < \langle \varphi_i, h_I \rangle \leq (1 + \epsilon)^{-q+1} |I|. \tag{5.3.3}$$

By definition, the collections $\mathcal{B}_{(i,q)}$ are pairwise disjoint and for i fixed, $\mathcal{B}_{(i,q)}$ is contained in the Haar support of φ_i. Let,

$$\begin{aligned} b_{(i,q)} &= \sum \{h_J : J \in \mathcal{B}_{(i,q)}\}, \\ a_q &= (1 + \epsilon)^{-q}. \end{aligned}$$

By (5.3.3) for $I \in \mathcal{B}_{(i,q)}$ we obtain that $|a_q|I| - \langle \varphi_i, h_I \rangle| \leq \epsilon a_q |I|$, and also $a_q |I| \leq \langle \varphi_i, h_I \rangle$. Hence we get

$$|a_q - \langle \varphi_i, \frac{h_I}{|I|} \rangle| \leq \epsilon \langle \varphi_i, \frac{h_I}{|I|} \rangle.$$

As $\|\varphi_i\|_{\text{BMO}} \leq 1$, the following approximation holds,

$$||\varphi_i - \sum_{q=1}^{M} a_q b_{(i,q)}||_{\text{BMO}} \leq \epsilon.$$

Recall that $\mathcal{B}_{(i,q)}$ is contained in the Haar support of φ_i. Hence it satisfies the γ^{-2}-Carleson condition. Now we apply Theorem 3.2.3 of Chapter 3 to the disjoint collections of dyadic intervals $\{\mathcal{B}_{(i,q)} : i \leq N, \quad \text{and} \quad q \leq M\}$. It asserts that there exists an orthogonal projection P such that

$$\begin{aligned} ||P||_{\mathrm{BMO}} &\leq 4, \\ ||P(\varphi_i) - \varphi_i||_{\mathrm{BMO}} &\leq 2\epsilon||\varphi_i||_{\mathrm{BMO}} \quad \text{for} \quad i \leq N, \\ \text{rank} \quad (P) &\leq f(N, \epsilon, \gamma). \end{aligned}$$

□

UAP data with small Haar coefficients

We will now present the proof of Theorem 5.3.2 under the additional hypothesis that the functions φ_j have disjoint Haar support, and that the Haar coefficients of φ_j are sufficiently small. For $N \in \mathbb{N}$ and $\epsilon > 0$ we define the rather small number

$$\gamma(N, \epsilon) = \frac{\epsilon^3}{N^3} \left(\log(\frac{N^2}{\epsilon}) \right)^{-N}.$$

Theorem 5.3.5. *Let $\varphi_1, \ldots \varphi_N \in BMO$, let $\epsilon > 0$. Assume that $\varphi_1, \ldots \varphi_N$ are disjointly supported over the Haar system, and that each Haar coefficient satisfies*

$$|\langle \varphi_j, h_I \rangle| \leq \gamma |I| ||\varphi_j||_{BMO},$$

where $\gamma < \gamma(N, \epsilon)$. Then there exists an operator T on BMO such that

$$\begin{aligned} ||T||_{BMO} &\leq 10, \\ ||T\varphi_j - \varphi_j||_{BMO} &\leq \epsilon ||\varphi_j||_{BMO}, \\ \text{rank}(T) &\leq f(N, \epsilon). \end{aligned}$$

There are two independent components in the proof of Theorem 5.3.5. The first component is the construction of a remarkable operator R that satisfies the following three conditions,

$$\begin{aligned} ||R|| &\leq 10, \\ ||R\varphi_j - \varphi_j|| &\leq \epsilon ||\varphi_j||, \end{aligned}$$

and R can be factored through ℓ^∞,

$$\begin{array}{ccc} \mathrm{BMO} & \xrightarrow{R} & \mathrm{BMO} \\ E \searrow & & \nearrow F \\ & \ell^\infty & \end{array} \quad \text{with} \quad ||E|| \cdot ||F|| \leq f(N, \epsilon). \tag{5.3.4}$$

In the second part we are guided by the proof of Theorem 5.3.4. There we observed that the deep combinatorial methods developed for the proof of Theorem 3.2.3 enable us to transform a factorization diagram like (5.3.4) into a finite rank condition with good upper bounds.

We recall the definitions for the localized square function and for blocks of dyadic intervals in $\mathcal{D}$. Let $f \in L^2$, with Haar expansion $f = \sum a_I h_I$. Given a collection of dyadic intervals $\mathcal{H}$, then the localized square function $S(f|\mathcal{H})$ is defined by the equation

$$S^2(f|\mathcal{H}) = \sum_{I \in \mathcal{H}} a_I^2 h_I^2.$$

For a given dyadic interval I we define $I \cap \mathcal{H}$ to be the collection of dyadic intervals

$$I \cap \mathcal{H} = \{J \in \mathcal{H} : J \subseteq I\}.$$

Recall that a collection $\mathcal{L}(I) \subseteq \mathcal{D}$ is a block of dyadic intervals if I is the only maximal interval of $\mathcal{L}(I)$ and if the following connectedness property holds true. Let $J \in \mathcal{L}(I)$ and let $K \in \mathcal{D}$, then

$$J \subseteq K \subseteq I \text{ implies } K \in \mathcal{L}(I).$$

Repeatedly in the course of proving Theorem 5.3.5 we use the following simple and general observation relating blocks of dyadic intervals, the Carleson packing condition, the localized square function and the norm in BMO through a simple inequality. Let $\{\mathcal{L}(I) : I \in \Omega\}$ be a decomposition of $\mathcal{D}$ into disjoint blocks of dyadic intervals so that the index set Ω satisfies the Carleson packing condition. Let $f \in L^2$. Assume that for the localized square functions $S(f|\mathcal{L}(I))$ there hold bounds as follows,

$$\|S^2(f|\mathcal{L}(I))\|_\infty \le a \quad \text{for} \quad I \in \Omega,$$

and

$$\int_I S^2(f|\mathcal{L}(I)) \le b|I| \quad \text{for} \quad I \in \Omega.$$

Then $f \in \mathrm{BMO}$ and the norm of f in BMO satisfies an upper bound in terms of a, b and $[\![\Omega]\!]$,

$$\|f\|_{\mathrm{BMO}}^2 \le a + b[\![\Omega]\!]. \tag{5.3.5}$$

This estimate is useful when $b << a$. We verify (5.3.5) by showing that for $J \in \mathcal{D}$,

$$\int_J S^2(f|Q(J)) \le a|J| + b[\![\Omega]\!]|J|,$$

where $Q(J) = J \cap \mathcal{D}$. Given J, choose $I_0 \in \Omega$ such that $J \in \mathcal{L}(I_0)$. Let $\{I_p\}$ be an enumeration of the maximal intervals in $J \cap \Omega$, (and assume that $\max J \cap \Omega$ is infinite). Put

$$\mathcal{F} = J \cap \mathcal{L}(I_0).$$

Then

$$Q(J) = \mathcal{F} \cup \bigcup_{p=1}^{\infty} \bigcup_{I \in I_p \cap \Omega} \mathcal{L}(I)$$

and

$$S^2(f|Q(J)) = S^2(f|\mathcal{F}) + \sum_{p=1}^{\infty} \sum_{I \in I_p \cap \Omega} S^2(f|\mathcal{L}(I)).$$

Integrating over J gives that

$$\begin{aligned} \int_J S^2(f|Q(J)) &\leq a|I| + \sum_{p=1}^{\infty} \sum_{I \in I_p \cap \Omega} b|I| \\ &\leq a|J| + b[\![\Omega]\!] \sum_{p=1}^{\infty} |I_p|. \end{aligned}$$

Since $\sum_{p=1}^{\infty} |I_p| \leq |J|$ the claim (5.3.5) is verified. We close by remarking the preceding observation is the model for several proofs in this section.

Proof of Theorem 5.3.5. Part 1.

In the following diagram we summarize the logical dependence of the parameters we will encounter in the construction of the resolving operator. They will appear in the estimates for the error terms and in upper bounds for the rank of the resolving operator.

$$\begin{array}{ccccc} & & L & \xrightarrow{(5.3.31)} & \gamma \\ & & {\scriptstyle Prop.\ 5.3.7}\uparrow & & \uparrow{\scriptstyle (5.3.31)} \\ N, \epsilon & \xrightarrow{(5.3.18)} & \eta & \xrightarrow{(5.3.51)} & A \end{array} \tag{5.3.6}$$

For $\epsilon > 0$ and $N \in \mathbb{N}$ we choose η, L, A, M, by the formulas

$$\eta = \frac{\epsilon}{16N^2}, \quad L = |\log \eta|^N, \quad A = \frac{N}{\epsilon\eta}, \quad M = AL. \tag{5.3.7}$$

All the above determine $\gamma = \gamma(\epsilon, N)$ by the relation

$$\gamma \leq \frac{\epsilon}{2 + N\eta + 2M}. \tag{5.3.8}$$

We are given the functions $\varphi_1, \ldots, \varphi_N$, disjointly supported over the Haar system. We assume that they are normalized by $||\varphi_j||_{\mathrm{BMO}} = 1$, and satisfy

$$|\langle \varphi_j, h_I \rangle| \leq \gamma |I|,$$

for every dyadic interval I. We begin by analyzing their sum

$$\varphi = \sum_{i=1}^{N} \varphi_i.$$

Here we use again a standard stopping time procedure introduced in Chapter 1.

Proposition 5.3.6. *The dyadic intervals $\mathcal{D}$ can be grouped into disjoint blocks of dyadic intervals $\{\mathcal{L}(I) : I \in \Omega\}$ so that the following conditions hold:*

1. *The index set Ω satisfies the $N\gamma^{-1} + 1$ Carleson condition.*
2. *The square function $S^2(\varphi|\mathcal{L}(I))$ satisfies the pointwise estimate*

$$||S^2(\varphi|\mathcal{L}(I))||_\infty \leq \gamma + \gamma^2. \tag{5.3.9}$$

3. *If $\mathcal{B}(I)$ is any block of dyadic intervals, strictly larger than $\mathcal{L}(I)$, then the square function $S^2(\varphi|\mathcal{B}(I))$ satisfies*

$$||S^2(\varphi|\mathcal{B}(I)||_\infty \geq \gamma + \gamma^2.$$

Proof. We use stopping time arguments on the square function of φ. We let $\mathcal{L}_0 = \mathcal{L}([0,1])$ be the largest block of dyadic intervals such that for any $x \in [0,1]$,

$$S^2(\varphi|\mathcal{L}_0)(x) \leq \gamma + \gamma^2.$$

Now let $\mathcal{K}^1 = \mathcal{D} \setminus \mathcal{L}_0$ and let $\{I_1^1, \ldots, I_j^1, \ldots\}$ be the maximal intervals of $\mathcal{K}^1$. For each I_j^1 we define $\mathcal{L}(I_j^1)$, to be the largest block of dyadic intervals satisfying

$$S^2(\varphi|\mathcal{L}(I_j^1))(x) \leq \gamma + \gamma^2.$$

Then we put $\mathcal{K}^2 = \mathcal{K}^1 \setminus \bigcup_{j=1}^\infty \mathcal{L}(I_j^1)$ and let $\{I_1^2, \ldots, I_j^2, \ldots\}$ be an enumeration of the maximal intervals of $\mathcal{K}^2$. For each of the intervals I_j^2 we define $\mathcal{L}(I_j^2)$ to be the largest block of dyadic intervals such that

$$S^2(\varphi|\mathcal{L}(I_j^2))(x) \leq \gamma + \gamma^2.$$

Continuing in this fashion we obtain the index set Ω as

$$\Omega = [0,1] \cup \bigcup_{n=1}^\infty \bigcup_{j=1}^\infty I_j^n.$$

The decomposition of $\mathcal{D}$ is consequently given by

$$\mathcal{D} = \bigcup\{\mathcal{L}(I) : I \in \Omega\},$$

where $\mathcal{L}(I)$ is the largest block of dyadic intervals satisfying the conditions

$$S^2(\varphi \,|\, \mathcal{L}(I))(x) \leq \gamma + \gamma^2.$$

By the stopping time definition of the blocks $\mathcal{L}(I)$ it is evident that for every block $\mathcal{B}(I)$ strictly larger than $\mathcal{L}(I)$ we obtain

$$\|S^2(\varphi \,|\, \mathcal{B}(I))\|_\infty \geq \gamma + \gamma^2.$$

Hence condition 3) of Proposition 5.3.6 holds true.

Next we verify that Ω satisfies the Carleson packing condition and $[\![\Omega]\!] \leq N\gamma^{-1} + 1$. To this end we fix $I \in \Omega$ and consider the collection

$$\mathcal{B}(I) = \mathcal{L}(I) \cup G_1(I, \Omega).$$

A moment's reflection yields that $\mathcal{B}(I)$ is a block of dyadic intervals, and clearly $\mathcal{B}(I)$ is strictly larger than $\mathcal{L}(I)$. Hence $S^2(\varphi \,|\, \mathcal{B}(I))$ satisfies a lower bound on the set $G_1^*(I, \Omega)$. We have,

$$S^2(\varphi \,|\, \mathcal{B}(I))(x) \geq \gamma + \gamma^2 \quad \text{for} \quad x \in G_1^*(I, \Omega). \tag{5.3.10}$$

Observe that $G_1(I, \Omega)$ is a collection of pairwise disjoint dyadic intervals, and by hypothesis the Haar coefficients of φ are so small that

$$\langle \varphi, \frac{h_J}{|J|} \rangle^2 \leq \gamma^2 \quad \text{for} \quad J \in G_1(I, \Omega). \tag{5.3.11}$$

Combining (5.3.10) and (5.3.11) yields a crucial lower bound for $S^2(\varphi \,|\, \mathcal{L}(I))$ on the set $G_1^*(I, \Omega)$,

$$S^2(\varphi \,|\, \mathcal{L}(I))(x) \geq \gamma \quad \text{for} \quad x \in G_1^*(I, \Omega). \tag{5.3.12}$$

By integrating (5.3.12) over I we find that

$$\int_I S^2(\varphi \,|\, \mathcal{L}(I)) \geq \gamma \sum_{K \in G_1(I, \Omega)} |K|. \tag{5.3.13}$$

Recall that $\varphi = \varphi_1 + \cdots + \varphi_N$ where the functions φ_i are disjointly supported over the Haar system and satisfy $\|\varphi_i\|_{\mathrm{BMO}} \leq 1$. Hence, by Proposition 5.1.1, $\|\varphi\|_{\mathrm{BMO}} \leq \sqrt{N}$. It follows that for $J \in \Omega$,

$$\int_J S^2(\varphi \,|\, Q(J)) \leq N|J|, \tag{5.3.14}$$

where $Q(J) = \{L \in \mathcal{D} : L \subseteq J\}$. By exploiting (5.3.13) and (5.3.14) we find the upper estimate for the Carleson constant of Ω as follows.

$$\begin{aligned}
\gamma \sum_{I \in \Omega \cap J} |I| &= \gamma |J| + \sum_{I \in \Omega \cap J} \gamma \sum_{K \in G_1(I, \Omega)} |K| \\
&\leq \gamma |J| + \sum_{I \in \Omega \cap J} \int_I S^2(\varphi \,|\, \mathcal{L}(I)) \\
&\leq \gamma |J| + \int_J S^2(\varphi \,|\, Q(J)) \\
&\leq \gamma |J| + N|J|.
\end{aligned}$$

$\square$

Our next aim is to display more of the structure of the index set Ω. We extract a sub-collection $\mathcal{E}_0$ of Ω by defining that $I \in \Omega$ belongs to $\mathcal{E}_0$ if

$$\int S^2(\varphi|\mathcal{L}(I)) \leq \frac{\gamma}{2}|I|. \tag{5.3.15}$$

We fix $j \in \{1, \ldots, N\}$ and define $\mathcal{E}_j$ to be the collection of dyadic intervals $I \in \Omega \backslash \mathcal{E}_0$ such that

$$\int S^2(\varphi_j|\mathcal{L}(I)) \leq \eta\gamma|I|. \tag{5.3.16}$$

Finally we define a perturbation ψ_j of φ_j by subtracting the contribution coming from $\mathcal{E}_0 \cup \mathcal{E}_j$,

$$\psi_j = \varphi_j - \sum_{I \in \mathcal{E}_0 \cup \mathcal{E}_j} \sum_{J \in \mathcal{L}(I)} \langle \varphi_j, h_J \rangle \frac{h_J}{|J|}. \tag{5.3.17}$$

In the next proposition we show several things. First we prove that ψ_j is close to φ_j. Second we decompose the intervals $\Omega \setminus \mathcal{E}_0$ into groups $\{\mathcal{C}_i\}$ satisfying the following homogeneity conditions: Up to small relative discretisation errors, the averages

$$\frac{1}{|I|} \int S^2(\psi_j|\mathcal{L}(I)), \quad I \in \mathcal{C}_i$$

depend only on the indices i and j, and **not** on the particular choice of interval $I \in \mathcal{C}_i$.

Recall that the constants $L = L(N, \epsilon), \eta = \eta(N, \epsilon)$ and $\gamma = \gamma(N, \epsilon)$ appearing in the proposition below have been defined earlier in (5.3.7).

Proposition 5.3.7. *$\mathcal{E}_0$ satisfies the 2-Carleson condition. For $1 \leq j \leq N$,*

$$||\varphi_j - \psi_j||_{BMO} \leq \sqrt{4\gamma + 2\eta N}. \tag{5.3.18}$$

The index set $\Omega \setminus \mathcal{E}_0$ can be grouped as

$$\Omega \setminus \mathcal{E}_0 = \mathcal{C}_1 \cup \cdots \cup \mathcal{C}_L$$

so that the following holds.

1. *For $i \leq L$ and $j \leq N$ there exist coefficients $a_{ij} \geq 0$ such that*

$$\frac{a_{ij}}{2} \leq \frac{1}{|I|} \int S^2(\psi_j|\mathcal{L}(I)) \leq a_{ij}, \quad \text{for} \quad I \in \mathcal{C}_i. \tag{5.3.19}$$

2. *If a coefficient a_{ij} is non-zero, then it satisfies the inequalities*

$$\eta\gamma \leq a_{ij} \leq 2\gamma. \tag{5.3.20}$$

Proof. We start by showing that the collection $\mathcal{E}_0$ satisfies the 2-Carleson condition. Let $I_0 \in \mathcal{E}_0$ and let $\{I_k\}$ be an enumeration of the intervals in the first generation $G_1(I_0, \mathcal{E}_0)$. So $\{I_k\}$ is the list of the maximal intervals of $\mathcal{E}_0$ which are contained in I_0. In the course of proving Proposition 5.3.6 we established (5.3.13) asserting that the following lower bound holds,

$$\int_{I_0} S^2(\varphi|\mathcal{L}(I_0)) \geq \gamma \sum_{k=1}^{\infty} |I_k|.$$

On the other hand as $I_0 \in \mathcal{E}_0$ we have

$$\int S^2(\varphi|\mathcal{L}(I_0)) \leq \frac{\gamma}{2}|I_0|.$$

The lower estimate and the upper estimate are matching so that γ cancels and

$$\sum |I_k| \leq \frac{1}{2}|I_0|.$$

By summing a geometric series $\sum_{n=0}^{\infty} q^n$ with $q = \frac{1}{2}$ we obtain the upper bound 2 for the Carleson constant of $\mathcal{E}_0$. Hence, by the square function estimate (5.3.9), the norm in BMO of

$$\sum_{I \in \mathcal{E}_0} \sum_{J \in \mathcal{L}(I)} \langle \varphi_j, h_J \rangle \frac{h_J}{|J|} \tag{5.3.21}$$

is bounded by $\leq \sqrt{2(\gamma + \gamma^2)}$.

Let $j \leq N$, and fix $I \in \mathcal{E}_j$. Let $J \in \mathcal{E}_j \cap I$. Recall that $\mathcal{E}_j \subseteq \Omega \setminus \mathcal{E}_0$. Thus $J \notin \mathcal{E}_0$. Consequently the following two inequalities hold for $\mathcal{L}(J)$,

$$\int S^2(\varphi|\mathcal{L}(J)) \geq \frac{\gamma}{2}|J| \quad \text{and} \quad \int S^2(\varphi_j|\mathcal{L}(J)) \leq \eta\gamma|J|.$$

We combine the two inequalities into one single estimate, namely

$$2\eta \int S^2(\varphi|\mathcal{L}(J)) \geq \int S^2(\varphi_j|\mathcal{L}(J)).$$

Next sum over $J \in \mathcal{E}_j \cap I$ and take into account that the BMO norm of φ is bounded by $\sqrt{N}$.

$$\begin{aligned} \sum_{J \in \mathcal{E}_j \cap I} \int S^2(\varphi_j|\mathcal{L}(J)) &\leq 2\eta \sum_{J \in \mathcal{E}_j \cap I} \int S^2(\varphi|\mathcal{L}(J)) \\ &\leq 2\eta \int S^2(\varphi|Q(I)) \\ &\leq 2\eta N|I|. \end{aligned} \tag{5.3.22}$$

Combining the global estimates (5.3.22) with the local square function estimate (5.3.9) proves that the norm in BMO of the term

$$\sum_{I\in\mathcal{E}_j}\sum_{J\in\mathcal{L}(I)}\langle\varphi_j,h_J\rangle\frac{h_J}{|J|} \tag{5.3.23}$$

is bounded by $\sqrt{\gamma+\gamma^2+2\eta N}$. The estimates for (5.3.23) and (5.3.21) imply (5.3.18).

In the next step of the proof we group the intervals in $\Omega\setminus\mathcal{E}_0$. Fix $j\in\{1,\dots,N\}$. Then recall that we produced ψ_j from φ_j by removing from the Haar support of φ_j the following blocks of intervals $\{\mathcal{L}(I): I\in\mathcal{E}_0\cup\mathcal{E}_j\}$. Recall that $I\in\mathcal{E}_j$, if the condition (5.3.16) holds. The following dichotomy is the result of removing the blocks $\{\mathcal{L}(I): I\in\mathcal{E}_j\}$. Either

$$\int S^2(\psi_j|\mathcal{L}(I))>\eta\gamma|I|,$$

or

$$\int S^2(\psi_j|\mathcal{L}(I))=0.$$

Hence for every $I\in\Omega\setminus\mathcal{E}_0$ there exist $n_j\in\{-1,0,1,2,\dots,|\log_2\eta|,\infty\}$ such that

$$\frac{2^{-n_j}}{2}\gamma<\frac{1}{|I|}\int S^2(\psi_j|\mathcal{L}(I))\le 2^{-n_j}\gamma.$$

Now we change our point of view and fix a vector

$$(n_1,\dots,n_N),$$

with $n_j\in\{-1,0,1,2,1,\dots,-\log_2\eta,\infty\}$. Then we define the collection $\mathcal{C}_{(n_1,\dots,n_N)}$ to consist of all dyadic intervals $I\in\Omega\setminus\mathcal{E}_0$ which satisfy the set of conditions

$$\frac{2^{-n_j}}{2}\gamma<\frac{1}{|I|}\int S^2(\psi_j|\mathcal{L}(I))\le 2^{-n_j}\gamma,$$

for any $j\in\{1,\dots,N\}$. Hence for any given interval $I\in\mathcal{C}_{(n_1,\dots,n_N)}$ the index

$$(n_1,\dots,n_N)$$

of the collection stores the discretized values of the integrals

$$\frac{1}{|I|}\int S^2(\psi_j|\mathcal{L}(I)),$$

$j\in\{1,\dots,N\}$. The relative discretization error is bounded by a factor of 2. By counting all possible choices of indices $(n_1,\dots,n_N)$, we see that there are at most $L = L(N,\eta)$ different collections $\mathcal{C}_{(n_1,\ldots,n_N)}$. For simplicity of notation we enumerate them as

$$\mathcal{C}_1, \ldots, \mathcal{C}_L.$$

By construction these collections satisfy the conclusions (5.3.19) and (5.3.20) of the Proposition 5.3.7. Finally we remark that no interval in $\Omega \setminus \mathcal{E}_0$ has been attributed to more than one of the groups $\mathcal{C}_i$, and that none of the intervals has been overlooked by our construction. □

Below, we ignore the contribution to φ_j coming from the blocks indexed by intervals in $\mathcal{E}_0$ or $\mathcal{E}_j$. This strategy is justified by (5.3.18) showing that the error terms resulting from discarding the index sets $\mathcal{E}_0$ and $\mathcal{E}_j$, are sufficiently small in BMO. Now we continue with a more detailed study of the index sets $\mathcal{C}_i$ and with the analysis of the functions

$$\sum_{J \in \mathcal{L}(I)} \langle \psi_j, h_J \rangle \frac{h_J}{|J|}, \quad I \in \mathcal{C}_i.$$

Fix the constant A and one of the index sets $\mathcal{C}_i$. Define $\mathcal{M}_i$ to be the collection of all intervals $I \in \mathcal{C}_i$, which satisfy,

$$\sum_{J \in G_A(I,\mathcal{C}_i)} |J| \geq \frac{1}{4}|I|. \tag{5.3.24}$$

Proposition 5.3.8. *The collection of dyadic intervals $\mathcal{C}_i \setminus \mathcal{M}_i$ satisfies the $2A$-Carleson packing condition. For each $I \in \mathcal{M}_i$, the collection*

$$\mathcal{R}(I) = \bigcup_{n=0}^{A} \bigcup_{J \in G_n(I,\mathcal{C}_i)} \mathcal{L}(J) \tag{5.3.25}$$

gives the estimate

$$\int S^2(\psi_j|\mathcal{R}(I)) \geq \frac{A}{16} \int S^2(\psi_j|\mathcal{L}(I)) \quad \textit{for} \quad 1 \leq j \leq N.$$

Proof. The inclusion $\{\mathcal{C}_i \setminus \mathcal{M}_i\} \subseteq \mathcal{C}_i$ implies that

$$\sum_{J \in G_A(I,\mathcal{C}_i \setminus \mathcal{M}_i)} |J| \leq \sum_{J \in G_A(I,\mathcal{C}_i)} |J|,$$

for every dyadic interval I. Hence by definition of $\mathcal{M}_i$ the following estimate holds for $I \in \mathcal{C}_i \setminus \mathcal{M}_i$,

$$\sum_{J \in G_A(I,\mathcal{C}_i \setminus \mathcal{M}_i)} |J| \leq \frac{1}{4}|I|. \tag{5.3.26}$$

An immediate consequence of (5.3.26) is that each of the collections $\mathcal{C}_i \setminus \mathcal{M}_i$ satisfies the $2A$-Carleson packing condition.

Now we fix $I \in \mathcal{M}_i$ and turn to the relation between the collections $\mathcal{R}(I)$ and $\mathcal{L}(I)$. Let $j \leq N$. The left-hand side estimate of the homogeneity conditions (5.3.19) in Proposition 5.3.7, and the defining condition (5.3.24) give the lower bound

$$\begin{aligned}\int S^2(\psi_j|\mathcal{R}(I)) &= \sum_{n=0}^{A} \sum_{J \in G_n(I,\mathcal{C}_i)} \int S^2(\psi_j|\mathcal{L}(J)) \\ &\geq \frac{1}{2} \sum_{n=0}^{A} \sum_{J \in G_n(I,\mathcal{C}_i)} a_{i,j}|J| \\ &\geq \frac{Aa_{i,j}}{8}|I|.\end{aligned} \tag{5.3.27}$$

Conversely the right-hand side of the homogeneity conditions (5.3.19) in Proposition 5.3.7 is just the estimate

$$\int S^2(\psi_j|\mathcal{L}(I)) \leq a_{i,j}|I|. \tag{5.3.28}$$

Combining the inequalities (5.3.27) and (5.3.28) gives the desired estimate between $\mathcal{R}(I)$ and $\mathcal{L}(I)$. Namely we obtain that

$$\int S^2(\psi_j|\mathcal{R}(I)) \geq \frac{A}{16} \int S^2(\psi_j|\mathcal{L}(I)). \qquad \square$$

For $i \leq L$ the collection $\mathcal{M}_i \subseteq \mathcal{C}_i$ was introduced by (5.3.24). Put

$$\mathcal{K} = \mathcal{M}_1 \cup \cdots \cup \mathcal{M}_L. \tag{5.3.29}$$

Recall that $\mathcal{M}_i \subseteq \Omega$. Hence by Proposition 5.3.6, $\mathcal{K}$ satisfies the Carleson packing condition with

$$[\![\mathcal{K}]\!] \leq N\gamma^{-1} + 1.$$

We define the error function E_j by the equation

$$\varphi_j = \sum_{I \in \mathcal{K}} \sum_{J \in \mathcal{L}(I)} \langle \psi_j, h_J \rangle \frac{h_J}{|J|} + E_j. \tag{5.3.30}$$

Next we estimate the norm in BMO of the error term E_j (this clears the road for part two of the proof of Theorem 5.3.5). Proposition 5.3.6 and the first assertion of Proposition 5.3.8 are the main ingredients in showing that the error term E_j is small.

Proposition 5.3.9. *Let $j \leq N$, then E_j satisfies the estimate*

$$||E_j||_{BMO} \leq \sqrt{6\gamma + 2\eta N + \gamma 2AL}. \tag{5.3.31}$$

Proof. First we identify the components which constitute the error term E_j. Let

$$\mathcal{E} = \mathcal{C}_1 \setminus \mathcal{M}_1 \cup \cdots \cup \mathcal{C}_L \setminus \mathcal{M}_L, \tag{5.3.32}$$

and define

$$z_j = \sum_{I\in\mathcal{E}} \sum_{J\in\mathcal{L}(I)} \langle \psi_j, h_J \rangle \frac{h_J}{|J|}.$$

Then the following identity holds,

$$E_j = \varphi_j - \psi_j + z_j. \tag{5.3.33}$$

Note that $\varphi_j - \psi_j$ and z_j, are disjointly supported over the Haar system. Hence by Proposition 5.1.1,

$$\|E_j\|^2_{\mathrm{BMO}} \leq \|\varphi_j - \psi_j\|^2_{\mathrm{BMO}} + \|z_j\|^2_{\mathrm{BMO}}.$$

Proposition 5.3.7 gives that $\|\varphi_j - \psi_j\|^2_{\mathrm{BMO}} \leq 4\gamma + 2\eta N$. We turn to estimating z_j. Each of the groups $\mathcal{C}_i \setminus \mathcal{M}_i$ satisfies the $2A$-Carleson condition. Hence by definition (5.3.32) the collection $\mathcal{E}$ satisfies a $2AL$-Carleson condition. Recall that by Proposition 5.3.6 for each $J \in \mathcal{E}$, we have the estimates

$$\begin{aligned}\int S^2(z_j|\mathcal{L}(J)) &\leq \int S^2(\varphi_j|\mathcal{L}(J)) \\ &\leq 2\gamma|J|.\end{aligned}$$

Let $I \in \mathcal{E}$ be given and let $\{J_k\}$ be an enumeration of $I \cap \mathcal{E}$. Then we estimate as follows,

$$\begin{aligned}\int S^2(z_j|Q(I)) &= \sum_{k=1}^{\infty} \int S^2(z_j|\mathcal{L}(J_k)) \\ &\leq \sum_{k=1}^{\infty} 2\gamma|J_k| \\ &\leq 4\gamma AL|I|.\end{aligned} \tag{5.3.34}$$

Combine (5.3.34) with (5.3.9) to obtain that $||z_j||^2_{\mathrm{BMO}} \leq \gamma + \gamma^2 + 4\gamma AL$, as claimed. □

A Provisional Appraisal.

Let us pause here to better see where we stand on our way to proving the uniform approximation property. We form now an operator R which almost reproduces the functions φ_j and whose norm is bounded independent of N. Unfortunately R is missing a good estimate on the algebraic dimension of its range. However the range of R satisfies a strong geometric bound expressed by the Carleson packing condition of the index set $\mathcal{K}$. Given ψ_j and $J \in \mathcal{K}$ we form the functions

$$x_{j,J} = \sum_{I \in \mathcal{L}(J)} \langle \psi_j, h_I \rangle \frac{h_I}{|I|} \quad \text{and} \quad y_{j,J} = \sum_{I \in \mathcal{R}(J)} \langle \psi_j, h_I \rangle \frac{h_I}{|I|}.$$

Note that we used $\mathcal{L}(J)$ to form $x_{j,J}$ and $\mathcal{R}(J)$ (defined in Lemma 5.3.8) for $y_{j,J}$. Define the operator R by the equation

$$Rf = \sum_{j=1}^{N} \sum_{J \in \mathcal{K}} \langle f, y_{j,J} \rangle \frac{x_{j,J}}{||y_{j,J}||_2^2}. \tag{5.3.35}$$

A direct calculation yields that

$$R\psi_j = \psi_j - \sum_{I \in \mathcal{E}} \sum_{J \in \mathcal{L}(I)} \langle \psi_j, h_J \rangle \frac{h_J}{|J|}.$$

Next observe that $R\psi_j = R\varphi_j$, hence

$$R\varphi_j = \varphi_j - E_j,$$

where E_j is the error function defined and estimated in Proposition 5.3.9. Thus we know already that $||R\varphi_j - \varphi_j||_{\mathrm{BMO}} \leq \epsilon$. We show now that $||R||$ is independent of N, and that R can be factored through the space ℓ^∞.

Proposition 5.3.10.

$$||R||_{BMO} \leq 100, \tag{5.3.36}$$

and $R : BMO \to BMO$ admits the following factorization through ℓ^∞,

$$\begin{array}{ccc} BMO & \stackrel{R}{\longrightarrow} & BMO \\ E \searrow & & \nearrow F \\ & \ell^\infty & \end{array} \qquad \text{with} \quad ||E|| \cdot ||F|| \leq f(N, \epsilon). \tag{5.3.37}$$

Proof. We show first the norm estimate for R. Later we define the operators E and F that provide the factorization through ℓ^∞ and prove that the product of their norm can be estimated by a function of ϵ and N.

The estimate for $\|R\|$. Let $f \in$ BMO. Let I be a dyadic interval in the Haar support of Rf. There exists $J_0 \in \mathcal{K}$ such that $I \in \mathcal{L}(J_0)$. Let

$$\mathcal{F} = \{K \in \mathcal{L}(J_0) : K \subseteq I\}. \tag{5.3.38}$$

Let $\{I_p\}$ be an enumeration of the maximal intervals of $\mathcal{K}$ which are strictly contained in I. The following identity holds then for the square function,

$$S^2(Rf|Q(I)) = S^2(Rf|\mathcal{F}) + \sum_{p=1}^{\infty} S^2(Rf|Q(I_p)). \tag{5.3.39}$$

Estimates for the sum over p result from the following three simple observations. First note that the square function taken over $Q(I_p)$ has a very simple form. It is

$$\int S^2(Rf|Q(I_p)) = \sum_{j=1}^{N} \sum_{J\in I_p\cap\mathcal{K}} \langle f, y_{j,J}\rangle^2 \frac{||x_{j,J}||_2^2}{||y_{j,J}||_2^4}. \tag{5.3.40}$$

Second note that the collections $\{\mathcal{R}(J) : J \in \mathcal{K}\}$ are **not** pairwise disjoint. However each dyadic interval is contained in at most A of the collections $\{\mathcal{R}(J) : J \in \mathcal{K}\}$. Hence applying Bessel's inequality gives

$$\sum_{j=1}^{N} \sum_{J\in I_p\cap\mathcal{K}} \langle f, y_{j,J}\rangle^2 ||y_{j,J}||_2^{-2} \le A|I_p| \cdot ||f||_{\mathrm{BMO}}^2. \tag{5.3.41}$$

Third and last, we observe that the following identities hold,

$$\begin{aligned} ||x_{j,J}||_2^2 &= \int S^2(\psi_j|\mathcal{L}(J)), \\ ||y_{j,J}||_2^2 &= \int S^2(\psi_j|\mathcal{R}(J)). \end{aligned} \tag{5.3.42}$$

The integrals appearing on the right in the above equations (5.3.42) are related by the estimate in Proposition 5.3.8. So we find that

$$||x_{j,J}||_2^2 \le \frac{16}{A} ||y_{j,J}||_2^2. \tag{5.3.43}$$

Now we insert the estimate (5.3.43) into the identity (5.3.40). After that we use (5.3.41). This gives

$$\int S^2(Rf|Q(I_p)) \le 32|I_p| \cdot ||f||_{\mathrm{BMO}}^2. \tag{5.3.44}$$

We recall that the intervals $\{I_p\}$ are an enumeration of pairwise disjoint intervals. Indeed the collection of intervals $\{I_p\}$ is just $G_1(I, \mathcal{K})$. Hence we may take the sum over p in the estimates (5.3.44) and insert the result into (5.3.39). This gives the following estimate summing up the first half of the proof.

$$\int S^2(Rf|Q(I)) \le \int S^2(Rf|\mathcal{F}) + 32|I| \cdot \|f\|_{\mathrm{BMO}}^2. \tag{5.3.45}$$

Now we turn to the estimate for $S^2(Rf|\mathcal{F})$. This requires more subtle considerations. First recall that $\mathcal{F} \subseteq \mathcal{L}(J_0)$. Thus we may rewrite $S^2(Rf|\mathcal{F})$ as

$$S^2(Rf|\mathcal{F}) = \sum_{j=1}^{N} \langle f, y_{j,J_0}\rangle^2 \frac{S^2(x_{j,J_0}|\mathcal{F})}{||y_{j,J_0}||_2^4}. \tag{5.3.46}$$

Note that we have a pointwise upper estimate for $S^2(x_{j,J_0}|\mathcal{F})$. Indeed, the second assertion of Proposition 5.3.6 shows that for every $j \leq N$,

$$S^2(x_{j,J_0}|\mathcal{F}) \leq 2\gamma. \tag{5.3.47}$$

Conversely by (5.3.20) the following lower integral estimate holds true,

$$\int S^2(x_{j,J_0}|\mathcal{L}(J_0)) \geq \eta\gamma|J_0|. \tag{5.3.48}$$

We combine next the inequalities (5.3.47) and (5.3.48). After that we use the relation (5.3.43). This gives

$$\begin{aligned}\int S^2(x_{j,J_0}|\mathcal{F}) &\leq \frac{2|I|}{\eta|J_0|}||x_{j,J_0}||_2^2 \\ &\leq \frac{32|I|}{A\eta|J_0|}||y_{j,J_0}||_2^2.\end{aligned} \tag{5.3.49}$$

Note that we found a highly non-trivial estimate for $\int S^2(x_{j,J_0}|\mathcal{F})$. By contrast, the estimates for the coefficients $\langle f, y_{j,J_0}\rangle^2$ appearing in (5.3.46) are relatively direct. We simply apply Bessel's inequality and find that

$$\langle f, y_{j,J_0}\rangle^2 \leq |J_0|\,||f||_{\mathrm{BMO}}^2\,||y_{j,J_0}||_2^2. \tag{5.3.50}$$

Insert (5.3.49) and (5.3.50) into the identity (5.3.46). This gives

$$\begin{aligned}\int S^2(Rf|\mathcal{F}) &\leq \frac{32|I|}{A\eta|J_0|}\sum_{j=1}^{N}\langle f, y_{j,J_0}\rangle^2||y_{j,J_0}||_2^{-2} \\ &\leq \frac{32N}{A\eta}|I|\cdot||f||_{\mathrm{BMO}}^2.\end{aligned} \tag{5.3.51}$$

With (5.3.45) we found that

$$\int S^2(Rf|Q(I)) \leq |I|\left(32 + \frac{32N}{A\eta}\right)||f||_{\mathrm{BMO}}^2.$$

The choice of A and η gives that

$$||R||_{\mathrm{BMO}} \leq 100.$$

The factorization through ℓ^∞. Now we turn to the proof of the factorization (5.3.37). Recall that $\mathcal{K} \subseteq \Omega$ where Ω is defined in Proposition 5.3.6. Hence $\mathcal{K}$ satisfies the Carleson packing condition with

$$[\![\mathcal{K}]\!] \leq N\gamma^{-1} + 1.$$

This puts a strong restriction on the range of R which leads to the claimed factorization. Define the operators E, F as follows. Let $g \in \mathrm{BMO}$. Then

$$E(g) = \left\{ \langle g, \frac{y_{j,J}}{\|y_{j,J}\|_2^2} \rangle : j \leq N,\ J \in \mathcal{K} \right\}.$$

Let $(a_{j,J}), j \leq N,\ J \in \mathcal{K}$ be a bounded sequence of scalars. Then

$$F\left((a_{j,J})\right) = \sum \{ a_{j,J} x_{j,J} : j \leq N,\ J \in \mathcal{K} \}.$$

Next we show that operator E is bounded. First by the Cauchy–Schwarz inequality we have that

$$\int S(y_{j,J}) \leq |J|^{1/2} \left(\int S^2(y_{j,J}) \right)^{1/2}. \tag{5.3.52}$$

Observe next that

$$S^2(y_{j,J}) = S^2(\psi_j | \mathcal{R}(J)),$$

Combining the lower bounds of Proposition 5.3.8 with (5.3.19) and (5.3.20) we obtain

$$\int S^2(y_{j,J}) \geq |J| \frac{A\gamma\eta}{32}. \tag{5.3.53}$$

View (5.3.53) as an upper estimate for $|J|$, and insert it into (5.3.52). This gives

$$\int S(y_{j,J}) \leq \sqrt{32 A^{-1}\gamma^{-1}\eta^{-1}} \int S^2(y_{j,J}).$$

Hence by Fefferman's inequality we obtain

$$\begin{aligned} |\langle g, \frac{y_{j,J}}{\|y_{j,J}\|_2^2} \rangle| &\leq 2\sqrt{2} \|g\|_{\mathrm{BMO}} \frac{\|y_{j,J}\|_{H^1}}{\|y_{j,J}\|_2^2} \\ &\leq 16 \sqrt{A^{-1}\gamma^{-1}\eta^{-1}} \|g\|_{\mathrm{BMO}}. \end{aligned} \tag{5.3.54}$$

From (5.3.54) we get $\|E : \mathrm{BMO} \longrightarrow \ell^\infty\| \leq 16\sqrt{A^{-1}\gamma^{-1}\eta^{-1}}$. Upper bounds for the operator norm of F follow from the upper estimates for the Carleson constant $[\![\mathcal{K}]\!]$ and $\|x_{j,J}\|_{\mathrm{BMO}}$. We have $\|F : \ell^\infty \longrightarrow \mathrm{BMO}\| \leq [\![\mathcal{K}]\!]^{1/2} \sup_{j,J} \|x_{j,J}\|_{\mathrm{BMO}}$. In summary we obtain the following estimate for the factoring operators,

$$\|E : \mathrm{BMO} \longrightarrow \ell^\infty\| \cdot \|F : \ell^\infty \longrightarrow \mathrm{BMO}\| \leq f(N, \epsilon).$$ □

The rank of R is of course infinite, in general, since we cannot guarantee that the cardinality of $\mathcal{K}$ is finite. For the uniform approximation property we have to replace $\mathcal{K}$ in the above operator by a finite set. The Carleson packing condition $[\![\mathcal{K}]\!] \leq N\gamma^{-1} + 1$, and the factorization of R suggest that we do this by applying the combinatorial methods developed in Chapter 3.

Proof of Theorem 5.3.5. Part 2.

The reader is advised to review in Section 1.5 (page 109) the glueing process and the pigeon hole principle, and in Section 3.2 the statement of the combinatorial Theorem 3.2.2 on colored intervals and orthogonal projections (page 191).

We attach colors to the intervals of $\mathcal{K} = \mathcal{M}_1 \cup \cdots \cup \mathcal{M}_L$. Let $0 \le k \le A-1$, and put

$$\mathcal{M}_{i,k} = \bigcup_{m=0}^{\infty} G_{mA+k}(\mathcal{M}_i).$$

Note that the collections $\{\mathcal{M}_{i,k} : 1 \le i \le L, \text{ and } 0 \le k \le A-1\}$ form a decomposition of $\mathcal{K}$. We use it to attach colors to the intervals of $\mathcal{K}$. For $I \in \mathcal{K}$ we define its color as follows:

$$\text{The color of } I \text{ is } (i,k) \text{ iff } I \in \mathcal{M}_{i,k}. \tag{5.3.55}$$

There are at most $M = LA$ different colors.

Thus we turned $\mathcal{K}$ into a collection of colored dyadic intervals. There are at most $M = M(\epsilon, N)$ different colors. Recall that $\mathcal{K}$ satisfies the Carleson packing condition and its Carleson constant admits an upper bound that depends only on ϵ and N. Theorem 3.2.2 applied to the colored collection $\mathcal{K}$ gives its decomposition into blocks as

$$\mathcal{K} = \bigcup_{J \in \mathcal{J}} \mathcal{B}(J), \tag{5.3.56}$$

and for $J \in \mathcal{J}$ it gives collections $\mathcal{F}_1(J), \ldots, \mathcal{F}_{L(J)}(J)$, so that the following conditions hold.

$$\text{For } m \le L(J),\ \mathcal{F}_m(J) \ \text{ is monochromatic with respect to (5.3.55)}\ . \tag{5.3.57}$$

$$\mathcal{F}_1(J), \ldots, \mathcal{F}_{L(J)}(J) \text{ satisfies Jones's compatibility condition,} \tag{5.3.58}$$

$$\mathcal{B}(J) = \mathcal{F}_1(J) \cup \cdots \cup \mathcal{F}_{L(J)}(J), \tag{5.3.59}$$

and

$$\text{For } m \le L(J), \ \sum_{I \in \mathcal{F}_m(J)} |I| > \delta |J|. \tag{5.3.60}$$

The index set $\mathcal{J}$ satisfies the Carleson packing condition with constant ≤ 2. This is a relevant piece of information since $J \in \mathcal{J}$ is the unique maximal interval in $\mathcal{B}(J)$. Theorem 3.2.2 puts an upper bound on $L(I)$ and a lower bound on δ. We have

$$L(I) \le L(\epsilon, N) \quad \text{and} \quad \delta \ge \delta(\epsilon, N). \tag{5.3.61}$$

For each of the collections $\mathcal{F}_m(J)$ we use the abbreviation

$$F_m(J) = \bigcup_{I \in \mathcal{F}_m(J)} I.$$

Thus $F_m(J)$ is the point-set covered by the collection $\mathcal{F}_m(J)$. Recall that for each J fixed the families $\{F_m(J) : m \le L(J)\}$ form a nested collection of sets. By (5.3.61) the intersection pattern of the nested families $\{F_m(J) : m \le L(J)\}$ is rather limited as J ranges in $\mathcal{J}$. Hence applying the pigeon hole principle discussed after Theorem 1.5.11 in Section 1.5 gives the following: There exist $K = K(\epsilon, N)$ and disjoint subcollections $\mathcal{A}_l \subseteq \mathcal{J}$, so that

$$\mathcal{J} = \mathcal{A}_1 \cup \cdots \cup \mathcal{A}_K, \tag{5.3.62}$$

and so that for each fixed collection $\mathcal{A}_l$ the families

$$\{F_m(J) : m \le L(J)\} \quad \text{where} \quad J \in \mathcal{A}_l,$$

satisfy the so-called rescaling property. It consists of the three conditions stated below. First, there exists $L(l)$ such that for $J \in \mathcal{A}_l$,

$$L(J) = L(l). \tag{5.3.63}$$

Second if $I, J \in \mathcal{A}_l$ and $m, n \le L(l)$, then

$$F_m(I) \subseteq F_n(I) \quad \text{iff} \quad F_m(J) \subseteq F_n(J), \tag{5.3.64}$$

and

$$|F_m(J)| \,/\, |F_n(J)| \ \le\ 2|F_m(I)| \,/\, |F_n(I)|. \tag{5.3.65}$$

The third rescaling condition states that if $I, J \in \mathcal{A}_l$ and $m \le L(l)$ are fixed, then the monochromatic collections $\mathcal{F}_m(I)$ and $\mathcal{F}_m(J)$ satisfy

$$\text{color}\left(\mathcal{F}_m(I)\right) = \text{color}\left(\mathcal{F}_m(J)\right), \tag{5.3.66}$$

where the coloring is given by (5.3.55).

Next we fix $l \le K$ and $m \le L(l)$. Then we define the collections of dyadic intervals

$$\mathcal{F}_{l,m} = \bigcup \{\mathcal{F}_m(J) : J \in \mathcal{A}_l\}.$$

In this way we define disjoint and monochromatic families of intervals which form a decomposition of $\mathcal{K}$. The coloring of $\mathcal{K}$ is defined in (5.3.55). Hence the fact that the family $\mathcal{F}_{l,m}$ is a monochromatic collection of dyadic intervals implies that there exists $i \le L$, and $k \le A$ such that

$$\mathcal{F}_{l,m} \subseteq \mathcal{M}_{i,k}.$$

Consequently the decomposition of $\mathcal{K}$ given by

$$\{\mathcal{F}_{l,m} : 1 \le l \le K \text{ and } 1 \le m \le L(l)\}, \tag{5.3.67}$$

is a refinement of the initial decomposition

$$\{\mathcal{M}_{i,k} : 1 \le i \le L, \text{ and } 0 \le k \le A-1\} \tag{5.3.68}$$

with which we started.

Defining the resolving Operator

In Chapter 1 we introduced a glueing procedure to generate new operators from old ones, with the aim of reducing the rank. We determined a condition ensuring that the new operator is bounded on BMO. The examples we considered in Section 1.5 were somewhat simpler than the ones we are facing right now. Nevertheless the basic glueing mechanism remains the same. In particular we exploit the rescaling property and the pigeon hole principle.

Recall that in (5.3.29) we defined the collection $\mathcal{K}$ of dyadic intervals. We obtained a decomposition of $\mathcal{K}$ in (5.3.67). Now we fix $1 \leq j \leq N$ and $J \in \mathcal{K}$. As before we form the functions

$$x_{j,J} = \sum_{I \in \mathcal{L}(J)} \langle \psi_j, h_I \rangle \frac{h_I}{|I|}, \quad \text{and} \quad y_{j,J} = \sum_{I \in \mathcal{R}(J)} \langle \psi_j, h_I \rangle \frac{h_I}{|I|}.$$

Let $1 \leq l \leq K$, and $1 \leq m \leq L(l)$. Let $\{J_p\}$ be an enumeration of the intervals in $\mathcal{A}_l$. Recall that

$$\mathcal{F}_{l,m} = \bigcup_{p=1}^{\infty} \mathcal{F}_m(J_p). \tag{5.3.69}$$

We use the collection $\mathcal{F}_{l,m}$ to define the glueing procedure as follows.

$$x_{j,l,m} = \sum_{J \in \mathcal{F}_{l,m}} x_{j,J} \quad \text{and} \quad y_{j,l,m} = \sum_{J \in \mathcal{F}_{l,m}} y_{j,J}. \tag{5.3.70}$$

The rescaling property expressed in (5.3.63)–(5.3.65) and the Jones compatibility condition (5.3.58) allow us to show that the operator T defined below is bounded on BMO. With the rescaling property we are able to modify the proof of Proposition 5.3.10 so that it gives a good norm estimate for the operator T.

Proposition 5.3.11. *The operator T defined by*

$$Tf = \sum_{j=1}^{N} \sum_{l=1}^{K} \sum_{m=1}^{L(l)} \langle f, y_{j,l,m} \rangle \frac{x_{j,l,m}}{||y_{j,l,m}||_2^2}, \tag{5.3.71}$$

satisfies the conditions

$$||T||_{BMO} \leq 100,$$
$$||T\varphi_j - \varphi_j|| \leq \epsilon, \text{ for } j \leq N.$$
$$\text{Rank}\,(T) \leq f(N, \epsilon).$$

Proof. Note that by construction the operator T almost leaves invariant the functions φ_j. More precisely, for $j \in \{1, \ldots, N\}$,

$$T\varphi_j = \varphi_j - E_j.$$

With the error estimates in Proposition 5.3.9 we obtain

$$||T\varphi_j - \varphi_j|| \le \epsilon.$$

The rank of T satisfies the estimate

$$\operatorname{Rank} T \le NK \max_l L(l).$$

By (5.3.62) and (5.3.63) the factors K and $\max_l L(l)$ have upper bounds depending on N and ϵ. Thus $\operatorname{Rank} T \le f(N, \epsilon)$ as claimed.

Now we turn to proving the norm estimate of T. Let I be a dyadic interval in the Haar support of Tf. Then there exists $J \in \mathcal{K}$, such that

$$I \in \mathcal{L}(J).$$

We let $\{I_p : p \in \mathbb{N}\}$ be the sequence of maximal intervals in $\mathcal{K}$, which are strictly contained in I. This is clearly a sequence of pairwise disjoint dyadic intervals and $\sum |I_p| \le |I|$. Next we define the collection

$$\mathcal{F} = I \cap \mathcal{L}(J).$$

Then we have the following identity for the square function of Tf,

$$S^2(Tf|Q(I)) = S^2(Tf|\mathcal{F}) + \sum_{p\in\mathbb{N}} S^2(Tf|Q(I_p)). \tag{5.3.72}$$

At this point we need to apply two auxiliary estimates which we prove later. First we use Lemma 5.3.12 to estimate the term $S^2(Tf|\mathcal{F})$ appearing on the right-hand side of (5.3.72). It states that for $I \in \mathcal{K}$ and $\mathcal{F} \subseteq \mathcal{L}(I)$ we have

$$\int S^2(Tf|\mathcal{F}) \le |\mathcal{F}^*| \frac{2^9 N}{A\eta} ||f||^2_{\mathrm{BMO}}. \tag{5.3.73}$$

Second, Lemma 5.3.13 is used to obtain a good upper estimate in the sum over p in (5.3.72). It gives

$$\int S^2(Tf|Q(I)) \le 3 \cdot 2^{10} |I| \, ||f||^2_{\mathrm{BMO}}, \tag{5.3.74}$$

for $I \in \mathcal{K}$. Integrating the identity (5.3.72) and invoking (5.3.73) and (5.3.74) we find that

$$\begin{aligned}
\int S^2(Tf|Q(I)) &= \int S^2(Tf|\mathcal{F}) + \sum_{p\in\mathbb{N}} \int S^2(Tf|Q(I_p)) \\
&\le ||f||^2_{\mathrm{BMO}} \left(\frac{2^9 N}{A\eta} |I| + 3 \cdot 2^{10} \sum_{p\in\mathbb{N}} |I_p| \right) \\
&\le C ||f||^2_{\mathrm{BMO}} |I|,
\end{aligned}$$

where $C = (3 \cdot 2^{10} + \frac{2^9 N}{A\eta})$. This completes the proof of Proposition 5.3.11. □

It remains to verify the two crucial lemmata used in the above proof showing the boundedness of the operator T. We first establish that (5.3.73) holds true. Consider the diagram

$$I \xrightarrow{I\in\mathcal{L}(J)} J \xrightarrow{\mathcal{F}=I\cap\mathcal{L}(J)} \mathcal{F} \xrightarrow{\mathcal{F}\subseteq\mathcal{F}_{(\ell_0,m_0)}} \alpha=(l_0,m_0) \xrightarrow{\mathcal{F}_{(\ell_0,m_0)}\subseteq\mathcal{C}_{i_0}} i_0.$$

It displays the logical dependence of the intervals appearing in the proof above and their relation to the indices chosen in the course of Lemma 5.3.12.

Lemma 5.3.12. *Let $J \in \mathcal{K}$ and let $\mathcal{F}$ be a sub-collection of $\mathcal{L}(J)$, then*

$$\int S^2(Tf|\mathcal{F}) \leq |\mathcal{F}^*|\,\frac{2^9 N}{A\eta}||f||^2_{BMO}.$$

Proof. Fix $J \in \mathcal{K}$ and $\mathcal{F} \subseteq \mathcal{L}(J)$. There exists at most one pair

$$\alpha = (l_0, m_0) \tag{5.3.75}$$

with $l_0 \leq K$ and $m_0 \leq L(l)$ so that for some $j \leq N$,

$$\int S^2(x_{j,l_0,m_0}|\mathcal{F}) \neq 0. \tag{5.3.76}$$

Note however that for a given choice of $\alpha = (l_0, m_0)$ there is no restriction on the set of $j \in \{1,\dots,N\}$ such that (5.3.76) holds. Consequently the square function $S^2(Tf|\mathcal{F})$ assumes the simplified form

$$S^2(Tf|\mathcal{F}) = \sum_{j=1}^{N}\langle f, y_{j,\alpha}\rangle^2\frac{S^2(x_{j,\alpha}|\mathcal{F})}{||y_{j,\alpha}||_2^4}. \tag{5.3.77}$$

Recall the pointwise estimates of Proposition 5.3.6 which imply that

$$\int S^2(x_{j,\alpha}|\mathcal{F}) \leq 2\gamma|\mathcal{F}^*|. \tag{5.3.78}$$

Proposition 5.3.7 gives a matching lower bound for the above integral. We have

$$\int S^2(x_{j,\alpha}|\mathcal{L}(J)) \geq \gamma\eta|J|. \tag{5.3.79}$$

Combining the estimates (5.3.79) and (5.3.78) we find that

$$\int S^2(x_{j,\alpha}|\mathcal{F}) \leq \frac{2|\mathcal{F}^*|}{\eta|J|}\int S^2(x_{j,\alpha}|\mathcal{L}(J)). \tag{5.3.80}$$

Next we relate the L^2 norm of the local square function $S(x_{j,\alpha}|\mathcal{L}(J))$ appearing in (5.3.80) to the global L^2 norm of $y_{j,\alpha}$. Specifically we claim that the following estimate holds,

$$\frac{1}{|J|}\int S^2(x_{j,\alpha}|\mathcal{L}(J)) \leq \frac{32||y_{j,\alpha}||_2^2}{A}\left[\sum_{k=1}^{\infty}|J_k|\right]^{-1}, \tag{5.3.81}$$

where $\{J_k\}$ is an enumeration of the intervals in $\mathcal{F}_\alpha$.

Verification of (5.3.81). Here we exploit the rescaling property formulated in (5.3.65). Recall that we defined $\alpha = (l_0, m_0)$. We let $\{J_k\}$ be an enumeration of the intervals in $\mathcal{F}_{l_0,m_0}$. By construction, for $J_k \in \mathcal{F}_{l_0,m_0}$ we have

$$S^2(x_{j,\alpha}|\mathcal{L}(J_k)) = S^2(\psi_j|\mathcal{L}(J_k)). \tag{5.3.82}$$

Recall that the pigeon hole principle defines $\mathcal{F}_{l_0,m_0}$ to be monochromatic with respect to the coloring given by the decomposition (5.3.68). Hence there exists i_0 so that the following inclusion holds,

$$\mathcal{F}_{l_0,m_0} \subseteq \mathcal{M}_{i_0,k_0} \subseteq \mathcal{C}_{i_0}. \tag{5.3.83}$$

Combining the observation (5.3.82) with (5.3.83) and applying Proposition 5.3.7 gives

$$\frac{a_{i_0,j}}{2} \le \frac{1}{|J_k|}\int S^2(x_{j,\alpha}|\mathcal{L}(J_k)) \le a_{i_0,j}. \tag{5.3.84}$$

By definition of $\alpha = (l_0, m_0)$ in (5.3.75) it follows that $J \in \mathcal{F}_{l_0,m_0}$. Hence

$$\frac{a_{i_0,j}}{2} \le \frac{1}{|J|}\int S^2(x_{j,\alpha}|\mathcal{L}(J)) \le a_{i_0,j}. \tag{5.3.85}$$

Next we merge the estimates (5.3.84) and (5.3.85), and apply Proposition 5.3.8. Thus we obtain,

$$\begin{aligned} \frac{1}{|J|}\int S^2(x_{j,\alpha}|\mathcal{L}(J)) &\le \frac{2}{|J_k|}\int S^2(x_{j,\alpha}|\mathcal{L}(J_k)) \\ &\le \frac{32}{A}\frac{1}{|J_k|}\int S^2(y_{j,\alpha}|\mathcal{R}(J_k)). \end{aligned} \tag{5.3.86}$$

Multiply the inequality (5.3.86) by $|J_k|$, and take the sum over k. Finally divide by $\sum_{k=1}^{\infty} |J_k|$. This gives

$$\begin{aligned} \frac{1}{|J|}\int S^2(x_{j,\alpha}|\mathcal{L}(J)) &\le \frac{32}{A}\frac{\sum_{k=1}^{\infty}\int S^2(y_{j,\alpha}|\mathcal{R}(J_k))}{\sum_{k=1}^{\infty}|J_k|} \\ &= \frac{32}{A}\frac{||y_{j,\alpha}||_2^2}{\sum_{k=1}^{\infty}|J_k|}. \end{aligned} \tag{5.3.87}$$

An L^2 bound for $||y_{j,\alpha}||_{H^1}$. Next we compare the H^1 norm of $y_{j,\alpha}$ with its L^2 norm. Simply applying the Cauchy–Schwarz inequality gives the estimate

$$\begin{aligned} ||y_{j,\alpha}||_{H^1}^2 &\le \sum_{k=1}^{\infty}|J_k| \sum_{k'=1}^{\infty}\int S^2(y_{j,\alpha}|\mathcal{R}(J_{k'})) \\ &\le \sum_{k=1}^{\infty}|J_k|\ ||y_{j,\alpha}||_2^2. \end{aligned} \tag{5.3.88}$$

Note that the factor $\sum_{k=1}^{\infty}|J_k|$ appeared before in the denominator of the second line in (5.3.87).

Conclusion. We finish now the proof as follows. First we integrate over J the identity (5.3.77). Next fix $j \in \{1, \ldots, N\}$ and consider separately each of the resulting summands

$$\langle f, y_{j,\alpha}\rangle^2 \frac{\int S^2(x_{j,\alpha}|\mathcal{F})}{||y_{j,\alpha}||_2^4}.$$

To the scalar product we apply Fefferman's inequality,

$$|\langle f, y_{j,\alpha}\rangle| \leq 2\sqrt{2}||y_{j,\alpha}||_{H^1}||f||_{\mathrm{BMO}}.$$

To the integral of the square function $S^2(x_{j,\alpha}|\mathcal{F})$ we apply (5.3.80). Then using the estimates of (5.3.87) we replace the integral of the local square function by the global L^2 norm of $y_{j,\alpha}$. Finally we use (5.3.88) to replace the H^1 norm of $y_{j,\alpha}$ by its L^2 norm. Doing this gives that

$$\langle f, y_{j,\alpha}\rangle^2 \frac{\int S^2(x_{j,\alpha}|\mathcal{F})}{||y_{j,\alpha}||_2^4} \leq 2^9||f||^2_{\mathrm{BMO}}\frac{|\mathcal{F}^*|}{\eta A}.$$

Summing this estimate over $j \in \{1, \ldots, N\}$ we obtain the claimed estimate of Lemma 5.3.12. □

Lemma 5.3.13. *For $I_0 \in \mathcal{K}$,*

$$\int S^2(Tf|Q(I_0)) \leq 3 \cdot 2^{10}|I_0|\, ||f||^2_{BMO}.$$

Proof. We start with an outline of the proof, and show that it suffices to establish the crucial estimate (5.3.90). Its verification requires a detailed analysis of the glueing process and exploits the rescaling properties obtained by applying the pigeon hole principle.

The outline of the proof. Let $I_0 \in \mathcal{K}$. There exists $J_0 \in \mathcal{J}$ such that $I_0 \in \mathcal{B}(J_0)$. We rewrite the collection $\mathcal{K} \cap I_0$ as

$$I_0 \cap \mathcal{K} = \{I_0 \cap \mathcal{B}(J_0)\} \cup \bigcup_{J \in \mathcal{J} \cap J_0} \mathcal{B}(J).$$

For convenience we enumerate the intervals $\{J \in \mathcal{J} : J \subseteq J_0\}$ as $\{J_k : k \in \mathbb{N}\}$. Next we introduce the following blocks of dyadic intervals,

$$\mathcal{B}_0 = \bigcup \{\mathcal{L}(I) : I \in I_0 \cap \mathcal{B}(J_0)\}, \tag{5.3.89}$$

and

$$\mathcal{B}_k = \bigcup \{\mathcal{L}(I) : I \in \mathcal{B}(J_k)\}.$$

They induce a corresponding splitting of the square function of Tf. We obtain that

$$S^2(Tf|Q(I_0)) = S^2(Tf|\mathcal{B}_0) + \sum_{k=1}^{\infty} S^2(Tf|\mathcal{B}_k).$$

Below we give estimates for each of the summands appearing in the above sum. First we show that

$$\int S^2(Tf|\mathcal{B}_0) \leq C\|f\|^2_{\mathrm{BMO}}|I_0|. \tag{5.3.90}$$

After that we remark that the proof of (5.3.90) actually shows for any $J_k \in J_0 \cap \mathcal{J}$ the bound

$$\int S^2(Tf|\mathcal{B}_k) \leq C\|f\|^2_{\mathrm{BMO}}|J_k|. \tag{5.3.91}$$

Recall that $\mathcal{J}$ satisfies the Carleson packing condition with constant ≤ 2. Then taking the sum over $k \geq 1$ in (5.3.91) and adding (5.3.90) proves the global estimate of Lemma 5.3.13. In the argument below the logical dependence of the intervals and indices is

$$\begin{array}{ccccccc} l_0 & & & & & & \\ \uparrow{\scriptstyle J_0\in\mathcal{A}_{l_0}} & & & & & & \\ J_0 & \xrightarrow{I_0\in\mathcal{F}_{m_0}(J_0)} & m_0 & \xrightarrow{F_m(J_0)\subseteq F_{m_0}(J_0)} & m & \quad\text{and}\quad & (l_0,m) \xrightarrow{\mathcal{F}_{(l_0,m)}\subseteq\mathcal{C}_{i_0}} i_0. \\ \uparrow{\scriptstyle I_0\in\mathcal{B}(J_0)} & & & & & & \\ I_0 & & & & & & \end{array}$$

Proof of (5.3.90)**. Step 1.** The first step of the proof consists in verifying the identity (5.3.94) below, which gives a convenient representation for the square function $S^2(T(f)|\mathcal{B}_0)$. Recall that $I_0 \in \mathcal{B}(J_0)$, where $J_0 \in \mathcal{K}$. By (5.3.59) the block $\mathcal{B}(J_0)$ is decomposed as

$$\mathcal{B}(J_0) = \bigcup_{m=1}^{L(J_0)} \mathcal{F}_m(J_0),$$

where the collections $\{\mathcal{F}_m(J_0) : m \leq L(J_0)\}$, satisfy Jones's compatibility condition. As $I_0 \in \mathcal{B}(J_0)$, there exists a unique $m_0 \leq L(J_0)$ such that

$$I_0 \in \mathcal{F}_{m_0}(J_0). \tag{5.3.92}$$

Next we define the index set

$$\mathcal{U}_{m_0} = \{m : F_m(J_0) \subseteq F_{m_0}(J_0)\}. \tag{5.3.93}$$

Recall that the collections $\{\mathcal{F}_m(J_0) : m \leq L(J_0)\}$ satisfy the compatibility condition (5.3.58). Hence we may rewrite $\mathcal{B}_0$, using the index set $\mathcal{U}_{m_0}$ as follows,

$$\mathcal{B}_0 = \bigcup_{m\in\mathcal{U}_{m_0}} \{\mathcal{L}(I) : I \in I_0 \cap F_m(J_0)\}.$$

Let $l_0 \in \{1, \ldots, K\}$ be determined by the relation $J_0 \in \mathcal{A}_{l_0}$, where the collection $\mathcal{A}_{l_0}$ appears in the decomposition (5.3.56) of $\mathcal{J}$. Note that only for the index l_0 do there exist j and m such that

$$\int S^2(x_{j,l_0,m}|\mathcal{B}_0) \neq 0.$$

We have thus observed that the square function of Tf simplifies considerably in the presence of $\mathcal{B}_0$. Indeed the following identity holds,

$$S^2(Tf|\mathcal{B}_0) = \sum_{j=1}^{N} \sum_{m\in\mathcal{U}_{m_0}} \langle f, y_{j,l_0,m}\rangle^2 ||y_{j,l_0,m}||_2^{-4} S^2(x_{j,l_0,m}|\mathcal{B}_0). \tag{5.3.94}$$

Proof of (5.3.90)**. Step 2.** The second step in the proof of (5.3.90) consists in establishing the two estimates

$$\left[\sum_{J\in\mathcal{A}_{l_0}} |F_{m_0}(J)|\right] \int S^2(x_{j,l_0,m}|\mathcal{B}_0) \leq 8|I_0| \cdot ||x_{j,l_0,m}||_2^2 \tag{5.3.95}$$

and

$$||x_{j,l_0,m}||_2^2 \leq \frac{16}{A} ||y_{j,l_0,m}||_2^2. \tag{5.3.96}$$

Once (5.3.95) and (5.3.96) are established the proof of (5.3.90) can be completed as follows. First we remark that the Jones compatibility condition implies that, for $J \in \mathcal{A}_{l_0}$,

$$\sum_{m\in\mathcal{U}_{m_0}} \sum_{j=1}^{N} \sum_{I\in\mathcal{F}_m(J)} \langle f, y_{j,I}\rangle^2 ||y_{j,l_0,m}||_2^{-2} \leq 8A||f||_{\mathrm{BMO}}^2 |F_{m_0}(J)|.$$

Hence taking the sum over $J \in \mathcal{A}_{l_0}$ and applying Bessel's inequality we obtain that

$$\sum_{j=1}^{N} \sum_{m\in\mathcal{U}_{m_0}} \langle f, y_{j,l_0,m}\rangle^2 ||y_{j,l_0,m}||_2^{-2} \leq 8A||f||_{\mathrm{BMO}}^2 \sum_{J\in\mathcal{A}_{l_0}} |F_{m_0}(J)|. \tag{5.3.97}$$

Next we integrate the equation (5.3.94) and insert (5.3.95) and (5.3.96). Then use (5.3.97). This gives

$$\begin{aligned}
\frac{1}{|I_0|} \int S^2(Tf|\mathcal{B}_0) &= \frac{1}{|I_0|} \sum_{j=1}^{N} \sum_{m\in\mathcal{U}_{m_0}} \langle f, y_{j,l_0,m}\rangle^2 ||y_{j,l_0,m}||_2^{-4} \int S^2(x_{j,l_0,m}|\mathcal{B}_0) \\
&\leq \frac{2^7 A^{-1}}{\sum_{J\in\mathcal{A}_{l_0}} |F_{m_0}(J)|} \sum_{j=1}^{N} \sum_{m\in\mathcal{U}_{m_0}} \langle f, y_{j,l_0,m}\rangle^2 ||y_{j,l_0,m}||_2^{-2} \\
&\leq 2^{10} ||f||_{\mathrm{BMO}}^2.
\end{aligned}$$

Thus using (5.3.95) and (5.3.96) we established (5.3.90).

Proof of (5.3.95). We turn now to the proof of the first central estimate, which is (5.3.95). We begin by expanding the square function as

$$\int S^2(x_{j,l_0,m}|\mathcal{B}_0) = \sum_{I\in I_0\cap\mathcal{F}_m(J_0)} \int S^2(x_{j,l_0,m}|\mathcal{L}(I)). \tag{5.3.98}$$

Next we estimate the integrals appearing on the right-hand side of (5.3.98). By the assertions of the pigeon hole principle the collection $\mathcal{F}_{l_0,m}$ is monochromatic with respect to the coloring given by (5.3.68). Hence there exists i_0 and k_0 so that

$$\mathcal{F}_{l_0,m} \subseteq \mathcal{M}_{i_0,k_0} \subseteq \mathcal{C}_{i_0}. \tag{5.3.99}$$

Observe that for $I \in \mathcal{F}_{l_0,m}$, we have the identity

$$S^2(x_{j,l_0,m}|\mathcal{L}(I)) = S^2(\psi_j|\mathcal{L}(I)).$$

Hence Proposition 5.3.7 gives

$$\frac{a_{i_0,j}}{2} \leq \frac{1}{|I|}\int S^2(x_{j,l_0,m}|\mathcal{L}(I)) \leq a_{i_0,j}, \tag{5.3.100}$$

for every $I \in \mathcal{F}_{l_0,m}$. Taking the sum of (5.3.100) over the intervals $I \in I_0\cap\mathcal{F}_m(J_0)$ we obtain that

$$\int S^2(x_{j,l_0,m}|\mathcal{B}_0) \leq a_{i_0,j} \sum_{I\in I_0\cap\mathcal{F}_m(J_0)} |I|. \tag{5.3.101}$$

The right-hand side of (5.3.101) is just $a_{l_0,j}|I_0\cap F_m(J_0)|$. The collections $\{\mathcal{F}_m(J_0) : m \leq L(l_0)\}$, satisfy the compatibility condition (J), so that the following uniform estimate holds true,

$$\frac{|I_0 \cap F_m(J_0)|}{|I_0|} \leq 2\frac{|F_m(J_0)|}{|F_{m_0}(J_0)|}. \tag{5.3.102}$$

Now we eliminate the dependence on J_0, from the right-hand side of (5.3.102). The rescaling condition (5.3.65) asserts that the ratio $|F_m(J)|/|F_{m_0}(J)|$ is independent of $J \in \mathcal{A}_{l_0}$. By averaging we obtain the inequality

$$\frac{|F_m(J_0)|}{|F_{m_0}(J_0)|} \leq 2\frac{\sum_{J\in\mathcal{A}_{l_0}}|F_m(J)|}{\sum_{J\in\mathcal{A}_{l_0}}|F_{m_0}(J)|}. \tag{5.3.103}$$

Now combine (5.3.101) with the estimates (5.3.102) and (5.3.103). This proves the inequality

$$\frac{1}{|I_0|}\int S^2(x_{j,l_0,m}|\mathcal{B}_0) \leq 4a_{i_0,j}\frac{\sum_{J\in\mathcal{A}_{l_0}}|F_m(J)|}{\sum_{J\in\mathcal{A}_{l_0}}|F_{m_0}(J)|}. \tag{5.3.104}$$

Next we give an upper estimate for the constant $a_{i_0,j}$ appearing on the right-hand side of (5.3.104). Here we exploit the left-hand side of estimate (5.3.100). For $J \in \mathcal{A}_{l_0}$ the lower bound of the homogeneity relation (5.3.100) implies that

$$\frac{a_{i_0,j}}{2}|F_m(J)| \le \sum_{I \in \mathcal{F}_m(J)} \int S^2(x_{j,l_0,m}|\mathcal{L}(I)).$$

Summing over $J \in \mathcal{A}_{l_0}$ we obtain that

$$\begin{aligned}\frac{a_{i_0,j}}{2} \sum_{J \in \mathcal{A}_{l_0}} |F_m(J)| &\le \sum_{J \in \mathcal{A}_{l_0}} \sum_{I \in \mathcal{F}_m(J)} \int S^2(x_{j,l_0,m}|\mathcal{L}(I)) \\ &= ||x_{j,l_0,m}||_2^2.\end{aligned} \tag{5.3.105}$$

View (5.3.105) as an upper bound for $a_{i_0,j}$. Inserting it into (5.3.104) gives (5.3.95) by arithmetic.

Proof of (5.3.96). It remains to prove (5.3.96) relating the L^2 norms of $x_{j,l_0,m}$ and $y_{j,l_0,m}$ by the inequality

$$||x_{j,l_0,m}||_2^2 \le (16/A)||y_{j,l_0,m}||_2^2.$$

Start by recalling Proposition 5.3.8. It asserts that for every $J \in \mathcal{A}_{l_0}$ and $I \in \mathcal{F}(J)$,

$$\int S^2(\psi_j|\mathcal{L}(I)) \le \frac{16}{A} \int S^2(\psi_j|\mathcal{R}(I)). \tag{5.3.106}$$

Invoking the definitions of $x_{j,l_0,m}$ and $y_{j,l_0,m}$ the estimate (5.3.106) translates into

$$\int S^2(x_{j,l_0,m}|\mathcal{L}(I)) \le \frac{16}{A} \int S^2(y_{j,l_0,m}|\mathcal{R}(I)). \tag{5.3.107}$$

Next we sum the estimate (5.3.107) over $J \in \mathcal{A}_{l_0}$ and $I \in \mathcal{F}(J)$. This gives that

$$\sum_{J \in \mathcal{A}_{l_0}} \sum_{I \in \mathcal{F}(J)} \int S^2(x_{j,l_0,m}|\mathcal{L}(I)) \le \frac{16}{A} \sum_{J \in \mathcal{A}_{l_0}} \sum_{I \in \mathcal{F}(J)} \int S^2(y_{j,l_0,m}|\mathcal{R}(I)). \tag{5.3.108}$$

Observe that the left-hand side of (5.3.108) coincides with $||x_{j,l_0,m}||_2^2$ and that the right-hand side (5.3.108) equals $||y_{j,l_0,m}||_2^2$. Thus (5.3.96) follows from (5.3.108).

Conclusion. We have presented a complete proof of (5.3.90), that is the estimate

$$\int S^2(T(f)|\mathcal{B}_0) \le 2^{10}||f||^2_{\mathrm{BMO}}|I_0|.$$

Next we specialize it to the case when $J_0 = I_0$. This gives that for any $k \in \mathbb{N}_0$ the following holds,

$$\int S^2(T(f)|\mathcal{B}_k) \le 2^{10}||f||^2_{\mathrm{BMO}}|J_k|.$$

Recall that $\mathcal{J}$ satisfies the 2-Carleson constant. Hence

$$\begin{aligned}\int S^2(T(f)|Q(I_0)) &= \sum_{k=0}^{\infty} \int S^2(T(f)|\mathcal{B}_k)\\ &\leq 2^{10}||f||^2_{\mathrm{BMO}}|I_0| + 2^{10}||f||^2_{\mathrm{BMO}} \sum_{k=1}^{\infty} |J_k|\\ &\leq 3 \cdot 2^{10}||f||^2_{\mathrm{BMO}}|I_0|.\end{aligned}$$

□

General UAP data

To complete the proof that BMO satisfies the uniform approximation property we combine Proposition 5.3.3 with Theorem 5.3.4 and Theorem 5.3.5.

Proof of Theorem 5.3.2. Let $n \in \mathbb{N}$ and $\epsilon > 0$. Let $x_1, \ldots, x_n \in \mathrm{BMO}$. Apply Proposition 5.3.3 to $x_1, \ldots, x_n$ to obtain $\varphi_1, \ldots, \varphi_N$ so that the following conditions hold:

(a) For $i \leq n$ $\mathrm{dist}_{\mathrm{BMO}}(x_i, \mathrm{span}\{\varphi_1, \ldots, \varphi_N\}) \leq \epsilon \|x_i\|_{\mathrm{BMO}}$.

(b) $\varphi_1, \ldots, \varphi_N$ are disjointly supported over the Haar system.

(c)
$$N \leq \left(\frac{2n^3}{\epsilon}\right)^n.$$

Fix $j \leq N$ and let $\mathcal{F}_j$ be the Haar support of φ_j. For $\gamma = \epsilon^3 N^{-3} \left(\log(N^2\epsilon^{-1})\right)^{-N}$, define

$$\mathcal{B}_j = \{I \in \mathcal{F}_j : |\langle \varphi_j, h_I \rangle| > \gamma \|\varphi_j\|_{\mathrm{BMO}}\}.$$

Then put

$$\varphi_j^1 = \sum_{I \in \mathcal{B}_j} \langle \varphi_j, \frac{h_I}{|I|} \rangle h_I$$

and $\varphi_j^2 = \varphi_j - \varphi_j^1$. Thus $\{\varphi_1^1, \ldots, \varphi_N^1\}$ satisfy the hypothesis of Theorem 5.3.4 and $\{\varphi_1^2, \ldots, \varphi_N^2\}$ satisfy the hypothesis of Theorem 5.3.5. Moreover the Haar support of $\{\varphi_1^1, \ldots, \varphi_N^1\}$ and $\{\varphi_1^2, \ldots, \varphi_N^2\}$ are disjoint collections of dyadic intervals.

Apply Theorem 5.3.4 to $\{\varphi_1^1, \ldots, \varphi_N^1\}$. This gives a resolving operator T_1 so that

$$\begin{gathered}||T_1||_{\mathrm{BMO}} \leq 4,\\ ||T_1(\varphi_i^1) - \varphi_i^1||_{\mathrm{BMO}} \leq \epsilon ||\varphi_i^1||_{\mathrm{BMO}} \quad \text{for} \quad i \leq N,\\ \mathrm{rank}(T_1) \leq f_1(N, \epsilon, \gamma).\end{gathered}$$

If J is a dyadic interval not contained in the Haar support of $\{\varphi_1^1, \ldots, \varphi_N^1\}$, then $T_1(h_J) = 0$. Next apply Theorem 5.3.5 to $\{\varphi_1^2, \ldots, \varphi_N^2\}$. This gives a resolving operator T_2 so that

$$\begin{aligned} \|T_2\|_{\mathrm{BMO}} &\leq 10, \\ \|T_2(\varphi_i^2) - \varphi_i^2\|_{\mathrm{BMO}} &\leq \epsilon \|\varphi_i^2\|_{\mathrm{BMO}} \quad \text{for } i \leq N, \\ \operatorname{rank}(T_2) &\leq f_2(N, \epsilon). \end{aligned}$$

If J is not contained in the Haar support of $\{\varphi_1^2, \ldots, \varphi_N^2\}$ then $T_2(h_J) = 0$. Hence the operator $T = T_1 + T_2$ satisfies

$$\begin{aligned} \|T\|_{\mathrm{BMO}} &\leq \sqrt{10^2 + 4^2}, \\ \|T(x_i) - x_i\|_{\mathrm{BMO}} &\leq 4\epsilon \|x_i\|_{\mathrm{BMO}} \quad \text{for } i \leq N, \\ \operatorname{rank}(T) &\leq f_1(N, \epsilon, \gamma) + f_2(N, \epsilon) \\ &= f(\epsilon, n). \end{aligned}$$

□

We have now presented a complete proof of P. W. Jones's theorem that BMO satisfies the uniform approximation property. By a result of S. Heinrich [83] the uniform approximation property is a self-dual isomorphic invariant. Hence H^1 satisfies the uniform approximation property, since BMO does. In addition to the spaces L^p $(1 \leq p \leq \infty)$, H^1 and BMO the uniform approximation property has been investigated for the following classical Banach spaces.

(a) The reflexive Orlicz spaces satisfy the UAP (theorem of J. Lindenstrauss and L. Tzafriri [129]).

(b) $L(\ell^2)$, the space of bounded linear operators on Hilbert space does not satisfy the approximation property, hence it fails to satisfy the UAP (theorem of A. Szankowski [198]).

(c) The Schatten class S^r consists of compact operators on Hilbert space for which the singular values are in ℓ^r. (Singular values of a compact operator T are the eigenvalues of $(T^*T)^{1/2}$).

The Schatten class S^r *fails* the UAP for $r > 80$ and $1 \leq r < 80/79$ (theorem of A. Szankowski [199]).

Many years ago A. Pełczyński [171] posed the problem whether $H^\infty(\mathbb{D})$, the space of bounded analytic functions on the unit disk, satisfies the UAP. Now, there are available two extremely deep results pertaining to this problem: On the one hand there is the theorem of P. W. Jones asserting that BMO satisfies the UAP, and there are many connections between $H^\infty(\mathbb{D})$ and BMO, as discussed by J. Garnett [72]. This seems to support the conjecture that $H^\infty(\mathbb{D})$ satisfies the uniform approximation property. On the other hand we have a representation of $H^\infty(\mathbb{D})$ as the sub-space of $L(\ell^2)$, that is formed by the Toeplitz operators (see [120]) and the theorem of A. Szankowski [198] to the effect that $L(\ell^2)$ does not satisfy the approximation property. This suggests that $H^\infty(\mathbb{D})$ is likely to *fail* the uniform approximation property. However, for special cases of UAP data in $H^\infty(\mathbb{D})$, resolving operators were found in [28] and [23].

We conclude this section with a list of Banach spaces for which the verification of the uniform approximation property is outstanding. The following spaces are conjectured to satisfy the uniform approximation property.

1. $W^{k,\infty}(\mathbb{R}^n)$, the Sobolev space of functions on $\mathbb{R}^n$ $(n \geq 2)$ with bounded weak derivatives of order $\leq k$ $(k \geq 1)$, (conjecture of P. W. Jones).
2. $L^{p,q}(\mathbb{R})(p \neq q)$, the Lorentz space consisting of measurable functions on $\mathbb{R}$ for which
$$\|f\|_{L^{p,q}(\mathbb{R})} = \left(\int_0^\infty [t^{1/p} f^*(t)]^q \frac{dt}{t}\right)^{1/q} < \infty,$$
where f^* denotes the non-decreasing rearrangement of f (conjecture of W. B. Johnson).
3. S^r, the Schatten class of operators, *provided that* $|1/2 - 1/r|$ is small enough (conjecture of A. Szankowski).

5.4 Notes

In 1976 S. Kwapien and A. Pełczyński obtained Theorem 5.1.2 for $H^1(\mathbb{T})$ (the space of integrable functions on the unit circle for which the harmonic extension to the disk is analytic). It appeared in [120]. Historically the work of S. Kwapien and A. Pełczyński establishes the first *isomorphic invariant* for the space $H^1(\mathbb{T})$ which strongly suggests that $H^1(\mathbb{T})$ is a space with an unconditional basis. The proof given in the text is based on ideas of P. Wojtaszczyk [211]. The dichotomy of Theorem 5.1.3 is from [158]. In the context of L^p, W. B. Johnson [91] established the factorization method used in the proof of Theorem 5.1.3. The H^1 version of Johnson's factorization is in [158]. The dichotomy analogous to Problem 5.1.5 is established for the complemented subspaces of $(\sum L^2)_1$. This is the content of Proposition 8.1 in the memoir [27] by J. Bourgain, P. Cassaza, J. Lindenstrauss and L. Tzafriri.

Proposition 5.1.10, the estimate for the square function of ℓ_n^2 unit vectors in H^1, is due to J. Bourgain [17].

In the famous paper [169] A. Pełczinski establishes the decomposition method and proves that the spaces ℓ^p $(1 \leq p < \infty)$ are primary by showing that every infinite-dimensional, complemented subspace of ℓ^p is isomorphic to ℓ^p. The complemented subspaces of the non-separable space ℓ^∞ are determined by J. Lindenstrauss [127]. The result is analogous to the separable case: Every infinite-dimensional, complemented subspace of ℓ^∞ is isomorphic to ℓ^∞. In [20] J. Bourgain proved that $H^\infty(\mathbb{D})$, the space of bounded analytic functions in the unit disk is a primary space. In [20], J. Bourgain not only solves a particular problem about

$H^\infty(\mathbb{D})$, but also introduces a general method capable of treating several classes of non-separable Banach spaces. It is applied in [153] to BMO. Theorem 5.2.1 and its proof is from [153].

Using Bourgain's method G. Blower [13] proved that the (non-separable) Banach space of bounded linear operators on Hilbert space is primary. Applications of Bourgain's method to projective tensor products of ℓ^p spaces were given by A. Arias and J. Farmer [4] who investigate their isomorphic invariants, and use combinatorial methods (Ramsey's theorem) to study the problem of primarity for these spaces. Consult J. Lindenstrauss and L. Tzafriri [130] for a detailed presentation of the primary spaces among the classical Banach spaces. In particular [130] contains P. Enflo's result that the L^p spaces are primary for $1 < p < \infty$. P. Enflo's original proof is recorded by B. Maurey in [138]. P. Enflo's method extends to a large class of separable function spaces with an unconditional basis (see [93] by W. Johnson, B. Maurey, G. Schechtman, L. Tzafriri.)

A second proof of P. Enflo's result is due to D. Alspach, P. Enflo and E. Odell [2], who base their argument on the fact that L^p and $L^p(\ell^2)$ are isomorphic Banach spaces for $1 < p < \infty$. By Bourgain's Theorem 5.2.8 the spaces H^1 and $H^1(\ell^2)$ are not isomorphic Banach spaces. Hence the approach of D. Alspach, P. Enflo and E. Odell [2] does not extend to the H^1 case. Nevertheless H^1 is primary, see [153]. P. Enflo and T. Starbird [63] prove that the space L^1 is primary. The space of bounded continuous functions on the unit interval is shown by J. Lindenstrauss and A. Pełczyński [128] to be a primary space. For spaces of continuous functions over countable compact spaces the analogous result is due to P. Billard [11].

Theorem 5.2.7 presents J. Bourgain's description of complemented copies of H^1_n. Its main application is Theorem 5.2.8, hence the fact that H^1 is not isomorphic to $H^1(\ell^2)$. It should have many more applications, in particular it should be useful in connection with the conjecture of P. Wojtaszczyk discussed in Section 4.3.

Tent spaces are introduced by R.R Coifman, Y. Meyer and E.M. Stein in [54] who show that H^1 is isomorphic to a complemented subspace of the tent space T_α. The isomorphic identification of tent spaces with $H^1(\ell^2)$ is presented in Theorem 5.2.9. It is due to N. Kalton and P. Wojtaszczyk [109].

The concept of the uniform approximation property was introduced by A. Pełczyński and H. Rosenthal in [172]. They show in Theorem 5.3.1 that the L^p spaces satisfy the uniform approximation property. The proof given in the text is due to S. Kwapien. Lower estimates for the uniformity function on L^p are obtained by T. Figiel, W. Johnson and G. Schechtman [67] for the values $p = 1$ and $p = \infty$, and by J. Bourgain [25] for the remaining values $1 < p < \infty$. For UAP data consisting of independent random variables there exist upper estimates for the uniformity function due to W. Johnson and G. Schechtman [95], which are much lower than the estimates for general UAP data in [25].

J. Lindenstrauss and L. Tzafriri [129] show that a large class of Orlicz spaces satisfies the uniform approximation property. For an overview on approximation properties see [130] and [47]. Solving a problem of A. Pełczyński [171], P. W. Jones established the uniform approximation property for H^1 and BMO in [102].

S. Heinrich [83] proved that a Banach space satisfies the uniform approximation property if and only if its dual space does. V. Mascioni [137] gives a constructive proof of S. Heinrich's result, that could in principle be useful in simplifying the proof that BMO satisfies the uniform approximation property. In particular V. Mascioni's ideas in [137] should be helpful in passing from operators factoring through ℓ^∞ (Proposition 5.3.10) to the resolving operator of finite rank.

Executing the original argument of P. W. Jones in the context of tent spaces should result in some simplifications in the proof of the UAP for BMO.

The work of A. Szankowski [198] and [199] introduces extremely refined methods to establish that a given Banach space does not satisfy the uniform approximation property. Thereby the space of linear operators on Hilbert space and the Schatten classes S^r (for $r > 80$ or $r < 80/79$) are shown to fail the UAP.

Chapter 6

Atomic H^1 Spaces

This chapter contains the presentation of atomic H^1 spaces and their relation to dyadic H^1. We treat the example of H^1_{at} in considerable detail, obtaining equivalent norms using the Hilbert transform, the Lusin area function and the non-tangential maximal function. We investigate L. Carleson's biorthogonal system and show that it is an unconditional basis for H^1_{at}. Thereby we prove B. Maurey's theorem that H^1_{at} and H^1 are isomorphic Banach spaces. We close the chapter with the classification theorem for the Banach spaces $H^1(X, d, \mu)$.

6.1 Basic similarities between H^1 and H^1_{at}

In this section we give three important characterizations of H^1_{at}. We prove two-sided estimates relating the norm of $u \in H^1_{at}$ to the Lusin function, the non-tangential maximal function and the Hilbert transform, respectively.

We will now define the Lusin function and the non-tangential maximal function. Let $t \in \mathbb{R}$, then define Γ_t to be the convex hull of e^{it}, and the disk $\{z \in \mathbb{D} : |z| \leq \frac{1}{\sqrt{2}}\}$. Thus defined Γ_t is a sub domain of the unit disk, it is called the Stolz domain with vertex at e^{it}. In a neighborhood of its vertex the Stolz domain has the shape of a triangle with a right angle at e^{it}. Let $F : \mathbb{D} \to \mathbb{C}$ have continuous partial derivatives. The Lusin function $S_L(F)$ is defined by the equation

$$S_L(F)(t) = \left(\iint_{\Gamma_t} |\text{grad}\, F(z)|^2 dA(z) \right)^{1/2}, \tag{6.1.1}$$

where $t \in \mathbb{R}$, and where $dA(z)$ denotes the area measure in the complex plane. Note that $S_L(F)$ is a periodic function satisfying $S_L(F)(t+2\pi) = S_L(F)(t)$. Hence we may regard it as a well-defined function on the boundary of the unit disk.

Next we define the non-tangential maximal function of $F : \mathbb{D} \to \mathbb{C}$. As before we let $t \in \mathbb{R}$. Then we put

$$F^*(t) = \sup_{z \in \Gamma_t} |F(z)|. \tag{6.1.2}$$

The non-tangential maximal function of F satisfies $F^*(t + 2\pi) = F^*(t)$, it is therefore well defined on the boundary of the unit disk.

Next we fix an integrable function $u : \mathbb{T} \to \mathbb{C}$ defined initially on the boundary of the unit disk. To form its Lusin function and its non-tangential maximal function we need to extend u to the unit disk. For that purpose we use the harmonic extension. Recall the Poisson kernel for the unit disk,

$$P_t(z) = \frac{1 - |z|^2}{|e^{it} - z|^2},$$

where $z \in \mathbb{D}$, and $t \in \mathbb{R}$. The harmonic extension of $u : \mathbb{T} \to \mathbb{C}$ is obtained by integrating u against the Poisson kernel, thus

$$U(z) = \frac{1}{2\pi} \int_{-\pi}^{\pi} u(e^{it}) P_t(z) dt. \tag{6.1.3}$$

For $U : \mathbb{D} \to \mathbb{C}$ the Lusin function $S_L(U)(t)$ and the non-tangential maximal function $U^*(t)$ are defined by (6.1.1), respectively (6.1.2). Now we define $S_L(u)$ and u^* by putting

$$S_L(u)(t) = S_L(U)(t) \quad \text{and} \quad u^*(t) = U^*(t).$$

We emphasize that for a function u defined on the boundary of the unit disk the Lusin function $S_L(u)$ and the non-tangential maximal function u^* are given only through the harmonic extension of u.

Next we recall a basic theorem of Hardy and Littlewood concerning the non-tangential maximal function F^* when F is *analytic* in the unit disk. Precisely let $F : \mathbb{T} \to \mathbb{C}$ be integrable. Assume that the harmonic extension of F to the unit disk, is *analytic* in $\mathbb{D}$. Then, by the theorem of Hardy and Littlewood [82] for every $p > 0$ the following estimate holds,

$$\int_{-\pi}^{\pi} |F^*(e^{it})|^p dt \leq C \int_{-\pi}^{\pi} |F(e^{it})|^p dt. \tag{6.1.4}$$

The analyticity of F is responsible for (6.1.4) to hold for every $p > 0$. Without analyticity the validity of (6.1.4) is limited to the range of p strictly larger than 1. In fact if $u : \mathbb{T} \to \mathbb{C}$ is integrable, and $1 < p < \infty$, then

$$\int_{-\pi}^{\pi} |u^*(e^{it})|^p dt \leq C_p^p \int_{-\pi}^{\pi} |u(e^{it})|^p dt,$$

where $C_p \to \infty$ as $p \to 1$. By the maximum principle for harmonic functions the above estimate remains true in the limiting case where $p \to \infty$. In particular C_p

stays uniformly bounded as $p \to \infty$. A proof of the Hardy–Littlewood theorem is presented in [72], Theorem II.3.1.

The Lusin area function and the non-tangential maximal function show similar boundedness properties in the reflexive L^p spaces. Indeed, a result of J. Marcinkiewicz and A. Zygmund [134] asserts that for $1 < p < \infty$,

$$\int_{-\pi}^{\pi} |S_L(u)(e^{it})|^p dt \leq A_p^p \int_{-\pi}^{\pi} |u(e^{it})|^p dt. \tag{6.1.5}$$

As $p \to 1$ we have $A_p \to \infty$. For large values of p the constants in the theorem of J. Marcinkiewicz and A. Zygmund are not uniformly bounded. They satisfy $A_p \sim \sqrt{p}$, as $p \to \infty$. From a modern point of view, originating with L. Hörmander [86], the Lusin area function is regarded as an average of singular integrals. Thus (6.1.5) can be obtained from the L^p boundedness of singular integral operators, see [192].

Finally we recall the definition of the Hilbert transform. Let u be an integrable function defined on the boundary of the unit disk. Equivalently we may regard u as a 2π periodic function on the real line. Then for $t \in]-\pi, \pi[$ we define

$$H(u)(t) = \lim_{\epsilon \to 0} \frac{1}{2\pi} \left\{ \int_{-\pi}^{t-\epsilon} + \int_{t+\epsilon}^{\pi} u(y) \cot \frac{y-t}{2} dy \right\}. \tag{6.1.6}$$

N. Lusin showed that $Hu(t)$ exists for almost every $t \in]-\pi, \pi[$ provided that $u \in L^2$. In the case where u is merely integrable the existence (almost everywhere) of the limit (6.1.6) is a result of I. Privalov (the book of P. Koosis [118] contains detailed proofs and an interesting historic discussion). Extended to the real line 2π periodically, the Hilbert transform gives a function on the boundary of the unit disk.

Let $u : \mathbb{T} \to \mathbb{R}$ be integrable and assume that also the Hilbert transform $H(u) : \mathbb{T} \to \mathbb{R}$ is integrable. For $z \in \mathbb{D}$ the equation

$$F(z) = \frac{1}{2\pi} \int_{\pi}^{\pi} \{u(e^{it}) + iH(u)(e^{it})\} P_t(z) dt$$

defines an analytic function $F(z)$ in $\mathbb{D}$. Thus the harmonic extension of $u + iH(u)$ is *analytic* in the unit disk.

We equip the unit circle $\mathbb{T} = \{z \in \mathbb{D} : |z| = 1\}$ with the metric given by the Euclidean distance, and with linear Lebesgue measure. Thus defined the triple $(\mathbb{T}, \mathrm{dt}, |\cdot|)$ is a space of homogeneous type (for definitions see Section 2.1 in Chapter 2). An atom for $(\mathbb{T}, \mathrm{dt}, |\cdot|)$ is either a constant function on $\mathbb{T}$ or a function $a : \mathbb{T} \to \mathbb{R}$ such that

$$\int_{\mathbb{T}} a dt = 0 \quad \text{and} \quad \int_{\mathbb{T}} a^2 dt \leq |I|^{-1},$$

where I is an interval in $\mathbb{T}$ such that

$$\operatorname{supp} a \subseteq I.$$

In Section 2.1 we defined the associated atomic H^1 space to be the space of functions $f : \mathbb{T} \to \mathbb{R}$ which admit a decomposition into atoms,

$$f = \sum c_i a_i,$$

where a_i are atoms for $(\mathbb{T}, \mathrm{dt}, |\cdot|)$ and where

$$\sum |c_i| < \infty.$$

We denote this space by H^1_{at}. The norm of $f \in H^1_{at}$ is

$$||f||_{H^1_{at}} = \inf\{\sum |c_i|\},$$

where the infimum is taken over all decompositions of f into atoms. In this section we prove three different characterizations for a function $u : \mathbb{T} \to \mathbb{R}$ to be in the space H^1_{at}. We show that $u \in H^1_{at}$ is equivalent to $(u + iHu) \in L^1(\mathbb{T})$, as well as to $u^* \in L^1(\mathbb{T})$, and to $S_L(u) \in L^1(\mathbb{T})$. We summarize now the inequalities by which we obtain these characterizations.

1. The equivalence between the conditions $u^* \in L^1(\mathbb{T})$, and $u \in H^1_{at}$, is expressed by the two-sided estimates

$$c\|u\|_{H^1_{at}} \leq \|u^*\|_{L^1(\mathbb{T})} \leq C\|u\|_{H^1_{at}}. \tag{6.1.7}$$

The proof of (6.1.7) uses the following three results. First, a function in H^1_{at} has an integrable Hilbert transform. Second, a function with an integrable non-tangential maximal function belongs to H^1_{at}. These properties of H^1_{at} are established in Proposition 6.1.3 and Proposition 6.1.2. They give

$$c\|Hu\|_{L^1(\mathbb{T})} \leq \|u\|_{H^1_{at}} \leq C\|u^*\|_{L^1(\mathbb{T})}. \tag{6.1.8}$$

The result of Hardy and Littlewood (6.1.4) provides the third component in the proof of the maximal function characterization of H^1_{at}. We use it to show that (6.1.8) implies (6.1.7).

2. The Lusin area function and the Hilbert transform are connected by the following result of A. Calderón [39],

$$c\|S_L(u)\|_{L^1(\mathbb{T})} \leq \|(u + iHu)\|_{L^1(\mathbb{T})} \leq C\|S_L(u)\|_{L^1(\mathbb{T})}. \tag{6.1.9}$$

We present A. Calderón's original proof which is an impressive example of elegance and power. Its key is a remarkable integral inequality, the so-called square-duality relation. Let $F : \mathbb{D} \to \mathbb{C}$ denote the harmonic extension of $u + iH(u)$. Then the Lusin function of $|F|^{1/3} : \mathbb{D} \to \mathbb{R}$ satisfies the estimate

$$\begin{aligned} &c\int_{\mathbb{T}} S_L^2\left(|F|^{1/3}\right) G dt \\ &\leq \int_{\mathbb{T}} |F|^{2/3} G dt + \int_{\mathbb{T}} \left(|F|^{1/3}\right)^* S_L\left(|F|^{1/3}\right) S_L(G) dt, \end{aligned}$$

whenever $G \in L^3(\mathbb{T})$ is non-negative. We obtain important applications of the square-duality relation by exploiting that the homogeneities on its right-hand side and left-hand side are different. Indeed we have $S^2(|F|^{1/3})$ on the left, where as $S(|F|^{1/3})$ appears on the right-hand side.

Characterization by maximal functions

We start with the maximal function characterization of H^1_{at}.

Theorem 6.1.1. *Let $u : \mathbb{T} \to \mathbb{R}$ be integrable, then the following a priori estimates hold,*

$$c\|u\|_{H^1_{at}} \leq \|u^*\|_{L^1(\mathbb{T})} \leq C\|u\|_{H^1_{at}}.$$

The starting point is the following proposition which shows that the Hilbert transform is well behaved on the atoms of $\mathbb{T}$.

Proposition 6.1.2. *Let a be an atom for $\mathbb{T}$. Then the Hilbert transform of a is integrable, satisfying*

$$\int_{-\pi}^{\pi} |H(a)|(t)dt \leq C,$$

where C is a universal constant

Proof. Let $I \subseteq \mathbb{T}$ be an interval. Let $a : \mathbb{T} \to \mathbb{R}$ satisfy $\operatorname{supp} a \subseteq I$,

$$||a||_2 \leq |I|^{-1/2},$$

and

$$\int_{\mathbb{T}} a = 0.$$

Assume that there exists $N \in \mathbb{N}$ such that $|I| = 2^{-N}$. For $0 \leq n \leq N$ we define $(2^n I) \subseteq \mathbb{T}$ to be the interval with the same midpoint as I satisfying $|(2^n I)| = 2^n |I|$. Next put

$$J_n = \{t \in [-\pi, \pi[: e^{it} \in (2^n I)\}.$$

Note that while $(2^n I)$ is an interval in $\mathbb{T}$ by definition, it is not necessarily true that J_n is an interval in $[-\pi, \pi[$. (That happens when $e^{i\pi} \in (2^n I)$ and $(2^n I) \neq \mathbb{T}$.)

To estimate the L^1 norm of $H(a)$ we treat separately the integrals

$$\int_{J_1} |H(a)|(t)dt, \tag{6.1.10}$$

and

$$\int_{[-\pi,\pi[\setminus J_1} |H(a)|(t)dt. \tag{6.1.11}$$

First we estimate (6.1.10). By the Cauchy–Schwarz inequality and the L^2 boundedness of the Hilbert transform we have that

$$\begin{aligned}\int_{J_1} |H(a)|(t)dt &\leq |J_1|^{1/2}\|H(a)\|_2 \\ &\leq C|I|^{1/2}\|a\|_2 \\ &\leq C.\end{aligned}$$

Now we turn to the corresponding estimate for (6.1.11). For $t \notin J_1$ we have

$$H(a)(t) = \frac{1}{2\pi}\int_{J_0} a(y)\cot\frac{y-t}{2}dy,$$

where the integral on the right-hand side is absolutely converging. Let e^{it_I} denote the midpoint of I. Next use that $\int_{\mathbb{T}} a = 0$, to find the identity

$$H(a)(t) = \int_{J_0} a(y)\left\{\cot\frac{y-t}{2} - \cot\frac{t_I-t}{2}\right\}dy.$$

Let $n \leq N$, let $t \in J_n \setminus J_{n-1}$, and let $y \in J_0$, then we have the following pointwise estimate, resulting from the periodicity of the cotangent function

$$\left|\cot\frac{y-t}{2} - \cot\frac{t_I-t}{2}\right| \leq \frac{C|I|}{|(2^nI)|^2}.$$

Invoking this estimate we obtain that

$$\int_{J_n\setminus J_{n-1}} |H(a)(t)|dt \leq C2^{-n}\int_{-\pi}^{\pi}|a(t)|dt.$$

Summing up we showed the estimate

$$\begin{aligned}\int_{[-\pi,\pi[\setminus J_1} |H(a)|(t)dt &= \sum_{n=1}^{N}\int_{J_n\setminus J_{n-1}} |H(a)(t)|dt \\ &\leq C\int_{-\pi}^{\pi}|a(t)|dt.\end{aligned}$$

The Cauchy–Schwarz inequality implies that $\int_{\mathbb{T}}|a(t)|dt \leq 1$. $\square$

If $u \in H^1_{at}$, then $u = \sum\lambda_I a_I$ where the a_I are atoms and where $\lambda_I \in \mathbb{R}$ satisfies

$$\sum|\lambda_I| \leq C\|u\|_{H^1_{at}}.$$

Hence Proposition 6.1.2 implies that $H(u)$ is integrable and

$$\int_{-\pi}^{\pi}|u + iH(u)|dt \leq C\|u\|_{H^1_{at}}. \tag{6.1.12}$$

Next we prove that functions with integrable non-tangential maximal function are contained in H^1_{at}. We show, if u^* is integrable, then u can be decomposed into an absolutely convergent series of atoms.

Proposition 6.1.3. *Let $u : \mathbb{T} \to \mathbb{R}$ be integrable and assume that also u^*, its non-tangential maximal function, is integrable. Then $u \in H^1_{at}$, and*

$$\|u\|_{H^1_{at}} \le C \int_{-\pi}^{\pi} u^* dt. \tag{6.1.13}$$

Proof. Let $k \in \mathbb{Z}$ and define

$$E_k = \{e^{it} \in \mathbb{T} : u^*(t) > 2^k\}.$$

As u^* is the supremum of continuous functions, E_k is an open set in $\mathbb{T}$. We let $\mathcal{C}_k$ be the collection of disjoint open intervals such that

$$E_k = \bigcup_{I \in \mathcal{C}_k} I.$$

Let $I \in \mathcal{C}_k$. Denote by $m_I(u)$ the average of u over the interval I. We claim that for $I \in \mathcal{C}_k$ the mean value of u over I admits the upper bound

$$|m_I(u)| \le C2^k. \tag{6.1.14}$$

Skipping over the details, we remark that (6.1.14) can be verified using the following version of the mean value property: For a smooth harmonic function defined on a square, the average over the sides of the boundary is equal to the average over the diagonals. For details consult the book of C. Bennett and R. Sharpley [8] p. 370.

Next we define

$$a_I = 2^{-k}|I|^{-1}\left[(u - m_I(u))1_I - \sum_{\{J \in \mathcal{C}_{k+1} : J \subseteq I\}} (u - m_J(u))1_J\right].$$

Observe that $\operatorname{supp} a_I \subseteq I$, and

$$\int_{\mathbb{T}} a_I dt = 0.$$

The claimed estimate (6.1.14) implies

$$\|a_I\|_\infty \le C|I|^{-1}.$$

Putting $\lambda_I = 2^k|I|$ we obtain

$$u = \sum_{k=-\infty}^{\infty} \sum_{I \in \mathcal{C}_k} \lambda_I a_I.$$

Next we claim that the series on the right-hand side represents an atomic decomposition of u which gives correct estimates for $\|u\|_{H^1_{at}}$. To verify the claim we recall the definition of the sequence λ_I and estimate

$$\begin{aligned}\sum_{k=-\infty}^{\infty} \sum_{I\in\mathcal{C}_k} \lambda_I &= \sum_{k=-\infty}^{\infty} 2^k \sum_{I\in\mathcal{C}_k} |I| \\ &= \sum_{k=-\infty}^{\infty} 2^k |E_k| \\ &\le C \int_{-\pi}^{\pi} u^*\, dt.\end{aligned}$$

□

Proof of Theorem 6.1.1. By (6.1.12) for every $u \in H^1_{at}$ we have

$$\int_{-\pi}^{\pi} |H(u)|\, dt \le C\|u\|_{H^1_{at}}.$$

Write

$$F = u + iH(u).$$

The harmonic extension of F to the unit disk defines an analytic function. Hence (6.1.4) — the estimate of Hardy and Littlewood — is applicable. It implies that F^* is integrable and satisfies

$$\int_{-\pi}^{\pi} F^* dt \le C \int_{\pi}^{\pi} |F| dt.$$

Taking real parts in the definition of F^* and using (6.1.12) we obtain the following estimate for $u \in H^1_{at}$,

$$\int_{-\pi}^{\pi} u^* dt \le C\|u\|_{H^1_{at}}. \tag{6.1.15}$$

Conversely by Proposition 6.1.3, if $u^* \in L^1(\mathbb{T})$, then u admits an atomic decomposition showing that

$$\|u\|_{H^1_{at}} \le C \int_{-\pi}^{\pi} u^* dt. \tag{6.1.16}$$

Combining (6.1.15) and (6.1.16) gives the maximal function characterization as claimed. □

Observe that by proving Theorem 6.1.1 we showed simultaneously that

$$c\|(u + iH(u)\|_{L^1(\mathbb{T})} \le \|u\|_{H^1_{at}} \le C\|(u + iH(u)\|_{L^1(\mathbb{T})}.$$

Characterization by square functions

Now we study the connection between the Lusin square function and the Hilbert transform. The next result provides the square function characterization for H^1_{at} due to A. Calderón.

Theorem 6.1.4. *Let $u : \mathbb{T} \to \mathbb{R}$ be integrable, then the following a priori estimates hold,*

$$c\|S_L(u)\|_{L^1(\mathbb{T})} \leq \|(u + iHu)\|_{L^1(\mathbb{T})} \leq C\|S_L(u)\|_{L^1(\mathbb{T})}.$$

Before we start proving Calderón's theorem we should single out some of the ingredients that are used in the proof. We will repeatedly apply Green's theorem in the following form. Let u be a smooth function defined in a neighborhood of the unit disk, then

$$\frac{1}{\pi}\iint_{\mathbb{D}} \Delta u(z) \log \frac{1}{|z|} dA(z) = \frac{1}{2\pi}\int_{-\pi}^{\pi} u(e^{it})dt - u(0). \tag{6.1.17}$$

We also use a very interesting set of identities holding for the modulus of analytic functions: Let F be analytic, $p \geq 1$ and $q > 0$, then

$$\left|\text{grad}|F|^{p/q}\right| = p|F|^{(p-1)/q}\left|\text{grad}|F|^{1/q}\right|. \tag{6.1.18}$$

This identity relates first order differentiation of various powers of $|F|$. The next identity is between the Laplacian, thus second order differentiation, and the gradient of $|F|$. Let $q > 0$ and let F be analytic. Then

$$\Delta(|F|^{2/q}) = 4\left|\text{grad}|F|^{1/q}\right|^2. \tag{6.1.19}$$

As a first application of the identity (6.1.19) and Green's Theorem (6.1.17) we prove the following lemma. It will be used several times below.

Lemma 6.1.5. *Let $F : \mathbb{T} \to \mathbb{C}$ be integrable. Assume that the harmonic extension of F to the unit disk is analytic, and that $F(0) = 0$. Then*

$$c\|F\|_{L^1(\mathbb{T})} \leq \left\|S_L^2\left(|F|^{1/2}\right)\right\|_{L^1(\mathbb{T})} \leq C\|F\|_{L^1(\mathbb{T})},$$

where

$$S_L^2\left(|F|^{1/2}\right)(e^{it}) = \iint_{\Gamma_t}\left|\text{grad}|F(z)|^{1/2}\right|^2 dA(z).$$

Proof. The starting point is Green's theorem. It expresses the $L^1(\mathbb{T})$ norm as an area integral. Recall that $F(0) = 0$, hence (6.1.17) gives that

$$\frac{1}{2\pi}\int_{-\pi}^{\pi}|F|(e^{it})dt = \frac{1}{\pi}\iint_{\mathbb{D}}\Delta(|F|)\log\frac{1}{|z|}dA(z).$$

Next by (6.1.19) we may replace the Laplacian in the above integral as follows,

$$\frac{1}{\pi}\iint_{\mathbb{D}} \Delta(|F|)\log\frac{1}{|z|}dA(z) = \frac{4}{\pi}\iint_{\mathbb{D}} \left|\mathrm{grad}|F|^{1/2}\right|^2 \log\frac{1}{|z|}dA(z). \tag{6.1.20}$$

Next we replace the factor $\log\frac{1}{|z|}$ by $1-|z|$. This is possible since the following three simple facts hold. First for $|z| \geq 1/4$ we have that $\log\frac{1}{|z|} \sim 1-|z|$. Second $\log\frac{1}{|z|}$ is integrable over the unit disk with respect to the area measure. Third $\left|\mathrm{grad}|F|^{1/2}\right|^2$ is a subharmonic function. Combined these facts imply that the integral on the right-hand side of (6.1.20) is bounded above and below by (a constant multiple of) the integral

$$\iint_{\mathbb{D}} \left|\mathrm{grad}|F|^{1/2}\right|^2 (1-|z|)dA(z). \tag{6.1.21}$$

Applying Fubini's theorem shows that the integral in (6.1.21) is (up to universal constants) the same as

$$\int_{-\pi}^{\pi} S_L^2(|F|^{1/2})(e^{it})dt.$$ □

The next proposition contains the first part of the square function characterization of H^1_{at}.

Proposition 6.1.6. *Let $F : \mathbb{T} \to \mathbb{C}$ be integrable. Assume that the harmonic extension of F to the unit disk is analytic. Then*

$$||S_L(F)||_{L^1(\mathbb{T})} \leq C||F||_{L^1(\mathbb{T})}.$$

Proof. We start with the identity (6.1.18). Putting $p = q = 2$ we obtain that

$$|\mathrm{grad}\,|F|| = 2|F|^{1/2}\left|\mathrm{grad}\,|F|^{1/2}\right|.$$

Taking the supremum of $|F|^{1/2}$ over the Stolz domain Γ_t, and then integrating the square of $\left|\mathrm{grad}\,|F|^{1/2}\right|$ over Γ_t, gives the following estimate between the Lusin function and the non-tangential maximal function.

$$S_L(|F|) \leq 2\left(|F|^{1/2}\right)^* S_L\left(|F|^{1/2}\right).$$

Integrating and applying the Cauchy–Schwarz inequality gives that

$$\begin{aligned} \|S_L(F)\|_{L^1(\mathbb{T})} &\leq 2\left\|\left(|F|^{1/2}\right)^* S_L\left(|F|^{1/2}\right)\right\|_{L^1(\mathbb{T})} \\ &\leq \left\|\left(|F|^{1/2}\right)^*\right\|_{L^2(\mathbb{T})}\left\|S_L\left(|F|^{1/2}\right)\right\|_{L^2(\mathbb{T})}. \end{aligned} \tag{6.1.22}$$

Next we show that both factors appearing in the second line of (6.1.22) are bounded by $C||F||^{1/2}_{L^1(\mathbb{T})}$. The estimate for the factor involving $S_L\left(|F|^{1/2}\right)$ is the content of Lemma 6.1.5.

Observe that

$$\left(|F|^{1/2}\right)^* = (|F|^*)^{1/2}.$$

Hence the following inequality for the non-tangential maximal function holds true, by (6.1.4),

$$\left\|\left(|F|^{1/2}\right)^*\right\|_{L^2(\mathbb{T})} \leq C\|F\|^{1/2}_{L^1(\mathbb{T})}. \tag{6.1.23}$$

Thus with (6.1.23) and Lemma 6.1.5 the conclusion of Proposition 6.1.6 follows from (6.1.22). □

Proposition 6.1.7. *Let $F : \mathbb{T} \to \mathbb{C}$ be integrable with $\int F = 0$. Assume that the harmonic extension of F to the unit disk is analytic. Then*

$$||F||_{L^1(\mathbb{T})} \leq C||S_L(F)||_{L^1(\mathbb{T})}.$$

Proof. We evaluate the identities (6.1.18) with $p = 2$ and $q = 2$. Thus

$$|\mathrm{grad}\,|F|| = 2|F|^{1/2}\left|\mathrm{grad}\,|F|^{1/2}\right|.$$

Next we take $p = 3/2$ and $q = 3$ and we obtain

$$\left|\mathrm{grad}\,|F|^{1/2}\right| = \frac{3}{2}|F|^{1/6}\left|\mathrm{grad}\,|F|^{1/3}\right|.$$

It follows that

$$\left|\mathrm{grad}\,|F|^{1/2}\right| = \left(3/2\left|\mathrm{grad}\,|F|^{1/3}\right|\right)^{3/4}(1/2\,|\mathrm{grad}\,|F||)^{1/4}. \tag{6.1.24}$$

Next we take the square of (6.1.24), then we integrate over the cone Γ_t and finally we apply Hölder's inequality with the parameters 4/3 and 4. This gives the following pointwise estimate for the Lusin function,

$$S_L^2(|F|^{1/2}) \leq \left(\frac{3}{2}S_L\left(|F|^{1/3}\right)\right)^{3/2}\left(\frac{1}{2}S_L\left(|F|\right)\right)^{1/2}. \tag{6.1.25}$$

Next we apply Lemma 6.1.5 followed by the estimate in (6.1.25) and the Cauchy–Schwarz inequality. This gives that

$$\begin{aligned} ||F||_{L^1(\mathbb{T})} &\leq C\left\|S_L^2\left(|F|^{1/2}\right)\right\|_{L^1(\mathbb{T})} \\ &\leq C\left\|S_L^{3/2}\left(|F|^{1/3}\right)\right\|_{L^2(\mathbb{T})}\left\|S_L^{1/2}\left(|F|\right)\right\|_{L^2(\mathbb{T})}. \end{aligned} \tag{6.1.26}$$

Note that the first factor in the second line of (6.1.26) is just $\|S_L(|F|^{1/3})\|^{3/2}_{L^3(\mathbb{T})}$. Similarly we may rewrite the second factor as $\|S_L\left(|F|\right)\|^{1/2}_{L^1(\mathbb{T})}$. It is also true that $\mathrm{grad}\,|F| = |F'|$, hence the following identity holds for the Lusin function,

$$S_L\left(|F|\right) = S_L\left(F\right).$$

In summary we showed that

$$||F||_{L^1(\mathbb{T})} \leq C \left\| S_L \left(|F|^{1/3} \right) \right\|_{L^3(\mathbb{T})}^{3/2} ||S_L(F)||_{L^1(\mathbb{T})}^{1/2}. \tag{6.1.27}$$

In Proposition 6.1.8 we will prove that the following crucial inequality holds,

$$\left\| S_L \left(|F|^{1/3} \right) \right\|_{L^3(\mathbb{T})}^{3/2} \leq C \, \|F\|_{L^1(\mathbb{T})}^{1/2}. \tag{6.1.28}$$

Inserting (6.1.28) into (6.1.27) gives that $||F||_{L^1(\mathbb{T})} \leq C||S_L(F)||_{L^1(\mathbb{T})}$. □

To complete the proof of Proposition 6.1.7 we show now that (6.1.28) holds.

Proposition 6.1.8.

$$\left\| S_L \left(|F|^{1/3} \right) \right\|_{L^3(\mathbb{T})} \leq C \, \|F\|_{L^1(\mathbb{T})}^{1/3}.$$

Proof. Let $G \in L^3(\mathbb{T})$ be non-negative. We denote its harmonic extension to the unit disk again by G. We claim that the following integral inequality holds,

$$\int_{-\pi}^{\pi} S_L \left(|F|^{1/3} \right)^2 G \, dt \leq C \int_{-\pi}^{\pi} |F|^{2/3} G \, dt + C \int_{-\pi}^{\pi} \left(|F|^{1/3} \right)^* S_L \left(|F|^{1/3} \right) S_L(G) dt.$$

From this claim we derive the conclusion of Proposition 6.1.8 as follows. We estimate the first integral on the right-hand side using Hölder's inequality with the parameters $1/3 + 2/3 = 1$. For the second integral use Hölder's inequality with $1/3 + 1/3 + 1/3 = 1$.

$$\begin{aligned} \int_{-\pi}^{\pi} S_L^2 \left(|F|^{1/3} \right) G \leq{}& C \left\| |F|^{2/3} \right\|_{L^{3/2}(\mathbb{T})} \|G\|_{L^3(\mathbb{T})} \\ &+ C \left\| \left(|F|^{1/3} \right)^* \right\|_{L^3(\mathbb{T})} \left\| S_L \left(|F|^{1/3} \right) \right\|_{L^3(\mathbb{T})} \|S_L \left(G \right) \|_{L^3(\mathbb{T})}. \end{aligned} \tag{6.1.29}$$

Let us now comment on estimates for the third and the first factor appearing in the second line of (6.1.29). Recall that by (6.1.5) a harmonic function G in the unit disk satisfies $\|S_L \left(G \right) \|_{L^3(\mathbb{T})} \leq C\|G\|_{L^3(\mathbb{T})}$. As F is analytic, the estimate of Hardy and Littlewood (6.1.4) implies that taking the non-tangential maximal function is a bounded operation in $L^1(\mathbb{T})$. Rewriting (6.1.4) gives that

$$\left\| \left(|F|^{1/3} \right)^* \right\|_{L^3(\mathbb{T})} \leq C \left\| F^{1/3} \right\|_{L^3(\mathbb{T})}.$$

Notice also that clearly $\left\| F^{1/3} \right\|_3 = \|F\|_1^{1/3}$. Hence taking the supremum in (6.1.29) over all non-negative G with $\|G\|_3 \leq 1$, we find that

$$\left\| S_L^2 \left(|F|^{1/3} \right) \right\|_{L^{3/2}(\mathbb{T})} \leq \|F\|_{L^1(\mathbb{T})}^{2/3} + C \, \|F\|_{L^1(\mathbb{T})}^{1/3} \left\| S_L \left(|F|^{1/3} \right) \right\|_{L^3(\mathbb{T})}.$$

By simple arithmetic, the homogeneities of this inequality give the estimate

$$\left\| S_L\left(|F|^{1/3}\right)\right\|_{L^3(\mathbb{T})} \leq C\,\|F\|^{1/3}_{L^1(\mathbb{T})}.$$

Now we turn to the proof of the square-duality relation,

$$\int_{-\pi}^{\pi} S_L^2\left(|F|^{1/3}\right)G \leq C\int_{-\pi}^{\pi}|F|^{2/3}G + C\int_{-\pi}^{\pi}\left(|F|^{1/3}\right)^* S_L\left(|F|^{1/3}\right)S_L(G). \tag{6.1.30}$$

Recalling first the definition of the Lusin function and then that G is non-negative we find with Fubini's theorem the estimate

$$\int_{-\pi}^{\pi} S_L^2\left(|F|^{1/3}\right)Gdt \leq C\iint_{\mathbb{D}}|\mathrm{grad}(|F|^{1/3})|^2(w)\int_{-(1-|w|)\pi}^{(1-|w|)\pi}G(e^{it}\frac{w}{|w|})dtdA(w).$$

Next by pointwise estimates for the Poisson kernel and by $(1-|w|)\leq 2\log\frac{1}{|w|}$ we obtain the following estimate for the non-negative function G,

$$\int_{-(1-|w|)\pi}^{(1-|w|)\pi}G(e^{it}\frac{w}{|w|})dt \leq C\log(\frac{1}{|w|})G(w). \tag{6.1.31}$$

Inserting the estimate (6.1.31) we obtain the inequality

$$\int_{-\pi}^{\pi} S_L\left(|F|^{1/3}\right)^2 Gdt \leq C\iint_{\mathbb{D}}4G\left|\mathrm{grad}\left(|F|^{1/3}\right)\right|^2\log(\frac{1}{|z|})dA(z). \tag{6.1.32}$$

We thus replaced the integral over $\mathbb{T}$ by an integral over the unit disk $\mathbb{D}$, and thereby we made a first step towards applying Green's theorem below. The second step consists in finding pointwise upper bounds for the integrand on the right-hand side of (6.1.32). That is, we seek estimates for $G|\mathrm{grad}(|F|^{1/3})|^2$. We start by calculating

$$\Delta(|F|^{2/3}G) = G\Delta(|F|^{2/3}) + 2\mathrm{grad}(|F|^{2/3})\cdot\mathrm{grad}(G). \tag{6.1.33}$$

We rewrite the identity (6.1.33) using that F is an analytic function. By (6.1.19) with $q=3$ we have

$$\Delta(|F|^{2/3}) = 4\left|\mathrm{grad}(|F|^{1/3})\right|^2.$$

Equation (6.1.18) with $p=2$ and $q=3$ gives that

$$\left|\mathrm{grad}(|F|^{2/3})\right| = 2|F|^{1/3}\left|\mathrm{grad}(|F|^{1/3})\right|.$$

Inserting back into equation (6.1.33) and rearranging terms gives the inequality

$$4G\left|\mathrm{grad}(|F|^{1/3})\right|^2 \leq \Delta(G|F|^{2/3}) + 4|F|^{1/3}\left|\mathrm{grad}(|F|^{1/3})\right|\cdot|\mathrm{grad}(G)|. \tag{6.1.34}$$

We return to estimating integrals. Multiply the inequality (6.1.34) by $\log(1/|z|)$ and integrate over the unit disk. This gives that

$$\begin{aligned}&\int_{\mathbb{D}} 4G\left|\operatorname{grad}(|F|^{1/3})\right|^2 \log\frac{1}{|z|}dA(z)\\ &\leq \int_{\mathbb{D}} \Delta(|F|^{2/3}G)\log\frac{1}{|z|}dA(z) + 4\int_{\mathbb{D}} |F|^{1/3}|\operatorname{grad}(|F|^{1/3})|\,|\operatorname{grad}(G)|\log\frac{1}{|z|}dA(z).\end{aligned} \tag{6.1.35}$$

Applying Green's theorem we rewrite the first integral appearing in the second line of (6.1.35). By Theorem 6.1.5 we obtain that

$$2\int_{\mathbb{D}} \Delta(|F|^{2/3}G)\log\frac{1}{|z|}dA(z) = \int_{-\pi}^{\pi} |F|^{2/3}G dt. \tag{6.1.36}$$

Finally we estimate the second integral using Fubini's theorem and the Cauchy Schwarz inequality. This gives

$$\int_{\mathbb{D}} |F|^{1/3}\left|\operatorname{grad}(|F|^{1/3})\right|\cdot|\operatorname{grad}(G)|\log\frac{1}{|z|}dA(z) \leq C\int_{-\pi}^{\pi} (|F|^{1/3})^* S_L(|F|^{1/3})S_L(G). \tag{6.1.37}$$

Summing up we showed that (6.1.32) followed by (6.1.35)– (6.1.37) implies the inequality

$$\int_{-\pi}^{\pi} S_L\left(|F|^{1/3}\right)^2 G \leq C\int_{-\pi}^{\pi} |F|^{2/3}G + C\int_{-\pi}^{\pi} \left(|F|^{1/3}\right)^* S_L\left(|F|^{1/3}\right) S_L(G). \qquad \square$$

For a power series $F(z) = \sum a_n z^n$ in the unit disk, R. E. A. C. Paley [168] shows that the sequence of lacunary coefficients is square summable provided that the boundary values are integrable on the unit circle. Paley's classic inequality asserts that

$$\left(\sum |a|_{2^n}^2\right)^{1/2} \leq \frac{1}{\pi}\int_{-\pi}^{\pi} |F(e^{it})|dt.$$

The square function estimate of Proposition 6.1.7 reveals the connection between Paley's classic inequality and J. Bourgain's estimate (1.3.32) for Rademacher projections on H^1. Indeed, by Proposition 6.1.7 and Paley's inequality, it follows that

$$\sum |a|_{2^n}^2 \leq C\int_{-\pi}^{\pi} |F(e^{it})|dt \int_{-\pi}^{\pi} S_L(F)(e^{it})|dt.$$

6.2 Carleson's biorthogonal system

In this section we prove B. Maurey's result that H^1 and H^1_{at} are isomorphic Banach spaces, and that there exists an unconditional basis in H^1_{at}. We present L. Carleson's approach to B. Maurey's theorem since it determines explicitly the unconditional basis in H^1_{at} and gives a simple characterization of its dual space using biorthogonal coefficients.

Recall that H^1_{at} is the atomic H^1 space defined over the space of homogeneous type $(\mathbb{T}, |\cdot|, dt)$. In Chapter 2, on page 124, we showed that the dual space to H^1_{at} can be represented as the space of integrable functions $h : \mathbb{T} \to \mathbb{R}$ satisfying

$$\sup_I \int_I |h - m_I(h)|^2 \frac{dt}{|I|} < \infty,$$

where the supremum is extended over all intervals $I \subseteq \mathbb{T}$. We find it convenient to identify functions on the unit circle with one-periodic functions on the real line. Intervals on $\mathbb{T}$ correspond to sets of the form $A + \mathbb{Z}$ where A is an interval on the real line satisfying $|A| < 1$. Having identified functions on $\mathbb{T}$ with one-periodic functions on $\mathbb{R}$ we denote the dual space of H^1_{at} by $\mathrm{BMO}([0,1))$.

This section is devoted to the analysis of Carleson's system $\{1_{[0,1[}\} \cup \{\varphi_J : J \in \mathcal{D}\}$, defined below in (6.2.3), and its relation to the spaces H^1_{at} and $\mathrm{BMO}([0,1))$. Carleson's system consists of continuous and piecewise linear functions which form an unconditional basis in L^2. Moreover it is equivalent to the L^2 normalized Haar system. Hence there exists a biorthogonal system $\{\psi_I : I \in \mathcal{D}\}$ defined uniquely by the equations

$$\int \psi_I = 0 \quad \text{and} \quad \langle \varphi_J, \psi_I \rangle = \delta_{I,J}.$$

In this section we establish the following results of L. Carleson [45].

1. Let $f \in L^2$, then the biorthogonal coefficients $\{\langle f, \psi_I \rangle : I \in \mathcal{D}\}$ determine the norm in L^2,

$$c\|f\|_2^2 \le \langle f, 1_{[0,1]} \rangle^2 + \sum_{I \in \mathcal{D}} \langle f, \psi_I \rangle^2 \le C\|f\|_2^2.$$

2. Let $f \in \mathrm{BMO}([0,1))$, then the biorthogonal coefficients satisfy the upper estimate

$$|\langle f, \psi_I \rangle| \le C\|f\|_{\mathrm{BMO}([0,1))} |I|^{1/2}.$$

3. For $f \in \mathrm{BMO}([0,1))$ the coefficients $\{\langle f, \psi_I \rangle : I \in \mathcal{D}\}$ determine its norm in the following way,

$$c\|f\|^2_{\mathrm{BMO}([0,1))} \le \sup_{I \in \mathcal{D}} \frac{1}{|I|} \sum_{J \subseteq I} \langle f, \psi_J \rangle^2 \le C\|f\|^2_{\mathrm{BMO}([0,1))}.$$

4. The biorthogonal functionals $\{\psi_I\}$ form an unconditional basis for H^1_{at}. Specifically, for $g \in H^1_{at}$, with $\int_0^1 g = 0$,

$$c\|g\|_{H^1_{at}} \le \sup\{\sum |\langle g, \varphi_I \rangle \langle \psi_I, f \rangle| : \|f\|_{\mathrm{BMO}([0,1))} \le 1\} \le C\|g\|_{H^1_{at}}.$$

We present a detailed exposition of Carleson's original work. Central to this study is the verification of the coefficient estimates

$$|\langle f, \psi_I \rangle| \le C \|f\|_{\mathrm{BMO}([0,1))} |I|^{1/2}, \tag{6.2.1}$$

where C is independent of I or f. The starting point is the equation

$$\langle f, h_I \rangle = \langle f, \psi_I \rangle \langle \varphi_I, h_I \rangle + \sum_{\{J : J \ne I\}} \langle f, \psi_J \rangle \langle \varphi_J, h_I \rangle.$$

We then proceed by distinguishing between two cases. In the case where

$$|\langle f, \psi_I \rangle \langle \varphi_I, h_I \rangle| \le C |\langle f, h_I \rangle|,$$

the bound (6.2.1) for $|\langle f, \psi_I \rangle|$ follows quite easily. The converse case is characterized by the estimate

$$|\langle f, \psi_I \rangle \langle \varphi_I, h_I \rangle| \gg |\langle f, h_I \rangle|. \tag{6.2.2}$$

For (6.2.2) to hold the value of $\langle f, \varphi_I \rangle \langle \varphi_I, h_I \rangle$ must be matched with opposite sign and with precision of order $|I|$ by the value of

$$\sum_{\{J : J \ne I\}} \langle f, \psi_J \rangle \langle \varphi_J, h_I \rangle.$$

With beautiful combinatorial reasoning L. Carleson exploits this observation and turns it into a proof of the coefficient estimates (6.2.1). The core of L. Carleson's proof is formed by the so-called *compensation argument.* Starting from (6.2.2) its main objective is to establish the *compensation inequalities,* and to show that they imply (6.2.1).

The definition of Carleson's biorthogonal system

We give now the definition of the Carleson system. Let $k \in \mathbb{N}$ with $k \ge 10$. Let $\varphi : \mathbb{R} \to [-1, 1]$ be continuous, and piecewise linear with nodes at the following eight points, listed in increasing order,

$$-2^{-k+1},\ -2^{-k},\ \frac{1}{2} - 2^{-k+1},\ \frac{1}{2} - 2^{-k},\ \frac{1}{2} + 2^{-k},\ \frac{1}{2} + 2^{-k+1},\ 1 + 2^{-k},\ 1 + 2^{-k+1}.$$

Hence φ is determined by its values at these points. We put

$$\begin{array}{lll} \varphi(-2^{-k+1}) = 0, & \varphi(-2^{-k}) = +1, & \varphi(\frac{1}{2} - 2^{-k+1}) = +1, \\ \varphi(\frac{1}{2} - 2^{-k}) = 0, & \varphi(\frac{1}{2} + 2^{-k}) = 0, & \varphi(\frac{1}{2} + 2^{-k+1}) = -1, \\ \varphi(1 + 2^{-k}) = -1 & \text{and} & \varphi(1 + 2^{-k+1}) = 0. \end{array}$$

The value of the parameter 2^{-k} determines how close φ is to the first Haar function. We have

$$\int_{\mathbb{R}} |\varphi - h_{[0,1]}|^2 = 4 \cdot 2^{-k} + 4 \frac{2^{-k}}{3}.$$

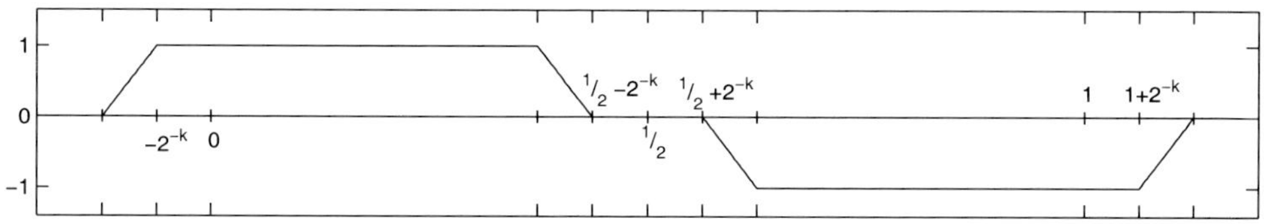

Figure 6.1: The graph of φ.

It also determines the Lipschitz constant of φ. For $x, y \in \mathbb{R}$,

$$|\varphi(x) - \varphi(y)| \leq 2^k |x - y|.$$

We define the entire system by rescaling using the single function φ, as follows. Given a dyadic interval $I \in \mathcal{D}$, let $l(I)$ be its left endpoint. For $|I| \leq 1/2$ put

$$\widetilde{\varphi}_I(x) = |I|^{-1/2} \varphi\left(\frac{x - l(I)}{|I|}\right). \tag{6.2.3}$$

Define φ_I to be the 1-periodic extension of $\widetilde{\varphi}_I$. Thus $\varphi_I \in \mathrm{BMO}([0,1))$.

For $\varphi_{[0,1]}$ (and only for $\varphi_{[0,1]}$) a minor adjustment is required; we proceed as follows. Let $a(x)$ be the affine function given by the conditions

$$a(0) = -2^{-k+1} \quad \text{and} \quad a(1) = 1 + 2^{-k+1}.$$

Then put

$$\widetilde{\varphi}_{[0,1]}(x) = \varphi(a(x)) \quad x \in \mathbb{R}.$$

Define $\varphi_{[0,1]}$ to be the 1-periodic extension of $\widetilde{\varphi}_{[0,1]}$.

We begin the investigation of the Carleson system by comparing it to the L^2 normalized Haar system. Unwinding the definition of φ_I or equivalently glancing at its graph shows that

$$\langle \varphi_I, \frac{h_I}{|I|^{1/2}} \rangle = 1 - 3 \cdot 2^{-k}, \qquad |I| \leq 1/2. \tag{6.2.4}$$

Next expand φ_I (with $|I| \leq 1/2$) using the L^2 normalized Haar system,

$$\varphi_I = (1 - 3 \cdot 2^{-k}) \frac{h_I}{|I|^{1/2}} + \sum_{\{J : J \neq I\}} H(I, J) \frac{h_J}{|J|^{1/2}}, \tag{6.2.5}$$

where we defined

$$H(I, J) = \langle \varphi_I, \frac{h_J}{|J|^{1/2}} \rangle.$$

Clearly the size of the off-diagonal coefficients $\{H(I,J) : I \neq J\}$ determines how close φ_I is to $h_I/|I|^{1/2}$. Our next proposition gives a quantitative estimate for the assertion that the first term in the Haar expansion (6.2.5) dominates the sum over $\{J : J \neq I\}$.

Proposition 6.2.1. *For every $I \in \mathcal{D}$,*

$$\sum_{\{J:J\neq I\}} |H(I,J)| + |H(J,I)| \leq 2^{-k/2+8}.$$

Proof. Let I, J be fixed dyadic intervals such that $J \neq I$. Assume without loss of generality that $J \neq [0,1]$. The relation between $|I|, |J|$ and the smoothing parameter 2^{-k} determines the value of $H(I,J)$. Let $|I| = 2^{-i}$ and $|J| = 2^{-j}$. We will distinguish between the following three cases.

1. $j \geq i + k$.
2. $i + k > j > i$.
3. $j \leq i$.

Let $j \geq i + k$. Let $m(J)$ denote the midpoint of J. Then, using that h_J has mean zero, we find

$$\begin{aligned} |H(I,J)| &\leq \left| \int (\varphi_I - \varphi_I(m(J))) \frac{h_J}{|J|^{1/2}} \right| \\ &\leq 2^{k-2} \frac{|J|^{3/2}}{|I|^{3/2}}. \end{aligned} \tag{6.2.6}$$

Let $i + k > j > i$. Then by comparing the graph of φ_I with that of h_J, we find

$$\begin{aligned} |H(I,J)| &\leq 2^{-k} \frac{3}{2} \frac{|I|}{|I|^{1/2}|J|^{1/2}} \\ &= 2^{-k} \frac{3}{2} \frac{|I|^{1/2}}{|J|^{1/2}}. \end{aligned} \tag{6.2.7}$$

Let $j \leq i$. If $I \neq J$ then we have again the estimate

$$|H(I,J)| \leq 2^{-k} \frac{3}{2} \frac{|I|^{1/2}}{|J|^{1/2}}. \tag{6.2.8}$$

With the above estimates for the coefficients $H(I,J)$ we will now prove that

$$\sum_{\{J:J\neq I\}} |H(I,J)| \leq 2^{-k/2} 2^6. \tag{6.2.9}$$

Let $j \leq i$. Then the cardinality of the collection $\{J : H(I,J) \neq 0, |J| = 2^{-j}\}$ is bounded by 2. By (6.2.8) we obtain that

$$\begin{aligned} \sum_{j=0}^{i} \sum_{\{J:|J|=2^{-j}\}} |H(I,J)| &\leq 3 \cdot 2^{-k} \sum_{j=0}^{i} 2^{-i/2+j/2} \\ &\leq 3\sqrt{2}(\sqrt{2}+1) \cdot 2^{-k}. \end{aligned} \tag{6.2.10}$$

Let $i+k>j>i$. The cardinality of the collection $\{J : J \neq I,\ H(I,J) \neq 0,\ |J| = 2^{-j}\}$ is now bounded by 4. Thus by (6.2.7),

$$\begin{aligned}\sum_{j=i+1}^{i+k-1} \sum_{\{J:|J|=2^{-j}\}} |H(I,J)| &\leq 6 \cdot 2^{-k} \sum_{j=i+1}^{i+k-1} 2^{-i/2+j/2} \\ &\leq 6(\sqrt{2}+1) \cdot 2^{-k/2}.\end{aligned} \tag{6.2.11}$$

Let $j \geq i+k$. The cardinality of the collection $\{J : H(I,J) \neq 0, |J| = 2^{-j}\}$ equals $4 \cdot 2^{j-i-k}$. Hence by (6.2.6),

$$\begin{aligned}\sum_{j=i+k}^{\infty} \sum_{\{J:|J|=2^{-j}\}} |H(I,J)| &\leq 2 \sum_{j=i+k}^{\infty} (2^{j-i-k})(2^k 2^{3i/2-3j/2}) \\ &\leq 2 \sum_{j=i+k}^{\infty} 2^{i/2-j/2} \\ &\leq 2\sqrt{2}(\sqrt{2}+1) \cdot 2^{-k/2}.\end{aligned} \tag{6.2.12}$$

Taking the sum of the estimates (6.2.10)–(6.2.12) gives (6.2.9). Parallel considerations show that

$$\sum_{\{J:J\neq I\}} |H(J,I)| \leq 2^{-k/2} 2^7. \qquad \square$$

The estimate of Proposition 6.2.1 is used to show that the Carleson system is equivalent to the L^2 normalized Haar system in L^2. It allows us to use a perturbation argument based on Neumann series and Riesz–Thorin interpolation.

Theorem 6.2.2. *Carleson's system is equivalent to the L^2 normalized Haar system. The operator $T : L^2 \to L^2$, defined by the relations $h_I/|I|^{1/2} \mapsto \varphi_I$ and $1_{[0,1]} \mapsto 1_{[0,1]}$, is close to the identity*

$$|| \operatorname{Id} - T||_{L^2,L^2} \leq 2^{-k/2+8}.$$

For $k \geq 17$ the operator T is an isomorphism, and the Carleson system is an unconditional basis in L^2.

Proof. Let I be a dyadic interval. For convenience assume that $|I| \leq 1/2$. We expand $T(h_I/|I|^{1/2}) = \varphi_I$ in its Haar series using the the matrix $H = (H(I,J))$.

$$\varphi_I = (1 - 3 \cdot 2^{-k}) \frac{h_I}{|I|^{1/2}} + \sum_{\{J:J\neq I\}} H(I,J) \frac{h_J}{|J|^{1/2}}.$$

Next, for any sequence of scalars $\{c_I\}$,

$$\left\|(1-3\cdot 2^{-k})\sum_I c_I h_I/|I|^{1/2} - \sum_I c_I\varphi_I\right\|_2^2 = \left\|\sum_{I\in\mathcal{D}}\sum_{\{J:J\neq I\}} c_I H(I,J)h_J/|J|^{1/2}\right\|_2^2$$

$$= \sum_{J\in\mathcal{D}}\left[\sum_{\{I:J\neq I\}} c_I H(I,J)\right]^2 .$$

By Proposition 6.2.1 we have

$$\sup_I \sum_{\{J:J\neq I\}} |H(I,J)| + |H(J,I)| \leq 2^{-k/2+8}.$$

Combined with the M. Riesz interpolation theorem this estimate gives the following conclusion,

$$\sum_{J\in\mathcal{D}}\left[\sum_{\{I:J\neq I\}} c_I H(I,J)\right]^2 \leq 2^{-k+16}\sum_I c_I^2. \qquad \square$$

By Theorem 6.2.2, in the space L^2, the Carleson system is equivalent to the L^2 normalized Haar system. In particular $\{\varphi_I : I\in\mathcal{D}\}\cup\{1_{[0,1]}\}$ is an unconditional basis in L^2. It has therefore associated to it a biorthogonal system. We denote it by $\{\psi_I : I\in\mathcal{D}\}\cup\{1_{[0,1]}\}$. The equations defining the biorthogonal system are

$$\int \psi_J = 0 \quad\text{and}\quad \langle\varphi_I,\psi_J\rangle = \delta_{I,J},$$

where $\delta_{I,J} = 0$ for $I\neq J$, and $\delta_{I,I} = 1$. It follows that every $f\in L^2$ admits a unique expansion along the system, $\{\varphi_I : I\in\mathcal{D}\}\cup\{1_{[0,1]}\}$, where the coefficients are computed by integration against the biorthogonal functionals, $\{\psi_I : I\in\mathcal{D}\}\cup\{1_{[0,1]}\}$. Let $f\in L^2$, then

$$f = \langle f, 1_{[0,1]}\rangle + \sum\langle f,\psi_I\rangle\varphi_I,$$

and the convergence of the above series is in L^2. The isomorphism T of Theorem 6.2.2 established the equivalence between the L^2 normalized Haar system and $\{\varphi_I : I\in\mathcal{D}\}$. It gives that

$$c\|f\|_2^2 \leq \langle f, 1_{[0,1]}\rangle^2 + \sum\langle f,\psi_I\rangle^2 \leq C\|f\|_2^2.$$

Carleson coefficients versus Haar coefficients

In the previous subsection we showed that $\{\varphi_J\}$ has associated to it a system of biorthogonal functionals with a uniform bound on their L^2 norm. As a result every $f\in L^2$ can be expanded as

$$f = \langle f, 1_{[0,1]}\rangle + \sum\langle f,\psi_I\rangle\varphi_I. \tag{6.2.13}$$

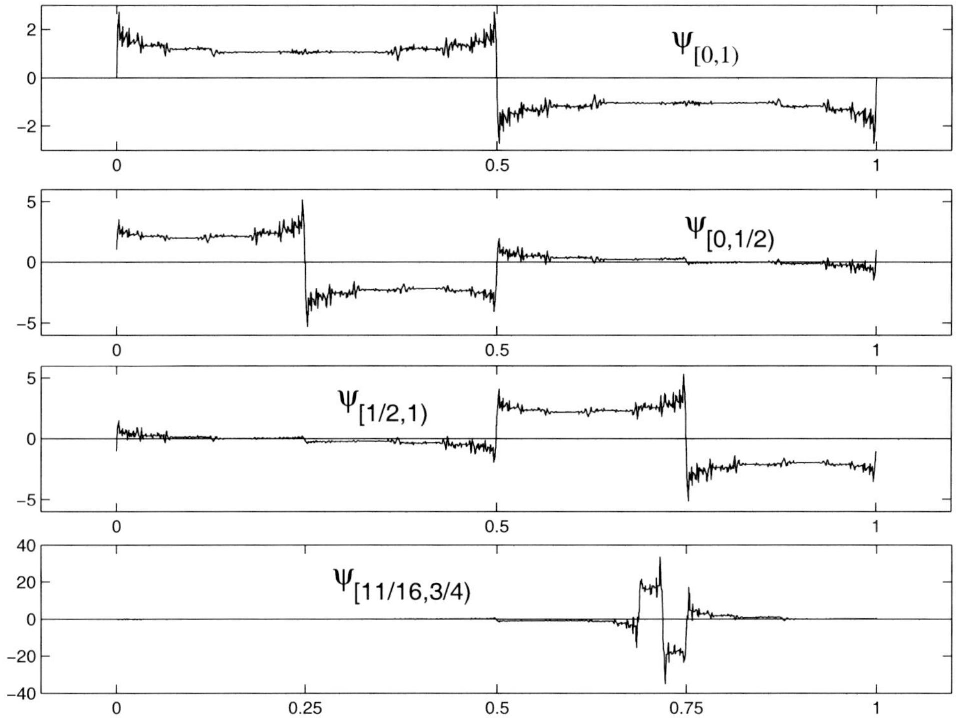

Figure 6.2: The graph of $\psi_{[0,1)}, \psi_{[0,1/2)}, \psi_{[1/2,1)}$ and $\psi_{[11/16,3/4)}$.

We obtained (6.2.13) by comparing $\{\varphi_J\}$ to the L^2 normalized Haar functions. Now we continue the comparison of the Carleson system and the Haar system. Eventually we will learn how to *exploit* the expansion on the right-hand side of the equation

$$\langle f, h_I \rangle = \sum \langle f, \psi_J \rangle \langle \varphi_J, h_I \rangle,$$

to obtain specific and non-trivial information about the size of the Haar coefficient $\langle f, h_I \rangle$. The first step consists in evaluating precisely the scalar product $\langle \varphi_J, h_I \rangle$, where $I, J \in \mathcal{D}$. This is the purpose of the following discussion which provides a thorough preparation for the most delicate part of Carleson's proof.

The values of the scalar product $\langle \varphi_J, h_I \rangle$. We fix numbers n, $q \in \mathbb{N}$ such that $n \geq q+1$. Consider I_{q-1} a dyadic interval of length 2^{-q+1}. Let μ be the midpoint of I_{q-1}. For each $j \geq q$ we define the dyadic intervals

$$J_j = [\mu - 2^{-j}, \mu), \quad \text{and} \quad I_j = [\mu, \mu + 2^{-j}). \tag{6.2.14}$$

Observe that for $j \geq q$, J_j and I_j are adjacent dyadic intervals and that

$$J_q \cup I_q = I_{q-1}.$$

The interval I_{q-1} is then the smallest dyadic interval containing I_n and J_n. Below we also need to consider the dyadic interval $R_n = I_{n-1} \setminus I_n$. Thus R_n is adjacent to the right of I_n and satisfies $|R_n| = 2^{-n}$. The intervals we listed in (6.2.14) and their relation to I_n will play a central role in this section.

Fix now $j \in \{1, \ldots, n\}$ and a dyadic interval J with

$$|J| = 2^{-j}.$$

We will evaluate the Haar coefficients $\langle \varphi_J, h_{I_n} \rangle$. In the discussion below we will see that $\langle \varphi_J, h_{I_n} \rangle$ depends on the size and the relative location of the dyadic intervals J, and I_n. In particular the value depends on the relation between n, j, q, and k. We proceed therefore very carefully. Below we consider separately the following cases for j.

1. $n \geq j \geq q$.
2. $j = q - 1$.
3. $q - 2 \geq j \geq q - k + 2$.
4. $q - k + 1 \geq j \geq 0$.

Case 1. Let $n \geq j \geq q$. In this case the only intervals J for which $\langle \varphi_J, h_{I_n} \rangle \neq 0$ are the intervals J_j defined by (6.2.14), together with I_n, I_{n-1} and $R_n = I_{n-1} \setminus I_n$. First we note that

$$\langle \varphi_{R_n}, h_{I_n} \rangle = -\frac{3}{2} 2^{-k} 2^{-n/2}, \quad \text{and} \quad \langle \varphi_{I_{n-1}}, h_{I_n} \rangle = \frac{3}{2} \sqrt{2} \cdot 2^{-k} 2^{-n/2}. \tag{6.2.15}$$

We determine now the Haar coefficients $\langle \varphi_{J_j}, h_{I_n} \rangle$. Aiming at precise values we distinguish between the following three sub-cases.

$$\text{If } j \leq n - k, \text{ then } \langle \varphi_{J_j}, h_{I_n} \rangle = 0. \tag{6.2.16}$$

$$\text{If } j = n - k + 1, \text{ then } \langle \varphi_{J_j}, h_{I_n} \rangle = -\frac{1}{2} 2^{-k} 2^{-j/2}. \tag{6.2.17}$$

$$\text{If } j \geq n - k + 2, \text{ then } \langle \varphi_{J_j}, h_{I_n} \rangle = -\frac{3}{2} 2^{-k} 2^{-j/2}. \tag{6.2.18}$$

Case 2. Let $j = q - 1$. Define $J_{q-1} = I_{q-1}$. In this case J_{q-1} is the only interval J with $|J| = 2^{-q+1}$ for which $\langle \varphi_J, h_{I_n} \rangle \neq 0$. We consider again three sub-cases as above. Consequently we find a similar set of values for the Haar coefficients, however in the present case we observe opposite signs. Precisely, the following holds.

$$\text{If } q - 1 \leq n - k, \text{ then } \langle \varphi_{J_{q-1}}, h_{I_n} \rangle = 0. \tag{6.2.19}$$

$$\text{If } q-1=n-k+1, \text{ then } \langle \varphi_{J_{q-1}}, h_{I_n} \rangle = +\frac{1}{2}2^{-k}2^{-(q-1)/2}. \tag{6.2.20}$$

$$\text{If } q-1 \geq n-k+2, \text{ then } \langle \varphi_{J_{q-1}}, h_{I_n} \rangle = +\frac{3}{2}2^{-k}2^{-(q-1)/2}. \tag{6.2.21}$$

Case 3. Let $q-2 \geq j \geq q-k+2$. In this case it is easy to see that the geometry of the graph of φ causes the Haar coefficient at I_n to vanish. Thus, for any dyadic interval K satisfying $|K| = 2^{-j}$, we obtain that

$$\langle \varphi_K, h_{I_n} \rangle = 0.$$

Case 4. Finally let $q-k+1 \geq j \geq 0$. In this case there exists exactly one dyadic interval K_j such that $|K_j| = 2^{-j}$ and $\langle \varphi_{K_j}, h_{I_n} \rangle \neq 0$. Recall that φ_{K_j} is a piecewise linear function and that its Lipschitz constant is $2^{k+3j/2}$. Moreover, when restricted to the interval I_n the function φ_{K_j} is just affine with slope $\leq 2^{k+3j/2}$. Using that the mean value of h_{I_n} vanishes, we estimate now the Haar coefficient. Let m be the midpoint of I_n. Then the following estimate holds.

$$\begin{aligned} |\langle \varphi_{K_j}, h_{I_n} \rangle| &\leq \int_{I_n} |\varphi_{K_j} - \varphi_{K_j}(m)| \\ &= 2^{+k-2}2^{+3j/2-2n}. \end{aligned}$$

We complete the preparatory considerations by collecting information about the dyadic intervals

$$\mathcal{N}_{I_n} = \{J : |J| \geq |I_n|, J \neq I_n, \langle \varphi_J, h_{I_n} \rangle \neq 0\} \setminus \{R_n, I_{n-1}\}.$$

The discussion above allows us to obtain a representation of the collection $\mathcal{N}_{I_n}$. Of course it is the relation of n to k and q that determines the structure of $\mathcal{N}_{I_n}$. We distinguish between the following two cases,

(a) $n \leq q+k-3$.

(b) $n \geq q+k-2$.

Case a. Let $n \leq q+k-3$. Then $n-k+1 \leq q-2$ and the interval J_{q-1} is the largest interval $J_j \in \{J_n, \ldots, J_{q-1}\}$ for which

$$\langle \varphi_{J_j}, h_{I_n} \rangle \neq 0.$$

It follows that

$$\mathcal{N}_{I_n} \subseteq \{J_n, \ldots, J_{q-1}, K_{q-k+1}, \ldots, K_0\}, \tag{6.2.22}$$

where for $0 \leq j \leq q-k+1$, we denote by K_j a dyadic interval of length 2^{-j}.

Case b. Let $n \geq q + k - 2$. Then $n - k + 1 \leq q - 1$ (in the limiting case $n = q+k-2$ we have $n-k+1 = q-1$). The interval J_{n-k+1} is the largest interval $J_j \in \{J_n, \ldots, J_{q-1}\}$ for which

$$\langle \varphi_{J_j}, h_{I_n} \rangle \neq 0.$$

This gives the representation

$$\mathcal{N}_{I_n} \subseteq \{J_n, \ldots, J_{n-k+1}, K_{q-k+1}, \ldots, K_0\}. \tag{6.2.23}$$

The information collected about the Haar coefficients $\langle \varphi_J, h_{I_n} \rangle$ will be needed in our analysis of Carleson's biorthogonal system.

Two helpful examples. Before we approach the main results in this section we will work out two examples in detail. We define a very simple function f and we will compare the coefficients coming from the Carleson system with its Haar coefficients. Recall that the Haar functions are normalized in L^∞ whereas the functions in the Carleson system are approximately normalized in L^2. Consequently we will be comparing the following two numbers, $\langle f, h_{I_n} \rangle$ and $\langle f, \psi_{I_n} 2^{-n/2} \rangle$. The numerical values appearing in the conclusion of our example are difficult to interpret at first sight. It may be helpful to apply a change of scale and use the following two rules of thumb: First, think of 2^{-n} as being the unit size. Second, consider 2^{-n-k} as being so small that it cannot be distinguished from zero. At the same time $2^{-n}k$ should be viewed as a very large number.

Example 6.2.3. *Let $n \geq q + k - 1$, and define*

$$f = \sum_{j=n-k+1}^{n} 2^{-j/2} \left(\varphi_{I_j} + \varphi_{J_j} \right).$$

For this function the ratio between the Haar coefficient at I_n and the coefficient with respect to the Carleson system is of order 2^{-k}. The Haar coefficient at I_n is given as

$$\langle f, h_{I_n} \rangle = \frac{3}{2} 2^{-n-k}. \tag{6.2.24}$$

The coefficient with respect to the Carleson system satisfies

$$\langle f, \psi_{I_n} 2^{-n/2} \rangle = 2^{-n}. \tag{6.2.25}$$

Moreover the integrals of f over I_n and J_n behave differently in the following sense,

$$\left| \int_{I_n} f - \int_{J_n} f \right| \geq 2^{-n}(2k - 6). \tag{6.2.26}$$

Remark. We move to the front some comments concerning the estimate (6.2.26). Dividing (6.2.26) by 2^{-n} gives a lower bound for the BMO norm of f. Indeed it follows from (6.2.26) that $\|f\|_{\mathrm{BMO}} \geq (k-3)$. Next observe that the norm of $2^{-j/2}\left(\varphi_{I_j}+\varphi_{J_j}\right)$ in BMO is bounded by 2. Hence we obtained a lower bound for $\|f\|_{\mathrm{BMO}}$ that is of the same order as the upper bound given by the triangle inequality. It follows that very little cancellation takes place in the sum defining f.

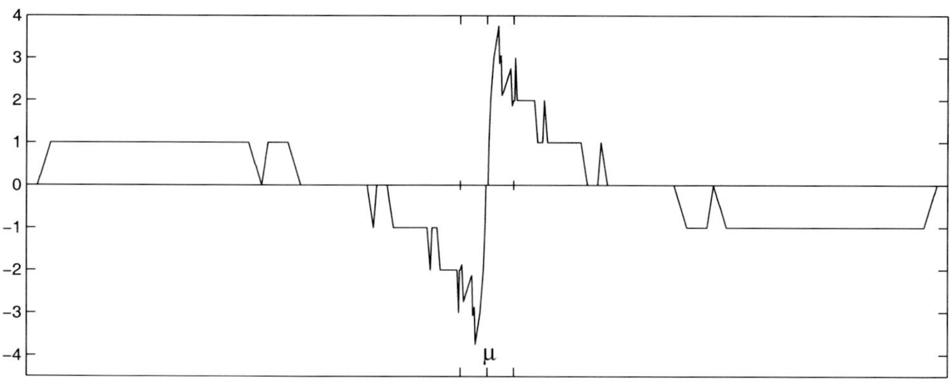

Figure 6.3: $f=\sum_{j=n-k+1}^{n} 2^{-j/2}(\varphi_{I_j}+\varphi_{J_j}), k=5, n=10, \mu=1/2.$

Proof of (6.2.24)–(6.2.26). By the biorthogonality of the functions φ_I and ψ_I we obtain immediately that $\langle f, \psi_{I_n}\rangle = 2^{-n/2}$. Hence (6.2.25) follows.

Next we evaluate the Haar coefficient of f at I_n. First observe that we already know the values of the coefficients appearing in the first line of (6.2.27) below. Indeed $\langle \varphi_{I_n}, h_{I_n}\rangle$ is determined in (6.2.4). The remaining coefficients correspond to I_{n-1} and J_j satisfying $n-k+1 \leq j \leq n$. The values of the coefficients are recorded in (6.2.15) and in the equations (6.2.17) and (6.2.18).

$$\begin{aligned}\langle f, h_{I_n}\rangle &= 2^{-n/2}\langle \varphi_{I_n}, h_{I_n}\rangle + 2^{-(n-1)/2}\langle \varphi_{I_{n-1}}, h_{I_n}\rangle + \sum_{j=n-k+1}^{n} 2^{-j/2}\langle \varphi_{J_j}, h_{I_n}\rangle \\ &= 2^{-n}(1-2^{-k}3) + 2^{-n-k}3 - \frac{1}{2}2^{-n-1} - \frac{3}{2}\sum_{j=n-k+2}^{n} 2^{-k-j}.\end{aligned} \tag{6.2.27}$$

To simplify the above expression we introduce the abbreviation

$$X = 2^k - \frac{1}{2}2^{k-1} - \frac{3}{2}\sum_{j=n-k+2}^{n} 2^{-j+n}.$$

Then collecting terms and rewriting the second line of (6.2.27) using X we find that

$$\langle f, h_{I_n}\rangle = 2^{-k-n}X.$$

Below we will calculate that $X = \frac{3}{2}$. Hence the above equation implies (6.2.24). The value of X is determined as follows. First we evaluate the geometric sum and obtain

$$\sum_{j=n-k+2}^{n} 2^{-j+n} = 2^{k-1} - 1.$$

Then we insert this value into the defining equation for X. This gives

$$X = 2^k - \frac{1}{2}2^{k-1} - \frac{3}{2}2^{k-1} + \frac{3}{2} = \frac{3}{2}.$$

Next we turn to the verification of (6.2.26). For a dyadic interval J we introduce the notation

$$\Delta(J) = \int_{I_n} \varphi_J - \int_{J_n} \varphi_J.$$

Then by the definition of f we obtain that

$$\int_{I_n} f - \int_{J_n} f = \sum_{j=n-k+1}^{n} 2^{-j/2}\left(\Delta(J_j) + \Delta(I_j)\right).$$

Next by drawing the graph of φ_{J_j} and φ_{Ij} and by taking into account the position of the intervals J_j and I_j relative to I_n we determine the values of $\Delta(J_j)$, $\Delta(I_j)$ for $n \geq j \geq n-k+1$. First notice that throughout we have $\Delta(J_j) = \Delta(I_j)$. Next consider separately the cases $j = n$, $j = n-1$ and $n-2 \geq j \geq n-k+1$. We get

$$\Delta(I_n) = \Delta(J_n) = -\frac{3}{2}2^{-k}2^{-n/2}, \tag{6.2.28}$$

$$\Delta(I_{n-1}) = \Delta(J_{n-1}) = \frac{2^{-n/2}}{\sqrt{2}} - 3\sqrt{2}\cdot 2^{-k}2^{-n/2}, \tag{6.2.29}$$

and

$$\Delta(J_j) = \Delta(I_j) = 2^{-n}2^{j/2} - \frac{3}{2}2^{-k}2^{-j/2}, \quad \text{for} \quad n-2 \geq j \geq n-k+1. \tag{6.2.30}$$

These values will now be inserted into the above equation for the difference. This gives

$$\int_{I_n} f - \int_{J_n} f = 2\cdot 2^{-n} - 15\cdot 2^{-k-n} + 2\sum_{j=n-k+1}^{n-2}(2^{-n} - \frac{3}{2}2^{-k-j}).$$

Summing a geometric series we determine the value of the above expression. We have that

$$\sum_{j=n-k+1}^{n-2} 2^{-k-j} = 2^{-n}(1 - 4 \cdot 2^{-k}).$$

Thus by inserting we find that

$$\left| \int_{I_n} f\,dt - \int_{J_n} f dt \right| \geq 2^{-n}(2k - 6). \qquad \square$$

In the following example we define a function g which appears to be very similar to the function f considered above. In fact the only difference is that the defining sum for g starts with the index $j = n - k + 2$, as opposed to $j = n - k + 1$ in the definition of f. Despite the similarities between f and g we will find that the Haar coefficient $\langle g, h_{I_n} \rangle$ is much larger than $\langle f, h_{I_n} \rangle$. It turns out that the ratio between the two Haar coefficients is of order 2^k. As we will see below it is just the absence of J_{n-k+1} which is responsible for this fact.

Example 6.2.4. *Let $n \geq q + k - 1$, and define*

$$g = \sum_{j=n-k+2}^{n} 2^{-j/2} \left(\varphi_{I_j} + \varphi_{J_j} \right).$$

Then at I_n the coefficient with respect to the Carleson system and the Haar coefficient are comparable within a factor of 4. Precisely we have $\langle g, \psi_{I_n} 2^{-n/2} \rangle = 2^{-n}$, and the Haar coefficient is

$$\langle g, h_{I_n} \rangle = \frac{1}{4} 2^{-n} + \frac{3}{2} 2^{-n-k}. \tag{6.2.31}$$

Proof. The biorthogonality of φ_I and ψ_I implies that $\langle g, \psi_{I_n} \rangle = 2^{-n/2}$. This gives the value for the coefficient with respect to the Carleson system. We evaluate the Haar coefficient of g at I_n following the pattern of the previous example. The definition of g gives

$$\begin{aligned} \langle g, h_{I_n} \rangle &= 2^{-n/2} \langle \varphi_{I_n}, h_{I_n} \rangle + 2^{-(n-1)/2} \langle \varphi_{I_{n-1}}, h_{I_n} \rangle + \sum_{j=n-k+2}^{n} 2^{-j/2} \langle \varphi_{J_j}, h_{I_n} \rangle \\ &= 2^{-n}(1 - 2^{-k}3) + 2^{-k-n}3 - \frac{3}{2} \sum_{j=n-k+2}^{n} 2^{-k-j}. \end{aligned} \tag{6.2.32}$$

To simplify the above expression we introduce the abbreviation

$$Y = 2^k - \frac{3}{2} \sum_{j=n-k+2}^{n} 2^{-j+n}.$$

We use Y to rewrite the second line of (6.2.32). Collecting terms gives

$$\langle g, h_{I_n} \rangle = +2^{-k-n} Y. \tag{6.2.33}$$

Below we will calculate that

$$Y = \frac{1}{4} 2^k + \frac{3}{2}.$$

Hence the equation (6.2.33) implies (6.2.31). The value of Y is determined quite simply as follows. As in the previous example we write down the value of the geometric sum

$$\sum_{j=n-k+2}^{n} 2^{-j+n} = 2^{k-1} - 1.$$

Then we insert into the defining equation for Y. This gives

$$\begin{aligned} Y &= 2^k - \frac{3}{2} 2^{k-1} + \frac{3}{2} \\ &= \frac{1}{4} 2^k + \frac{3}{2}. \end{aligned}$$

□

The compensation argument

The following theorem provides an upper estimate for the H^1_{at} norm of the biorthogonal functionals $\{\psi_I : I \in \mathcal{D}\}$.

Theorem 6.2.5. *The biorthogonal coefficients satisfy the inequality*

$$|\langle f, \psi_I \rangle| \le A ||f||_{BMO([0,1))} |I|^{1/2}, \tag{6.2.34}$$

where $A = A(k)$ is independent of $I \in \mathcal{D}$ or $f \in BMO([0,1))$.

To help organizing the set of coefficients $\{\langle f, \psi_I \rangle : I \in \mathcal{D}\}$ we introduce a simple and useful device. Let $m \in \mathbb{N}$, then define

$$M_m = \sup\{|\langle f, \psi_I \rangle| \cdot |I|^{-1/2} : |I| \ge 2^{-m}\}.$$

The assertion of Theorem 6.2.5 is equivalent to saying that the sequence $\{M_m\}$ is bounded by $A||f||_{\mathrm{BMO}([0,1))}$. Clearly $\{M_m\}$ is an increasing sequence. Thus it suffices to show that there exists a subsequence $n(j) \uparrow \infty$, such that

$$M_{n(j)} \le A||f||_{\mathrm{BMO}([0,1))}.$$

To get started we show the following short proposition.

Proposition 6.2.6. *There exists a strictly increasing sequence $n(j) \uparrow \infty$, such that for each $n(j)$ the following holds,*

$$M_{n(j)+p} \le 2^{p/2} M_{n(j)}, \quad \forall p \in \mathbb{N}. \tag{6.2.35}$$

Proof. Assume to the contrary that there exists $K \in \mathbb{N}$ such that for all $n \geq K$ there exists $p \geq 1$ satisfying

$$M_{n+p} \geq 2^{p/2} M_n.$$

Iterating this assumption we find a sequence $n(j) \in \mathbb{N}$, such that

$$M_{n(j)} \geq M_{n(j-1)} 2^{(n(j)-n(j-1))/2},$$

and $n(j) - n(j-1) \geq 1$. Equivalently we have for every $j \geq 1$, that

$$M_{n(j)} \geq 2^{n(j)/2} M_{n(1)} 2^{-n(1)/2}.$$

By definition of $M_{n(j)}$ there exists a dyadic interval I_j such that $|I_j| \geq 2^{-n(j)}$ and

$$\langle f, \psi_{I_j} \rangle^2 \geq M_{n(j)}^2 |I_j|.$$

Inserting the growth condition of the sequence $M_{n(j)}$ we find that for every $j \in \mathbb{N}$,

$$\langle f, \psi_{I_j} \rangle^2 \geq M_{n(1)}^2 2^{-n(1)}.$$

Passing to an infinite subcollection of the intervals $\{I_j\}$, if necessary we obtained

$$\langle f, \psi_{I_j} \rangle^2 \geq M_{n(1)}^2 2^{-n(1)}.$$

Clearly this contradicts the fact that the Carleson system is an unconditional basis in L^2, and that

$$\sum_j \langle f, \psi_{I_j} \rangle^2 \leq 4 \| f - \int f \|_2^2 \leq 4 \| f \|_{\mathrm{BMO}([0,1))}^2. \qquad \square$$

With the above proposition the inequality (6.2.34) for the biorthogonal functionals is a result of the next theorem.

Theorem 6.2.7. *Let $n \in \mathbb{N}$ such that M_n satisfies* (6.2.35). *Then*

$$M_n \leq A \| f \|_{BMO([0,1))},$$

where $A = A(k)$ is independent of n or $f \in BMO([0,1))$.

Proof. We show how Theorem 6.2.7 follows from Theorem 6.2.8 below. Fix $n \in \mathbb{N}$. Select a dyadic interval I_n for which

$$|\langle f, \psi_{I_n} \rangle| \cdot |I_n|^{-1/2} = M_n.$$

Without losing any generality we assume that I_n satisfies also the following conditions.

(a) $|I_n| = 2^{-n}$.

(b) $\langle f, \psi_{I_n} \rangle \geq 0$.

(c) I_n is the left half of its dyadic predecessor.

Let us now describe the interval I_n, its position within the tree of all dyadic intervals and some intervals around I_n, which play an important role in the proof to come. Let μ denote the left endpoint of I_n. Then $I_n = [\mu, \mu + 2^{-n})$. We define J_n to be the dyadic interval of length 2^{-n} such that μ is the right endpoint of J_n. Thus $J_n = [\mu - 2^{-n}, \mu)$. We denote the dyadic predecessors of I_n respectively J_n by $I_{n-1} \supset \cdots \supset I_0 = [0,1)$, respectively $J_{n-1} \supset \cdots \supset J_0 = [0,1)$. Clearly the chains of dyadic intervals I_j, J_j will have to merge somewhere. We encode this fact as follows. Define $q \in \mathbb{N}$ by the following two conditions.

1. $I_q \cap J_q = \emptyset$.

2. $I_q \cup J_q = I_{q-1} = J_{q-1}$.

Notice that I_{q-1} is the smallest dyadic interval containing both I_n and J_n. Hence the assumption that I_n is the left half of its dyadic predecessor implies that $q \leq n-1$. At this point we have reduced the proof of Theorem 6.2.7 to the following critical implication.

Theorem 6.2.8. *Assume that* $\log_2 k \geq 88$. *If*

$$|\langle f, h_{I_n} \rangle| \leq 2^{-k-n} M_n,$$

and if also

$$|\langle f, h_{J_m} \rangle| \leq 2^{-k-m} M_n, \quad \textit{for} \quad n-k \leq m \leq n,$$

then

$$\left| \int_{I_n} f - \int_{J_n} f \right| \geq c \log k \cdot 2^{-n} M_n,$$

where $c > 0$ *is a universal constant independent of* k, *or* n.

We obtain Theorem 6.2.8 by combining Lemma 6.2.11 and Lemma 6.2.12.

Theorem 6.2.8 is the central piece in Carleson's solution to the unconditional basis problem for H^1_{at}. Its proof introduces a new method, called by Carleson, the compensation argument. Quoting from Carleson's introduction [45], the compensation argument is the *delicate* part of a proof that is *not quite easy*.

Before we proceed we will make some remarks about the structure of the proof of Theorem 6.2.8. To this end we introduce the following notation. Let $0 \leq i \leq n$, and $i \neq q-1$. Then select non-negative numbers $0 \leq \alpha_i \leq 2$, and $0 \leq \beta_i \leq 2$, such that

$$\begin{aligned} \langle f, \psi_{J_i} \rangle |J_i|^{-1/2} &= M_n(1 - \alpha_i), \\ \langle f, \psi_{I_i} \rangle |I_i|^{-1/2} &= M_n(1 - \beta_i). \end{aligned} \tag{6.2.36}$$

For the remaining case where $i = q-1$ we choose $0 \le \alpha_{q-1} \le 2$, and $0 \le \beta_{q-1} \le 2$, such that

$$\begin{aligned} \langle f, \psi_{J_{q-1}} \rangle |J_{q-1}|^{-1/2} &= M_n(\alpha_{q-1} - 1), \\ \langle f, \psi_{I_{q-1}} \rangle |I_{q-1}|^{-1/2} &= M_n(\beta_{q-1} - 1). \end{aligned} \tag{6.2.37}$$

We treat the case $i = q-1$ and $i \neq q-1$ differently to take into account the plus sign in (6.2.20) and (6.2.21) and the minus sign in the corresponding equations (6.2.17) and (6.2.18). Doing this allows us to write the equations (6.2.61) and (6.2.62) below in closed form.

The main lemma in the proof of Theorem 6.2.8 is Lemma 6.2.9 below. It asserts that the size of the numbers α_i is intimately related to the hypothesis that at I_n the Haar coefficient of f satisfies $|\langle f, h_{I_n} \rangle| \le 2^{-k-n} M_n$. Denote

$$P(I_n) = \frac{3}{2} \sum_{i=n-k+2}^{n} \alpha_i 2^{n-i}. \tag{6.2.38}$$

Lemma 6.2.9 shows that $|\langle f, h_{I_n} \rangle| \le 2^{-k-n} M_n$ implies that

$$P(I_n) \le 44. \tag{6.2.39}$$

This inequality is called the compensation inequality for the interval I_n, its proof is called the compensation argument. In Lemma 6.2.10 we apply the compensation argument to the intervals J_m with $n-k \le m \le n$. Using that $P(I_n) \le 44$ we obtain from Lemma 6.2.10 a set of compensation inequalities showing that

$$\sum_{i=n-k+2}^{n-k+A} \beta_i \le 88, \tag{6.2.40}$$

where $A = \log(k-1)$. In Lemma 6.2.12 which is the final step in the proof of Theorem 6.2.8 we exploit that the non-negative coefficients α_i, β_i, are small in the sense that (6.2.39) and (6.2.40) hold true. We show there the following lower bound for the integral of f over the adjacent intervals J_n, I_n,

$$\left| \int_{I_n} f - \int_{J_n} f \right| \ge c M_n \log k |I_n|. \tag{6.2.41}$$

The above estimate implies an upper bound for M_n. Indeed since

$$\left| \int_{I_n} f - \int_{J_n} f \right| \le C \|f\|_{\mathrm{BMO}([0,1))} |I_n|,$$

we obtain from (6.2.41) that

$$M_n \le C (\log k)^{-1} \|f\|_{\mathrm{BMO}([0,1))}.$$

Now we turn to the main step of the proof of Theorem 6.2.8. It shows that

$$P(I_n) = \frac{3}{2} \sum_{i=n-k+2}^{n} \alpha_i 2^{n-i}$$

is strongly related to lower estimates for the Haar coefficient $\langle f, h_{I_n}\rangle$. Recall that the definition of the non-negative numbers α_i involves only the coefficients $\langle f, \psi_J\rangle$ formed by the Carleson system. Hence their important role in linking the Carleson system to the Haar system and BMO.

Lemma 6.2.9. *If $n \le q + k - 3$, then*

$$|\langle f, h_{I_n}\rangle| \ge 2^{-n} M_n/16. \tag{6.2.42}$$

If $n \ge q + k - 2$, then

$$|\langle f, h_{I_n}\rangle| \ge 2^{-n-k} M_n P(I_n) - 2^{-n-k} M_n 43. \tag{6.2.43}$$

The essence of Lemma 6.2.9 could be formulated briefly as follows: If $|\langle f, h_{I_n}\rangle| \le 2^{-n-k} M_n$, then $n \ge q + k - 2$, and

$$P(I_n) \le 44.$$

Observe that the lower estimate (6.2.42) holding for $n \le q+k-3$, is already quite satisfactory. After dividing by 2^{-n} we see that the first assertion of Lemma 6.2.9 gives

$$M_n \le 2^{n+4} |\langle f, h_{I_n}\rangle| \le C ||f||_{\mathrm{BMO}([0,1))}.$$

In the case when $n \ge q + k - 2$, we have identified the value of the parameter $P(I_n)$ as the crucial quantity: A lower estimate for $P(I_n)$ gives a lower estimate for $|\langle f, h_{I_n}\rangle|$, and thus an upper bound for M_n. If however $|\langle f, h_{I_n}\rangle| \le 2^{-n-k} M_n$, then the conclusion of Lemma 6.2.9 states that

$$\frac{3}{2} \sum_{i=n-k+2}^{n} \alpha_i 2^{n-i} \le 44,$$

which means in a very strong sense that the non-negative coefficients α_i satisfy $\alpha_i \approx 0$.

Proof of Lemma 6.2.9. Part 1. The first part of the proof establishes the basic estimate (6.2.52). We simplify notation and put

$$M = M_n.$$

Using that the Carleson system is an unconditional basis for L^2 we expand the Haar coefficient $\langle f, h_{I_n}\rangle$ as follows.

$$\langle f, h_{I_n}\rangle = \sum_{I \in \mathcal{D}} \langle f, \psi_I\rangle \langle \varphi_I, h_{I_n}\rangle.$$

Actually contributing to the above expansion are the following dyadic intervals $\{I \in \mathcal{D} : \langle \varphi_I, h_{I_n} \rangle \neq 0\}$. Next split this collection of dyadic intervals into the following two parts (the notation should remind us of "north of I_n" and "south of I_n"). We put

$$\mathcal{N}_{I_n} = \{I : |I| \geq |I_n|,\, I_n \neq I,\, \langle \varphi_I, h_{I_n} \rangle \neq 0\} \setminus \{R_n, I_{n-1}\},$$

and

$$\mathcal{S}_{I_n} = \{I : |I| < |I_n|, \;\; \langle \varphi_I, h_{I_n} \rangle \neq 0\} \cup \{R_n, I_{n-1}\}.$$

Rewrite the equation for $\langle f, h_{I_n} \rangle$ using $\mathcal{N}_{I_n}$ and $\mathcal{S}_{I_n}$,

$$\langle f, h_{I_n} \rangle = \langle f, \psi_{I_n} \rangle \langle \varphi_{I_n}, h_{I_n} \rangle + \sum_{I \in \mathcal{N}_{I_n} \cup \mathcal{S}_{I_n}} \langle f, \psi_I \rangle \langle \varphi_I, h_{I_n} \rangle. \tag{6.2.44}$$

Clearly our aim is to understand better the sum appearing on the right-hand side of the above equation.

We start by showing that the sum over $\mathcal{S}_{I_n}$ represents a negligible quantity for the problem at hand. At this point of the proof we are using the hypothesis that

$$M_{n+p} \leq 2^{p/2} M, \;\; \forall p \in \mathbb{N}.$$

We claim the following upper bound for the sum over $\mathcal{S}_{I_n}$,

$$\sum_{I \in \mathcal{S}_{I_n}} |\langle f, \psi_I \rangle \langle \varphi_I, h_{I_n} \rangle| \leq 34 \cdot M 2^{-n-k}. \tag{6.2.45}$$

To prove the claim, fix $l \in \mathbb{N}$ with $l \geq n+1$. Observe that the collection

$$\{J \in \mathcal{S}_{I_n} : |J| = 2^{-l}\}$$

consists of six intervals. Indeed, if $I \in \{J \in \mathcal{S}_{I_n} : |J| = 2^{-l}\}$, then either I is adjacent to the midpoint of I_n, to the right endpoint of I_n, or to the left endpoint of I_n. Each of these cases accounts for two intervals in $\{J \in \mathcal{S}_{I_n} : |J| = 2^{-l}\}$. Next let $I \in \{J \in \mathcal{S}_{I_n} : |J| = 2^{-l}\}$. We determine the value of the scalar product $\langle \varphi_I, h_{I_n} \rangle$ as follows. If I is adjacent to the midpoint of I_n, then

$$|\langle \varphi_I, h_{I_n} \rangle| = 3 \left(2^{-k} |I|^{1/2}\right). \tag{6.2.46}$$

In the remaining cases where I is adjacent to the right or to the left endpoint of I_n, we have that

$$|\langle \varphi_I, h_{I_n} \rangle| = \frac{3}{2} \left(2^{-k} |I|^{1/2}\right). \tag{6.2.47}$$

Moreover as $M_l \leq 2^{(l-n)/2} M$, we have that

$$\begin{aligned} |\langle f, \psi_I \rangle| &\leq 2^{-l/2} M_l \\ &\leq 2^{-n/2} M. \end{aligned} \tag{6.2.48}$$

To summarize we have four intervals in $\{J \in \mathcal{S}_{I_n} : |J| = 2^{-l}\}$ for which (6.2.47) holds and the remaining two intervals satisfy (6.2.46). Combining this with (6.2.48) gives satisfactory estimates for the products $\langle f, \psi_I\rangle\langle\varphi_I, h_{I_n}\rangle$ and their sum

$$\begin{aligned}\sum_{l=n+1}^{\infty} \sum_{\{I\in\mathcal{S}_{I_n}:|I|=2^{-l}\}} |\langle f, \psi_I\rangle\langle\varphi_I, h_{I_n}\rangle| &\leq 12M2^{-k}\sum_{l=n+1}^{\infty} 2^{-(n+l)/2} \\ &= 12(\sqrt{2}+1)M2^{-k-n}.\end{aligned} \tag{6.2.49}$$

Next recall that the contribution of the intervals $R_n = I_{n-1}\setminus I_n$ and I_{n-1} is given by

$$\langle\varphi_{R_n}, h_{I_n}\rangle = -\frac{3}{2}2^{-k}2^{-n/2}, \quad \text{and} \quad \langle\varphi_{I_{n-1}}, h_{I_n}\rangle = \frac{3}{2}\sqrt{2}\cdot 2^{-k}2^{-n/2}. \tag{6.2.50}$$

This proves (6.2.45), the claimed estimate for the sum over $\mathcal{S}_{I_n}$.

Below we insert (6.2.45) into (6.2.44). We also use that

$$\langle f, \psi_{I_n}\rangle\langle\varphi_{I_n}, h_{I_n}\rangle = M2^{-n}(1-2^{-k}3). \tag{6.2.51}$$

Thus we obtain the lower bound

$$|\langle f, h_{I_n}\rangle| \geq \left| M2^{-n} + \sum_{I\in\mathcal{N}_{I_n}} \langle f, \psi_I\rangle\langle\varphi_I, h_{I_n}\rangle\right| - \{34+3\}M2^{-n-k}. \tag{6.2.52}$$

Next distinguish between the cases $n \leq q+k-3$ and $n \geq q+k-2$. In both cases the preceding estimate (6.2.52) is the starting point of further analysis.

Part 2. Now we prove (6.2.42). Let $n \leq q+k-3$. This case is the easier of the two. We claim that if $n \leq q+k-3$, then

$$\sum_{I\in\mathcal{N}_{I_n}} |\langle f, \psi_I\rangle\langle\varphi_I, h_{I_n}\rangle| \leq \frac{7}{8}M2^{-n}. \tag{6.2.53}$$

What counts here is that $1-\frac{7}{8}$ is large compared to 2^{-k}. We start proving (6.2.53) by recalling the following inclusion for the index set $\mathcal{N}_{I_n}$. It is valid when $n \leq q+k-3$.

$$\mathcal{N}_{I_n} \subseteq \{J_n, \ldots, J_{q-1}, K_{q-k+1}, \ldots, K_0\}. \tag{6.2.54}$$

Here, for $0 \leq j \leq q-k+1$ the intervals K_j satisfy $|K_j| = 2^{-j}$. Fix $0 \leq j \leq q-k$. Then

$$|\langle f, \psi_{K_j}\rangle| \leq M|K_j|^{1/2},$$

and

$$|\langle\varphi_{K_j}, h_{I_n}\rangle| = 2^{+k-2}|I_n|^2|K_j|^{-3/2}.$$

Inserting these values and taking the product gives

$$|\langle f, \psi_{K_j}\rangle\langle\varphi_{K_j}, h_{I_n}\rangle| \leq M2^{+k-2}2^{-2n}2^j.$$

Summing over the index set $0 \leq j \leq q-k+1$, we find that

$$\begin{aligned}\sum_{j=0}^{q-k+1} |\langle f, \psi_{K_j}\rangle\langle\varphi_{K_j}, h_{I_n}\rangle| &\leq M2^{k-2n-2}\sum_{j=0}^{q-k+1} 2^j \\ &\leq M2^{q-2n}.\end{aligned} \tag{6.2.55}$$

Next fix $q-1 \leq i \leq n$. Then we have

$$|\langle f, \psi_{J_i}\rangle| \leq M|J_i|^{1/2} \quad \text{and} \quad |\langle\varphi_{J_i}, h_{I_n}\rangle| = 3(2^{-k-1})|J_i|^{1/2}.$$

This gives

$$\begin{aligned}\sum_{i=q-1}^{n} |\langle f, \psi_{J_i}\rangle\langle\varphi_{J_i}, h_{I_n}\rangle| &\leq 3M\cdot 2^{-k-1}\sum_{i=q-1}^{n} 2^{-i} \\ &\leq 6M\cdot 2^{-k-q}.\end{aligned} \tag{6.2.56}$$

Adding the estimates (6.2.55) and (6.2.56) we find that

$$\sum_{I\in\mathcal{N}_{I_n}} |\langle f, \psi_I\rangle\langle\varphi_I, h_{I_n}\rangle| \leq M(6\cdot 2^{-k-q} + 2^{-2n+q}). \tag{6.2.57}$$

The assumption that $n \leq q+k-3$, and the fact that $n \geq q+1$ gives

$$M(6\cdot 2^{-k-q} + 2^{-2n+q}) \leq \frac{7}{8}M2^{-n}. \tag{6.2.58}$$

Indeed, verify first that (6.2.58) holds for $n = q+k-3$ and $n = q+1$. Then estimate in a straightforward way using the improved hypothesis $q+2 \leq n \leq q+k-4$. Insert the estimate (6.2.57) into the previously established inequality (6.2.52). Using that $1 - 7/8 = 1/8$, and that $k \geq 10$ this gives the following lower bound for the Haar coefficients $\langle f, h_{I_n}\rangle$ (it is valid when $n \leq q+k-3$).

$$\begin{aligned}|\langle f, h_{I_n}\rangle| &\geq \left|M2^{-n} + \sum_{I\in\mathcal{N}_{I_n}} \langle f, \psi_I\rangle\langle\varphi_I, h_{I_n}\rangle\right| - \{34+3\}\, M2^{-k-n} \\ &\geq \frac{1}{8}M2^{-n} - \{34+3\}\, M2^{-k-n} \\ &\geq \frac{1}{16}M2^{-n}.\end{aligned} \tag{6.2.59}$$

Part 3. Next we prove (6.2.43). Let $n \geq q+k-2$. For $n \geq q+k-2$ we have the following representation for $\mathcal{N}_{I_n}$,

$$\mathcal{N}_{I_n} \subseteq \{J_n, \cdots, J_{n-k+1}, K_{q-k+1}, \cdots, K_0\}.$$

Rewrite the inequality $n \geq q+k-2$, as $q-2n \leq -k-n+2$. Inserting this into (6.2.55) we obtain

$$\begin{aligned}\sum_{j=0}^{q-k+1} |\langle f, \psi_{K_j}\rangle\langle\varphi_{K_j}, h_{I_n}\rangle| &\leq M2^{q-2n} \\ &\leq 4M2^{-n-k}.\end{aligned}$$

We insert this estimate into (6.2.52), the lower bound for $|\langle f, h_{I_n}\rangle|$ obtained in the first part of the proof. This gives that

$$|\langle f, h_{I_n}\rangle| \geq \left| M2^{-n} + \sum_{i=n-k+1}^{n} \langle f, \psi_{J_i}\rangle\langle\varphi_{J_i}, h_{I_n}\rangle \right| - \{34+7\}\, M2^{-n-k}. \qquad (6.2.60)$$

We continue by recalling what is known about the values of $\langle f, \psi_{J_i}\rangle\langle\varphi_{J_i}, h_{I_n}\rangle$ appearing in the above sum. We distinguish between the cases $n-k+2 \leq i \leq n$, and $i = n-k+1$. This distinction is necessary in view of the different factors $-3/2$ and $-1/2$ appearing in the formulas below: It follows from

1. If $n-k+2 \leq i \leq n$, then

$$\langle f, \psi_{J_i}\rangle\langle\varphi_{J_i}, h_{I_n}\rangle = -\frac{3}{2} 2^{-i-k} M(1-\alpha_i). \qquad (6.2.61)$$

   The equation (6.2.61) follows from (6.2.18) and (6.2.21) in combination with (6.2.36) and (6.2.37).

2. If $i = n-k+1$, then

$$\langle f, \psi_{J_i}\rangle\langle\varphi_{J_i}, h_{I_n}\rangle = -\frac{1}{2} 2^{-i-k} M(1-\alpha_i). \qquad (6.2.62)$$

   The equation (6.2.62) follows from (6.2.17) and (6.2.20) in combination with (6.2.36) and (6.2.37).

We displayed the above identities with the purpose of inserting them into (6.2.60). Before we do that we introduce, for sake of convenience, the abbreviation

$$X = 2^{+k} - \frac{1}{2} 2^{+k-1} - \frac{3}{2} \sum_{i=n-k+2}^{n} 2^{-i+n}.$$

Later we will calculate the value of X and find that $X = 3/2$. However for the time being it is really more instructive to write it in just the form we used above.

We insert now into (6.2.60) the identities we obtained in (6.2.61) and (6.2.62). Next we pull out the common factor $2^{-n-k}M$. This gives the equations

$$\begin{aligned}\left| M2^{-n} + \sum_{i=n-k+1}^{n} \langle f, \psi_{J_i}\rangle\langle\varphi_{J_i}, h_{I_n}\rangle \right| &= 2^{-n-k}M \left| 2^{+k} - \frac{1}{2}2^{+k-1}(1-\alpha_{n-k+1}) - \frac{3}{2}\sum_{i=n-k+2}^{n} 2^{-i+n}(1-\alpha_i) \right| \\ &= 2^{-n-k}M \left| X + \frac{1}{2}2^{+k-1}\alpha_{n-k+1} + \frac{3}{2}\sum_{i=n-k+2}^{n} 2^{-i+n}\alpha_i \right|.\end{aligned}$$

Next we determine the value of X. First we re-index the sum on the right of its definition. This gives

$$\sum_{i=n-k+2}^{n} 2^{-i+n} = \sum_{j=0}^{k-2} 2^j = 2^{k-1} - 1.$$

Then we insert the value for the geometric sum and find that

$$X = \frac{3}{2}.$$

Below it is *important* that we have a universal upper bound for X. By (6.2.60) the resulting lower bound for $|\langle f, h_{I_n}\rangle|$ is now

$$|\langle f, h_{I_n}\rangle| \geq 2^{-n-k}M \left[\frac{3}{2}\sum_{i=n-k+2}^{n} 2^{-i+n}\alpha_i \right] - 2^{-n-k}M\{34+9\}.$$

As $34 + 9 = 43$ this completes the proof of Lemma 6.2.9. □

The proof given above constitutes the compensation argument. We applied it to the interval I_n. Next we apply it to the intervals J_m, $n-k \leq m \leq n$. We consider the Haar coefficients, $\langle f, h_{J_m}\rangle$ where $n-k \leq m \leq n$ and try to obtain lower bounds for $|\langle f, h_{J_m}\rangle|$. We will therefore repeat the proof of the previous lemma. We execute Carleson's compensation argument and uncover again the associated compensation inequalities. It turns out that now the value of the following parameter controls the lower estimates for the Haar coefficient $|\langle f, h_{J_m}\rangle|$,

$$P(J_m) = \left| -\alpha_m 2^{+k} + \frac{1}{4}2^{+k}\beta_{m-k+1} + \frac{3}{2}\sum_{i=m-k+2}^{m} 2^{-i+m}\beta_i \right|.$$

In its formulation and in its proof the next lemma is an "affine image" of Lemma 6.2.9. Its presentation here is motivated only by the desire to be definite and complete. The proof below uses once more the compensation argument — shifted and rescaled to the interval J_m. We will apply Lemma 6.2.10 in the course of proving Lemma 6.2.11, where we will consider only those m for which $\alpha_m \leq 2^{-5}$.

Lemma 6.2.10. *If $m \le q+k-3$, then*

$$|\langle f, h_{J_m}\rangle| \ge 2^{-m} M \left(|1-\alpha_m| - \frac{15}{16} \right).$$

If $m \ge q+k-2$, then

$$|\langle f, h_{J_m}\rangle| \ge 2^{-m-k} M P(J_m) - 43 \cdot 2^{-m-k} M.$$

Proof. We define first the following two collections of dyadic intervals,

$$\mathcal{N}_{J_m} = \{I : |I| \ge |J_m|,\ J_m \ne I,\ \langle \varphi_I, h_{J_m}\rangle \ne 0\} \setminus \{J_{m-1}, J_{m-1} \setminus J_m\},$$

and

$$\mathcal{S}_{J_m} = \{I : |I| < |J_m|,\ \ \langle \varphi_I, h_{J_m}\rangle \ne 0\} \cup \{J_{m-1}, J_{m-1} \setminus J_m\}.$$

With this notation we expand $\langle f, h_{J_m}\rangle$. The expansion is naturally centered around the interval J_m.

$$\langle f, h_{J_m}\rangle = \langle f, \psi_{J_m}\rangle \langle \varphi_{J_m}, h_{J_m}\rangle + \sum_{I \in \mathcal{N}_{J_m} \cup \mathcal{S}_{J_m}} \langle f, \psi_I\rangle \langle \varphi_I, h_{J_m}\rangle. \tag{6.2.63}$$

The intervals contained in $\mathcal{S}_{J_m}$ contribute relatively little to the above sum. Indeed the proof of (6.2.45) shows that the sum over the index set $\mathcal{S}_{J_m}$ is bounded as follows,

$$\sum_{I \in \mathcal{S}_{J_m}} |\langle f, \psi_I\rangle \langle \varphi_I, h_{J_m}\rangle| \le 2^{-m-k} 34 M. \tag{6.2.64}$$

To obtain (6.2.64) we use the additional hypothesis that $M_{n+p} \le 2^{p/2} M_n$. Next recall that $\langle f, \psi_{J_m}\rangle = M(1-\alpha_m)2^{-m/2}$, and that $\langle \varphi_{J_m}, h_{J_m}\rangle = 2^{-m/2}(1-3\cdot 2^{-k})$. Multiplying allows us to rewrite the leading term in (6.2.63) as

$$\langle f, \psi_{J_m}\rangle \langle \varphi_{J_m}, h_{J_m}\rangle = M(1-\alpha_m)2^{-m}(1-3\cdot 2^{-k}).$$

Then with the estimate (6.2.64) for the sum over $\mathcal{S}_{J_m}$ we obtain the lower bound

$$|\langle f, h_{J_m}\rangle| \ge \left| M(1-\alpha_m)2^{-m} + \sum_{I \in \mathcal{N}_{J_m}} \langle f, \psi_I\rangle \langle \varphi_I, h_{J_m}\rangle \right| - 37 M 2^{-m-k}. \tag{6.2.65}$$

Next distinguish between the cases $m \le q+k-3$ and $m \ge q+k-2$. In both cases the preceding estimate is the place from which we start developing the proof.

Let $m \le q+k-3$. As in the proof of Lemma 6.2.9 the following estimate holds,

$$\sum_{I \in \mathcal{N}_{J_m}} |\langle f, \psi_I\rangle \langle \varphi_I, h_{J_m}\rangle| \le \frac{7}{8} M 2^{-m}. \tag{6.2.66}$$

Inserting the estimate (6.2.66) into the lower bound (6.2.65) gives that

$$|\langle f, h_{J_m}\rangle| \geq M|1-\alpha_m|2^{-m} - \frac{7}{8}M2^{-m} - 37M2^{-k-m}.$$

With $k \geq 10$ we obtain therefore that

$$|\langle f, h_{J_m}\rangle| \geq 2^{-m}M\left(|1-\alpha_m| - \frac{15}{16}\right).$$

This completes the proof for the case $m \leq q+k-3$.

Next we consider $m \geq q+k-2$. We return to the lower estimate obtained in (6.2.65). Recall that for $m \geq q+k-2$ the following representation holds,

$$\mathcal{N}_{J_m} \subseteq \{I_m, \ldots, I_{m-k+1}, K_{q-k+1}, \ldots, K_0\}.$$

We estimate now the contribution to the sum on the right-hand side of (6.2.65) which comes from of the intervals $K_{q-k+1}, \ldots, K_0$. Recall (6.2.55), where we proved that

$$\sum_{j=0}^{q-k+1} |\langle f, \psi_{K_j}\rangle\langle\varphi_{K_j}, h_{J_m}\rangle| \leq M2^{q-2m}.$$

As $m \geq q+k-2$, we have $q-2m-1 \leq -k-m+1$, hence

$$\sum_{j=0}^{q-k+1} |\langle f, \psi_{K_j}\rangle\langle\varphi_{K_j}, h_{J_m}\rangle| \leq 4M2^{-m-k}.$$

Inserting this estimate into the lower bound (6.2.65) shows that,

$$|\langle f, h_{J_m}\rangle| \geq \left| M(1-\alpha_m)2^{-m} + \sum_{i=m-k+1}^{m} \langle f, \psi_{I_i}\rangle\langle\varphi_{I_i}, h_{J_m}\rangle \right| - 41M2^{-m-k}. \tag{6.2.67}$$

The preceding analysis showed that the crucial contribution to the right-hand side of (6.2.65) comes from the intervals $I_m, \ldots, I_{m-k+1}$. Indeed all the rest contributes an amount that is less than $41M2^{-m-k}$. We will now calculate the value of the sum

$$\sum_{i=m-k+1}^{m} \langle f, \psi_{I_i}\rangle\langle\varphi_{I_i}, h_{J_m}\rangle,$$

with as much precision as possible. We have to distinguish between the following two cases. The case where $m-k+2 \leq i \leq m$, and $i = m-k+1$.

1. If $m-k+2 \leq i \leq m$, then

$$\langle f, \psi_{I_i}\rangle\langle\varphi_{I_i}, h_{J_m}\rangle = -\frac{3}{2}M(1-\beta_i)2^{-i-k}.$$

2. If $i = m - k + 1$, then

$$\langle f, \psi_{I_i}\rangle\langle\varphi_{I_i}, h_{J_m}\rangle = -\frac{1}{2}M(1-\beta_i)2^{-i-k}.$$

We use again the somewhat artificial abbreviation

$$X = 2^{+k} - \frac{1}{2}2^{+k-1} - \frac{3}{2}\sum_{i=m-k+2}^{m} 2^{-i+m}.$$

Recall that $X = 3/2$. Recall further that we put

$$P(J_m) = \left| -\alpha_m 2^{+k} + \frac{1}{4}2^{+k}\beta_{m-k+1} + \frac{3}{2}\sum_{i=m-k+2}^{m} 2^{-i+m}\beta_i \right|.$$

With this notation and with the values for $\langle f, \psi_{I_i}\rangle\langle\varphi_{I_i}, h_{J_m}\rangle$ we obtain the following identities.

$$\begin{aligned}
&\left| M(1-\alpha_m)2^{-m} + \sum_{i=m-k+1}^{m} \langle f, \psi_{I_i}\rangle\langle\varphi_{I_i}, h_{J_m}\rangle \right| \\
&= 2^{-m-k}M\left| (1-\alpha_m)2^{+k} - \frac{1}{2}2^{+k-1}(1-\beta_{m-k+1}) - \frac{3}{2}\sum_{i=m-k+2}^{m} 2^{-i+m}(1-\beta_i) \right| \\
&= 2^{-m-k}M\left| X - P(J_m)\right|.
\end{aligned}$$

Thus we showed that the Haar coefficient $\langle f, h_{J_m}\rangle$ is bounded from below as follows,

$$|\langle f, h_{J_m}\rangle| \geq 2^{-m-k}MP(J_m) - 43M2^{-m-k}. \qquad \square$$

The following lemma is a simple bookkeeping device. We use it to merge the conclusions of Lemma 6.2.9 and Lemma 6.2.10.

Lemma 6.2.11. *Assume that*

$$|\langle f, h_{I_n}\rangle| < 2^{-k-n}M, \tag{6.2.68}$$

and that for each m satisfying $n-k \leq m < n$,

$$|\langle f, h_{J_m}\rangle| \leq 2^{-k-m}M. \tag{6.2.69}$$

Then the following compensation inequalities hold true,

$$\sum_{i=n-k+2}^{n} \alpha_i \leq 44, \quad \textit{and} \quad \sum_{i=n-k+2}^{n-k+A} \beta_i \leq 88, \tag{6.2.70}$$

where $A = \log_2(k-1)$.

Proof. By Lemma 6.2.9 the hypothesis (6.2.68) implies that $n \geq q + k - 2$, and that $P(I_n) \leq 44$. By definition of $P(I_n)$ this gives

$$\sum_{i=n-k+2}^{n} \alpha_i \leq 44.$$

Let $A = \log_2(k-1)$ and assume that it is a natural number. The compensation inequality $P(I_n) \leq 44$ implies that $0 \leq \alpha_m \leq 1/32$, for $n-k+1 \leq m \leq n-k+A$. Hence

$$|1 - \alpha_m| \geq \frac{31}{32}. \tag{6.2.71}$$

With (6.2.71) and the hypothesis (6.2.69) the conclusion of Lemma 6.2.10 gives that $m \geq q + k - 2$ and $P(J_m) \leq 44$ for $n - k + 1 \leq m \leq n - k + A$. Writing out the inequality $P(J_m) \leq 44$ and taking into account that $3/2 \geq 1$, we find that

$$\sum_{i=m-k+2}^{m} 2^{-i+m}\beta_i \leq 44 + \alpha_m 2^k. \tag{6.2.72}$$

We would like to take advantage of the compensation inequality for I_n. So we multiply the inequalities (6.2.72) with the factor 2^{-k-m+n}. This gives

$$\sum_{i=m-k+2}^{m} 2^{-i-k+n}\beta_i \leq 44 \cdot 2^{-k-m+n} + \alpha_m 2^{-m+n}, \tag{6.2.73}$$

where $n-k+2 \leq m \leq n-k+A$. In this way the coefficient in front of α_m has the same form as in the compensation inequality. We take the sum of the estimates (6.2.73) over the range $n - k + 2 \leq m \leq n - k + A$. This gives

$$\sum_{m=n-k+2}^{n-k+A} \sum_{i=m-k+2}^{m} 2^{-i-k+n}\beta_i \leq 44 \sum_{m=n-k+2}^{n-k+A} 2^{-k-m+n} + \sum_{m=n-k+2}^{n-k+A} \alpha_m 2^{-m+n}. \tag{6.2.74}$$

The geometric sum appearing on the right-hand side of (6.2.74) is easily evaluated. Its value is ≤ 1. The second sum is bounded by $P(I_n)$. Hence, by (6.2.74) we obtain that

$$\sum_{m=n-k+2}^{n-k+A} \sum_{i=m-k+2}^{m} 2^{-i-k+n}\beta_i \leq 88. \tag{6.2.75}$$

We now change the order of summation on the left-hand side of (6.2.75). Doing this allows us to exploit that the summands on the left do not depend on m. We find thus the identity

$$\sum_{m=n-k+2}^{n-k+A} \sum_{i=m-k+2}^{m} 2^{-i-k+n}\beta_i = (k-1) \sum_{i=n-2k+4}^{n-k+A} 2^{-i-k+n}\beta_i.$$

Insert now the above line into (6.2.75). Observe that for $n-2k+4 \le i \le n-k+A$ we obtain that $2^{-i-k+n} \ge 2^{-A}$. Hence, the non-negative numbers β_i satisfy the simple estimate

$$2^{-A}(k-1) \sum_{i=n-2k+4}^{n-k+A} \beta_i \le 88. \tag{6.2.76}$$

We observe that by our choice for A, the factor $2^{-A}(k-1)$ equals 1. Hence, dividing (6.2.76) by $2^{-A}(k-1)$ shows that (6.2.70) holds. □

Next we prove that the compensation inequalities (6.2.70) imply an upper bound for $M = M_n$. The idea of the proof was present already in Example 6.2.3. Note that by Lemma 6.2.11 the hypothesis of the following lemma is satisfied with $A = \log(k-1)$, $B = 44$ and $C = 88$.

Lemma 6.2.12. *Let $A, B, C \in \mathbb{N}$, be fixed. Assume that there hold compensation inequalities in the form*

$$\sum_{i=n-k+2}^{n-k+A} \beta_i \le C \quad \text{and} \quad \sum_{i=n-k+2}^{n} \alpha_i \le B. \tag{6.2.77}$$

Then

$$\left| \int_{I_n} f - \int_{J_n} f \right| \ge 2^{-n} M(2A - B - C - 12).$$

Proof. For any dyadic interval J we denote

$$\Delta(J) = \int_{I_n} \varphi_J - \int_{J_n} \varphi_J.$$

We define the collection of dyadic intervals

$$\mathcal{N} = \{J : |J| \ge 2^{-n+2}, \Delta(J) \ne 0\} \quad \text{and} \quad \mathcal{S} = \{J : |J| < 2^{-n+2}, \Delta(J) \ne 0\}.$$

With this notation we have

$$\int_{I_n} f - \int_{J_n} f = \sum_{J \in \mathcal{N} \cup \mathcal{S}} \langle f, \psi_J \rangle \Delta(J).$$

We easily observe that the proof of (6.2.45) gives

$$\sum_{J \in \mathcal{S}} |\langle f, \psi_J \rangle \Delta(J)| \le 34 M 2^{-n-k}. \tag{6.2.78}$$

Now we show the lower estimate

$$\sum_{J \in \mathcal{N}} \langle f, \psi_J \rangle \Delta(J) \ge 2^{-n} M(2A - B - C - 11), \tag{6.2.79}$$

where A, B are the constants appearing in the hypothesis of the lemma. Clearly the estimates (6.2.78) and (6.2.79) imply the conclusion of the lemma.

We now turn to the verification of (6.2.79). We start the proof by analyzing the index set $\mathcal{N}$. For $\mathcal{N}$ we have the representation

$$\mathcal{N} \subseteq \{I_{n-2}, J_{n-2}, \ldots, I_{n-k+1}, J_{n-k+1}, K_{q-k+1}, \ldots, K_0\}.$$

Correspondingly, we have that

$$\begin{aligned}\sum_{J\in\mathcal{N}} \langle f, \psi_J\rangle \Delta(J) = & \sum_{j=n-k+1}^{n-2} \langle f, \psi_{J_j}\rangle \Delta(J_j) + \langle f, \psi_{I_j}\rangle \Delta(I_j) \\ & + \sum_{i=0}^{q-k+1} \langle f, \psi_{K_i}\rangle \Delta(K_i).\end{aligned} \tag{6.2.80}$$

Next we show that the sum in the second line of (6.2.80) is bounded by $M2^{-n+3}$. Fix $i \le q-k+1$. Recall that $\operatorname{Lip}(\varphi_{K_i}) \le 2^{+3i/2+k}$. As $I_n \cup J_n$ is an interval of diameter $\le 2^{-n+1}$, we obtain that

$$\left| \int_{I_n} \varphi_{K_i} - \int_{J_n} \varphi_{K_i} \right| \le 2^{-2n+3i/2+k+1}.$$

Next recall that $|\langle f, \psi_{K_i}\rangle| \le M2^{-i/2}$ by definition. Hence for $i \le q-k+1$,

$$|\langle f, \psi_{K_i}\rangle \Delta(K_i)| \le M2^{-2n+k+i+1},$$

and by summing the geometric series we find

$$\sum_{i=0}^{q-k+1} |\langle f, \psi_{K_i}\rangle \Delta(K_i)| \le M2^{q-2n+3}. \tag{6.2.81}$$

Using that $n \ge q+1$ we obtain from (6.2.80) and (6.2.81) that

$$\sum_{J\in\mathcal{N}} \langle f, \psi_J\rangle \Delta(J) \ge -M2^{-n+2} + \sum_{j=n-k+1}^{n-2} \langle f, \psi_{J_j}\rangle \Delta(J_j) + \langle f, \psi_{I_j}\rangle \Delta(I_j). \tag{6.2.82}$$

Hence to verify (6.2.79) it remains to find a lower bound for the sum appearing in (6.2.82). To this end we let $n-k+1 \le j \le n-2$ and $I_j, J_j \in \mathcal{N}$. The values of the differences $\Delta(I_j)$, $\Delta(J_j)$ are calculated most easily by looking at the graph of φ_{J_j} respectively φ_{I_j}. We obtained in (6.2.30) that

$$\Delta(I_j) = \Delta(J_j) = 2^{-n}2^{+j/2} - \frac{3}{2}2^{-k}2^{-j/2} \quad \text{for} \quad n-k+1 \le j \le n-2.$$

Now we rewrite the sum appearing on the right-hand side (6.2.82) using some of the notation we introduced earlier. We use $\langle f, \psi_{J_j}\rangle 2^{+j/2} = M(1-\alpha_j)$ and $\langle f, \psi_{I_j}\rangle 2^{+j/2} = M(1-\beta_j)$. Thus we have that

$$\langle f, \psi_{J_j}\rangle \Delta(J_j) = M(1-\alpha_j)\{2^{-n} - \frac{3}{2}2^{-k}2^{-j}\},$$

and also

$$\langle f, \psi_{I_j}\rangle \Delta(I_j) = M(1-\beta_j)\{2^{-n} - \frac{3}{2}2^{-k}2^{-j}\}.$$

Now we sum the above identities over $n-k+1 \le j \le n-2$. This gives a new formula for the sum (6.2.79). Most interestingly, the right-hand side of (6.2.83) below contains only terms about which we have information available due to the hypothesis of the lemma.

$$\begin{aligned} &\sum_{j=n-k+1}^{n-2} \langle f, \psi_{J_j}\rangle \Delta(J_j) + \langle f, \psi_{I_j}\rangle \Delta(I_j) \\ &\ge -2^{-n+1}M + \sum_{j=n-k+2}^{n-2} (2^{-n} - \frac{3}{2}2^{-k-j})\,(M(1-\alpha_j) + M(1-\beta_j))\,. \end{aligned} \tag{6.2.83}$$

Our aim is to find a lower bound for the right-hand side of (6.2.83). To this end we consider separately the summands containing $M(1-\alpha_j)$, and $M(1-\beta_j)$. We will first obtain the lower bound for the sum involving the terms $M(1-\alpha_j)$. We claim that the following estimate holds true.

$$\sum_{j=n-k+2}^{n-2} \left(2^{-n} - \frac{3}{2}2^{-k-j}\right) M(1-\alpha_j) \ge 2^{-n}M(k-B-4). \tag{6.2.84}$$

To verify (6.2.84) it suffices to combine the following three simple observations. First counting the number of summands gives $\sum_{j=n-k+2}^{n-2} 2^{-n} = 2^{-n}(k-3)$. Second we sum the geometric series

$$\sum_{j=n-k+2}^{n-2} 2^{-k-j} = 2^{-n-1}(1-2^{-k+3}). \tag{6.2.85}$$

Third we invoke the compensation inequality as stated in the hypothesis, that is

$$\sum_{j=n-k+2}^{n-2} \alpha_j \le B.$$

Combining these three remarks gives the lower bound stated in (6.2.84). Now we turn to the remaining sum in (6.2.83) where the summands contain $M(1-\beta_j)$.

We claim that the compensation inequalities for β_j stated in the hypothesis give the lower bound

$$\sum_{j=n-k+2}^{n-2}\left(2^{-n}-\frac{3}{2}2^{-k-j}\right)M(1-\beta_j)\geq 2^{-n}M(A-k-1). \tag{6.2.86}$$

To verify (6.2.86) we split the range of summation into two groups. We consider separately the cases $n-k+2\leq j\leq n-k+A$ and $n-k+A<j\leq n-2$. For the first case where $n-k+2\leq j\leq n-k+A$, we invoke the hypothesis stating that

$$\sum_{j=n-k+2}^{n-k+A}\beta_j\leq C. \tag{6.2.87}$$

Using (6.2.85) and (6.2.87) gives that

$$\sum_{j=n-k+2}^{n-k+A}(1-\beta_j)\left(2^{-n}-\frac{3}{2}2^{-k-j}\right)\geq 2^{-n}(A-C-3). \tag{6.2.88}$$

On the other hand when $n-k+A+1\leq j\leq n-2$ we just use the trivial estimate

$$(1-\beta_j)(2^{-n}-\frac{3}{2}2^{-k-j})\geq -2^{-n}-\frac{3}{2}2^{-k-j}.$$

Summing over j and counting the number of summands gives that trivially

$$\sum_{j=n-k+A+1}^{n-2}(1-\beta_j)\left(2^{-n}-\frac{3}{2}2^{-k-j}\right)\geq 2^{-n}(A-k-1). \tag{6.2.89}$$

It remains to take the sum of (6.2.88) and (6.2.89) to see that (6.2.86) holds.

Compare now the lower bound (6.2.86) with the right-hand side of (6.2.84). Note that the term $-k$ appearing in (6.2.86) is matched by $+k$ in (6.2.84). Hence the combination of (6.2.83) with (6.2.84) and (6.2.86) gives that

$$\sum_{j=n-k+1}^{n-2}\langle f,\psi_{J_j}\rangle\Delta(J_j)+\langle f,\psi_{I_j}\rangle\Delta(I_j)\geq 2^{-n}M(2A-B-C-7). \tag{6.2.90}$$

By (6.2.82) the estimate (6.2.90) implies (6.2.79). □

We have now completed the proof of Theorem 6.2.8. We deduce it by putting together Lemma 6.2.11 and Lemma 6.2.12.

Carleson's unconditional basis for H^1_{at}

In this subsection we prove that the Carleson system is an unconditional basis in H^1_{at}. We begin by proving first the corresponding dual estimate in BMO([0, 1)).

Theorem 6.2.13. *For $f \in BMO([0,1))$ the biorthogonal coefficients $\{\langle f, \psi_I\rangle : I \in \mathcal{D}\}$ determine the norm as*

$$A^{-1}||f||^2_{BMO([0,1))} \le \sup_{J\in\mathcal{D}} \frac{1}{|J|}\sum_{I\subseteq J}\langle f,\psi_I\rangle^2 \le A||f||^2_{BMO([0,1))}. \tag{6.2.91}$$

The constant $A = A(k)$ depends only on the smoothing parameter k.

Proof. Let $f \in \mathrm{BMO}([0,1))$, and normalize such that $||f||_{\mathrm{BMO}([0,1))} = 1$. We start proving the right-hand side estimate. Viewing $f \in \mathrm{BMO}([0,1))$ as an element of L^2 and assuming that $\int f = 0$, we expand f using Carleson's biorthogonal system,

$$f = \sum_{I\in\mathcal{D}}\langle f,\psi_I\rangle\varphi_I. \tag{6.2.92}$$

By Theorem 6.2.2 the convergence in (6.2.92) is unconditional in L^2. Let J be a fixed dyadic interval. Depending on J we split the collection of all dyadic intervals $\mathcal{D}$ into the following three families. We let

$$\begin{aligned}\mathcal{A}_1 &= \{I\in\mathcal{D} : I\subseteq J\},\\ \mathcal{A}_2 &= \{I\notin\mathcal{A}_1 : |I|\le|J|\},\\ \mathcal{A}_3 &= \{I\in\mathcal{D} : |I|>|J|\}.\end{aligned}$$

Accordingly we split $f = f_1 + f_2 + f_3$ where for $i \in \{1,2,3\}$, we put

$$f_i = \sum_{I\in\mathcal{A}_i}\langle f,\psi_I\rangle\varphi_I.$$

Our normalizing assumption that $||f||_{\mathrm{BMO}([0,1))} = 1$, the definition of the norm in $\mathrm{BMO}([0,1))$ and the triangle inequality, give the lower bounds

$$\begin{aligned}1 &\ge \left(\int_J |f-m_J(f)|^2\frac{dt}{|J|}\right)^{1/2}\\ &\ge \left(\int_J |f_1|^2\frac{dt}{|J|}\right)^{1/2} - \left(\int_J |f_2|^2\frac{dt}{|J|}\right)^{1/2} - \left(\int_J |f_3-m_J(f)|^2\frac{dt}{|J|}\right)^{1/2}.\end{aligned} \tag{6.2.93}$$

Now we analyze the three terms appearing above. We find a lower bound for the first term and upper bounds for the remaining ones. From now on, the coefficient estimates

$$|\langle f,\psi_I\rangle| \le C|I|^{1/2},$$

established in Theorem 6.2.5 play a decisive role. First we estimate $\int_J |f_1|^2$ from below. Let $|J| = 2^{-j}$. Let $\mathcal{R}$ denote the intervals in $\mathcal{A}_1$ which are adjacent to the right endpoint of J. Similarly let $\mathcal{L}$ denote the intervals in $\mathcal{A}_1$ which are adjacent to the left endpoint of J. Then with $\mathcal{E} = \mathcal{R}\cup\mathcal{L}$ we define

$$h = \sum_{I\in\mathcal{E}}\langle f,\psi_I\rangle\varphi_I, \quad \text{and} \quad g = f_1 - h. \tag{6.2.94}$$

Recall that $|J| = 2^{-j}$. For $\ell \geq 0$ let $\mathcal{E}_\ell = \{I \in \mathcal{E} : |I| = 2^{-j-\ell}\}$. Note that $\mathcal{E}_0 = \{J\}$, and that for $\ell \geq 1$ the collection $\mathcal{E}_\ell$ consists of two intervals. By Theorem 6.2.2 it follows that

$$\int_J |h|^2 \leq 4 \sum_{I \in \mathcal{E}} \langle f, \psi_I \rangle^2. \tag{6.2.95}$$

Now estimate the sum on the right-hand side. For $I \in \mathcal{E}_\ell$, we have $|I| = 2^{-j-\ell}$. Hence Theorem 6.2.5 gives that

$$\langle f, \psi_I \rangle^2 \leq C 2^{-j-\ell}. \tag{6.2.96}$$

Inserting the estimate (6.2.96) into (6.2.95) and summing the geometric series gives that

$$\sum_{I \in \mathcal{E}} \langle f, \psi_I \rangle^2 \leq C|J|. \tag{6.2.97}$$

Next, observe that for $I \in \mathcal{A}_1 \setminus \mathcal{E}$ the support of φ_I is contained in the interval J. Now Theorem 6.2.2 implies that the L^2 norm of g can be expressed through its coefficients as

$$\begin{aligned} \sum_{I \in \mathcal{A}_1 \setminus \mathcal{E}} \langle f, \psi_I \rangle^2 &\leq 4 \int_J |g|^2 \\ &\leq 8 \int_J |f_1|^2 + 8 \int_J |h|^2. \end{aligned} \tag{6.2.98}$$

We combine (6.2.98) with the upper bounds obtained in (6.2.97) and (6.2.95). This gives

$$\sum_{I \in \mathcal{A}_1} \langle f, \psi_I \rangle^2 \leq C|J| + C \int_J |f_1|^2.$$

Next use the upper bound for $\int_J |f_1|^2$ given by (6.2.93). Thus we obtain that

$$\sum_{I \in \mathcal{A}_1} \langle f, \psi_I \rangle^2 \leq C|J| + C \int_J |f_3 - m_J(f)|^2 dt + C \int_J |f_2|^2 dt. \tag{6.2.99}$$

Now we turn to the estimate for the second term appearing on the right-hand side of (6.2.99). We first rewrite $m_J(f) = m_J(f_1) + m_J(f_2) + m_J(f_3)$. Then we estimate the integrand as

$$|f_3(x) - m_J(f)| \leq |f_3(x) - m_J(f_3)| + |m_J(f_2)| + |m_J(f_1)|. \tag{6.2.100}$$

We continue finding an estimate for $|m_J(f_1)|$ appearing in the last line. Observe with h defined in (6.2.94),

$$m_J(f_1) = m_J(h).$$

Hence by using again Theorem 6.2.5 and the estimate $|m_J(\varphi_I)| \le 2^{-k}(3|I|)/(2|J|)$ holding for $I \in \mathcal{E}_\ell$, we find that

$$|\langle f, \psi_I\rangle m_J(\varphi_I)| \le C2^{-\ell}, \quad \text{for} \quad I \in \mathcal{E}_\ell.$$

Consequently the following estimate holds for $m_J(f_1)$.

$$\begin{aligned} |m_J(f_1)| &\le \sum_{\ell=0}^{\infty} \sum_{I\in\mathcal{E}_\ell} |\langle f, \psi_I\rangle m_J(\varphi_I)| \\ &\le C. \end{aligned} \tag{6.2.101}$$

Next we prove a pointwise upper bound for $|f_3(x) - m_J(f_3)|$. Recall that the definition of $\varphi_I(x)$ gives this Lipschitz estimate $\mathrm{Lip}(\varphi_I) \le 2^{+k}|I|^{-3/2}$. Hence by Theorem 6.2.5 for $I \in \mathcal{A}_3$ we find that for $x \in J$,

$$|\langle f, \psi_I\rangle(\varphi_I(x) - m_J(\varphi_I))| \le C\frac{2^k \mathrm{diam}(J)}{|I|}.$$

Let $(2 \cdot I)$ denote the interval with the same midpoint as I and twice its diameter. The support of φ_I is (clearly) contained in the interval $(2 \cdot I)$. In summary for $x \in J$,

$$\begin{aligned} |f_3(x) - m_J(f_3)| &\le \sum_{I\in\mathcal{A}_3} |\langle f, \psi_I\rangle(\varphi_I(x) - m_J(\varphi_I))| \\ &\le C \sum_{I\in\mathcal{A}_3} \frac{2^k \mathrm{diam}(J)}{|I|} 1_{(2\cdot I)}(x) \\ &\le C2^k. \end{aligned} \tag{6.2.102}$$

It remains to find good estimates for $m_J(f_2)$ and $\int_J |f_2|^2$. Clearly it suffices to estimate the latter integral since

$$|m_J(f_2)| \le \left(\int_J |f_2|^2 \frac{dt}{|J|}\right)^{1/2}.$$

Now estimates for $|f_2|^2$ over the interval J are obtained just like the estimates for $|h|^2$ in (6.2.95) and (6.2.97). We find

$$\int_J |f_2|^2 dt \le C|J|. \tag{6.2.103}$$

Summing up we combine the estimates (6.2.101), (6.2.102) and (6.2.103) and insert them first into (6.2.100) and then into (6.2.99). This proves that

$$\sum_{I\in\mathcal{A}_1} \langle f, \psi_I\rangle^2 \le C|J|.$$

Now we prove the left-hand side estimate of Theorem 6.2.13. We start by fixing an interval $D \subseteq [0,1)$. The interval D is not necessarily dyadic. Note that it suffices to show that there exists $m = m(D)$ so that

$$\frac{1}{|D|}\int_D |f(x)-m|^2 dx \le 2^k C \sup_{J\in\mathcal{D}} \frac{1}{|J|}\sum_{I\subseteq J}\langle f,\psi_I\rangle^2.$$

Choose two (adjacent) dyadic intervals J_1, J_2 such that

$$D \subseteq J_1 \cup J_2,$$

and such that every $y \notin J_1 \cup J_2$ satisfies

$$\operatorname{dist}(y, D) \ge \frac{|D|}{2}.$$

Next let I be a dyadic interval such that $I \cap J_1 = \emptyset$, $I \cap J_2 = \emptyset$, and $|I| \le |D|$. For this choice of I our definition of J_1, J_2, gives that $\varphi_I(x) = 0$, for $x \in D$. Now define

$$g(x) = \sum_{I\subseteq J_1\cup J_2} \langle f,\psi_I\rangle\varphi_I(x), \quad \text{and} \quad h(x) = f(x) - g(x).$$

Fix $x \in D$ and $x_0 \in D$. Then using that $\operatorname{Lip}(\varphi_I) \le 2^k|I|^{-3/2}$ we obtain for $x \in D$, the following pointwise estimate,

$$|h(x) - h(x_0)| \le 2^k C \sup_I |I|^{-1/2}|\langle f,\psi_I\rangle|. \tag{6.2.104}$$

Next write $f(x) - h(x_0) = g(x) + h(x) - h(x_0)$. Combining Theorem 6.2.2 and (6.2.104) gives that

$$\begin{aligned}\int_D |f(x)-h(x_0)|^2 dx &\le C\int_D |g|^2 + 2^k C|D| \sup_I \frac{\langle f,\psi_I\rangle^2}{|I|}\\ &\le 2^k C|D| \sup_{J\in\mathcal{D}} \frac{1}{|J|}\sum_{I\subseteq J}\langle f,\psi_I\rangle^2.\end{aligned}$$

□

Next we prove that Carleson's system is an unconditional basis in H^1_{at}.

Theorem 6.2.14. *The biorthogonal functionals $\{\psi_I : I \in \mathcal{D}\}$ form an unconditional basis in H^1_{at}. The Banach spaces H^1_{at} and H^1 are isomorphic. Specifically, the linear extension of the map*

$$T : \psi_I \to \frac{h_I}{|I|^{1/2}}, \quad \textit{and} \quad 1_{[0,1]} \to 1_{[0,1]}$$

defines an isomorphism between H^1_{at} and H^1.

Proof. Recall first that by Theorem 6.2.2 the map T defines an isomorphism on L^2. Hence, as $H^1_{at} \cap L^2$ is dense in H^1_{at}, is suffices to prove that

$$c\|g\|_{H^1_{at}} \leq \|Tg\|_{H^1} \leq C\|g\|_{H^1_{at}}, \tag{6.2.105}$$

for every $g \in H^1_{at} \cap L^2$. Theorem 6.2.2 asserts that for $g \in H^1_{at} \cap L^2$ and $f \in \mathrm{BMO}([0,1))$,

$$\langle g, f\rangle = \sum \langle g, \varphi_I\rangle\langle f, \psi_I\rangle,$$

where the sum on the right-hand side is absolutely converging. Hence by duality,

$$||g||_{H^1_{at}} \sim \sup\left\{\sum \langle g, \varphi_I\rangle\langle f, \psi_I\rangle : ||f||_{\mathrm{BMO}([0,1))} \leq 1\right\}.$$

By Theorem 6.2.13 we obtain that

$$||g||_{H^1_{at}} \sim \sup\left\{\sum \langle g, \varphi_I\rangle b_I\right\}, \tag{6.2.106}$$

where the supremum is taken over all sequences (b_I) satisfying

$$\sup_J \frac{1}{|J|} \sum_{I \subseteq J} b_I^2 \leq 1. \tag{6.2.107}$$

Clearly if (b_I) satisfies (6.2.107), then the same holds for $(\pm b_I)$. Consequently

$$||g||_{H^1_{at}} \sim \sup\left\{\sum |\langle g, \varphi_I\rangle b_I|\right\},$$

where the supremum is taken over all sequences (b_I) satisfying (6.2.107). In particular

$$\|\sum \pm\langle g, \varphi_I\rangle\psi_I\|_{H^1_{at}} \leq C||g||_{H^1_{at}}.$$

Let us now compare the H^1_{at} norm of g with the norm of the function in (dyadic) H^1,

$$Tg = G = \sum \langle g, \varphi_I\rangle \frac{h_I}{|I|^{1/2}}.$$

By the duality between the dyadic H^1 and dyadic BMO we obtain that

$$||G||_{H^1} \sim \sup\left\{\sum |\langle g, \varphi_I\rangle b_I| : \sup_J \frac{1}{|J|} \sum_{I \subseteq J} b_I^2 \leq 1\right\}. \tag{6.2.108}$$

A direct comparison between (6.2.106) and (6.2.108) gives that (6.2.105) holds. □

6.3 Spaces of homogeneous type

The remaining sections of this chapter are devoted to proving the classification theorem for the Banach spaces $H^1(X, d, \mu)$. The style of the exposition below will be more cursory than in the previous chapters of this book and I refer to [155] for additional details.

On a set $X \times X$ let d be a quasi-metric, that is, d is symmetric, positive definite and satisfies the following so-called quasi-triangle inequality,

$$d(x,z) \leq K(d(x,y) + d(y,z)), \tag{6.3.1}$$

for $x, y, z, \in X$. Let μ be a probability measure on X. We say that μ satisfies the doubling condition, if there exists $A > 0$ so that for all $x \in X$, and $r > 0$,

$$\mu(B(x, 2r)) \leq A\mu(B(x,r)). \tag{6.3.2}$$

If for d the quasi-triangle inequality (6.3.1) holds and if the probability measure μ satisfies (6.3.2), then the triple (X, d, μ) is called a *space of homogeneous type.* We consider quasi-metrics only which satisfy the following Lipschitz condition. There exists $0 < \beta \leq 1$ and $C > 0$ so that for each $x, y, z \in X$,

$$|d(x,y) - d(y,z)| \leq Cd(x,z)^{\beta} \left(d(x,y) + d(y,z)\right)^{1-\beta}. \tag{6.3.3}$$

Finally we assume that the measure μ and the quasi-metric d are related as follows. If $x \in X$ and $\mu(\{x\}) \leq r \leq 1$, then the measure of a ball centered at x with radius r is bounded above and below by its radius. Thus,

$$cr \leq \mu(B(x,r)) \leq Cr, \quad \text{for} \quad \mu(\{x\}) \leq r \leq 1, \tag{6.3.4}$$

where $c > 0, C > 0$. For the remaining values of r where $r < \mu(\{x\})$ we assume that the ball $B(x,r)$ contains only one point, hence

$$B(x,r) \setminus \{x\} = \emptyset, \quad \text{for} \quad r < \mu(\{x\}). \tag{6.3.5}$$

Clearly the conditions (6.3.3)–(6.3.5) imply (6.3.1) and (6.3.2). Conversely there exists a theorem of R. Macias and C. Segovia [131] and [132] to the effect that for every space of homogeneous type there exists a metric d' and $\alpha > 0$ so that (6.3.3)–(6.3.5) are satisfied and

$$C_1 d'(x,y) \leq d^{\alpha}(x,y) \leq C_2 d'(x,y), \quad \text{for} \quad x, y \in X.$$

On a space of homogeneous type we define the following H^1 space of functions. An atom for (X, d, μ) is either a constant or a real-valued function a on X satisfying

$$\int_X a d\mu = 0,$$

and

$$\int_X a^2 d\mu \le \mu(B)^{-1},$$

where B is a ball in X containing the support of a. We define the associated atomic H^1 space to be the space of functions f admitting a decomposition into atoms a_i,

$$f = \sum c_i a_i, \tag{6.3.6}$$

where $c_i \in \mathbb{R}$ and

$$\sum |c_i| < \infty.$$

The right-hand side of (6.3.6) we call atomic decomposition of f. The space of functions defined that way is denoted $H^1(X, d, \mu)$. The norm of $f \in H^1(X, d, \mu)$ is given by

$$||f||_{H^1(X,d,\mu)} = \inf\{\sum |c_i|\},$$

where the infimum is taken over all decompositions (6.3.6). In this chapter we will present the isomorphic classification of these spaces. We will prove the following result.

Theorem 6.3.1. *Let (X, d, μ) be a space of homogeneous type. The Banach space $H^1(X, d, \mu)$, is isomorphic to one of the following spaces: dyadic H^1, $(\sum H_n^1)_1$, or ℓ^1.*

We will isolate geometric conditions on (X, d, μ), which determine the isomorphic type of $H^1(X, d, \mu)$. The simplest of these conditions is

$$\mu\{x \in X : \mu(\{x\}) = 0\} > 0, \tag{6.3.7}$$

characterizing the case when $H^1(X, d, \mu)$ is isomorphic to the usual dyadic H^1. It is easy to see from (6.3.3) and (6.3.5) that in a space of homogeneous type a point $x \in X$ is an isolated point precisely when $\mu(\{x\}) > 0$. Hence, the mere absence of isolated points in X implies that $H^1(X, d, \mu)$ is isomorphic to dyadic H^1. The remaining possibilities for the isomorphic types of $H^1(X, d, \mu)$ are the spaces $(\sum H_n^1)_1$, or ℓ^1. These are possible only when $\mu\{x \in X : \mu(\{x\}) = 0\} = 0$, hence only when, up to a set of measure zero, X consists of isolated points.

We will use a quantitative measure, modeled after the Carleson constant, to define the isomorphic invariant by which we will distinguish the spaces $(\sum H_n^1)_1$ and ℓ^1. This requires a bit more preparation: Let $q \in \mathbb{N}$ be large enough so that

$$q^{-\beta} \le \frac{1}{2}, \tag{6.3.8}$$

where $\beta > 0$ is given by (6.3.3). In a space of homogeneous type there exist collections of balls $\mathcal{B}^n$ which serve as a substitute and generalization of dyadic intervals of length 2^{-n}. Their definition is as follows. First select any maximal set of points

$$x_1, \ldots, x_m,$$

satisfying

$$d(x_i, x_j) \geq q^{-n} \quad \text{for} \quad i \neq j. \tag{6.3.9}$$

Then form the balls $B_i = B(x_i, 4q^{-n})$, and define

$$\mathcal{B}^n = \{B_i : 1 \leq i \leq m\}.$$

In Proposition 6.3.2 below we will show that $\mathcal{B}^n$ is a covering of X by balls with bounded overlap. The collections $\{\mathcal{B}^n : n \in \mathbb{N}\}$, determine an isomorphic invariant for the space $H^1(X, d, \mu)$. Indeed, fix $B \in \mathcal{B}^n$, and let

$$w(B) = \sum_{m=n}^{\infty} \sum_{\{Q \in \mathcal{B}^m, B \cap Q \neq \emptyset\}} \operatorname{diam}(Q).$$

Then define the Carleson constant of the collection $\mathcal{B} = \bigcup_n \mathcal{B}^n$ to be

$$[\![\mathcal{B}]\!] = \sup_n \sup_{B \in \mathcal{B}^n} \frac{w(B)}{\operatorname{diam}(2B)}. \tag{6.3.10}$$

We will now summarize how the underlying space of homogeneous type determines the isomorphic class of $H^1(X, d, \mu)$.

(a) $H^1(X, d, \mu)$ is isomorphic to dyadic H^1, if and only if

$$\mu\{x \in X : \mu(\{x\}) = 0\} > 0.$$

(b) $H^1(X, d, \mu)$ is isomorphic to $(\sum H^1_n)_1$, if and only if

$$\mu\{x \in X : \mu(\{x\}) = 0\} = 0, \qquad \text{and} \qquad [\![\mathcal{B}]\!] = \infty.$$

(b) $H^1(X, d, \mu)$ is isomorphic to ℓ^1, if and only if

$$[\![\mathcal{B}]\!] < \infty.$$

Next we present a detailed construction of the families $\{\mathcal{B}^n : n \in \mathbb{N}\}$, referred to above. They will serve as a substitute for dyadic intervals. The proof illustrates very well how the measure and the metric interact on spaces of homogeneous type. From now on we fix a large integer q so that (6.3.8) holds.

Proposition 6.3.2. *For $n \in \mathbb{N}$ there exists a collection of balls $\mathcal{B}^n$, so that the following conditions hold.*

1. $$X = \bigcup_{B \in \mathcal{B}^n} B.$$

2. *Each ball $B \in \mathcal{B}^n$ intersects at most $A_1 = A_1(K, A)$ different balls in $\mathcal{B}^n$.*

3. *For each* $B \in \mathcal{B}^n$,

$$q^{-n} \leq \text{ diam } B \leq Kq^{-n}.$$

Proof. Fix $n \in \mathbb{N}$. Then select a maximal set of points

$$x_1, \ldots, x_m,$$

such that

$$d(x_i, x_j) \geq q^{-n}, \quad \text{for} \quad i \neq j. \tag{6.3.11}$$

We claim that the union of the sets $B_i = B(x_i, 4q^{-n})$ covers X. Indeed suppose that there exists $w \in X$ which is not contained in the union $\bigcup_{i=1}^m B_i$. Then $d(w, x_i) \geq 4q^{-n}$, for each $i \leq m$. This contradicts the maximality condition (6.3.11) used to define the points $\{x_i\}$. Recall that $K > 0$ is defined by the quasi-triangle inequality (6.3.1). We claim that the balls

$$D_i = B(x_i, (4K)^{-1}q^{-n})$$

are pairwise disjoint. If not there exist $j \neq k$, and $w \in D_j \cap D_k$. Hence $d(w, x_j) \leq (4K)^{-1}q^{-n}$ and also $d(w, x_k) \leq (4K)^{-1}q^{-n}$. Consequently

$$\begin{aligned} d(x_j, x_k) &\leq K(d(w, x_j) + d(w, x_k)) \\ &\leq K(2K)^{-1}q^{-n} \\ &\leq q^{-n}/2. \end{aligned}$$

This contradicts the lower estimate (6.3.11).

Now let $A_1 \in \mathbb{N}$ and suppose that there exists a point $w \in X$ which intersects more than A_1 of the sets B_i. Next we show that there exists an upper bound for A_1. Here we exploit the doubling property (6.3.2) of the measure μ. Without loss of generality we may assume that w intersects $B_1, \ldots, B_{A_1}$. Then by triangle inequality the union $B_1 \cup \cdots \cup B_{A_1}$ is contained in $B(w, 8Kq^{-n})$. Hence

$$\mu(B_1 \cup \cdots \cup B_{A_1}) \leq CKq^{-n}, \tag{6.3.12}$$

for some universal constant C. On the other hand

$$B_j \supseteq D_j.$$

Note that we obtain B_j by doubling D_j approximately $\log K$ times. Thus we find that

$$\mu(D_j) \geq c_1\mu(B_j)/(A \log K).$$

Recall now that the sets D_j are pairwise disjoint. This allows us to estimate

$$\begin{aligned} \mu(B_1 \cup \cdots \cup B_{A_1}) &\geq \sum_{j=1}^{A_1} \mu(D_j) \\ &\geq c_1 \sum_{j=1}^{A_1} \mu(B_j)/(A \log K) \\ &\geq c_2 A_1 q^{-n}/(A \log K). \end{aligned} \tag{6.3.13}$$

Merging the estimates (6.3.12) and (6.3.13) and cancelling q^{-n} gives the following upper bound for A_1,

$$A_1 \leq CAK \log K. \qquad \square$$

With the following proposition we refine the properties of the collection $\mathcal{B}^n$.

Proposition 6.3.3. *Let $L \in \mathbb{N}$. The covering $\mathcal{B}^n$ can be split into $A_2 = A_2(L, A_1)$ disjoint collections*

$$\mathcal{B}_1^n, \ldots, \mathcal{B}_{A_2}^n,$$

such that,

$$\operatorname{dist}(B, B') \geq Lq^{-n}, \tag{6.3.14}$$

for $B, B' \in \mathcal{B}_i^n$.

Proof. Let $B \in \mathcal{B}^n$, and fix $x_B \in B$, then choose

$$R = 4LK^2. \tag{6.3.15}$$

Next we define the collection

$$\mathcal{A} = \{B(x_B, Rq^{-n}) : B \in \mathcal{B}^n\}.$$

This is a finite collection of overlapping balls in X. Note that each ball in $\mathcal{A}$ intersects at most $N = N(R)$ different elements of $\mathcal{A}$. Now we split $\mathcal{A}$ into $N+1$ collections $\{\mathcal{L}_i\}$, in such a way that the balls in $\mathcal{L}_i$ are pairwise disjoint. We let $\{U_1, \ldots, U_m\}$ be an enumeration of the sets in $\mathcal{A}$. The argument proceeds now inductively. Let $\ell \leq m$. Suppose that we have been able to obtain a splitting of $\{U_1, \ldots, U_\ell\}$ into $\mathcal{L}_1, \ldots, \mathcal{L}_{N+1}$ such that each $\mathcal{L}_i$ contains pairwise disjoint balls. Now consider $U_{\ell+1}$. Assume for the moment that for each $i \leq N+1$ there exists $L_i \in \mathcal{L}_i$, such that

$$U_{\ell+1} \cap L_i \neq \emptyset.$$

This means that $U_{\ell+1}$ intersects at least $N+1$ elements of $\mathcal{A}$, which can't happen. Therefore we found $i_0 \leq N+1$ so that for each $L \in \mathcal{L}_{i_0}$,

$$U_{\ell+1} \cap L = \emptyset.$$

In this way we obtained a proper splitting of $\mathcal{A} = \{U_1, \ldots, U_{\ell+1}\}$. This completes the proof that $\mathcal{A}$ may be split into collections $\mathcal{L}_1, \ldots, \mathcal{L}_{N+1}$ so that each $\mathcal{L}_i$ consists of pairwise disjoint sets.

Clearly the elements in $\mathcal{A}$ are indexed by the elements in $\mathcal{B}^n$. Thus the splitting of $\mathcal{A}$ into disjoint collections $\mathcal{L}_1, \ldots, \mathcal{L}_{N+1}$ induces a corresponding splitting of $\mathcal{B}^n$ into subcollections $\mathcal{B}_1^n, \ldots, \mathcal{B}_{N+1}^n$. We claim now that $\mathcal{B}_1^n, \ldots, \mathcal{B}_{N+1}^n$ satisfies the conclusion of our proposition. Indeed suppose that there exist $A, B \in \mathcal{B}_i^n$, $b \in B$, and $a \in A$ such that $d(a, b) \leq Lq^{-n}$. Then by the quasi-triangle inequality we obtain that

$$\begin{aligned} d(x_A, x_B) &\leq K(d(b, x_B) + K(d(b, a) + d(a, x_A,))) \\ &\leq (K^2L + K^2 + K)q^{-n}. \end{aligned} \tag{6.3.16}$$

With our choice of (6.3.15) the estimate (6.3.16) contradicts the fact that the collection $\mathcal{L}_i$ consists of pairwise disjoint sets. □

Let $M \in \mathbb{N}$ be a large integer to be specified below. Now we fix $1 \le i \le A_2$ and $1 \le k \le M$. Then for $j = j(k, i)$ we form

$$\mathcal{M}_j = \bigcup_{n=1}^{\infty} \mathcal{B}_i^{Mn+k}. \tag{6.3.17}$$

In that way we define

$$N = A_2 \times M \tag{6.3.18}$$

collections of balls. Next fix $j \le N$. We discard those balls $B \in \mathcal{M}_j$ for which $\mu(B)$ is much larger than $\operatorname{diam}(B)$. Note that if a ball B satisfies $\mu(B) \ge L \operatorname{diam} B$, then B consists of one isolated point. This is a consequence of the doubling property of the measure μ. We put

$$\mathcal{N}_j = \{B \in \mathcal{M}_j : L \cdot \operatorname{diam}(B) \ge \mu(B)\}, \quad \text{for } j \le N, \text{ and } \mathcal{N} = \mathcal{N}_1 \cup \cdots \cup \mathcal{N}_N. \tag{6.3.19}$$

Now we construct a nested family of measurable sets associated to $\mathcal{N}_j$.

Proposition 6.3.4. *For any $B \in \mathcal{N}_j$ there exists a set $T(B)$, such that*

$$B \subseteq T(B) \quad \text{and} \quad \mathit{diam}(T(B)) \le C\,\mathit{diam}(B),$$

and so that the entire collection

$$\mathcal{E}_j = \{T(B) : B \in \mathcal{N}_j\} \tag{6.3.20}$$

forms a nested family of measurable sets.

Proof. Let $n \in \mathbb{N}$, and let

$$\mathcal{L}_n = \mathcal{B}_i^{Mn+k} \cap \mathcal{N}_j$$

be the layers of $\mathcal{N}_j$. Now fix any $B \in \mathcal{L}_n$. We define the set $T(B)$ as follows. We put

$$x \in T(B),$$

if there exists a chain of sets $\{B_i : i \in \mathbb{N}\}$, so that

1. $B_i \in \mathcal{L}_n \cup \cdots \cup \mathcal{L}_{n+k} \cdots$,
2. $B_1 = B$,
3. $B_i \cap B_{i+1} \neq \emptyset$,

and

$$x \in \bigcup_{i=1}^{\infty} B_i.$$

This definition makes the collection

$$\mathcal{E}_j = \{T(B) : B \in \mathcal{N}_j\}, \quad j \leq N, \tag{6.3.21}$$

a nested collection of measurable sets (automatically!). Moreover with the separation property of Proposition 6.3.3 one verifies easily that the chain $\{B_i : i \geq 1\}$, does not extend "horizontally" by an amount larger than $C\mathrm{diam}(B)$. Thus,

$$\mathrm{diam}(T(B)) \leq C\mathrm{diam}(B). \qquad \square$$

The nested collections $\{\mathcal{E}_j : j \leq N\}$ defined by (6.3.20) reflect the geometry of (X, d, μ) as follows. The condition $\mu\{x \in X : \mu(\{x\}) = 0\} > 0$ is equivalent to

$$\max_{j \leq N} \mu(\limsup \mathcal{E}_j) > 0. \tag{6.3.22}$$

Recall that $\mathcal{B} = \bigcup_n \mathcal{B}^n$. We defined the Carleson constant of $\mathcal{B}$ in (6.3.10). It follows that $[\![\mathcal{B}]\!] = \infty$ if and only if

$$\max_{j \leq N} \sup_{A \in \mathcal{E}_j} \sum_{\{B \in \mathcal{E}_j \cap A\}} \frac{\mu(B)}{\mu(A)} = \infty. \tag{6.3.23}$$

We use the generations of $\mathcal{E}_j$ to define a sequence of increasing σ-algebras on X. We put $\mathcal{F}_0^j = \{X\}$, and let

$$\mathcal{A}_n = G_n(\mathcal{E}_j), \quad n \in \mathbb{N}.$$

Then we define inductively

$$\mathcal{F}_{n+1}^j = \sigma(\mathcal{F}_n^j, \mathcal{A}_{n+1}) \quad n \in \mathbb{N}. \tag{6.3.24}$$

Observe that by B. Maurey's isomorphism established in Theorem 4.1.3, the martingale space $H^1[(\mathcal{F}_n^j)_{n=1}^\infty]$ is isomorphic to $X[\mathcal{E}_j]$, the space spanned by three-valued martingale differences associated to the nested family $\mathcal{E}_j$.

In the following two sections we show that $H^1[(\mathcal{F}_n^1)] \oplus \cdots \oplus H^1[(\mathcal{F}_n^N)]$ is isomorphic to $H^1(X, d, \mu)$. We obtain this result by applying the decomposition principle of A. Pełczyński and the two theorems stated next. First in Section 6.4 we prove that

Theorem 6.3.5. *$X[\mathcal{E}_1] \oplus \cdots \oplus X[\mathcal{E}_N]$ is isomorphic to a complemented subspace of $H^1(X, d, \mu)$.*

Second, in Section 6.5 we show the converse to Theorem 6.3.5.

Theorem 6.3.6. *$H^1[(\mathcal{F}_n^1)] \oplus \cdots \oplus H^1[(\mathcal{F}_n^N)]$ contains a complemented copy of $H^1(X, d, \mu)$*

Thus the problem of identifying the isomorphic type of *atomic* H^1 spaces is reduced to the case of *martingale* H^1 spaces. The latter ones are classified by Theorem 4.1.3 and Theorem 4.2.1.

Lipschitz partitions of unity

The covering $\mathcal{B}^n$ defined in Proposition 6.3.3 induces the following Lipschitz partition of unity on (X, d, μ). We let g be a smooth and non-negative function on $\mathbb{R}$ such that g is equal to 1 on the interval $]-1, 1[$, and 0 outside of $]-2, 2[$. Then fix $B \in \mathcal{B}^n$, and define

$$r_B(x) = g\left(\frac{d(x, B)}{\operatorname{diam} B}\right). \tag{6.3.25}$$

Note that $r_B(x) = 1$ for $x \in B$, and that $r_B(x) = 0$, when $d(x, B) \geq 2\operatorname{diam} B$. Hence,

$$1 \leq \sum_{B \in \mathcal{B}^n} r_B(x) \leq K_1$$

where $K_1 = K_1(A, K, A_1)$. Then we define a partition of unity by the formula

$$s_B(x) = r_B(x) \left(\sum_{A \in \mathcal{B}^n} r_A(x) \right)^{-1}.$$

Note that s_B satisfies the Lipschitz estimate

$$|s_B(x) - s_B(y)| \leq C\mu(B)^{-\beta} d(x, y)^\beta, \tag{6.3.26}$$

and its support is centered around B,

$$B \subseteq \operatorname{supp} s_B \subseteq KB.$$

The partition of unity $\{s_B : B \in \mathcal{B}^n\}$ induces a positive and symmetric kernel

$$S_n(x, y) = \sum_{B \in \mathcal{B}^n} s_B(x) s_B(y) / ||s_B||_1. \tag{6.3.27}$$

The kernel satisfies a Lipschitz estimate

$$|S_n(x, y) - S_n(u, y)| \leq C d(x, u)^\beta q^{n + \beta n}, \tag{6.3.28}$$

and

$$\int S_n(x, y) d\mu(x) = 1. \tag{6.3.29}$$

Moreover $S_n(x, y)$ is supported around the diagonal as follows: If $d(x, y) \geq Kq^{-n}$, then

$$S_n(x, y) = 0.$$

For $\beta > 0$ the Lipschitz class Lip_β consists of functions $f : X \to \mathbb{R}$ for which there is $C > 0$ so that

$$|f(x) - f(y)| \leq C d(x, y)^\beta, \quad \text{for} \quad x, y \in X. \tag{6.3.30}$$

We let $\operatorname{Lip}_\beta(f)$ denote the infimum over all $C > 0$ for which (6.3.30) holds true.

$H^1(X,d,\mu)$ estimates for molecules

In this subsection we establish a general and widely used condition asserting that a given operator is bounded on $H^1(X,d,\mu)$.

Suppose that a linear operator T defined initially on $L^2(X,\mu)$ maps atoms into atoms. Then, of course, T admits a unique extension to a bounded operator on $H^1(X,d,\mu)$. The same conclusion holds true if the operator T has the property that it maps a single atom into a convex combination of atoms. This observation motivates the search for conditions implying that a given function in $L^2(X,\mu)$ can be written as a convex combination of atoms. Most frequently the following criterion is used.

We say that a function $m : X \to \mathbb{R}$ is a molecule centered at x_0 if there exists $\epsilon > 0$ such that

$$\left\{\int_X m^2(x)d\mu(x)\right\}\left\{\int_X m^2(x)d(x_0,x)^{1+\epsilon}d\mu(x)\right\}^{1/\epsilon} \le 1,$$

and

$$\int m d\mu = 0.$$

Next we will show that a molecule can be written as an absolutely convergent series of atoms, so that the sum of the absolute values of the coefficients depends only on $\epsilon > 0$ and a universal constant derived from the space of homogeneous type (X,d,μ). It follows that a linear operator mapping atoms into molecules (with fixed $\epsilon > 0$) admits an extension to a bounded operator on $H^1(X,d,\mu)$.

Theorem 6.3.7. *Any molecule admits an absolutely convergent decomposition into atoms, and hence belongs to $H^1(X,d,\mu)$.*

Proof. Let m be an ϵ-molecule centered at x_0. Then put $R = ||m||_2^{-2}$ and let $A_0 = B(x_0,R)$. Define the following sequence of annuli centered at x_0.

$$A_i = \{x : R2^{i-1} \le d(x,x_0) \le R2^i\}.$$

Then define coefficients

$$a_i = \int_{A_i} m d\mu,$$

and the following pieces of m,

$$m_i = m1_{A_i} - a_i\mu(A_i)^{-1}1_{A_i}.$$

Note that m_i is supported on a ball of radius $R2^i$, and $\int m_i d\mu = 0$. Now let

$$c_i = \sum_{k=i}^{\infty} a_k.$$

By assumption $c_0 = \int m d\mu = 0$, hence partial summation gives the identities

$$\begin{aligned} m &= \sum_{i=0}^{\infty} m_i + \sum_{i=0}^{\infty} a_i \mu(A_i)^{-1} 1_{A_i} \\ &= \sum_{i=0}^{\infty} m_i + \sum_{i=0}^{\infty} c_{i+1} (\mu(A_{i+1})^{-1} 1_{A_{i+1}} - \mu(A_i)^{-1} 1_{A_i}). \end{aligned}$$

Now we will show that this is indeed an atomic decomposition of m. Note that m_i is supported on a ball of radius $R2^i$, and $\int m_i d\mu = 0$. Analogously $(\mu(A_{i+1})^{-1} 1_{A_{i+1}} - \mu(A_i)^{-1} 1_{A_i})$ is supported on a ball of radius $R2^{i+1}$. Next we verify the L^2 estimates

$$\begin{aligned} 2^{i\epsilon} \int m_i^2 &\leq 4(R2^i)^{-1}, \\ \int (\mu(A_{i+1})^{-1} 1_{A_{i+1}} - \mu(A_i)^{-1} 1_{A_i})^2 &\leq C(R2^i)^{-1}, \end{aligned} \tag{6.3.31}$$

and the upper bound for the coefficients

$$c_i \leq 2^{-i\epsilon/2}. \tag{6.3.32}$$

It is worth pointing out that, in proving the first estimate in (6.3.31) we use that ϵ-molecules satisfy the upper bound

$$\int m^2(x) d(x, x_0)^{1+\epsilon} d\mu(x) \leq ||m||_2^{-2\epsilon}.$$

Fix $i \in \mathbb{N}$ and write

$$\begin{aligned} \int m_i^2(x) d\mu &\leq \int_{A_i} m^2(x) d(x, x_0)^{1+\epsilon} d(x, x_0)^{-1-\epsilon} d\mu \\ &\leq (R2^{i-1})^{(-1-\epsilon)} ||m||_2^{-2\epsilon} \\ &\leq 2^{1+\epsilon} (R2^i)^{-1} 2^{-i\epsilon}. \end{aligned}$$

For the second estimate in (6.3.31) observe that, pointwise, $|\mu(A_{i+1})^{-1} 1_{A_{i+1}} - \mu(A_i)^{-1} 1_{A_i}| \leq C(R2^i)^{-1}$. It remains to square this and integrate over its support. The support has measure $\leq CR2^i$.

Finally we prove the estimate (6.3.32) for the coefficients c_i. We begin with Cauchy–Schwarz,

$$|c_i| = \sum_{k=i}^{\infty} \int_{A_k} |m| \leq \sum_{k=i}^{\infty} \mu(A_k)^{1/2} \|m_k^2\|_2. \tag{6.3.33}$$

The bound for $\int m_k^2$, shows that the summands of the series on the right-hand side of (6.3.33) are dominated by a geometric progression,

$$\mu(A_k)^{1/2} (\int m_k^2)^{1/2} \leq C(R2^k)^{1/2} (R2^k)^{-1/2} 2^{-k\epsilon/2}.$$

Summing the resulting geometric series,

$$\sum_{k=i}^{\infty} 2^{-k\epsilon/2},$$

gives the upper estimate $|c_i| \leq C_\epsilon 2^{-i\epsilon/2}$. □

6.4 Orthogonal projections in atomic H^1 spaces

The previous section was devoted to the construction of the nested collections $\{\mathcal{E}_j : j \leq N\}$. See Proposition 6.3.4. Here we will show that $X[\mathcal{E}_j]$ is isomorphic to a complemented subspace of $H^1(X, d, \mu)$. Following is an outline of how this is done below. We start with a martingale difference sequence $\{h_A : A \in \mathcal{N}_j\}$ on a probability space (X, μ), generating a copy of $X[\mathcal{E}_j]$. Then we regularize the martingale difference sequence using the kernels $\{S_n : n \in \mathbb{N}\}$, defined by (6.3.27) in the previous section. We will show that for the resulting family of L^2 normalized functions $\{g_A : A \in \mathcal{N}_j\}$ the Gram matrix

$$G(A, B) = \int g_A g_B d\mu,$$

is a positive definite and almost diagonal matrix. Hence we may use the square root of its inverse to form a biorthogonal system defined by

$$f_A = \sum G^{-1/2}(A, B) g_B.$$

This system gives rise to an orthogonal projection

$$Pf = \sum \langle f, f_A \rangle f_A.$$

We will prove that P is bounded on $H^1(X, d, \mu)$ and that its range is isomorphic to $X[\mathcal{E}_j]$. In this way we will obtain an explicit form of the embedding and projection showing that $X[\mathcal{E}_j]$ is isomorphic to a complemented subspace of $H^1(X, d, \mu)$.

Given $\mathcal{E}_j = \{T(B) : B \in \mathcal{N}_j\}$, we start by constructing L^2 normalized martingale difference sequences, $\{h_B : B \in \mathcal{N}_j\}$ satisfying,

(a) $\operatorname{supp} h_B \subseteq T(B)$,

(b) $\int h_B d\mu = 0$,

(c) $\sup_{x \in B} |h_B(x)| \leq C \inf_{x \in B} |h_B(x)|$, for some universal constant C,

(d) h_B is constant when restricted to any $A \in G_1(T(B), \mathcal{E}_j)$.

Now we continue by regularizing the martingale differences defined above. Let $k_0 \in \mathbb{N}$ be a fixed, large integer depending on (X, d, μ). Let $n \in \mathbb{N}$ and $B \in \mathcal{B}^n \cap \mathcal{N}_j$.

$$\text{For } M(n) = n + k_0, \quad \text{define} \quad \phi_B(x) = \int S_{M(n)}(x, y) h_B(y) d\mu(y), \tag{6.4.1}$$

and renormalize in L^2,

$$g_B(x) = \phi_B(x)/||\phi_B||_2. \tag{6.4.2}$$

The Gram matrix $G = (G(A,B))$ of the system $\{g_B : B \in \mathcal{N}_j\}$ is defined to be the matrix whose entries are

$$G(A,B) = \int_X g_A g_B d\mu, \quad \text{where} \quad A, B \in \mathcal{N}_j. \tag{6.4.3}$$

With the next proposition we summarize the basic properties of the system $\{g_B : B \in \mathcal{N}_j\}$. We show that G is positive definite, and that its entries decay rapidly off the diagonal. Recall that the constant M appearing in the conclusion of Proposition 6.4.1 is defined in (6.3.17).

Proposition 6.4.1. *For any $A \in \mathcal{N}_j$, the function g_A, is supported in a constant multiple of the ball A. Let $A \in \mathcal{B}^n \cap \mathcal{N}_j$, and let $B \in \mathcal{B}^m \cap \mathcal{N}_j$. If $m \neq n$, then*

$$|G(A,B)| \leq C(k_0) \min\left\{\frac{\mu(A)}{\mu(B)}, \frac{\mu(B)}{\mu(A)}\right\}^{1/2+\beta}. \tag{6.4.4}$$

If $m = n$ and $A \neq B$, then

$$G(A,B) = 0. \tag{6.4.5}$$

There exists $0 < \epsilon(M) \leq C(k_0)q^{-M\beta}$ so that for any sequence of scalars $\{c_A : A \in \mathcal{N}_j\}$,

$$(1-\epsilon(M))\sum c_A^2 \leq ||\sum c_A g_A||_2^2 \leq (1+\epsilon(M))\sum c_A^2.$$

Hence, the Gram matrix G is invertible and satisfies

$$\|\operatorname{Id} - G\|_{\ell^2} \leq \epsilon(M) \quad \text{with} \quad 0 < \epsilon(M) \leq C(k_0)q^{-M\beta}. \tag{6.4.6}$$

Proof. Let $A, B \in \mathcal{B}^n \cap \mathcal{N}_j$, and $A \neq B$, then by Proposition 6.3.3, $\operatorname{supp} g_A \cap \operatorname{supp} g_B = \emptyset$. Hence $G(A,B) = 0$, and (6.4.5) holds.

Next we turn to the verification of (6.4.4). Let $A \in \mathcal{N}_j \cap \mathcal{B}^m$. We first obtain Lipschitz estimates for g_A directly from Lipschitz estimates of the kernel $S_{M(n)}$.

$$\begin{aligned}|g_A(x) - g_A(y)| &\leq C \int |S_{M(n)}(x,u) - S_{M(n)}(y,u) \cdot |h_A(u)| d\mu(u) \\ &\leq C(k_0)\mu(A)^{-1/2-\beta} d(x,y)^\beta.\end{aligned}$$

Fix $A, B \in \mathcal{N}_j$ with $\mu(B) \leq \mu(A)$, then fix $a \in A$. Now estimate using that $\int g_B d\mu = 0$,

$$\begin{aligned}|G(A,B)| &\leq \int |g_B(x)(g_A(x) - g_A(a))| d\mu \\ &\leq \int |g_B| d\mu \mathrm{Lip}_\beta(g_A)\mu(B)^\beta \\ &\leq C(k_0)(\mu(B)/\mu(A))^{1/2+\beta}.\end{aligned} \tag{6.4.7}$$

Thus we verified (6.4.4).

Let $\{c_A : A \in \mathcal{N}_j\}$ be given, then

$$||\sum c_A g_A||_2^2 = \sum c_A^2 + \sum_A \sum_{B \neq A} c_A c_B G(A,B).$$

Exploiting the estimate (6.4.7) we prove now good estimates for the off diagonal terms appearing on the right-hand side. We claim that these are bounded by a small fraction of $\sum c_A^2$. Fix $A \in \mathcal{N}_j$ and $l \geq 1$. Define $G_l(A)$ to be the l-th generation of $T(A)$ with respect to the nested collection $\mathcal{E}_j$. Then put

$$\mathcal{H}_l(A) = \{B : T(B) \in G_l(A)\}.$$

By (6.4.5) we have the identity

$$\sum_A \sum_{B \neq A} c_A c_B G(A,B) = \sum_{l=1}^{\infty} \sum_A c_A \sum_{B \in \mathcal{H}_l(A)} c_B G(A,B).$$

Note that for $l \geq 1$ and $B \in \mathcal{H}_l(A)$ we have that $\mu(B) \sim q^{-lM}\mu(A)$, hence by (6.4.7),

$$|G(A,B)| \leq C(k_0)(\mu(B)/\mu(A))^{1/2+\beta} \leq C(k_0)q^{-lM(1/2+\beta)}.$$

Using also the symmetry of the Gram matrix, $G(A,B) = G(B,A)$, we obtain that

$$\sum_A \sum_{B \neq A} |c_A c_B G(A,B)| \leq C(k_0) \sum_{l=1}^{\infty} \sum_A |c_A| \sum_{B \in \mathcal{H}_l(A)} |c_B| q^{-lM(1/2+\beta)}. \tag{6.4.8}$$

The cardinality of the index set $\mathcal{H}_l(A)$ is of order q^{lM}. Next we use the Cauchy–Schwarz inequality and invoke the estimate for the cardinality of $\mathcal{H}_l(A)$. Then the right-hand side of the inequality (6.4.8) can be dominated by

$$C(k_0)\left(\sum c_A^2\right)^{1/2} \sum_{l=1}^{\infty} q^{-lM\beta} \left(\sum c_B^2\right)^{1/2}. \tag{6.4.9}$$

Note that the geometric series in (6.4.9) starts with $l = 1$. Evaluating its sum we obtain finally the following bound for the off diagonal terms,

$$\sum_A \sum_{B \neq A} |c_A c_B G(A,B)| \leq C(k_0) q^{-M\beta} \sum_A c_A^2.$$

Putting $\epsilon(M) = C(k_0)q^{-M\beta}$ finishes the proof. □

Proposition 6.4.1 implies immediately that the Gram matrix G is positive definite. Its inverse G^{-1} has therefore a well-defined square root denoted by $G^{-1/2}$.

Its existence is guaranteed by the functional calculus of spectral theory. By (6.4.6) the functional calculus yields an expansion of $G^{-1/2}$ in an absolutely convergent power series of the form

$$G^{-1/2} = \sum_{k=0}^{\infty} c_k(\,\mathrm{Id} - G)^k,$$

where the coefficients satisfy $c_k \sim k^{-1/2}$. Below, we use the power series expansion of $G^{-1/2}$ to show that (6.4.4) and (6.4.6) imply that the entries of $G^{-1/2}$ (indexed by $A, B \in \mathcal{N}_j$) satisfy the estimate

$$|G^{-1/2}(A,B)| \le C \min\left\{\frac{\mu(A)}{\mu(B)}, \frac{\mu(B)}{\mu(A)}\right\}^{1/2+\beta/8} \left\{1 + \frac{d(A,B)}{\max\{\mu(A),\mu(B)\}}\right\}^{-1-\beta/8}.$$

See Theorem 6.4.6 and (6.4.45). The off-diagonal decay of $G^{-1/2}$ allows us to construct an orthonormal system which satisfies good localization and decay properties. This is the content of the next theorem.

Theorem 6.4.2. *There exists an orthonormal system of functions* $\{f_B : B \in \mathcal{N}_j\}$ *on* (X, d, μ), *so that the following conditions hold.*

1. $\mathrm{span}\{f_B : B \in \mathcal{N}_j\} = \mathrm{span}\{g_B : B \in \mathcal{N}_j\}$.
2. $\displaystyle\int f_B d\mu = 0$.
3. $\displaystyle |f_B(x)| \le \frac{C}{\mu(B)^{1/2}} \left(1 + \frac{\mathrm{dist}(x,B)}{\mu(B)}\right)^{-1-\beta/(16)}.$
4. $\displaystyle |f_B(x) - f_B(y)| \le C \frac{d(x,y)^{\beta/(16)}}{\mu(B)^{1/2+\beta/(16)}} \left(1 + \frac{\mathrm{dist}(x,B)}{\mu(B)}\right)^{-1-\beta/(16)}.$

Proof. By Theorem 6.4.1 the Gram matrix G is positive definite hence its inverse and the square root of its inverse are well defined. We obtain the system f_B by applying the matrix $G^{-1/2}$, to the functions $\{g_A\}$, that is, we let

$$f_B(x) = \sum G^{-1/2}(A,B) g_A(x).$$

Theorem 6.4.6 below describes the decay of the off diagonal coefficients in the matrix $G^{-1/2}$. It asserts that

$$G^{-1/2}(A,B) \le C \min\left\{\frac{\mu(A)}{\mu(B)}, \frac{\mu(B)}{\mu(A)}\right\}^{1/2+\beta/8} \left\{1 + \frac{d(A,B)}{\max\{\mu(A),\mu(B)\}}\right\}^{-1-\beta/8}.$$

For $x \in X$, fixed, the support of $g_A(x)$, is contained in a set of diameter $\le C\mu(A)$. Hence the pointwise estimate for $f_B(x)$,

$$|f_B(x)| \le C \sum_{\{A:\ \mathrm{dist}(x,A) \le C\mu(A)\}} \mu(A)^{-1/2} G^{-1/2}(A,B).$$

Now we split the index set $\{A \in \mathcal{N}_j : \operatorname{dist}(x, A) \leq C\mu(A)\}$ in three parts. Define $K_0 \in \mathbb{N}$ such that

$$\frac{K_0}{2} \leq 1 + \frac{\operatorname{dist}(x, B)}{\mu(B)} \leq K_0.$$

Depending on the value of K_0 we split the index set $\{A : \operatorname{dist}(x, A) \leq C\mu(A)\}$. Start by "drawing a disk" around x with radius K_0. Then let $\mathcal{A}$ be the part of the index set which is not contained in this disk. The remaining part of the index set is divided into balls with large measure and those with small measure. Precisely, we define

$$\begin{aligned} \mathcal{A} &= \{A \in \mathcal{N}_j : \operatorname{dist}(x, A) \leq C\mu(A) \text{ and } \mu(A) > K_0\mu(B)\}, \\ \mathcal{B} &= \{A \in \mathcal{N}_j : \operatorname{dist}(x, A) \leq C\mu(A) \text{ and } \mu(B) \leq \mu(A) \leq K_0\mu(B)\}, \\ \mathcal{C} &= \{A \in \mathcal{N}_j : \operatorname{dist}(x, A) \leq C\mu(A) \text{ and } \mu(A) < \mu(B)\}. \end{aligned} \tag{6.4.10}$$

We record now estimates for each of these cases. In the first case we have

$$\sum_{A \in \mathcal{A}} \mu(A)^{-1/2} G^{-1/2}(A, B) \leq K_0^{-1-\beta/8} \mu(B)^{-1/2}.$$

For the second case we obtain

$$\sum_{A \in \mathcal{B}} \mu(A)^{-1/2} G^{-1/2}(A, B) \leq K_0^{-1-\beta/8} (\log K_0) \mu(B)^{-1/2}.$$

Finally in the third case we consider two sub-cases. First if $\operatorname{dist}(x, B) \leq C\mu(B)$, then

$$\sum_{A \in \mathcal{C}} \mu(A)^{-1/2} G^{-1/2}(A, B) \leq \mu(B)^{-1/2}.$$

Second if $\operatorname{dist}(x, B) \geq C\mu(B)$, then we simply get

$$\sum_{A \in \mathcal{C}} \mu(A)^{-1/2} G^{-1/2}(A, B) = 0.$$

Combining the estimates of these three cases into one statement we formulate

$$\sum_{\{A:\ \operatorname{dist}(x,A) \leq C\mu(A)\}} \mu(A)^{-1/2} G^{-1/2}(A, B) \leq \mu(B)^{-1/2} \left(1 + \frac{\operatorname{dist}(x, B)}{\mu(B)}\right)^{-1-\beta/(16)}. \qquad \square$$

The $H^1(X, d, \mu)$ bounds for f_B appearing in the next theorem are a direct consequence of the pointwise decay established in Theorem 6.4.2.

Theorem 6.4.3. *The orthonormal functions $f_B, B \in \mathcal{N}_j$ defined by Theorem* 6.4.2 *satisfy the $H^1(X, d, \mu)$ estimates*

$$||f_B||_{H^1(X,d,\mu)} \leq C\mu(B)^{1/2}.$$

Proof. Let $B \in \mathcal{N}_j$. By Theorem 6.3.7 it suffices to check that $\mu(B)^{-1/2} f_B$ is a multiple of a $\beta/(16)$-molecule. Recall first that $\int f_B(x)d\mu(x) = 0$, and $\|f_B\|_2 = 1$. Now fix $x_B \in B$. We will show that

$$\int_X f_B^2(x)d(x_B, x)^{1+\beta/(16)} d\mu(x) \leq C\mu(B)^{1+\beta/(16)}.$$

Let $A_0 = B$. For $j \geq 1$, we form the annuli

$$A_j = \{x \in X : 2^{j-1}\mu(B) \leq d(x_B, x) \leq 2^j \mu(B)\}.$$

The pointwise estimate of Proposition 6.4.2 and the defining equation of the annuli A_j give that for $j \geq 1$,

$$\begin{aligned}\int_{A_j} & f_B^2(x)d(x_B, x)^{1+\beta/(16)} d\mu(x) \\ & \leq \mu(A_j) \sup_{x\in A_j} f_B^2(x) \sup_{x\in A_j} d(x_B, x)^{1+\beta/(16)} \\ & \leq C \left[2^j \mu(B)\right] \left[2^{-j(2+\beta/8)} \mu(B)^{-1}\right] \left[(\mu(B) 2^j)^{1+\beta/(16)}\right].\end{aligned}$$

Collecting terms in the second line of the above inequality we find that

$$\int_{A_j} f_B^2(x)d(x, B)^{1+\beta/(16)} d\mu(x) \leq C2^{-j\beta/(16)} \mu(B)^{1+\beta/(16)}. \tag{6.4.11}$$

Next we take the sum of the estimates (6.4.11). This gives

$$\begin{aligned}\int_{X\setminus B} f_B^2(x)d(x_B, x)^{1+\beta/(16)} d\mu(x) & \leq \sum_{j=1}^{\infty} \int_{A_j} f_B^2(x)d(x_B, x)^{1+\beta/(16)} d\mu(x) \\ & \leq C\mu(B)^{1+\beta/(16)} \sum_{j=1}^{\infty} 2^{-j\beta/(16)}.\end{aligned} \tag{6.4.12}$$

To complete the proof we observe that $\int_B d(x_B, x)^{1+\beta/(16)} d\mu(x) \leq C\mu(B)^{2+\beta/(16)}$. Hence Proposition 6.4.2 gives that

$$\int_B f_B^2(x)d(x_B, x)^{1+\beta/(16)} d\mu(x) \leq C\mu(B)^{1+\beta/(16)}. \tag{6.4.13}$$

Combining (6.4.12) and (6.4.13) completes the proof. □

Theorem 6.4.4. *The orthogonal projection given by*

$$P(f) = \sum_{B\in\mathcal{N}_j} \langle f, f_B\rangle f_B$$

extends to a bounded operator on $H^1(X, d, \mu)$.

Proof. The proof uses molecules, the norm estimates of Theorem 6.4.3 and a splitting of the index set $\mathcal{N}_j$, analogous to the one introduced in the course of proving Theorem 6.2.13.

Part 1. Let $a : X \to \mathbb{R}$ be an atom for (X, d, μ). Let I be a ball in X such that $\operatorname{supp} a \subseteq I$, and

$$\int a^2 \leq \mu(I)^{-1}.$$

We split the collection $\mathcal{N}_j$ into three subcollections which are determined by the support of the atom a. Define

$$\begin{aligned} \mathcal{A} &= \{B \in \mathcal{N}_j : \mu(B) \geq \mu(I)\}, \\ \mathcal{B} &= \{B \in \mathcal{N}_j : \mu(B) < \mu(I) \text{ and } \operatorname{dist}(B, I) \leq \mu(I)\}, \\ \mathcal{C} &= \{B \in \mathcal{N}_j : \mu(B) < \mu(I) \text{ and } \operatorname{dist}(B, I) > \mu(I)\}. \end{aligned} \tag{6.4.14}$$

Before we start estimating we will decompose these collections into levels consisting of sets of comparable measure. Recall that

$$\mathcal{N}_j \subseteq \bigcup_n \mathcal{B}_i^{Mn+k}, \tag{6.4.15}$$

for some fixed $k \leq M$ and $i \leq A_2$. Now we select the level n_0 which is determined by $\mu(I)$. We choose $n_0 \in \mathbb{N}$ by the following rule. If $B \in \mathcal{A}$, then there exists $n \leq n_0$ such that $B \in \mathcal{B}_i^{Mn+k}$; if $B \in \mathcal{B} \cup \mathcal{C}$, then there exists $n \geq n_0 + 1$ so that $B \in \mathcal{B}_i^{Mn+k}$.

Part 2. Here we estimate

$$\|\sum_{B \in \mathcal{A}} \langle a, f_B \rangle f_B\|_{H^1(X,d,\mu)}.$$

With the triangle inequality this sum is dominated by

$$\sum_{B \in \mathcal{A}} \|\langle a, f_B \rangle f_B\|_{H^1(X,d,\mu)}.$$

Recall that by Theorem 6.4.3, we have

$$\|f_B\| \leq C\mu(B)^{1/2}, \tag{6.4.16}$$

and $\int a d\mu = 0$. The Lipschitz estimate of Proposition 6.4.2 gives that

$$|\langle a, f_B \rangle| \leq \operatorname{diam}(I)^{\beta/(16)} \mu(B)^{-1/2-\beta/(16)} \left(1 + \frac{\operatorname{dist}(B, I)}{\mu(B)}\right)^{-1-\beta/(16)}. \tag{6.4.17}$$

Now we split the index set $\mathcal{A}$ into its levels. For $n \leq n_0$ we put

$$\mathcal{A}_n = \mathcal{A} \cap \mathcal{B}_i^{Mn+k}.$$

Observe that

$$\sum_{B \in \mathcal{A}_n} \left(1 + \frac{\operatorname{dist}(B, I)}{\mu(B)}\right)^{-1-\beta/(16)} \leq C, \tag{6.4.18}$$

where $C > 0$ is independent of the level n. Moreover for $B \in \mathcal{A}_n$ we have $\mu(B)^{-\beta/(16)} \sim q^{(Mn+k)\beta/(16)}$. Hence combining (6.4.16)–(6.4.18) we find that

$$\sum_{B\in\mathcal{A}} \|\langle a, f_B\rangle f_B\|_{H^1(X,d,\mu)} \leq C\text{diam}(I)^{\beta/(16)} \sum_{n=1}^{n_0} q^{(Mn+k)\beta/(16)}. \tag{6.4.19}$$

It remains to analyze the factor in (6.4.19) given by the geometric series. First note that

$$q^{(Mn_0+k)\beta/(16)} \sim \mu(I)^{-\beta/(16)} \sim \text{diam}(I)^{-\beta/(16)}.$$

Inserting this in the estimate (6.4.19) we obtain that

$$\sum_{B\in\mathcal{A}} \|\langle a, f_B\rangle f_B\|_{H^1(X,d,\mu)} \leq C.$$

Part 3. Now we turn to estimating

$$||\sum_{B\in\mathcal{C}} \langle a, f_B\rangle f_B||_{H^1(X,d,\mu)}. \tag{6.4.20}$$

Recall that the index set $\mathcal{C}$ contains those $B \in \mathcal{N}_j$ which are well separated from I and which satisfy $\mu(B) \leq \mu(I)$. Again the first move is to apply the triangle inequality. So we will actually find an estimate for the sum

$$\sum_{B\in\mathcal{C}} ||f_B||_{H^1(X,d,\mu)} \int_I |f_B(x)a(x)|d\mu(x). \tag{6.4.21}$$

By Theorem 6.4.2 for $B \in \mathcal{C}$ the following pointwise estimate holds for $x \in I$,

$$|f_B(x)| \leq \mu(B)^{+1/2+\beta/(16)} \text{ dist}(I, B)^{-1-\beta/(16)}.$$

We insert this, and recall that $||f_B|| \leq C\mu(B)^{1/2}$. Thus we obtain that

$$||f_B||_{H^1(X,d,\mu)} \int_I |f_B(x)a(x)|d\mu(x) \leq C\mu(B)^{1+\beta/(16)} \text{ dist}(I, B)^{-1-\beta/(16)}. \tag{6.4.22}$$

Next we split the index set $\mathcal{C}$ in its natural layers. For $n \geq n_0 + 1$, define

$$\mathcal{C}_n = \mathcal{C} \cap \mathcal{B}_i^{Mn+k},$$

where k, i are determined by (6.4.15). We thus arrived at the estimate,

$$\sum_{n=n_0+1}^{\infty} \sum_{B\in\mathcal{C}_n} \mu(B)^{1+\beta/(16)} \text{ dist}(I, B)^{-1-\beta/(16)}. \tag{6.4.23}$$

To this end we fix $n \geq n_0 + 1$. Then for $B \in \mathcal{C}_n$ we have $\operatorname{dist}(B, I) \geq (\mu(I) + m\mu(B))$ for some $m \in \mathbb{N}$. Moreover, the sets $B \in \mathcal{C}_n$ satisfy the measure estimate $\mu(B) \sim q^{-(Mn+k)}$. Now we abbreviate

$$A_0 = \mu(I) q^{(Mn+k)}.$$

Then the inner sum of (6.4.23), has the upper bound

$$\begin{aligned} \sum_{B \in \mathcal{C}_n} \mu(B)^{1+\beta/(16)} \operatorname{dist}(I, B)^{-1-\beta/(16)} &\leq C \sum_{m=1}^{\infty} (m + A_0)^{-1-\beta/(16)} \\ &\leq C A_0^{-\beta/(16)}. \end{aligned} \tag{6.4.24}$$

We have now reduced our task of estimating (6.4.20) to that of evaluating a geometric series. Indeed, as $A_0 = \mu(I) q^{(Mn+k)}$, we obtained with (6.4.22) and (6.4.24), that

$$\|\sum_{B \in \mathcal{C}} \langle a, f_B \rangle f_B\| \leq C \mu(I)^{-\beta/(16)} \sum_{n=n_0+1}^{\infty} q^{-(Mn+k)\beta/(16)}. \tag{6.4.25}$$

Note that by choice of n_0,

$$\sum_{n=n_0}^{\infty} q^{-(Mn+k)\beta/(16)} \leq C \mu(I)^{\beta/(16)}. \tag{6.4.26}$$

In (6.4.25) the factor $\mu(I)^{-\beta/(16)}$ is thus compensated by the value of the geometric sum in (6.4.26). Thus we find,

$$\|\sum_{B \in \mathcal{C}} \langle a, f_B \rangle f_B\| \leq C.$$

Part 4. Now we turn to the estimates for

$$b = \sum_{B \in \mathcal{B}} \langle a, f_B \rangle f_B.$$

Here the index set comprises the elements in $\mathcal{N}_j$ which are close to I, and for which $\mu(B) \leq C\mu(I)$. Therefore loosely speaking the sum b is strongly concentrated around I. This observation makes it plausible that b is a molecule, centered around (any point of) I. We will now prove that this is in fact the case. To begin with, we have by orthonormality of the system f_B,

$$\int b^2 d\mu \leq \int a^2 d\mu \leq \mu(I)^{-1}.$$

Hence in order for b to be a molecule, it suffices that

$$\int_X b(x)^2 d(x, x_I)^{1+\beta/(16)} d\mu(x) \leq \mu(I)^{\beta/(16)}.$$

Now we split the domain of integration. Consider first the integral over the region $4I$. Here we estimate simply by pulling out the factor $d(x,x_I)^{1+\beta/(16)}$, hence

$$\int_{4I} b(x)^2 d(x,x_I)^{1+\beta/(16)} d\mu(x) \le \mu(I)^{1+\beta/(16)} \int b^2(x) d\mu(x) \le \mu(I)^{\beta/(16)}.$$

Now we turn to the remaining integral over $X \setminus 4I$. We will show that

$$\int_{X\setminus 4I} b(x)^2 d(x,x_I)^{1+\beta/(16)} d\mu(x) \le \mu(I)^{\beta/(16)}. \tag{6.4.27}$$

Here we start with an unusual estimate, namely we observe that we have the upper bound

$$b^2(x) \le \Big(\sum_{B\in\mathcal{B}} \langle a, f_B\rangle^2\Big)\Big(\sum_{B\in\mathcal{B}} f_B^2(x)\Big). \tag{6.4.28}$$

An estimate like the one above is rarely used, mainly because it is almost always too crude. Here however it works pretty well, because the factor $\sum_{B\in\mathcal{B}} f_B^2(x)$ is concentrated around I and we are now interested in estimates over the set $X \setminus 4I$. Note that by orthonormality $(\sum_{B\in\mathcal{B}}\langle a, f_B\rangle^2) \le \mu(I)^{-1}$. Hence by (6.4.28)

$$\int_{X\setminus 4I} b(x)^2 d(x,x_I)^{1+\beta/(16)} d\mu(x) \le \mu(I)^{-1} \int_{X\setminus 4I} \sum_{B\in\mathcal{B}} f_B^2(x) d(x,x_I)^{1+\beta/(16)} d\mu(x). \tag{6.4.29}$$

Hence (6.4.27) is implied by the estimate

$$\sum_{B\in\mathcal{B}} \int_{X\setminus 4I} f_B^2(x) d(x,x_I)^{1+\beta/(16)} d\mu(x) \le \mu(I)^{1+\beta/(16)}. \tag{6.4.30}$$

Its verification requires again that we display the layers of the index set $\mathcal{B}$. Let $k \le M$, and $i \le A_2$ be determined by the inclusion (6.4.15). Recall that $n_0 \in \mathbb{N}$ is chosen in the first part of the proof. For $n \ge n_0 + 1$ we write

$$\mathcal{L}_n = \mathcal{B} \cap \mathcal{B}_i^{Mn+k}. \tag{6.4.31}$$

For each of these collections we will find that the following estimate holds,

$$\sum_{B\in\mathcal{L}_n} \int_{X\setminus 4I} f_B^2(x) d(x,x_I)^{1+\beta/(16)} d\mu(x) \le \mu(I)^{1-\beta/(16)} q^{-(Mn+k)\beta/8}. \tag{6.4.32}$$

Let us first observe how the estimate (6.4.32) implies (6.4.30). We take the right-hand side of (6.4.32) and sum over $n \ge n_0 + 1$. Evaluating the geometric series we find that

$$\mu(I)^{1-\beta/(16)} \sum_{n=n_0+1}^{\infty} q^{-(Mn+k)\beta/8} \le \mu(I)^{1-\beta/(16)} q^{-(Mn_0+k)\beta/8} \sim \mu(I)^{1+\beta/(16)}.$$

We now turn to the proof of (6.4.32). We begin with an estimate for the integrand. We claim that for $x \in X \setminus 4I$,

$$\sum_{B\in\mathcal{L}_n} f_B^2(x) \le C\mu(I)q^{-(Mn+k)\beta/8} \operatorname{dist}(x,I)^{-2-\beta/8}. \tag{6.4.33}$$

To prove the claim (6.4.33) we invoke the pointwise bounds established for the Franklin functions in Theorem 6.4.2,

$$f_B^2(x) \le \mu(B)^{1+\beta/8} \operatorname{dist}(x,B)^{-2-\beta/8}. \tag{6.4.34}$$

Note that for any $x \in X \setminus 4I$ and any $B \in \mathcal{L}_n$, we have the uniform estimate

$$\operatorname{dist}(x,I) \le \operatorname{dist}(x,B).$$

Hence the summands of (6.4.35) below, can be estimated independent of $B \in \mathcal{L}_n$. To find an upper bound of (6.4.35) we take out one of the summands and multiply it with the cardinality of the index set, (denoted here $|\mathcal{L}_n|$). This gives

$$\sum_{B\in\mathcal{L}_n} \operatorname{dist}(x,B)^{-2-\beta/8} \le |\mathcal{L}_n| \operatorname{dist}(x,I)^{-2-\beta/8}. \tag{6.4.35}$$

By (6.4.31) for $B \in \mathcal{L}_n$ we have $\mu(B) \sim q^{-(Mn+k)}$. Hence the cardinality of $\mathcal{L}_n$ satisfies the upper bound

$$|\mathcal{L}_n| \le C\mu(I)q^{(Mn+k)}. \tag{6.4.36}$$

Combining the estimates (6.4.34) – (6.4.36) proves (6.4.33). It follows from (6.4.33) that for $x \in X \setminus 4I$ the following estimate holds,

$$\sum_{B\in\mathcal{L}_n} f_B^2(x) \operatorname{dist}(x,I)^{1+\beta/(16)} \le C\mu(I)q^{-(Mn+k)\beta/8} \operatorname{dist}(x,I)^{-1-\beta/(16)}.$$

Integrating this bound and changing the order of summation and integration we obtain

$$\begin{aligned}&\sum_{B\in\mathcal{L}_n} \int_{X\setminus 4I} f_B^2(x) \operatorname{dist}(x,I)^{1+\beta/(16)} d\mu(x)\\ &\le \mu(I)q^{-(Mn+k)\beta/8} \int_{X\setminus 4I} \operatorname{dist}(x,I)^{-1-\beta/(16)} d\mu(x).\end{aligned} \tag{6.4.37}$$

It remains to find the value of the integral appearing in the second line of (6.4.37). A sufficiently good approximation is obtained as follows. First split $X \setminus 4I$ into dyadic annuli $A_i = 2^{i+1}I \setminus 2^iI$. Then for $x \in A_i$ the ratio $\operatorname{dist}(x,I)/\operatorname{dist}(A_i,I)$ is bounded above and below by a constant independent of i or I. Finally sum the values $\mu(A_i)\operatorname{dist}(A_i,I)^{-1-\beta/(16)}$. The resulting estimate for the integral in (6.4.37) is

$$\int_{X\setminus 4I} \operatorname{dist}(x,I)^{-1-\beta/(16)} d\mu(x) \le \mu(I)^{-\beta/(16)}.$$

This completes the proof of (6.4.30). □

By Theorem 6.4.4 the closed linear span of the orthonormal system $\{f_B : B \in \mathcal{N}_j\}$ is a complemented subspace of $H^1(X,d,\mu)$. Recall that the nested collection $\mathcal{E}_j$ is defined by Proposition 6.3.4. By a reasoning similar to the one appearing in the proof of Theorem 6.4.4 the space $X[\mathcal{E}_j]$ (spanned by three-valued martingale differences) is isomorphic to the closed linear span of the system $\{f_B : B \in \mathcal{N}_j\}$ in $H^1(X,d,\mu)$. For the details of the proof the reader is referred to [155]. In summary we proved that the martingale space $X[\mathcal{E}_j]$ is isomorphic to a complemented subspace of $H^1(X,d,\mu)$.

The square root of the Gram matrix

For $A, B \in \mathcal{N}_j$ and $\gamma > 0$ we define

$$\sigma(A,B,\gamma) = \min\left\{\frac{\mu A}{\mu B}, \frac{\mu B}{\mu A}\right\}^{1/2+\gamma}\left\{1 + \frac{d(A,B)}{\max\{\mu A, \mu B\}}\right\}^{-1-\gamma}.$$

A matrix $R = (R(A,B))$ is called an almost diagonal matrix if there exists $C > 0$ and $\gamma > 0$ such that

$$|R(A,B)| \leq C\sigma(A,B,\gamma), \quad A, B \in \mathcal{N}_j.$$

For instance the Gram matrix of the system $\{g_A, A \in \mathcal{N}_j\}$ is an almost diagonal matrix. Our first result states that the class of almost diagonal matrices is stable under the formation of products.

Proposition 6.4.5. *Let R be an infinite matrix with entries $R(A,B), A, B \in \mathcal{N}_j$, and assume that*

$$|R(A,B)| \leq M_1\sigma(A,B,\gamma),$$

for all $A, B \in \mathcal{N}_j$, and some M_1. Let R^k be the k-th power of R and let its entries be $R_k(A,B)$. Then there exists $M_2 \geq 0$ depending only on (X,d,μ) such that

$$|R_k(A,B)| \leq (M_1M_2)^k\sigma(A,B,\gamma/2),$$

for any $k \in \mathbb{N}$ and $A, B \in \mathcal{N}_j$.

Proof. The proof is inductive. We let $S = (S(A,B))$ be an almost diagonal matrix satisfying $|S(A,B)| \leq (M_1M_2)^{k-1}\sigma(A,B,\gamma/2)$. Explicitly this means that,

$$|S(A,B)| \leq (M_1M_2)^{k-1}\min\left\{\frac{\mu A}{\mu B}, \frac{\mu B}{\mu A}\right\}^{1/2+\gamma/2}\left\{1 + \frac{d(A,B)}{\max\{\mu A, \mu B\}}\right\}^{-1-\gamma/2}.$$

In this way the matrix S assumes the role of R^{k-1} and we are assuming that the theorem holds true for R^{k-1}. By hypothesis, R satisfies $|R(A,B)| \leq M_1\sigma(A,B,\gamma)$. Note that the entries of the product SR are given by

$$w(A,B) = \sum_{P \in \mathcal{N}_j} S(A,P)R(P,B). \tag{6.4.38}$$

Now we claim that the following estimate holds,

$$|w(A,B)| \le (M_1M_2)^k \min\left\{\frac{\mu A}{\mu B}, \frac{\mu B}{\mu A}\right\}^{1/2+\gamma/2} \left\{1 + \frac{d(A,B)}{\max\{\mu A, \mu B\}}\right\}^{-1-\gamma/2}.$$

We will only be sketching the proof of the claim, omitting the details. The idea is to split the sum $\sum_{P\in\mathcal{N}_j}$ in (6.4.38) into three pieces depending on the measure of P relative to the measure of A and B. This is done as follows. Let $\mu A \le \mu B$, then we have

$$\sum_{P\in\mathcal{N}_j} = \sum_{\{P\in\mathcal{N}_j:\mu P\le\mu A\}} + \sum_{\{P\in\mathcal{N}_j:\mu A\le\mu P\le\mu B\}} + \sum_{\{P\in\mathcal{N}_j:\mu B\le\mu P\}}. \tag{6.4.39}$$

Each of the sums in (6.4.39) is now estimated using the assumptions we made above. These are the induction hypothesis and the fact that R is an almost diagonal matrix,

$$|S(A,P)| \le (M_1M_2)^{k-1}\sigma(A,B,\gamma/2), \quad \text{and} \quad |R(P,B)| \le M_1\sigma(A,B,\gamma).$$

The calculations involved are quite long; we omit them and refer to [155] and [70] for the details. □

The constant M_2 appearing in Proposition 6.4.5 is an absolute constant that can be traced to the quasi-metric constant (6.3.1) and the doubling constant in (6.3.2). Under the additional hypothesis that the almost diagonal matrix G is positive definite and sufficiently close to the identity, we will show now that also the square root of the inverse $G^{-1/2}$ is an almost diagonal matrix.

Theorem 6.4.6. *Let G be a symmetric matrix with entries $G(A,B), A, B \in \mathcal{N}_j$. Assume that*

$$|G(A,B)| \le K_1\sigma(A,B,\gamma), \tag{6.4.40}$$

and that the induced operator satisfies

$$\|\operatorname{Id} - G\|_{\ell^2} \le 1/K_2 \quad \text{with} \quad K_2 \ge ((K_1+1)M_2)^{(4/\gamma)-1}. \tag{6.4.41}$$

(Here M_2 is the constant appearing in Proposition 6.4.5.) Then $G^{-1/2}$ is diagonally dominant and the entries of $G^{-1/2}$ satisfy

$$|G^{-1/2}(A,B)| \le K_3\sigma(A,B,\gamma/2)^{1-\gamma/4}. \tag{6.4.42}$$

Proof. Let $R = \operatorname{Id} - G$ and let R^k be the k-th power of R. The link between $G^{-1/2}$ and the powers R^k comes from elementary spectral theory of symmetric operators. With the spectral theorem $G^{-1/2}$ has a series representation as

$$G^{-1/2} = \sum_{k=1}^{\infty} c_k R^k,$$

where the coefficients c_k satisfy $c_k \le Ck^{-1/2}$.

Let the entries of R^k be $R_k(A,B)$. We will verify that for a fixed choice of $A, B \in \mathcal{N}_j$ the following estimate holds for the sum of the powers,

$$\sum_{k=1}^{\infty} |R_k(A,B)| \le \sigma(A,B,\gamma/2)^{1-\gamma/4}.$$

By Proposition 6.4.5, we have

$$|R_k(A,B)| \le (K_1+1)^k M_2^k \sigma(A,B,\gamma/2) \tag{6.4.43}$$

for $k \in \mathbb{N}$. Note that we also have the following inequality as a result of the operator estimate (6.4.41),

$$|R_k(A,B)| \le K_2^{-k}. \tag{6.4.44}$$

We will use the estimate (6.4.43) for small values of k, and (6.4.44) for large values of k. The transition from one regime to the next happens at the critical value k_0. Define k_0 by the relation

$$(K_1+1)^{k_0} M_2^{k_0} \le \sigma(A,B,\gamma/2)^{-\gamma/4} \le (K_1+1)^{k_0+1} M_2^{k_0+1}.$$

The following estimate holds for $k \le k_0$,

$$\begin{aligned}\sum_{k=1}^{k_0} |R_k(A,B)| &\le \sum_{k=1}^{k_0} (K_1+1)^k M_2^k \sigma(A,B,\gamma/2)\\ &\le 2(K_1+1)^{k_0} M_2^{k_0} \sigma(A,B,\gamma/2).\end{aligned}$$

The choice of k_0 implies that

$$\sum_{k=1}^{k_0} |R_k(A,B)| \le 2\sigma(A,B,\gamma/2)^{1-\gamma/4}.$$

Now we turn to the remainder. Estimating the geometric sum gives

$$\sum_{k=k_0+1}^{\infty} |R_k(A,B)| \le 2K_2^{-k_0}.$$

The assumption that $K_2 \ge ((K_1+1)M_2)^{(4/\gamma)-1}$ together with the choice of k_0 implies that

$$K_2^{-k_0} \le C\sigma(A,B,\gamma/2)^{1-\gamma/4}.$$

We thus verified that

$$\sum_{k=1}^{\infty} |R_k(A,B)| \le C\sigma(A,B,\gamma/2)^{1-\gamma/4}.$$

Finally,the spectral theorem implies that

$$G^{-1/2} = \sum_{k=1}^{\infty} c_k R^k,$$

where the coefficients satisfy $c_k \leq Ck^{-1/2}$. This gives

$$G^{-1/2}(A,B) \leq C\sigma(A,B,\gamma/2)^{1-\gamma/4}. \qquad \square$$

Remarks. 1. An elementary calculation shows that

$$\sigma(A,B,\gamma/2)^{1-\gamma/4} \leq C\sigma(A,B,\gamma/8). \tag{6.4.45}$$

2. The hypothesis of Theorem 6.4.6 is matched by the conclusion of Proposition 6.4.1. Hence by choosing the constant M large enough in (6.3.17) it follows that the inverse to the square root of $G(A,B), A,B \in \mathcal{N}_j$ is an almost diagonal matrix.

6.5 Martingale approximation in atomic H^1 spaces

In this section we will prove that $H^1(X,d,\mu)$ is isomorphic to a complemented subspace of the direct sum

$$H^1[(\mathcal{F}_n^1)] \oplus \cdots \oplus H^1[(\mathcal{F}_n^N)].$$

The theorem of B. Maurey implies then that $H^1(X,d,\mu)$ is isomorphic to a complemented subspace of dyadic H^1. More can be said by taking into account Theorem 6.3.5 and the classification theorems for martingale H^1 spaces (Theorem 4.1.3 and Theorem 4.2.1): Combining these results with Theorem 6.5.2 below gives the isomorphic classification of atomic H^1 spaces.

We use a remarkable family of symmetric kernels $\{k_A : A \in \mathcal{N}\}$, constructed by R. Coifman. Its properties are summarized as follows.

(a) For $f \in \mathrm{BMO}(X,d,\mu)$, respectively $f \in L^2(X,\mu)$, and

$$\lambda_A(f)(x) = \int f(y)k_A(x,y)d\mu(y), \tag{6.5.1}$$

the series

$$f_1 = \sum_{A\in\mathcal{N}} \lambda_A(f) \tag{6.5.2}$$

converges unconditionally in $\mathrm{BMO}(X,d,\mu)$, respectively in $L^2(X,\mu)$.

(b) The series (6.5.2) provides an approximation of f,

$$||f - f_1||_{\mathrm{BMO}(X)} \leq \frac{1}{2}||f||_{\mathrm{BMO}(X)}, \tag{6.5.3}$$

and simultaneously

$$||f - f_1||_{L^2(X)} \leq \frac{1}{2}||f||_{L^2(X)}. \tag{6.5.4}$$

(c) The operator

$$K_j(f) = \sum_{B \in \mathcal{N}_j} \lambda_B(f) \tag{6.5.5}$$

is bounded and linear on $\mathrm{BMO}(X, d, \mu)$ and on $H^1(X, d, \mu)$. By interpolation K_j is bounded on $L^p(X, \mu), 1 < p < \infty$.

We next define the space

$$Z_\infty = \{(K_1(f), \ldots, K_N(f)) : f \in \mathrm{BMO}(X, d, \mu)\},$$

equipped with the norm

$$\max_j ||K_j(f)||_{\mathrm{BMO}(X,d,\mu)}.$$

By (6.5.2), the space Z_∞ is naturally isomorphic to $\mathrm{BMO}(X, d, \mu)$ where the isomorphism is given by the map $f \to (K_1(f), \ldots, K_N(f))$.

By (6.5.2) we obtain a corresponding renorming of L^2. We define

$$Z_2 = \{(K_1(f), \ldots, K_N(f)) : f \in L^2(X)\},$$

and on Z_2 we define the norm

$$\left(\sum_j ||K_j(f)||_2^2\right)^{1/2}.$$

As a result the space Z_2 is a closed subspace of the N-fold direct sum $L^2(X, \mu) \oplus \cdots \oplus L^2(X, \mu)$. Hence the orthogonal projection P defined on $L^2(X, \mu) \oplus \cdots \oplus L^2(X, \mu)$ with range in Z_2 is bounded.

The article by Y. S. Han and G. Weiss [81] presents the unpublished results of R. Coifman referred to above. They are established by isolating and exploiting the singular integral representation for the operators K_j and applying a version of the $T(1)$-theorem valid in spaces of homogeneous type. The representation of K_j as a singular integral is possible since the kernels k_A, used in the definition of K_j, are of mean zero, $\int k_A(x, y)d\mu(x) = \int k_A(x, y)d\mu(y) = 0$, they satisfy a Lipschitz estimate,

$$|k_A(x, y) - k_A(x, z)| \leq C\mathrm{diam}(A)^{-1-\beta}d(y, z)^\beta,$$

and moreover, they are localized around a fixed multiple of $A \times A$. The $T(1)$ theorem asserts that for any subcollection $\mathcal{G} \subseteq \mathcal{N}$, the resulting kernel,

$$k(x, y) = \sum_{A \in \mathcal{G}} k_A(x, y),$$

defines a bounded singular integral operator an L^2. For details we refer to the book of S. Y. Han and E. T. Sawyer [80] or to A. Nahmod [161].

We will establish below that the operators K_j decompose any $f \in \mathrm{BMO}(X, d, \mu)$ into approximate martingales. The first step in this direction is the next proposition. It asserts that for a fixed j the function $K_j(f)$ defined by (6.5.5) is well adapted to the increasing sequence of σ-algebras $(\mathcal{F}_n^j)_{n=1}^\infty$ defined in (6.3.24).

Proposition 6.5.1. *The following a priori estimates hold for the linear operator* K_j,

$$||K_j(f)||_{BMO[(\mathcal{F}_n^j)]} \leq C||f||_{BMO(X,d,\mu)},$$

and

$$||K_j(f)||_{H^1[(\mathcal{F}_n^j)]} \leq C||f||_{H^1(X,d,\mu)}.$$

Proof. We first prove the BMO estimates. Let $h = K_j(f)$. Let D be an atom in the σ-algebra $\mathcal{F}_n^j$. We will show that

$$\int_D |h - \mathbb{E}(h|\mathcal{F}_{n-1}^j)|^2 d\mu \leq \mu(D)||f||^2_{\mathrm{BMO}(X,d,\mu)}. \tag{6.5.6}$$

We need to isolate the index set in the sum defining $h = K_j(f)$, which contributes to the integral on the left-hand side of (6.5.6). The index set consists of three parts as follows. First there is the collection

$$\mathcal{A} = \{A \in \mathcal{N}_j : T(A) \supseteq D \text{ and } T(A) \in \mathcal{F}_{n-2}^j\},$$

second

$$\mathcal{B} = \{A \in \mathcal{N}_j : T(A) \subseteq D\},$$

and third, there is $A_0 \in \mathcal{N}_j$ satisfying $T(A_0) \supseteq D$ and $T(A_0) \in \mathcal{F}_{n-1}^j \setminus \mathcal{F}_{n-2}^j$. Note that on the atom D, there holds the following identity,

$$\mathbb{E}(h|\mathcal{F}_{n-1}^j) = \sum_{A \in \mathcal{A}} \mathbb{E}(\lambda_A(f)|\mathcal{F}_{n-1}^j).$$

Thus we obtain a representation for the difference $h - \mathbb{E}(h|\mathcal{F}_{n-1}^j)$ appearing in (6.5.6). Indeed on D the following identity holds,

$$h - \mathbb{E}(h|\mathcal{F}_{n-1}^j) = \sum_{A \in \mathcal{A}} \lambda_A(f) - \mathbb{E}(\lambda_A(f)|\mathcal{F}_{n-1}^j) + \sum_{A \in \mathcal{B} \cup \{A_0\}} \lambda_A(f).$$

Next we give proper estimates for each of the above sums. We start with the second term. The integral representation for $\lambda_A(f)$, and the resulting singular integral representation for $\sum_{A\in\mathcal{B}\cup\{A_0\}}\lambda_A(f)$ justifies the application of the $T(1)$ theorem. Hence, the following L^2 estimate holds,

$$\int_D |\sum_{A\in\mathcal{B}\cup\{A_0\}}\lambda_A(f)|^2 d\mu \le \mu(D)||\lambda_{A_0}(f)||_\infty^2 + \sum_{A\in\mathcal{B}}||\lambda_A(f)||_2^2. \tag{6.5.7}$$

We continue with pointwise estimates for the term $\sum_{A\in\mathcal{A}}\lambda_A(f)-\mathbb{E}(\lambda_A(f)|\mathcal{F}_{n-1}^j)$. We start by applying the triangle inequality,

$$\sum_{A\in\mathcal{A}}|\lambda_A(f)-\mathbb{E}(\lambda_A(f)|\mathcal{F}_{n-1}^j)| \le \operatorname{diam} D^\beta \sum_{A\in\mathcal{A}}\operatorname{Lip}_\beta(\lambda_A(f)). \tag{6.5.8}$$

To analyze the right-hand side of (6.5.8), observe that for each $k\le n-1$, there exists at most one $A\in\mathcal{N}_j$ such that $T(A)\supseteq D$, and $T(A)\in\mathcal{F}_k^j\setminus\mathcal{F}_{k-1}^j$. Invoking the Lipschitz estimates $\operatorname{Lip}_\beta(\lambda_A(f))\le C\operatorname{diam}A^{-\beta}\|f\|_{\mathrm{BMO}(X)}$ we obtain from (6.5.7) and (6.5.8) the BMO estimate (6.5.6).

Now we prove the H^1 part of our proposition. Let $I\subseteq X$ be a ball and let $a: X\to\mathbb{R}$ be an atom supported on I such that

$$||a||_2^2\le\mu(I)^{-1}.$$

Now put $h=K_j(a)$. Next denote by S_j the martingale square function induced by the filtration $(\mathcal{F}_n^j)_{n=1}^\infty$. Thus

$$S_j(h)^2=\sum_{n=1}^\infty(\mathbb{E}(h|\mathcal{F}_n^j)-\mathbb{E}(h|\mathcal{F}_{n-1}^j))^2.$$

Now observe that the support of $S_j(h)^2$ is contained in the set

$$S=\bigcup\{T(A): A\in\mathcal{N}_j, A\cap I\ne\emptyset, \mu(A)\le\mu(I)\}.$$

Note that $\mu(S)\le C\mu(I)$. Hence we estimate using Cauchy–Schwarz, and $||h||_2\le C||a||_2\le C\mu(I)^{-1/2}$,

$$\begin{aligned}||S_j(h)||_1 &\le \mu(S)^{1/2}||S_j(h)||_2\\ &\le C\mu(S)^{1/2}||h||_2\\ &\le C.\end{aligned}$$

□

Let P be the orthogonal projection of $L^2(X,\mu)\oplus\cdots\oplus L^2(X,\mu)$ onto the closed subspace

$$Z_2=\{(K_1(f),\dots,K_N(f)): f\in L^2(X,\mu)\},$$

equipped with the norm

$$(\sum_j ||K_j(f)||_2^2)^{1/2}.$$

We write $P_j(f_1,\ldots,f_N)$ to denote the j -th component of the vector $P(f_1,\ldots,f_N)$. Now we are ready to prove the main theorem of this section which expresses an a priori estimate for the orthogonal projection P.

Theorem 6.5.2. *For every choice $h_j \in BMO[(\mathcal{F}_n^j)]$ the following a priori estimate holds for the components of the orthogonal projection P.*

$$\max_j ||P_j(h_1,\ldots,h_N)||_{BMO(X,d,\mu)} \le C(N) \max_j ||h_j||_{BMO[(\mathcal{F}_n^j)]}.$$

Hence, P extends to a bounded projection from $BMO[(\mathcal{F}_n^1)] \oplus \cdots \oplus BMO[(\mathcal{F}_n^N)]$ onto Z_∞.

With the duality theorem between $H^1(X,d,\mu)$ and $\mathrm{BMO}(X,d,\mu)$ we obtain the a priori bound of Theorem 6.5.2 from the following proposition.

Proposition 6.5.3. *Let $h_j \in BMO[(\mathcal{F}_n^j)]$. Then there exists $f \in BMO(X,d,\mu)$ such that*

$$P(h_1,\ldots,h_N) = (K_1(f),\ldots,K_N(f)),$$

and for any atom a_I in (X,d,μ), the following estimate holds,

$$\int f a_I d\mu \le C(N) \max ||h_j||_{BMO[(\mathcal{F}_n^j)]} + Cq^{-M\beta} ||f||_{BMO(X,d,\mu)},$$

where the constant C is a universal constant, independent of q, or M.

Proof. First we rewrite f as

$$f = f - f_1 + \sum_{i=1}^N K_i(f). \tag{6.5.9}$$

We let I be the ball in X supporting the atom a_I. For j we define the collection

$$\mathcal{N}_j(I) = \{B \in \mathcal{N}_j : \mu(B) \le q^{M\beta}\mu(I), I \cap B \ne \emptyset\}.$$

This defines

$$K_j^I(f) = \sum_{B \in \mathcal{N}_j(I)} \lambda_B(f).$$

We let P^I be the orthogonal projection of $L^2(X,\mu)\oplus\cdots\oplus L^2(X,\mu)$ onto the closed subspace

$$Z_2^I = \{(K_1^I(f),\ldots,K_N^I(f)) : f \in L^2(X,\mu)\}.$$

Next put $q_j = P_j^I(a_I,\ldots,a_I)$. We find $b \in L^2$ so that

$$q_j = \sum_{Q \in \mathcal{N}_j(I)} \lambda_Q(b),$$

moreover we may have that $\int b = 0, ||b||_2 \leq 2||a_I||_2$, and

$$\operatorname{supp} b \subseteq \bigcup\{(C \cdot Q) \in \mathcal{N}_j(I) : j \leq N\},$$

where $C > 0$ is an absolute constant. Note that the vector $(K_1^I(f), \ldots, K_N^I(f)) \in Z_2^I$ is contained in the range of the operator P^I. Hence by the orthogonality of the projection P^I, we find

$$\sum_{j=1}^N \int K_j^I(f) a_I = \sum_{j=1}^N \int K_j^I(f) q_j.$$

Thus we may rewrite

$$\begin{aligned} \sum_{j=1}^N \int K_j(f) a_I = & \sum_{j=1}^N \int K_j(f) q_j \\ & + \sum_{j=1}^N \int (K_j(f) - K_j^I(f)) a_I \\ & + \sum_{j=1}^N \int (K_j^I(f) - K_j(f)) q_j. \end{aligned} \tag{6.5.10}$$

Now we let $1 \leq j \leq N$ and put $r_j = \sum_{Q \notin \mathcal{N}_j(I)} \lambda_Q(b)$. By this choice of r_j, the vector $(q_j + r_j)$ is contained in Z_2. Hence the orthogonal projection leaves it invariant and we have $P((q_j + r_j)) = (q_j + r_j)$. Next recall that $P(h_1, \ldots, h_N) = (K_1(f), \ldots, K_N(f))$, so that by orthogonality we obtain

$$\sum_{j=1}^N \int K_j(f)(q_j + r_j) = \sum_{j=1}^N \int h_j(q_j + r_j).$$

Regrouping terms we obtain from the last identity that

$$\sum_{j=1}^N \int K_j(f) q_j = \sum_{j=1}^N \int h_j(q_j + r_j) - \sum_{j=1}^N \int K_j(f) r_j. \tag{6.5.11}$$

We add the equation (6.5.11) to the decomposition (6.5.10) and arrive thereby at the representation

$$\begin{aligned}\sum_{j=1}^{N}\int K_j(f)a_I &= \sum_{j=1}^{N}\int h_j(q_j+r_j)\\ &+\sum_{j=1}^{N}\int (K_j(f)-K_j^I(f))a_I\\ &+\sum_{j=1}^{N}\int (K_j^I(f)-K_j(f))q_j\\ &-\sum_{j=1}^{N}\int K_j(f)r_j.\end{aligned}$$

The first sum on the right-hand side is the leading term. We will estimate it using duality and Proposition 6.5.1. It will turn out that each of the remaining three sums is bounded by a small fraction of $||f||_{\mathrm{BMO}(X,d,\mu)}$. For our purposes the meaning of small is " small relative to $1/N$."

We now turn to estimating the integrals $|\int h_j(q_j+r_j)|$, using Fefferman's inequality and Proposition 6.5.1 . We have that,

$$\begin{aligned}|\int h_j(q_j+r_j)| &\leq C||h_j||_{\mathrm{BMO}[(\mathcal{F}_n^j)]}\cdot||q_j+r_j||_{H^1[(\mathcal{F}_n^j)]}\\ &\leq C||h_j||_{\mathrm{BMO}[(\mathcal{F}_n^j)]}\cdot||b||_{H^1(X,d,\mu)}\\ &\leq C(M,q)||h_j||_{\mathrm{BMO}[(\mathcal{F}_n^j)]}.\end{aligned}\tag{6.5.12}$$

We insert this estimate in the above representation. Thus we obtain the estimate

$$|\int fa_I|\leq||f-f_1||_{\mathrm{BMO}(X,d,\mu)}+C(M,N,q)\max||h_j||_{\mathrm{BMO}[(\mathcal{F}_n^j)]}+E_1+E_2+E_3,$$

where we used the abbreviation

$$\begin{aligned}E_1 &= \sum_{j=1}^{N}\left|\int (K_j(f)-K_j^I(f))a_I d\mu\right|,\\ E_2 &= \sum_{j=1}^{N}\left|\int (K_j(f)-K_j^I(f))q_j d\mu\right|,\\ E_3 &= ||f||_{\mathrm{BMO}(X,d,\mu)}\sum_{j=1}^{N}||r_j||_{H^1(X,d,\mu)}.\end{aligned}$$

We will now give estimates for the terms E_1, E_2, E_3, showing that they represent negligible error terms. We start estimating E_2.

$$\begin{aligned} E_2 &= \sum_{j=1}^{N} |\int K_j^I(f) - K_j(f))q_j|d\mu \\ &\leq \sum_{j=1}^{N} \sum_{B \notin \mathcal{N}_j(I)} |\int \lambda_B(f) q_j d\mu|. \end{aligned} \tag{6.5.13}$$

Now we fix $B \in \mathcal{N}_j$ with $B \notin \mathcal{N}_j(I)$, and seek estimates for the integrals appearing in the sum in the second line of (6.5.13). We let $\mathcal{U}$ be the collection of maximal sets in $\mathcal{N}_j(I)$. Our claim is that the following estimate holds,

$$|\int \lambda_B(f) q_j d\mu| \leq \sup\{\operatorname{diam}(P \cap B)^\beta : P \in \mathcal{U}\} \mu(B)^{-\beta} ||f||_{\mathrm{BMO}(X,d,\mu)}. \tag{6.5.14}$$

We prove the claim as follows. Let $b \in L^2$ such that

$$q_j = K_j^I(b).$$

Then for $R \in \mathcal{U}$ we put

$$b_R = \sum_{A \in \mathcal{N}_j, A \subseteq R} \lambda_A(b).$$

Using that the collection $\mathcal{U}$ consists of pairwise disjoint sets we obtain

$$||q_j||_2^2 = \sum_{R \in \mathcal{U}} ||b_R||_2^2 \leq \|a_I\|_2^2.$$

Next we fix $R \in \mathcal{U}$. As $\int b_R = 0$, we use below the Lipschitz continuity of $\lambda_B(f)$, and the fact that b_R is supported on R. Thus by Hölder's inequality we obtain that

$$|\int \lambda_B(f) b_R d\mu| \leq ||f||_{\mathrm{BMO}(X)} \operatorname{diam}(R \cap B)^\beta \mu(B)^{-\beta} \mu(R)^{1/2} ||b_R||_2. \tag{6.5.15}$$

Next we note that, by Hölder's inequality we find

$$\sum_{R \in \mathcal{U}} \mu(R)^{1-1/s} ||b_R||_{L^s} \leq q^{M\beta/2} \mu(I)^{1/2} ||a_I||_{L^2}. \tag{6.5.16}$$

Recall also that for any atom, $\mu(I)^{1/2}||a_I||_{L^2} \leq 1$. Hence by (6.5.15) and (6.5.16) we obtain the following estimate, by summing over $R \in \mathcal{U}$ and taking out the supremum of the terms $\operatorname{diam}(R \cap B)^\beta \mu(B)^{-\beta}$.

$$\sum_{R \in \mathcal{U}} |\int \lambda_B(f) b_R| \leq q^{M\beta/2} ||f||_{\mathrm{BMO}} \sup\{\operatorname{diam}(R \cap B)^\beta : R \in \mathcal{U}\} \mu(B)^{-\beta}.$$

To complete the proof of the claim (6.5.14) it remains to use the triangle inequality

$$|\int \lambda_B(f)q_j| \le \sum_{R\in\mathcal{U}} |\int \lambda_B(f)b_R|.$$

Now we return to estimating E_2. Recall that

$$E_2 \le \sum_{j=1}^{N} \sum_{B\notin\mathcal{N}_j(I)} |\int \lambda_B(f)q_j d\mu|.$$

With (6.5.14) we find that

$$E_2 \le Nq^{M\beta/2}||f||_{\mathrm{BMO}} \sum_{B\notin\mathcal{N}_j(I)} \mu(B)^{-\beta} \sup\{\mathrm{diam}(R\cap B)^\beta : R\in\mathcal{U}\}.$$

Now we consider the geometric series appearing on the right-hand side of the last line. Its sum is bounded by $Cq^{-M\beta}$, where $C>0$ is a universal constant. Thus we obtain that

$$E_2 \le CNq^{-M\beta/2}||f||_{\mathrm{BMO}}.$$

Note that choosing q large enough makes $q^{-M\beta/2}$ of order 2^{-M}. This is sufficiently small since $N = A_2M$, where A_2 is a universal constant.

Finally we turn to estimating E_3. We show that the norm in $H^1(X,d,\mu)$ of r_j is a small multiple of $1/N$. Recall that

$$r_j = \sum_{Q\notin\mathcal{N}_j(I)} \lambda_Q(b_I).$$

By the triangle inequality it suffices to obtain estimates for the norm of $\lambda_Q(b_I)$ in $H^1(X,d,\mu)$, when $Q\notin\mathcal{N}_j(I)$. First we note that the support of $\lambda_Q(b_I)$, is contained in a ball of diameter $\le C\mu(Q)$, and $\int\lambda_Q(b_I)=0$. Hence $\lambda_Q(b_I)$ is a multiple of an atom for (X,d,μ). We have that

$$||\lambda_Q(b_I)||_{H^1(X,d,\mu)} \le \mu(Q)^{1/2}||\lambda_Q(b_I)||_2.$$

Next examining the kernel representation of $\lambda_Q(b_I)$, we observe the following L^2 estimate,

$$||\lambda_Q(b_I)||_2 \le C\mu(Q)^{-1/2-\beta}(\mathrm{diam\,supp}\,b_I)^\beta.$$

Summing up we find that

$$\begin{aligned} ||r_j||_{H^1(X,d,\mu)} &\le C(\mathrm{diam\,supp}\,b_I)^\beta \sum_{Q\notin\mathcal{N}_j(I)} \{\mu(Q)^{-\beta} : Q\cap\mathrm{supp}\,b_I\ne\emptyset\} \\ &\le Cq^{-M\beta}. \end{aligned}$$

With our choice of M and q we obtain that $||r_j||_{H^1(X,d,\mu)} < C2^{-M} << 1/N$, as claimed. □

6.6 Notes

The construction giving the atomic decomposition in Proposition 6.1.3 is due to R. Coifman [55]. Our presentation is based on C. Bennett and R. Sharpley [8]. Theorem 6.1.1, the maximal function characterization of H^1_{at} follows from Proposition 6.1.3 and L^2 estimates for the Hilbert transform.

The square function characterization in Theorem 6.1.4 is a fundamental result of A. Calderón [39]. Extensions of Calderón's method were obtained by E. M. Stein, see [192] and [194].

Section 6.2 presents L. Carleson's work [45]. The biorthogonal system (φ_I, ψ_I) was introduced by L. Carleson with the aim of proving Theorem 6.2.13 and Theorem 6.2.14. The hardest component in Carleson's approach is the verification of the coefficient estimates in Theorem 6.2.5. In the text we explain L. Carleson's original proof. The figures in the text were produced by V. Pillwein using MATLAB routines. To obtain Figure 6.2, V. Pillwein started with $\{\varphi_I : |I| \geq 2^{-7}\}$ using $k = 5$. The truncated Gram matrix

$$G = (\langle \varphi_I, \varphi_J \rangle) \quad \text{with} \quad |J|, |I| \geq 2^{-7}$$

was inverted, and for $|J| \geq 2^{-7}$ the biorthogonal functional ψ_J was approximated by

$$\sum_{\{I:|I|\geq 2^{-7}\}} G^{-1}(I, J)\varphi_I,$$

where $G^{-1}(I, J)$ denotes the entries of G^{-1}, the inverse of the truncated Gram matrix.

Shortly after Carleson, P. Wojtaszczyk [210] proved that the biorthogonal Franklin system is an unconditional basis in H^1_{at}. J.O Stromberg [196] transferred the construction of the Franklin system from the boundary of the unit disk to $\mathbb{R}^n$, thereby obtaining an unconditional basis for $H^1(\mathbb{R}^n)$. Thus J. O. Stromberg found the first of the continuous wavelet systems that became an important tool in pure and applied mathematics. For an overview of the literature, see for instance the book by Y. Meyer and R. Coifman [147] or the introductory text by P. Wojtaszczyk [214].

Spaces of homogeneous type are systematically studied in [56]. A proof of the $T(1)$ theorem for singular integral operators on spaces of homogeneous type is presented in [50] where it is attributed to R. Coifman (unpublished). Also due to R. Coifman (unpublished) is the construction of smooth approximate identities and the reproduction formula of Calderón type. See [81] and [80].

The classification theorem for the spaces $H^1(X, d, \mu)$ is in [155]. The basic idea of using molecules to estimate the H^1 norm of Franklin functions is due to P. Wojtaszczyk [212]. The proof of the L^2 estimate in Proposition 6.4.1 follows A. Uchiyama [207].

Permanence properties of almost diagonal matrices such as Proposition 6.4.5 and Theorem 6.4.6 have a long history starting with the work of Z. Ciesielski [51], and J. Domsta [60]. For more recent results see S. Jaffard and Y. Meyer [89], P. Lemarie [125], M. Frazier and B. Jawerth [70], and the book by Y. Meyer and R. Coifman [147].

Bibliography

[1] D. J. Aldous. Unconditional bases and martingales in $L^p(F)$. *Math. Proc. Cambridge Philos. Soc.*, 85(1):117–123, 1979.

[2] D. Alspach, P. Enflo, and E. Odell. On the structure of separable $\mathcal{L}^p$ spaces $(1 < p < \infty)$. *Studia Math.*, 60(1):79–90, 1977.

[3] H. Arai. Degenerate elliptic operators, Hardy spaces and diffusions on strongly pseudoconvex domains. *Tohoku Math. J. (2)*, 46(4):469–498, 1994.

[4] A. Arias and J. D. Farmer. On the structure of tensor products of l^p-spaces. *Pacific J. Math.*, 175(1):13–37, 1996.

[5] K. Azuma. Weighted sums of certain dependent random variables. *Tôhoku Math. J. (2)*, 19:357–367, 1967.

[6] R. Bañuelos. A sharp good-λ inequality with an application to Riesz transforms. *Michigan Math. J.*, 35(1):117–125, 1988.

[7] R. Bañuelos and C. N. Moore. *Probabilistic behavior of harmonic functions*, volume 175 of *Progress in Mathematics*. Birkhäuser Verlag, Basel, 1999.

[8] C. Bennett and R. Sharpley. *Interpolation of Operators*, volume 129 of *Pure and Applied Mathematics*. Academic Press, New York, 1988.

[9] G. Bennett, L. E. Dor, V. Goodman, W. B. Johnson, and C. M. Newman. On uncomplemented subspaces of L^p, $1 < p < 2$. *Israel J. Math.*, 26(2):178–187, 1977.

[10] G. Beylkin, R. Coifman, and V. Rokhlin. Fast wavelet transforms and numerical algorithms. I. *Comm. Pure Appl. Math.*, 44(2):141–183, 1991.

[11] P. Billard. Sur la primarité des espaces $C(\alpha)$. *Studia Math.*, 62(2):143–162, 1978.

[12] C. J. Bishop and P. W. Jones. Harmonic measure, L^2 estimates and the Schwarzian derivative. *J. Anal. Math.*, 62:77–113, 1994.

[13] G. Blower. The Banach space $B(l^2)$ is primary. *Bull. London Math. Soc.*, 22(2):176–182, 1990.

[14] S. V. Bočkarev. Existence of a basis in the space of functions analytic in the disc, and some properties of Franklin's system. *Mat. Sb. (N.S.)*, 95(137):3–18, 159, 1974.

[15] A. Bonami. Ensembles $\Lambda(p)$ dans le dual de D^∞. *Ann. Inst. Fourier (Grenoble)*, 18(fasc. 2):193–204 (1969), 1968.

[16] J. Bourgain. *New classes of $\mathcal{L}^p$-spaces*, volume 889 of *Lecture Notes in Mathematics*. Springer-Verlag, Berlin, 1981.

[17] J. Bourgain. The nonisomorphism of H^1-spaces in one and several variables. *J. Funct. Anal.*, 46(1):45–57, 1982.

[18] J. Bourgain. Embedding L^1 in L^1/H^1. *Trans. Amer. Math. Soc.*, 278(2):689–702, 1983.

[19] J. Bourgain. The nonisomorphism of H^1-spaces in a different number of variables. *Bull. Soc. Math. Belg. Sér. B*, 35(2):127–136, 1983.

[20] J. Bourgain. On the primarity of H^∞-spaces. *Israel J. Math.*, 45(4):329–336, 1983.

[21] J. Bourgain. Some remarks on Banach spaces in which martingale difference sequences are unconditional. *Ark. Mat.*, 21(2):163–168, 1983.

[22] J. Bourgain. Extension of a result of Benedek, Calderón and Panzone. *Ark. Mat.*, 22(1):91–95, 1984.

[23] J. Bourgain. Sur l'approximation dans H^∞. In *Seminar on the geometry of Banach spaces, Vol. I, II (Paris, 1983)*, volume 18 of *Publ. Math. Univ. Paris VII*, pages 235–254. Univ. Paris VII, Paris, 1984.

[24] J. Bourgain. Vector valued singular integrals and the H^1-BMO duality. In *Israel seminar on geometrical aspects of functional analysis (1983/84)*, pages xvi, 23. Tel Aviv Univ., Tel Aviv, 1984.

[25] J. Bourgain. A remark on the behaviour of L^p-multipliers and the range of operators acting on L^p-spaces. *Israel J. Math.*, 79(2-3):193–206, 1992.

[26] J. Bourgain, H. Brezis, and P. Mironescu. Limiting embedding theorems for $W^{s,p}$ when $s \uparrow 1$ and applications. *J. Anal. Math.*, 87:77–101, 2002. Dedicated to the memory of Thomas H. Wolff.

[27] J. Bourgain, P. G. Casazza, J. Lindenstrauss, and L. Tzafriri. Banach spaces with a unique unconditional basis, up to permutation. *Mem. Amer. Math. Soc.*, 54(322):iv+111, 1985.

[28] J. Bourgain and O. Reinov. On the approximation properties for the space H^∞. *Math. Nachr.*, 122:19–27, 1985.

[29] J. Bourgain, H. P. Rosenthal, and G. Schechtman. An ordinal L^p-index for Banach spaces, with application to complemented subspaces of L^p. *Ann. of Math. (2)*, 114(2):193–228, 1981.

[30] H. Brezis. How to recognize constant functions. A connection with Sobolev spaces. *Uspekhi Mat. Nauk*, 57(4(346)):59–74, 2002.

[31] D. L. Burkholder. Distribution function inequalities for martingales. *Ann. Probability*, 1:19–42, 1973.

[32] D. L. Burkholder. Martingale transforms and the geometry of Banach spaces. In *Probability in Banach spaces, III (Medford, Mass., 1980)*, volume 860 of *Lecture Notes in Math.*, pages 35–50. Springer, Berlin, 1981.

[33] D. L. Burkholder. A nonlinear partial differential equation and the unconditional constant of the Haar system in L^p. *Bull. Amer. Math. Soc. (N.S.)*, 7(3):591–595, 1982.

[34] D. L. Burkholder. A geometric condition that implies the existence of certain singular integrals of Banach-space-valued functions. In *Conference on harmonic analysis in honor of Antoni Zygmund, Vol. I, II (Chicago, Ill., 1981)*, Wadsworth Math. Ser., pages 270–286. Wadsworth, Belmont, CA, 1983.

[35] D. L. Burkholder. An elementary proof of an inequality of R. E. A. C. Paley. *Bull. London Math. Soc.*, 17(5):474–478, 1985.

[36] D. L. Burkholder. A proof of Pełczyński's conjecture for the Haar system. *Studia Math.*, 91(1):79–83, 1988.

[37] D. L. Burkholder. Martingales and singular integrals in Banach spaces. In *Handbook of the geometry of Banach spaces, Vol. I*, pages 233–269. North-Holland, Amsterdam, 2001.

[38] A. Calderón. Intermediate spaces and interpolation, the complex method. *Studia Math.*, 24:113–190, 1964.

[39] A. Calderón. Commutators of Singular Integral Operators. *Proc. Nat. Acad. Sci. USA*, 53:1092–1099, 1965.

[40] M. Capon. Primarité de $l^p(L^1)$. *Math. Ann.*, 250(1):55–63, 1980.

[41] M. Capon. Primarité de $L^p(l^r)$, $1 < p, r < \infty$. *Israel J. Math.*, 36(3-4):346–364, 1980.

[42] M. Capon. Primarité de $L^p(L^r)$, $1 < p$, $r < \infty$. *Israel J. Math.*, 42(1-2):87–98, 1982.

[43] M. Capon. Primarité de $L^p(X)$. *Trans. Amer. Math. Soc.*, 276(2):431–487, 1983.

[44] L. Carleson. On convergence and growth of partial sums of Fourier series. *Acta Math.*, 116:135–157, 1966.

[45] L. Carleson. An explicit unconditional basis in H^1. *Bull. Sci. Math. (2)*, 104(4):405–416, 1980.

[46] L. Carleson and J. B. Garnett. Interpolating sequences and separation properties. *J. Analyse Math.*, 28:273–299, 1975.

[47] P. G. Casazza. Approximation properties. In *Handbook of the geometry of Banach spaces, Vol. I*, pages 271–316. North-Holland, Amsterdam, 2001.

[48] S.-Y. A. Chang and Z. Ciesielski. Spline characterizations of H^1. *Studia Math.*, 75(2):183–192, 1983.

[49] S.-Y. A. Chang, J. M. Wilson, and T. H. Wolff. Some weighted norm inequalities concerning the Schrödinger operators. *Comment. Math. Helv.*, 60(2):217–246, 1985.

[50] M. Christ. *Lectures on singular integral operators*, volume 77 of *CBMS Regional Conference Series in Mathematics.* Published for the Conference Board of the Mathematical Sciences, Washington, DC, 1990.

[51] Z. Ciesielski. Properties of the orthonormal Franklin system. *Studia Math.*, 23:141–157, 1963.

[52] Z. Ciesielski. Properties of the orthonormal Franklin system. II. *Studia Math.*, 27:289–323, 1966.

[53] Z. Ciesielski. Haar orthogonal functions in analysis and probability. In *A. Haar memorial conference, Vol. I, II (Budapest, 1985)*, volume 49 of *Colloq. Math. Soc. János Bolyai*, pages 25–56. North-Holland, Amsterdam, 1987.

[54] R. Coifman, Y. Meyer, and E. Stein. Some new Function Spaces and their Applications to Harmonic Analysis. *Jour. Funct. Anal*, 62:304–335, 1985.

[55] R. R. Coifman. A real variable characterization of H^p. *Studia Math.*, 51:269–274, 1974.

[56] R. R. Coifman and G. Weiss. Extensions of Hardy spaces and their use in analysis. *Bull. Amer. Math. Soc.*, 83(4):569–645, 1977.

[57] M. Cwikel, P. G. Nilsson, and G. Schechtman. Interpolation of weighted Banach lattices. A characterization of relatively decomposable Banach lattices. *Mem. Amer. Math. Soc.*, 165(787):vi+127, 2003.

[58] G. David. *Wavelets and singular integrals on curves and surfaces*, volume 1465 of *Lecture Notes in Mathematics.* Springer-Verlag, Berlin, 1991.

[59] G. David and J.-L. Journé. A boundedness criterion for generalized Calderón-Zygmund operators. *Ann. of Math. (2)*, 120(2):371–397, 1984.

[60] J. Domsta. A theorem on B-splines. *Studia Math.*, 41:291–314, 1972.

[61] R. Durrett. *Brownian motion and martingales in analysis.* Wadsworth Mathematics Series. Wadsworth International Group, Belmont, CA, 1984.

[62] P. Enflo. A counterexample to the approximation problem in Banach spaces. *Acta Math.*, 130:309–317, 1973.

[63] P. Enflo and T. W. Starbird. Subspaces of L^1 containing L^1. *Studia Math.*, 65(2):203–225, 1979.

[64] C. Fefferman and E. M. Stein. H^p spaces of several variables. *Acta Math.*, 129(3-4):137–193, 1972.

[65] T. Figiel. On equivalence of some bases to the Haar system in spaces of vector-valued functions. *Bull. Polish Acad. Sci. Math.*, 36(3-4):119–131 (1989), 1988.

[66] T. Figiel. Singular integral operators: a martingale approach. In *Geometry of Banach spaces (Strobl, 1989)*, volume 158 of *London Math. Soc. Lecture Note Ser.*, pages 95–110. Cambridge Univ. Press, Cambridge, 1990.

[67] T. Figiel, W. B. Johnson, and G. Schechtman. Factorizations of natural embeddings of l_n^p into L_r. I. *Studia Math.*, 89(1):79–103, 1988.

[68] T. Figiel, J. Lindenstrauss, and V. D. Milman. The dimension of almost spherical sections of convex bodies. *Acta Math.*, 139(1-2):53–94, 1977.

[69] T. Figiel and P. Wojtaszczyk. Special bases in function spaces. In *Handbook of the geometry of Banach spaces, Vol. I*, pages 561–597. North-Holland, Amsterdam, 2001.

[70] M. Frazier and B. Jawerth. A discrete transform and decompositions of distribution spaces. *J. Funct. Anal.*, 93(1):34–170, 1990.

[71] J. L. B. Gamlen and R. J. Gaudet. On subsequences of the Haar system in L^p $[0, 1](1 < p < \infty)$. *Israel J. Math.*, 15:404–413, 1973.

[72] J. B. Garnett. *Bounded analytic functions*, volume 96 of *Pure and Applied Mathematics*. Academic Press Inc.

[73] J. B. Garnett and P. W. Jones. The distance in BMO to L^∞. *Ann. of Math. (2)*, 108(2):373–393, 1978.

[74] J. B. Garnett and P. W. Jones. BMO from dyadic BMO. *Pacific J. Math.*, 99(2):351–371, 1982.

[75] A. M. Garsia. *Martingale inequalities: Seminar notes on recent progress.* W. A. Benjamin, Inc., Reading, Mass.-London-Amsterdam, 1973. Mathematics Lecture Notes Series.

[76] C. Goffman and G. Pedrick. *First course in functional analysis.* Prentice-Hall Inc., Englewood Cliffs, N.J., 1965.

[77] B. Golubov, A. Efimov, and V. Skvortsov. *Walsh series and transforms*, volume 64 of *Mathematics and its Applications (Soviet Series).* Kluwer Academic Publishers Group, Dordrecht, 1991. Theory and applications, Translated from the 1987 Russian original by W. R. Wade.

[78] V. I. Gurarii, M. Kadec, and V. I. Macaev. Distances between finite-dimensional analogs of the L^p-spaces. *Mat. Sb. (N.S.)*, 70 (112):481–489, 1966.

[79] A. Haar. Zur Theorie der orthogonalen Funktionensysteme (Erste Mitteilung). *Math. Ann.*, 69:311–371, 1910.

[80] Y. S. Han and E. T. Sawyer. Littlewood-Paley theory on spaces of homogeneous type and the classical function spaces. *Mem. Amer. Math. Soc.*, 110(530):vi+126, 1994.

[81] Y. S. Han and G. Weiss. Function spaces on spaces of homogeneous type. In *Essays on Fourier analysis in honor of Elias M. Stein (Princeton, NJ, 1991)*, volume 42 of *Princeton Math. Ser.*, pages 211–224. Princeton Univ. Press, Princeton, NJ, 1995.

[82] G. Hardy and J. E. Littlewood. A maximal theorem with function-theoretic applications. *Acta. Math.*, 54:81–116, 1930.

[83] S. Heinrich. Finite representability and super-ideals of operators. *Dissertationes Math. (Rozprawy Mat.)*, 172:37, 1980.

[84] S. Heinrich. Ultraproducts in Banach space theory. *J. Reine Angew. Math.*, 313:72–104, 1980.

[85] W. Hoeffding. Probability inequalities for sums of bounded random variables. *J. Amer. Statist. Assoc.*, 58:13–30, 1963.

[86] L. Hörmander. Estimates for translation invariant operators in L^p spaces. *Acta Math.*, 104:93–140, 1960.

[87] R. A. Hunt. Almost everywhere convergence of Walsh-Fourier series of L^2 functions. In *Actes du Congrès International des Mathématiciens (Nice, 1970), Tome 2*, pages 655–661. Gauthier-Villars, Paris, 1971.

[88] R. A. Hunt. Developments related to the a.e. convergence of Fourier series. In *Studies in harmonic analysis (Proc. Conf., DePaul Univ., Chicago, Ill., 1974)*, pages 20–37. MAA Stud. Math., Vol. 13. Math. Assoc. Amer., Washington, D.C., 1976.

[89] S. Jaffard and Y. Meyer. Bases d'ondelettes dans des ouverts de $\mathbf{R}^n$. *J. Math. Pures Appl. (9)*, 68(1):95–108, 1989.

[90] S. Janson and P. W. Jones. Interpolation between H^p spaces: the complex method. *J. Funct. Anal.*, 48(1):58–80, 1982.

[91] W. B. Johnson. Operators into L^p which factor through 1^p. *J. London Math. Soc. (2)*, 14(2):333–339, 1976.

[92] W. B. Johnson and J. Lindenstrauss. Basic concepts in the geometry of Banach spaces. In *Handbook of the geometry of Banach spaces, Vol. I*, pages 1–84. North-Holland, Amsterdam, 2001.

[93] W. B. Johnson, B. Maurey, G. Schechtman, and L. Tzafriri. Symmetric structures in Banach spaces. *Mem. Amer. Math. Soc.*, 19(217):v+298, 1979.

[94] W. B. Johnson and G. Pisier. The proportional UAP characterizes weak Hilbert spaces. *J. London Math. Soc. (2)*, 44(3):525–536, 1991.

[95] W. B. Johnson and G. Schechtman. Sums of independent random variables in rearrangement invariant function spaces. *Ann. Probab.*, 17(2):789–808, 1989.

[96] W. B. Johnson and G. Schechtman. On the distance of subspaces of l_p^n to l_p^k. *Trans. Amer. Math. Soc.*, 324(1):319–329, 1991.

[97] W. B. Johnson and G. Schechtman. Finite dimensional subspaces of L^p. In *Handbook of the geometry of Banach spaces, Vol. I*, pages 837–870. North-Holland, Amsterdam, 2001.

[98] P. W. Jones. Carleson measures and the Fefferman-Stein decomposition of BMO(**R**). *Ann. of Math. (2)*, 111(1):197–208, 1980.

[99] P. W. Jones. Interpolation between Hardy spaces. In *Conference on harmonic analysis in honor of Antoni Zygmund, Vol. I, II (Chicago, Ill., 1981)*, Wadsworth Math. Ser., pages 437–451. Wadsworth, Belmont, CA, 1983.

[100] P. W. Jones. On interpolation between H^1 and H^∞. In *Interpolation spaces and allied topics in analysis (Lund, 1983)*, volume 1070 of *Lecture Notes in Math.*, pages 143–151. Springer, Berlin, 1984.

[101] P. W. Jones. Recent advances in the theory of Hardy spaces. In *Proceedings of the International Congress of Mathematicians, Vol. 1, 2 (Warsaw, 1983)*, pages 829–838, Warsaw, 1984. PWN.

[102] P. W. Jones. BMO and the Banach space approximation problem. *Amer. J. Math.*, 107(4):853–893, 1985.

[103] P. W. Jones. Rectifiable sets and the traveling salesman problem. *Invent. Math.*, 102(1):1–15, 1990.

[104] P. W. Jones and P. F. X. Müller. Conditioned Brownian motion and multipliers into SL^∞. *GAFA, Geom. funct. anal.*, 14:319–379, 2004.

[105] N. J. Kalton. Differentials of complex interpolation processes for Köthe function spaces. *Trans. Amer. Math. Soc.*, 333(2):479–529, 1992.

[106] N. J. Kalton. Complex interpolation of Hardy-type subspaces. *Math. Nachr.*, 171:227–258, 1995.

[107] N. J. Kalton, C. Leránoz, and P. Wojtaszczyk. Uniqueness of unconditional bases in quasi-Banach spaces with applications to Hardy spaces. *Israel J. Math.*, 72(3):299–311 (1991), 1990.

[108] N. J. Kalton and S. Montgomery-Smith. Interpolation of Banach spaces. In *Handbook of the geometry of Banach spaces, Vol. 2*, pages 1131–1175. North-Holland, Amsterdam, 2003.

[109] N. J. Kalton and P. Wojtaszczyk. On nonatomic Banach lattices and Hardy spaces. *Proc. Amer. Math. Soc.*, 120(3):731–741, 1994.

[110] B. S. Kashin. The widths of certain finite-dimensional sets and classes of smooth functions. *Izv. Akad. Nauk SSSR Ser. Mat.*, 41(2):334–351, 478, 1977.

[111] B. S. Kashin and A. A. Saakyan. *Orthogonal series*, volume 75 of *Translations of Mathematical Monographs.* American Mathematical Society, Providence, RI, 1989. Translated from the Russian by Ralph P. Boas, Translation edited by Ben Silver.

[112] K. Kiener. *Über Produkte von quadratisch integrierbaren Funktionen endlicher Vielfalt.* Dissertation, Universität Innsbruck, 1969.

[113] K. Kiener. A problem concerning $\Lambda(p)$ sets of the Walsh-Paley-system. *Math. Balkanica*, 4:331–333, 1974. Papers presented at the Fifth Balkan Mathematical Congress (Belgrade, 1974).

[114] N. M. Kislovskaya and V. B. Osipov. Equivalence in L^1 of permutations of the haar system. *Mat. Zametki*, 29(6):877–885, 956, 1981.

[115] S. V. Kislyakov. Interpolation inequalities for Fourier multipliers and their applications. *Dokl. Akad. Nauk SSSR*, 292(1):29–33, 1987.

[116] S. V. Kislyakov. Fourier coefficients of continuous functions and a class of multipliers. *Ann. Inst. Fourier (Grenoble)*, 38(2):147–183, 1988.

[117] D. Kleper and G. Schechtman. Block bases of the Haar system as complemented subspaces of L^p, $2 < p < \infty$. *Proc. Amer. Math. Soc.*, 131(2):433–439, 2003.

[118] P. Koosis. *Introduction to H^p spaces*, volume 40 of *London Mathematical Society Lecture Note Series.* Cambridge University Press, Cambridge, 1980. With an appendix on Wolff's proof of the corona theorem.

[119] S. G. Krein, Y. I. Petunin, and E. M. Semënov. *Interpolation of linear operators*, volume 54 of *Translations of Mathematical Monographs.* American Mathematical Society, Providence, R.I., 1982. Translated from the Russian by J. Szűcs.

[120] S. Kwapień and A. Pełczyński. Some linear topological properties of the Hardy spaces H^p. *Compositio Math.*, 33(3):261–288, 1976.

[121] M. Lacey and C. Thiele. L^p estimates on the bilinear Hilbert transform for $2 < p < \infty$. *Ann. of Math. (2)*, 146(3):693–724, 1997.

[122] M. Lacey and C. Thiele. On Calderón's conjecture. *Ann. of Math. (2)*, 149(2):475–496, 1999.

[123] E. Landau. *Darstellung und Begründung einiger neuerer Ergebnisse der Funktionentheorie.* Springer-Verlag, Berlin, second edition, 1929.

[124] R. Latała and K. Oleszkiewicz. On the best constant in the Khinchin-Kahane inequality. *Studia Math.*, 109(1):101–104, 1994.

[125] P. G. Lemarié. Base d'ondelettes sur les groupes de Lie stratifiés. *Bull. Soc. Math. France*, 117(2):211–232, 1989.

[126] G. Lerman. Quantifying curvelike structures of measures by using L^2 Jones quantities. *Comm. Pure Appl. Math.*, 56(9):1294–1365, 2003.

[127] J. Lindenstrauss. On complemented subspaces of m. *Israel J. Math.*, 5:153–156, 1967.

[128] J. Lindenstrauss and A. Pełczyński. Contributions to the theory of the classical Banach spaces. *J. Functional Analysis*, 8:225–249, 1971.

[129] J. Lindenstrauss and L. Tzafriri. The uniform approximation property in Orlicz spaces. *Israel J. Math.*, 23(2):142–155, 1976.

[130] J. Lindenstrauss and L. Tzafriri. *Classical Banach Spaces. Vol. I and II.* Classics in Mathematics. Springer Verlag, Berlin Heidelberg New York, 1996.

[131] R. A. Macías and C. Segovia. A decomposition into atoms of distributions on spaces of homogeneous type. *Adv. in Math.*, 33(3):271–309, 1979.

[132] R. A. Macías and C. Segovia. Lipschitz functions on spaces of homogeneous type. *Adv. in Math.*, 33(3):257–270, 1979.

[133] N. G. Makarov. Probability methods in the theory of conformal mappings. *Algebra i Analiz*, 1(1):3–59, 1989.

[134] J. Marcinkiewicz and A. Zygmund. A theorem of Lusin. *Duke. Math. J.*, 4:473–485, 1938.

[135] V. Mascioni. s-numbers of projections in Banach spaces. *Israel J. Math.*, 67(1):82–94, 1989.

[136] V. Mascioni. Some remarks on the uniform approximation property in Banach spaces. *Studia Math.*, 96(3):243–253, 1990.

[137] V. Mascioni. On the duality of the uniform approximation property in Banach spaces. *Illinois J. Math.*, 35(2):191–197, 1991.

[138] B. Maurey. Sous-espaces complémentés de L^p, d'après P. Enflo. In *Séminaire Maurey-Schwartz 1974–1975: Espaces L^p, applications radonifiantes et géométrie des espaces de Banach, Exp. No. III*, page 15 pp. Centre Math., École Polytech., Paris, 1975.

[139] B. Maurey. Système de Haar. In *Séminaire Maurey-Schwartz 1974–1975: Espaces L^p, applications radonifiantes et géométrie des espaces de Banach, Exp. Nos. I et II*, pages 26 pp. (erratum, p. 1). Centre Math., École Polytech., Paris, 1975.

[140] B. Maurey. Plongement de H^1 dans un espace à base inconditionnelle. *C. R. Acad. Sci. Paris Sér. A-B*, 287(13):A865–A867, 1978.

[141] B. Maurey. Construction de suites symétriques. *C. R. Acad. Sci. Paris Sér. A-B*, 288(14):A679–A681, 1979.

[142] B. Maurey. Isomorphismes entre espaces H^1. In *Séminaire d'Analyse Fonctionnelle (1978–1979)*, pages Exp. No. 19–20, 7. École Polytech., Palaiseau, 1979.

[143] B. Maurey. Isomorphismes entre espaces H^1. *C. R. Acad. Sci. Paris Sér. A-B*, 288(4):A271–A273, 1979.

[144] B. Maurey. Isomorphismes entre espaces H^1. *Acta Math.*, 145(1-2):79–120, 1980.

[145] B. Maurey and G. Schechtman. Some remarks on symmetric basic sequences in L^1. *Compositio Math.*, 38(1):67–76, 1979.

[146] Y. Meyer. Wavelets and operators. In *Analysis at Urbana, Vol. I (Urbana, IL, 1986–1987)*, volume 137 of *London Math. Soc. Lecture Note Ser.*, pages 256–365. Cambridge Univ. Press, Cambridge, 1989.

[147] Y. Meyer and R. Coifman. *Wavelets*, volume 48 of *Cambridge Studies in Advanced Mathematics*. Cambridge University Press, Cambridge, 1997. Calderón-Zygmund and multilinear operators, Translated from the 1990 and 1991 French originals by David Salinger.

[148] V. D. Milman and G. Schechtman. *Asymptotic theory of finite-dimensional normed spaces*, volume 1200 of *Lecture Notes in Mathematics*. Springer-Verlag, Berlin, 1986. With an appendix by M. Gromov.

[149] P. F. X. Müller. On subsequences of the Haar basis in $H^1(\delta)$ and isomorphism between H^1-spaces. *Studia Math.*, 85(1):73–90 (1987), 1986.

[150] P. F. X. Müller. Classification of the isomorphic types of martingale-H^1 spaces. *Israel J. Math.*, 59(2):195–212, 1987.

[151] P. F. X. Müller. On the span of some three valued martingale difference sequences in L^p and H^1. *Israel J. Math.*, 60(1):39–53, 1987.

[152] P. F. X. Müller. A local version of a result of Gamlen and Gaudet. *Israel J. Math.*, 63(2):212–222, 1988.

[153] P. F. X. Müller. On projections in H^1 and BMO. *Studia Math.*, 89(2):145–158, 1988.

[154] P. F. X. Müller. Orthogonal projections on martingale H^1 spaces of two parameters. *Illinois J. Math.*, 38(4):554–573, 1994.

[155] P. F. X. Müller. The Banach space $H^1(X, d, \mu)$. I, II. *Math. Ann.*, 303(3):499–521, 523–544, 1995.

[156] P. F. X. Müller. Rearrangements of the Haar system that preserve BMO. *Proc. London Math. Soc. (3)*, 75(3):600–618, 1997.

[157] P. F. X. Müller. A family of complemented subspaces in VMO and its isomorphic classification. *Israel J. Math.*, 134:289–306, 2003.

[158] P. F. X. Müller and G. Schechtman. On complemented subspaces of H^1 and VMO. In *Geometric aspects of functional analysis (1987–88)*, volume 1376 of *Lecture Notes in Math.*, pages 113–125. Springer, Berlin, 1989.

[159] P. F. X. Müller and G. Schechtman. Several results concerning unconditionality in vector valued L^p and H^1 spaces. *Illinois J. Math.*, 35(2):220–233, 1991.

[160] P. F. X. Müller and G. Schechtman. A remarkable rearrangement of the Haar system in L^p. *Proc. Amer. Math. Soc.*, 125(8):2363–2371, 1997.

[161] A. R. Nahmod. Generalized uncertainty principles on spaces of homogeneous type. *J. Funct. Anal.*, 119(1):171–209, 1994.

[162] A. Naor and G. Schechtman. Remarks on non linear type and Pisier's inequality. *J. Reine Angew. Math.*, 552:213–236, 2002.

[163] Z. Nehari. *Conformal mapping.* Dover Publications Inc., New York, 1975. Reprinting of the 1952 edition.

[164] P. Nilsson. Interpolation of Banach lattices. *Studia Math.*, 82(2):135–154, 1985.

[165] I. Novikov and E. M. Semenov. *Haar series and linear operators*, volume 367 of *Mathematics and its Applications.* Kluwer Academic Publishers Group, Dordrecht, 1997.

[166] R. I. Ovsepian and A. Pełczyński. On the existence of a fundamental total and bounded biorthogonal sequence in every separable Banach space, and related constructions of uniformly bounded orthonormal systems in L^2. *Studia Math.*, 54(2):149–159, 1975.

[167] R. E. A. C. Paley. A remarkable series of orthogonal functions (I). *Proc. Lond. Math. Soc.*, 34:241–264, 1932.

[168] R. E. A. C. Paley. On the lacunary coefficients of power series. *Ann. of Math. (2)*, 34(3):615–616, 1933.

[169] A. Pełczyński. Projections in certain Banach Spaces. *Studia Math.*, 19(4):209–228, 1960.

[170] A. Pełczyński. All separable Banach spaces admit for every $\varepsilon > 0$ fundamental total and bounded by $1 + \varepsilon$ biorthogonal sequences. *Studia Math.*, 55(3):295–304, 1976.

[171] A. Pełczyński. *Banach spaces of analytic functions and absolutely summing operators.* American Mathematical Society, Providence, R.I., 1977. Expository lectures from the CBMS Regional Conference held at Kent State University, Kent, Ohio, July 11–16, 1976, Conference Board of the Mathematical Sciences Regional Conference Series in Mathematics, No. 30.

[172] A. Pełczyński and H. Rosenthal. Lokalisation techniques in L^p spaces. *Studia Math.*, 52:263–289, 1975.

[173] A. Pełczyński and C. Schütt. Factoring the natural injection $i^{(n)} : L_n^\infty \to L_n^1$ through finite-dimensional Banach spaces and geometry of finite-dimensional unitary ideals. In *Mathematical analysis and applications, Part B*, volume 7 of *Adv. in Math. Suppl. Stud.*, pages 653–683. Academic Press, New York, 1981.

[174] J. Pipher. A martingale inequality related to exponential square integrability. *Proc. Amer. Math. Soc.*, 118(2):541–546, 1993.

[175] G. Pisier. La méthode d'interpolation complexe: applications aux treillis de Banach. In *Séminaire d'Analyse Fonctionnelle (1978–1979)*, pages Exp. No. 17, 18. École Polytech., Palaiseau, 1979.

[176] G. Pisier. Some applications of the complex interpolation method to Banach lattices. *J. Analyse Math.*, 35:264–281, 1979.

[177] G. Pisier. Holomorphic semigroups and the geometry of Banach spaces. *Ann. of Math. (2)*, 115(2):375–392, 1982.

[178] G. Pisier. Probabilistic methods in the geometry of Banach spaces. In *Probability and analysis (Varenna, 1985)*, volume 1206 of *Lecture Notes in Math.*, pages 167–241. Springer, Berlin, 1986.

[179] G. Pisier. *The volume of convex bodies and Banach space geometry*, volume 94 of *Cambridge Tracts in Mathematics.* Cambridge University Press, Cambridge, 1989.

[180] A. N. Podkorytov and O. I. Reinov. On the Khinchin-Kahane inequality. *Algebra i Analiz*, 10(1):265–270, 1998.

[181] H. P. Rosenthal. On the subspaces of L^p $(p > 2)$ spanned by sequences of independent random variables. *Israel J. Math.*, 8:273–303, 1970.

[182] C. Samuel. Exemples d'espaces de Banach ayant la propriété de projection uniforme. In *Séminaire sur la Géométrie des Espaces de Banach (1977–1978)*, pages Exp. No. 27, 15. École Polytech., Palaiseau, 1978.

[183] C. Samuel. Primarité des produits d'espaces de suites. *Colloq. Math.*, 39(1):123–132, 1978.

[184] C. Samuel. Exemples d'espaces de Banach ayant la propriété de projection uniforme. *Colloq. Math.*, 43(1):117–126 (1981), 1980.

[185] J. P. Schauder. Eine Eigenschaft des Haarschen Orthogonalsystems. *Mathem. Zeitsch.*, 28:317–320, 1927.

[186] G. Schechtman. Random embeddings of Euclidean spaces in sequence spaces. *Israel J. Math.*, 40(2):187–192, 1981.

[187] G. Schechtman. Concentration results and applications. In *Handbook of the geometry of Banach spaces, Vol. 2*, pages 1603–1634. North-Holland, Amsterdam, 2003.

[188] F. Schipp. On equivalence of rearrangements of the Haar system in dyadic Hardy and BMO spaces. *Anal. Math.*, 16(2):135–141, 1990.

[189] C. Schütt. Unconditionality in tensor products. *Israel J. Math.*, 31(3-4):209–216, 1978.

[190] E. M. Semenov. Equivalence in L^p of permutations of the Haar system. *Dokl. Akad. Nauk SSSR*, 242(6):1258–1260, 1978.

[191] E. M. Semenov and B. Stöckert. Permutations of the Haar system in spaces L^p. *Anal. Math.*, 7(4):277–295, 1981.

[192] E. M. Stein. *Singular integrals and differentiability properties of functions.* Princeton Mathematical Series, No. 30. Princeton University Press, Princeton, N.J., 1970.

[193] E. M. Stein. *Topics in harmonic analysis related to the Littlewood-Paley theory.* Annals of Mathematics Studies, No. 63. Princeton University Press, Princeton, N.J., 1970.

[194] E. M. Stein. Variations on the Littlewood-Paley theme. In *Lectures in Modern Analysis and Applications, III*, pages 1–17. Lecture Notes in Mathematics, Vol. 170. Springer, Berlin, 1970.

[195] E. M. Stein and G. Weiss. *Introduction to Fourier analysis on Euclidean spaces.* Princeton University Press, Princeton, N.J., 1971. Princeton Mathematical Series, No. 32.

[196] J.-O. Strömberg. A modified Franklin system and higher-order spline systems on $\mathbf{R}^n$ as unconditional bases for Hardy spaces. In *Conference on harmonic analysis in honor of Antoni Zygmund, Vol. I, II (Chicago, Ill., 1981)*, Wadsworth Math. Ser., pages 475–494. Wadsworth, Belmont, CA, 1983.

[197] A. Szankowski. A Banach lattice without the approximation property. *Israel J. Math.*, 24(3-4):329–337, 1976.

[198] A. Szankowski. $B(\mathcal{H})$ does not have the approximation property. *Acta Math.*, 147(1-2):89–108, 1981.

[199] A. Szankowski. On the uniform approximation property in Banach spaces. *Israel J. Math.*, 49(4):343–359, 1984.

[200] S. J. Szarek. On the best constants in the Khinchin inequality. *Studia Math.*, 58(2):197–208, 1976.

[201] M. Talagrand. Isoperimetry, logarithmic Sobolev inequalities on the discrete cube and Margulis graph connectivity theorem. *Geom. Funct. Anal.*, 3:295–314, 1993.

[202] A. E. Taylor. A geometric theorem and its application to biorthogonal systems. *Bull. Amer. Math. Soc.*, 53:614–616, 1947.

[203] C. Thiele. *Time-frequency analysis in the discrete phase plane.* PhD thesis, Yale University, 1995.

[204] C. Thiele. The quartile operator and pointwise convergence of Walsh series. *Trans. Amer. Math. Soc.*, 352(12):5745–5766, 2000.

[205] A. Uchiyama. A maximal function characterization of H^p on the space of homogeneous type. *Trans. Amer. Math. Soc.*, 262(2):579–592, 1980.

[206] A. Uchiyama. The factorization of H^p on the space of homogeneous type. *Pacific J. Math.*, 92(2):453–468, 1981.

[207] A. Uchiyama. A constructive proof of the Fefferman-Stein decomposition of BMO $(\mathbf{R}^n)$. *Acta Math.*, 148:215–241, 1982.

[208] J. L. Walsh. A closed set of normal orthogonal functions. *Amer. J. Math.*, 55:5–24, 1923.

[209] J. L. Walsh. A property of Haar's system of orthogonal functions. *Math. Ann*, 90:38–45, 1923.

[210] P. Wojtaszczyk. The Franklin system is an unconditional basis in H^1. *Ark. Mat.*, 20(2):293–300, 1982.

[211] P. Wojtaszczyk. The Banach space H^1. In *Functional analysis: surveys and recent results, III (Paderborn, 1983)*, volume 90 of *North-Holland Math. Stud.*, pages 1–33. North-Holland, Amsterdam, 1984.

[212] P. Wojtaszczyk. H^p-spaces, $p \leq 1$, and spline systems. *Studia Math.*, 77(3):289–320, 1984.

[213] P. Wojtaszczyk. *Banach spaces for analysts*, volume 25 of *Cambridge Studies in Advanced Mathematics*. Cambridge University Press, Cambridge, 1991.

[214] P. Wojtaszczyk. *A mathematical introduction to wavelets*, volume 37 of *London Mathematical Society Student Texts*. Cambridge University Press, Cambridge, 1997.

[215] P. Wojtaszczyk. Uniqueness of unconditional bases in quasi-Banach spaces with applications to Hardy spaces. II. *Israel J. Math.*, 97:253–280, 1997.

[216] P. Wojtaszczyk. Spaces of analytic functions with integral norm. In *Handbook of the geometry of Banach spaces, Vol. 2*, pages 1671–1702. North-Holland, Amsterdam, 2003.

[217] T. M. Wolniewicz. On isomorphisms between Hardy spaces on complex balls. *Ark. Mat.*, 27(1):155–168, 1989.

List of Symbols

Index

Monografie Matematyczne

[1] S. Banach, *Théorie des opérations linéaires*, 1932
[2] S. Saks, *Théorie de l'integrale*, 1933
[3] C. Kuratowski, *Topologie* I, 1933
[4] W. Sierpiński, *Hypothèse de continu*, 1934
[5] A. Zygmund, *Trigonometrical Series*, 1935
[6] S. Kaczmarz, H. Steinhaus, *Theorie der Orthogonalreihen*, 1935
[7] S. Saks, *Theory of the integral*, 1937
[8] S. Banach, *Mechanika*, T. I, 1947
[9] S. Banach, *Mechanika*, T. II, 1947
[10] S. Saks, A. Zygmund, *Funkcje analityczne*, 1948
[11] W. Sierpiński, *Zasady algebry wyższej*, 1946
[12] K. Borsuk, *Geometria analityczna w n wymiarach*, 1950
[13] W. Sierpiński, *Działania nieskończone*, 1948
[14] W. Sierpiński, *Rachunek różniczkowy poprzedzony badaniem funkcji elementarnych*, 1947
[15] K. Kuratowski, *Wykłady rachunku różniczkowego i całkowego*, T. I, 1948
[16] E. Otto, *Geometria wykreślna*, 1950
[17] S. Banach, *Wstęp do teorii funkcji rzeczywistych*, 1951
[18] A. Mostowski, *Logika matematyczna*, 1948
[19] W. Sierpiński, *Teoria liczb*, 1950
[20] C. Kuratowski, *Topologie* I, 1948
[21] C. Kuratowski, *Topologie* II, 1950
[22] W. Rubinowicz, *Wektory i tensory*, 1950
[23] W. Sierpiński, *Algèbre des ensembles*, 1951
[24] S. Banach, *Mechanics*, 1951
[25] W. Nikliborc, *Równania różniczkowe*, Cz. I, 1951
[26] M. Stark, *Geometria analityczna*, 1951
[27] K. Kuratowski, A. Mostowski, *Teoria mnogości*, 1952
[28] S. Saks, A. Zygmund, *Analytic functions*, 1952
[29] F. Leja, *Funkcje analityczne i harmoniczne*, Cz. I, 1952
[30] J. Mikusiński, *Rachunek operatorów*, 1953
* [31] W. Ślebodziński, *Formes extérieures et leurs applications*, 1954
[32] S. Mazurkiewicz, *Podstawy rachunku prawdopodobieństwa*, 1956
[33] A. Walfisz, *Gitterpunkte in mehrdimensionalen Kugeln*, 1957
[34] W. Sierpiński, *Cardinal and ordinal numbers*, 1965
[35] R. Sikorski, *Funkcje rzeczywiste*, 1958
[36] K. Maurin, *Metody przestrzeni Hilberta*, 1959
[37] R. Sikorski, *Funkcje rzeczywiste*, T. II, 1959
[38] W. Sierpiński, *Teoria liczb* II, 1959

* [39] J. Aczél, S. Gołąb, *Funktionalgleichungen der Theorie der geometrischen Objekte*, 1960
[40] W. Ślebodziński, *Formes extérieures et leurs applications*, II, 1963
[41] H. Rasiowa, R. Sikorski, *The mathematics of metamathematics*, 1963
[42] W. Sierpiński, *Elementary theory of numbers*, 1964
* [43] J. Szarski, *Differential inequalities*, 1965
[44] K. Borsuk, *Theory of retracts*, 1967
[45] K. Maurin, *Methods of Hilbert spaces*, 1967
[46] M. Kuczma, *Functional equations in a single variable*, 1967
[47] D. Przeworska-Rolewicz, S. Rolewicz, *Equations in linear spaces*, 1968
[48] K. Maurin, *General eigenfunction expansions and unitary representations of topological groups*, 1968
[49] A. Alexiewicz, *Analiza funkcjonalna*, 1969
* [50] K. Borsuk, *Multidimensional analytic geometry*, 1969
* [51] R. Sikorski, *Advanced calculus. Functions of several variables*, 1969
[52] W. Ślebodziński, *Exterior forms and their applications*, 1971
[53] M. Krzyżański, *Partial differential equations of second order*, vol. I, 1971
[54] M. Krzyżański, *Partial differential equations of second order*, vol. II, 1971
[55] Z. Semadeni, *Banach spaces of continuous functions*, 1971
[56] S. Rolewicz, *Metric linear spaces*, 1972
[57] W. Narkiewicz, *Elementary and analytic theory of algebraic numbers*, 1974
[58] Cz. Bessaga, A. Pełczyński, *Selected topics in infinite dimensional topology*, 1975
* [59] K. Borsuk, *Theory of shape*, 1975
[60] R. Engelking, *General topology*, 1977
[61] J. Dugundji, A. Granas, *Fixed point theory*, 1982
* [62] W. Narkiewicz, *Classical problems in number theory*, 1986

The volumes marked with * are available at the exchange department of the library of the Institute of Mathematics, Polish Academy of Sciences.